Code Calculations

Based on the 2014 *National Electrical Code*®

$$I_{pri} = \frac{kVA \times 1,000}{E_{pri}}$$

$$\text{Max. } OCPD_{pri} = I_{pri} \times 125\%$$

NATIONAL JOINT APPRENTICESHIP
AND TRAINING COMMITTEE

NJATC Code Calculations is intended to be an educational resource for the user and contains procedures commonly practiced in industry and the trade. Specific procedures vary with each task and must be performed by a qualified person. For maximum safety, always refer to specific manufacturer recommendations, insurance regulations, specific job site and plant procedures, applicable federal, state, and local regulations, and any authority having jurisdiction. The *electrical training ALLIANCE* assumes no responsibility or liability in connection with this material or its use by any individual or organization.

3 4 5 6 7 8 9 – 14 – 9 8 7 6 5 4 3 2

Printed in the United States of America

Contents

Chapter 1 Conductor Allowable Ampacity Calculations i

Contents

Contents

Contents

Contents

Contents

Contents

Contents

Contents

Features

Problems provide application of *Code* requirements to real-world installations.

Calculator Icon indicates that a calculation using an industry standard calculator appears in the Appendix.

Code Excerpts are from NFPA 70®.

Figures, including photographs and artwork, clearly illustrate concepts from the text.

For additional information related to QR Codes, visit qr.njatcdb.org Item #1079

Quick Response Codes (QR Codes) create a link between the textbook and the Internet. They can be scanned using Smartphone applications to obtain additional information online. (To access the information without using a Smartphone, visit qr.njatc.org and enter the referenced Item #.)

Features

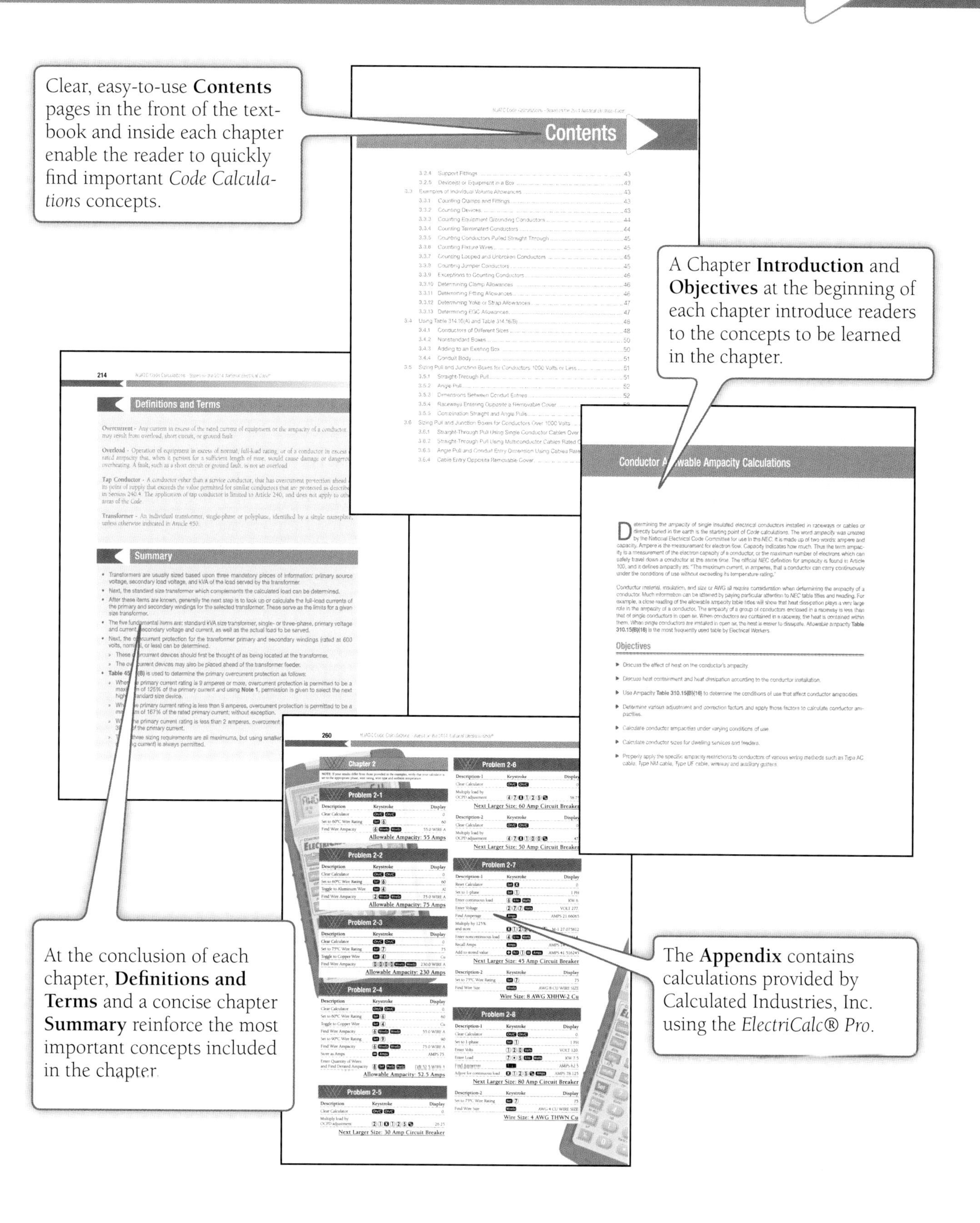

Clear, easy-to-use **Contents** pages in the front of the textbook and inside each chapter enable the reader to quickly find important *Code Calculations* concepts.

A Chapter **Introduction** and **Objectives** at the beginning of each chapter introduce readers to the concepts to be learned in the chapter.

At the conclusion of each chapter, **Definitions and Terms** and a concise chapter **Summary** reinforce the most important concepts included in the chapter.

The **Appendix** contains calculations provided by Calculated Industries, Inc. using the *ElectriCalc® Pro*.

Introduction

For additional information, visit qr.njatcdb.org Item #1534

The NJATC *Code Calculations* textbook was developed for our industry and by our industry, the IBEW/NECA industry. The NJATC continues to update this valuable *Code*-related applied mathematical approach to calculations based on the 2014 *National Electrical Code*.

The 2014 edition of *Code Calculations* focuses on solving math-related issues of electrical power, such as insulation, ampacity, terminations, continuous loads, box size, voltage drop, tap rules, motor loads, transformer loads, various building loads, and a host of other *Code*-related mathematical issues, in order to provide a *NEC* based electrically safe installation. This textbook is a solid foundation for a comprehensive training program for apprentices, Journeymen, foremen, Electrical Workers, estimators, electrical project managers, and electrical engineers. *Code Calculations* provides the necessary skills and knowledge to achieve a *Code*-compliant installation and is well-suited as *Code* exam preparation material.

For the 2011 and the 2014 edition of *Code Calculations*, select problems are solved using a state-of-the-art Construction Master® *ElectriCalc®Pro* electronic calculator. This provides the user with the experience of performing real-world calculations using an industry standard professional calculator.

About this Book

This textbook is primarily arranged according to the *National Electrical Code (NEC®)* and has been revised for 2014. Code calculations are calculated and determined using many of the wiring methods of *NEC* Chapter 3, beginning with ampacity calculations related to Article 310 and applied ampacity (with terminations related to Article 110). Next, ampacity calculations are applied to continuous and noncontinuous loads as well as fuses and circuit breakers including 100% rated circuit breakers to complete real-world ampacity applications. Additional updates include applied motor load calculations, appliance applications and straight-forward cable tray fill and ampacity calculations.

Acknowledgments

Technical information and assistance has been furnished by the following companies and organizations:

American Technical Publishers
Baldor Electric Company
Calculated Industries, Inc.
Eaton's Bussmann Business (formerly Cooper Bussmann)
Fluke
Honeywell/Salsibury
The Lincoln Electric Co.
NECA
Pass and Seymour LeGrand/Wiremold
Pass and Seymour LeGrand/Cabolfil
Philips Color Kinetics
Rigid
Raychem Quicknet, Tyco Thermal Controls
Schneider Electric
Thomas & Betts Corporation

Special thanks is extended to the National Fire Protection Association (NFPA) for allowing the educational use of the 2014 *NEC* related material.

QR Codes

Baldor Electric Company
Calculated Industries, Inc.
Eaton's Bussmann Business
General Electric Company
The Lincoln Electric Co.
National Fire Protection Association (NFPA)
Pass and Seymour LeGrand/Cabolfil
Schneider Electric
Southwire Company
Thomas & Betts Corporation
Tyco International Ltd.

Conductor Allowable Ampacity Calculations

Determining the *ampacity* of single insulated electrical conductors installed in raceways or cables or directly buried in the earth is the starting point of *Code* calculations. The word *ampacity* was created by the National Electrical Code Committee for use in the *NEC*. It is made up of two words: ampere and capacity. Ampere is the measurement for electron flow. Capacity indicates how much. Thus the term ampacity is a measurement of the electron capacity of a conductor, or the maximum number of electrons which can safely travel down a conductor at the same time. The official *NEC* definition for ampacity is found in Article 100, and it defines ampacity as: "The maximum current, in amperes, that a conductor can carry continuously under the conditions of use without exceeding its temperature rating."

Conductor material, insulation, and size or AWG all require consideration when determining the ampacity of a conductor. Much information can be attained by paying particular attention to *NEC* table titles and heading. For example, a close reading of the allowable ampacity table titles will show that heat dissipation plays a very large role in the ampacity of a conductor. The ampacity of a group of conductors enclosed in a raceway is less than that of single conductors in open air. When conductors are contained in a raceway, the heat is contained within them. When single conductors are installed in open air, the heat is easier to dissipate. Allowable ampacity **Table 310.15(B)(16)** is the most frequently used table by Electrical Workers.

Objectives

- Discuss the effect of heat on the conductor's ampacity.
- Discuss heat containment and heat dissipation according to the conductor installation.
- Use Ampacity **Table 310.15(B)(16)** to determine the conditions of use that affect conductor ampacities.
- Determine various adjustment and correction factors and apply those factors to calculate conductor ampacities.
- Calculate conductor ampacities under varying conditions of use.
- Calculate conductor sizes for dwelling services and feeders.
- Properly apply the specific ampacity restrictions to conductors of various wiring methods such as Type AC cable, Type NM cable, Type UF cable, wireway and auxiliary gutters.

Chapter 1

Table of Contents

1.1 General

Ampacities for conductors may be calculated by a complex mathematical formula or ampacities for conductors may be determined by using a combination of look-up tables and application factors. The determination of electrical conductor sizes for a given ampacity is a fundamental duty of a trained electrical worker. Often an electrical worker is called upon to make this selection for an installation. This chapter is an in-depth demonstration of using the *Code* look-up tables, applying the various application factors to those table values where necessary and finally, determining the appropriate conductor insulation and size for a wide selection of practical applications.

This chapter is devoted to using table values and to calculating the allowable ampacity of a conductor as presented in Parts II and III of **Article 310**. Within these parts of **Article 310**, particular attention is directed to the general rules, temperature limitations of conductors and the following:

310.15 Ampacities for Conductors Rated 0-2000 Volts, (A) General

310.15 Ampacities for Conductors Rated 0-2000 Volts, (B) Tables

- **Table 310.15(B)(16) through Table 310.15(B)(18) Allowable Ampacities for Conductors**

310.104 Conductor Constructions and Applications

- **Table 310.104(A) Conductor Applications and Insulations Rated 600 Volts**[1]

Do not be concerned with overcurrent protection for the conductors or the effects of connected loads at this time. These will be taken up in later chapters.

Before we begin determining actual ampacities, it is imperative to have a clear understanding of the term ampacity, the different electrical conductor insulations, and the temperature limitations of those insulated conductors.

1.2 Ampacity

The term *ampacity* means current-carrying capacity and is a coined word built from two common words, *ampere* and *capacity*, as shown in **Figure 1-1**.

AMP~~ERE CAP~~ACITY = AMPACITY

Figure 1-1. *Ampacity is derived from ampere and capacity.*

According to **Article 100 Definitions**, ampacity is defined as "the maximum current, in amperes, that a conductor can carry continuously under the conditions of use without exceeding its temperature rating." The four parts of this definition are as follows:

- Maximum current, in amperes
- Able to carry continuously
- Under the conditions of use
- Without exceeding its temperature rating

This definition identifies the current-carrying capacity of a conductor as the maximum amount of current that will raise the temperature of a particular conductor in a given environment to its rated temperature. In this chapter, generally and unless otherwise stated, a conductor is assumed to be an insulated conductor.

Insulated current-carrying conductors have two basic parts: the inner conducting metal and the outer covering of electrical insulation. Remembering that plastic melts and degrades at a much lower temperature than metal, the calculations of this book focus on the maximum operating temperature of electrical insulations as permitted in **Table 310.104(A)**. Therefore, the rated or maximum permitted temperature of a conductor is ultimately determined by the ability of the selected electrical insulating material (i.e. conductor insulation) to withstand those temperatures along their entire length without significant degradation.

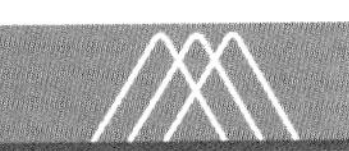

The U.S. is the fourth largest producer of copper in the world. The largest U.S. copper mine is found in Utah (Bingham Canyon). Other major mines are found in Arizona, New Mexico, and Nevada. In South America, Chile is the world's largest producer and along with Peru are major producers of copper. www.copper.org and others

Courtesy of The Copper Development

Conductor heating originates from many sources. These sources include the following:

1. *Area Heat.* This is the environment or ambient air temperature which may vary along the conductor's length or may vary during time of day.
2. *Load Current Heat.* This is the heat generated by the load current flowing in the metal conductor material due to the natural resistance of the conductor. This heat includes the heat generated by ordinary or fundamental current as well as any harmonic currents.
3. *Adjacent Conductor Heat.* This is the heat generated by current flowing through adjacent load-carrying conductors, which adds to the ambient heat and impedes the heat dissipation process within the conduit or cable assembly.
4. *Ability to Cool Down or Dissipate Heat.* This is the retained heat due to the inability of the surrounding media to dissipate heat. Surrounding media which may entrap heat include additional conductor insulation, outer covering or cable jacket, raceway or enclosure construction or placement, enclosure, soil temperature, wind velocity, and ambient air.

The temperature of a conductor, therefore, is reached by adding these four heat "sources." Since heat is the determining factor in the permitted or allowable ampacity of a conductor, **310.15(A)(3)** clearly prohibits a conductor from being used if the maximum operating temperature for its particular insulation is exceeded anywhere along the length of the conductor. Thus, the most important statement regarding conductors is: Do not permit conductors to be heated beyond their limit.

1.3 Tables

The ampacity tables used in this textbook are limited to **Table 310.15(B)(16)** and **Table 310.15(B)(17)**. Since conductors are placed in various applications and installed in various conditions, temperature corrections and adjustment factors may need to be applied to the ampacity of the selected conductor to ensure that the insulated conductor temperature is not exceeded. See **310.15(B)** for the exact text of these requirements and permissions. Other conditions (such as temperature rating of terminations and continuous duty loads) will be dealt with in later chapters.

The allowable ampacity of current-carrying conductors given in the *NEC* tables is not the true ampacity of the conductor as defined in **Article 100**. Rather, it is the allowable ampacity because the *Code* has established limiting parameters on the installation of conductors. Items of consideration when establishing the allowable ampacity tables are as follows:

1. Temperature compatibility with connected equipment, especially at the connection point
2. Coordination with circuit and system overcurrent protection
3. Compliance with the requirements of product listings
4. Preservation of the safety benefits of established industry practices and standardized procedures

Where tables are used to determine allowable ampacity, the table heading is used to determine whether the table applies in a certain situation or applies to specific conditions of use. The title of **Table 310.15(B)(16)** is as follows:

> **Allowable Ampacities of Insulated Conductors Rated Up to and Including 2000 Volts, 60°C Through 90°C (140°F Through 194°F), Not More Than Three Current-Carrying Conductors in Raceway, Cable, or Earth (Directly Buried), Based on Ambient Temperature of 30°C (86°F)**

The two important installation parameters or "conditions of use" for the current-carrying capacity of the conductor are stated in this title. They are:

1. Not more than three conductors in a raceway or cable
2. Ambient temperature not over 30°C (86°F)

Table 310.15(B)(16) expresses the allowable ampacity for a current-carrying conductor, provided there are not more than three conductors in the raceway or cable and provided that the ambient air temperature is 30°C. If either of the two basic factors is different, the allowable ampacity of the current-carrying conductor shown in **Table 310.15(B)(16)** must be changed.

Table 310.15(B)(17) changes the conditions of use from "... not more than three conductors in a raceway or cable" to "single-insulated conductors ... in free air." This is a major difference in the conditions of use. Therefore, the allowable ampacities of **Table 310.15(B)(16)** are much different from the allowable ampacities of **Table 310.15(B)(17)**. Make sure that the selected table matches the necessary conditions of use. For most of the work in this book, the conditions of use will dictate **Table 310.15(B)(16)**.

There are many types of conductor insulation found in the headings of allowable ampacity **Table 310.15(B)(16)** and **Table 310.15(B)(17)**. For an explanation of the type letters used in the tables and other important information concerning electrical conductor insulation, see **Table 310.104(A)**. Additional installation requirements are found in **310.10** through **310.15(A)(3)** and other articles of the *Code*. For the ampacities of flexible cords, see **Table 400.5(A)(1)**, and **Table 400.5(A)(2)**.

1.4 Factors Affecting Ampacity

There are two important factors which directly affect the ampacity of an insulated conductor. These factors are called ambient temperature correction factors and adjustment factors.

1.4.1 Ambient Temperature Correction Factors

Both **Table 310.15(B)(16)** and **Table 310.15(B)(17)** stipulate that the fundamental ambient temperature is 30°C (86°F). This is clearly stated in both of the table headings. An ambient temperature of 30°C (86°F) is considered as one of the conditions of use.

Where project conditions have ambient temperatures other than 30°C (86°F), the ampacity table must be corrected from 30°C (86°F) to the ambient temperature for the expected job conditions. This correction is made by using the correction factors found in **Table 310.15(B)(2)(a)**. Or, as an option, the correction may be made by using a simple ratio formula. Further information about using this ratio formula may be found in **310.15(B)(2)**.

For example, where the basic conditions of use or the ambient temperature is different from that shown in the table heading, additional correction is required to correct the allowable ampacity.

Referring to **Table 310.15(B)(16)**, a correction factor must be used where the ambient air temperature is different from 30°C (86°F), as stated in the table title. The correction factor is given as a percentage of the allowable ampacity value selected from **Table 310.15(B)(16)**. The correction factors are found in **Table 310.15(B)(2)(a)**. Correction factors, for the most part, are simple corrections for temperature once the ambient temperature is determined.

1.4.2 General Application of Adjustment Factors

Once the appropriate table is selected, other conditions of use may require adjustment. For example, again specifically referring to **Table 310.15(B)(16)**, an adjustment factor must be used where the number of current-carrying conductors in a raceway or cable exceeds three conductors, as stated in the table title. The adjustment factor is given as a percentage of the allowable ampacity value selected from **Table 310.15(B)(16)**. These adjustment factors for more than three current-carrying conductors in a raceway or cable are found in **Table 310.15(B)(3)(a)**.

Another adjustment factor may need to be applied where circular raceways are exposed to direct sunlight on rooftops. Again, the adjustment factor is given as a percentage of the allowable ampacity value selected from **Table 310.15(B)(16)**. These adjustment factors are found in **Table 310.15(B)(3)(c)**.

Adjustment factors to conductors are more difficult to determine, since many rules apply. **310.15(B)(3)** contains many parts that come into play as adjustment factors may need to be applied.

1.5 310.15(B) Ampacity Tables for Conductors Rated 0–2000 Volts

Section 310.15(B) applies to **Table 310.15(B)(16)** through **Table 310.15(B)(21)**. However, to see how it works, start out by applying it to **Table 310.15(B)(16)**. Ampacities found in **Table 310.15(B)(16)** are further explained in the following paragraphs.

1.5.1 General

Insulated conductors are defined in **Article 100** and recognized by **Section 310.104**. Therefore, only conductors described in **Table 310.104(A)** through **Table 310.104(E)** are considered insulated conductors suitable for general wiring. Notice that the subtle requirement of **Section 310.104** eliminates other insulated conductors for general wiring if they are not specifically recognized by this section and its tables.

Table 310.104(A) shows the physical properties of electrical insulation for general-use conductors specifically rated 600 volts or up to 1000 volts if listed and marked. **Table 310.104(A)** is used to properly apply the insulations listed in the column headings of **Table 310.15(B)(16)**. For example, in **Table 310.15(B)(16)**, the insulation type XHHW appears in both the 75°C and the 90°C columns for both copper and aluminum conductors. Knowing whether the location for the XHHW conductor is a wet or a dry location and using **Table 310.104(A)**, the correct maximum operating temperature is easily determined.

Specifically, **Table 310.104(A)** limits the maximum operating temperature of an XHHW to the ampacity of the 75°C column where that conductor is used in a wet location. Whereas, for a dry location, **Table 310.104(A)** limits the maximum operating temperature of Type XHHW to the 90°C ampacity column. Also notice that insulation Type XHHW-2 has the same maximum operating temperature of 90°C for both wet and dry locations.

1.5.2 General Requirements for Adjustment Factor Applications

To apply adjustment factors correctly, one needs to have a thorough understanding of which conductors within a circuit actually create enough heat to be labeled as generating "load current heat." (The term "load current heat" was previously explained in Section 1.2) There are many sections within the *Code* which assist us in making this current carrying or not current carrying determination. Some sections include conductors as always current carrying, some sections specifically exclude certain conductors as current carrying, and a few sections point out that a further determination must be made as to whether they are current carrying or not.

To correctly apply adjustment factors, each and every current-carrying conductor needs to be counted very accurately. Just as accurately, exempt conductors needed to be eliminated from the count.

Remember the spacing rule which permits raceways to adequately radiate some of their retained heat. Both **310.15(B)(3)(a)** and **310.15(B)(3)(b)** require physical spacing be provided and maintained between all raceways. There are no exceptions to these rules, but no dimension is specifically stated either.

1.5.3 Specific Requirements for Adjustment Factor Applications

The following are some of the adjustment factor requirements specific to **310.15(B)(3)(a)**:

1. Where the number of current-carrying conductors in a raceway or cable exceeds three, or where single conductor or multiconductor cables are installed without maintaining spacing for a continuous length longer than 24 inches and are not installed in raceways, the allowable ampacity of each conductor must be reduced by the factors of **Table 310.15(B)(3)(a)**.
2. Where current-carrying conductors are assembled as a set of parallel conductors, each conductor is counted separately.
3. Where conductors of different systems, such as control circuits and power circuits are installed in common raceways or cables, generally only the power and lighting conductors for branch circuits, feeders, and services need be counted.

See **366.23(A)** for adjustment factors for conductors in sheet metal auxiliary gutters and **376.22(B)** for adjustment factors for conductors in metal wireways.

Short sections of wireways may also be used as junction boxes to supplement raceway wiring methods.

The following adjustment factors are not required to be used, thus the standard *NEC* permissive text of **90.5(B)** "shall be permitted" is appropriate here. So the text reads: "The following adjustments shall be permitted for conductors and installation methods listed in **310.15(B)(3)(a)(1 through 5)**."

1. Where conductors are installed in cable trays, the provisions of **Section 392.80** shall apply.
2. Adjustment factors shall not apply to conductors in raceways having a length not exceeding 600 mm (24 in.).
3. Adjustment factors shall not apply to underground conductors entering or leaving an outdoor trench if those conductors have physical protection in the form of rigid metal conduit, intermediate metal conduit, rigid polyvinyl chloride conduit (PVC), or reinforced thermosetting resin conduit (RTRC) having a length not exceeding 3.05 m (10 ft), and if the number of conductors does not exceed four.
4. Adjustment factors shall not apply to certain Type AC cable or Type MC cable under the following conditions:
   a. The cables do not have an overall outer jacket.
   b. Each cable has not more than three current-carrying conductors.
   c. The conductors are 12 AWG copper.
   d. No more than 20 current-carrying conductors are installed without maintaining spacing, stacked, or are supported on "bridle rings."
5. An adjustment factor of 60% shall be applied to Type AC cable or Type MC cable under the following conditions:
   a. The cables do not have an overall outer jacket.
   b. The number of current-carrying conductors exceeds 20.
   c. The cables are stacked or bundled longer than 600 mm (24 in) without spacing being maintained.

1.5.4 Adjustment Factor for Circular Raceways Exposed to Sunlight on Rooftops

Where raceways and cables are exposed to direct sunlight on or above rooftops, the adjustments shown in **Table 310.15(B)(3)(c)** must be added to the outdoor ambient temperature to determine the applicable ambient temperature for application of the proper temperature correction factors of **Table 310.15(B)(2)(a)** or **Table 310.15(B)(2)(b)**. The informational note following the table points to only one possible source of average warmest outdoor ambient temperature. The average warmest outdoor ambient temperature for a particular geographical area is also available from many sources in weather almanacs as well as Internet-based weather sites. But, the final selected average warmest outdoor ambient temperature will need to meet with the approval of the authority having jurisdiction for the area in question, since no specific data appears in the *NEC*. New the 2014 edition of *Code Calculations*, type XHHW-2 conductors are not subject to these sunlight adjustment factors. This change permits the use of Type XHHW-2 copper or aluminum on a sunlight exposed roof top without further penalty.

1.5.5 Bare or Covered Conductors

Although bare and covered conductors have higher maximum operating temperatures than the insulated conductors of **Table 310.104(A)**, these conductors are not permitted to operate at a higher temperature than the adjacent insulated conductors. Bare and covered conductors operating at a higher temperature can cause irreparable harm to the insulated conductors if they are contained in the same raceway or cable. Therefore, the bare or covered conductor temperature may never exceed the lowest temperature rating of the adjacent insulated conductor for the purpose of determining ampacity.

1.5.6 Neutral Conductor

Both neutral conductor and neutral point are defined in **Article 100** of the *NEC*. The neutral conductor may count as a current-carrying conductor when applying the provisions of adjustment factors of **Table 310.15(B)(5)** according to the following statements:

1. A neutral that carries only the unbalanced current is not counted.
2. A neutral with two phases of a 3-phase system is counted.
3. Where the major portion of the load consists of nonlinear loads, harmonic currents are present in the neutral conductor of a 3-phase, 4-wire system. Therefore, the neutral conductor is counted.

1.5.7 Grounding and Bonding Conductors

Grounding and bonding conductors shall not be counted when applying the provisions of adjustment factors of **Table 310.15(B)(3)(a)**.

1.5.8 120/240-Volt, 3-Wire, Single-Phase Dwelling Services and Feeders

Conductor ampacities for 120/240-volt, single-phase dwelling services and feeders differ from those of **Table 310.15(B)(16)** and shall be permitted to be sized according to **Table 310.15(B)(7)**. But, if adjustment factors and correction factors have to be applied to these conductors, they are not required to be larger than the standard AWG sizes of **Table 310.15(B)(16)** allowed for service conductors. Finally, the grounded conductors for these services and feeders are permitted to be smaller than the ungrounded conductors according to the requirements of **Section 215.2**, **Section 220.61**, and **Section 230.42**.

1.6 Problems Using a Temperature Correction Factor

Any deviation from the 30°C ambient temperature designated for **Table 310.15(B)(16)** will cause a change in the allowable ampacity of the conductor. The correction factor for changing the allowable ampacity of the conductor is found in **Table 310.15(B)(2)(a)**. When the ambient temperature is hotter than 30°C or 85°F, the allowable ampacity of the conductor will decrease. When the ambient temperature is cooler than 30°C or 85°F, the allowable ampacity will increase. To use the temperature correction factor **Table 310.15(B)(2)(a)** properly, first determine the insulation Type letters, such as THW, THWN, etc. Knowing the Type letter, the user then proceeds to **Table 310.104(A)** to determine the temperature rating associated with that particular type of insulation. The temperature rating of a particular insulation is not affected by the type of metal conductor (copper or aluminum).

NOTE: The illustrated problems within this chapter are intended to show the basic allowable ampacity only. No other factors, such as overcurrent protection, are taken into consideration at this time.

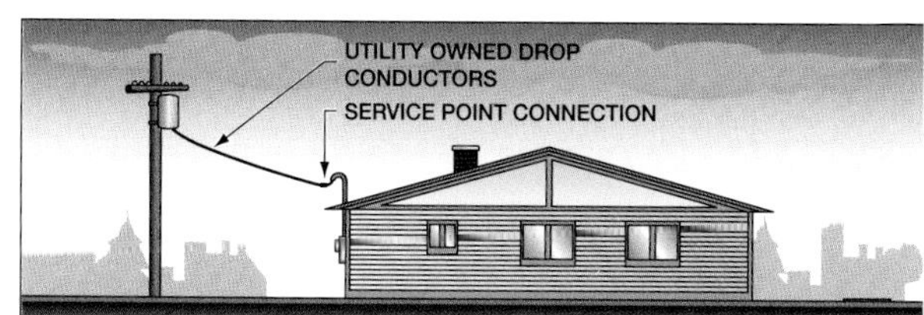

Utility owned drop conductors are not covered by the NEC according to 90.2(B)(5).

Problem 1-1

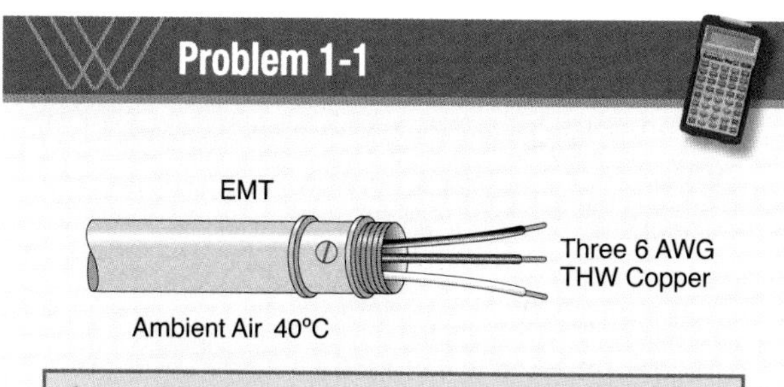

Three 6 AWG THW copper conductors are installed in electrical metallic tubing, Type EMT, in an ambient temperature of 40°C. What is the allowable ampacity of each of the three current-carrying conductors?

Solution

- Table 310.15(B)(16) Allowable Ampacity
  6 AWG THW copper = 65 amps
- Table 310.104(A), THW = 75°C
- Table 310.15(B)(2)(a) Temperature Correction Factors
  75°C (THW) @ 40°C ambient = 0.88

Ampacity at 40°C = Allowable Ampacity × Correction Factor
= 65 × 0.88
= 57.2

Answer: 57.2 amperes

Problem 1-2

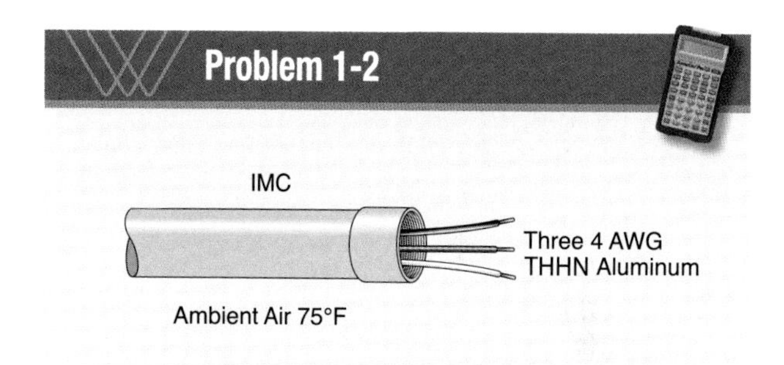

Three 4 AWG THHN aluminum conductors are installed in an intermediate metal conduit, Type IMC, in an ambient temperature of 75°F. What is the allowable ampacity of the current-carrying conductors?

Solution

Table 310.15(B)(16) Allowable Ampacity
4 AWG THHN aluminum = 75 amps
Table 310.15(B)(2)(a) Temperature Correction Factors
Table 310.104(A), THHN = 90°C
90°C (THHN) @ 75°F ambient = 1.04
75 amps × 1.04 = 78 amps
Answer: 78 amperes

NOTE: As the user becomes familiar with insulation temperature ratings, proceeding to **Table 310.104(A)** may become an unnecessary step. Simply using the column headings of **Table 310.15(B)(16)** to determine insulation temperature rating is also effective.

Problem 1-3

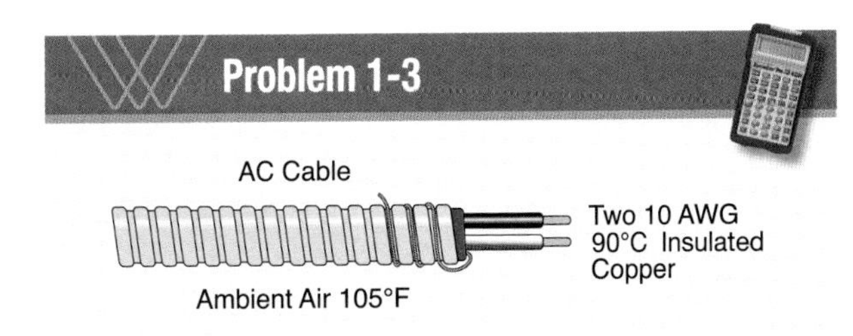

A 2-conductor armored cable, Type AC, with 10 AWG 90°C insulated copper conductors, is installed in open bar joist areas having an ambient temperature of 105°F. What is the allowable ampacity of the current-carrying conductors?

Solution
Table 310.15(B)(16) Allowable Ampacity
10 AWG 90°C insulated copper = 40 amps
Table 310.15(B)(2)(a) Temperature Correction Factors
90°C insulation @ 105°F ambient = 0.87
40 amps x 0.87 = 34.8 amps
Answer: 34.8 amperes

Problem 1-4

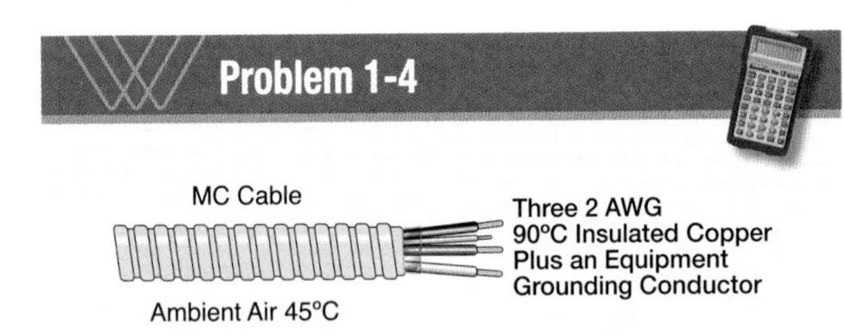

A 3-conductor metal-clad cable (with ground), Type MC, is installed in an area with an ambient temperature of 45°C. The conductors in the cable are 2 AWG copper with 90°C insulation. What is the allowable ampacity of the current-carrying conductors?

Solution
Table 310.15(B)(16) Allowable Ampacity
2 AWG copper with 90°C insulation = 130 amps
Table 310.15(B)(2)(a) Temperature Correction Factors
90°C conductor @ 45°C ambient = 0.87
130 amps x 0.87 = 113.1 amps
Answer: 113.1 amperes

For additional information, visit qr.njatcdb.org Item #1024

Courtesy of AFC Cable Systems®

Two examples of Type AC cable, 3-conductor with armor/bond wire ground.

Rigid Metal Conduit, Type RMC

1.7 Problems Using an Adjustment Factor

The title of the allowable ampacity **Table 310.15(B)(16)** indicates that the table applies only where not more than three current-carrying conductors are installed in any one raceway or cable. If there are more than three conductors in a raceway or cable, the allowable ampacity changes. The heat problem caused by adjacent conductors comes into play very prominently when there are many current-carrying conductors in the same raceway. **Table 310.15(B)(3)(a) Adjustment Factors for More Than Three Current-Carrying Conductors** gives the adjustment factors as a percent and is to be used to reduce the allowable ampacity listed in **Table 310.15(B)(16)** through **Table 310.15(B)(19)**.

The basic application of the adjustment factor table will be illustrated first. Some other specific factors, listed later in the subsection, will be looked at individually as they also affect the ampacity of the conductors.

Problem 1-5

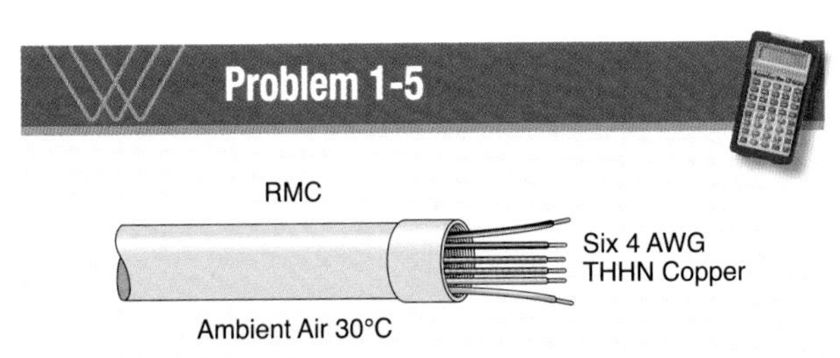

Two 3-phase motors are fed with six 4 AWG THHN copper conductors in the same rigid metal conduit, Type RMC, in an ambient temperature of 30°C. What is the allowable ampacity of each current-carrying conductor?

Solution
Table 310.15(B)(16) Allowable Ampacity
4 AWG THHN copper = 95 amps
Table 310.15(B)(3)(a) Adjustment Factors
6 conductors = 80%
95 x 0.80 = 76 amps
Answer: 76 amperes

Problem 1-6

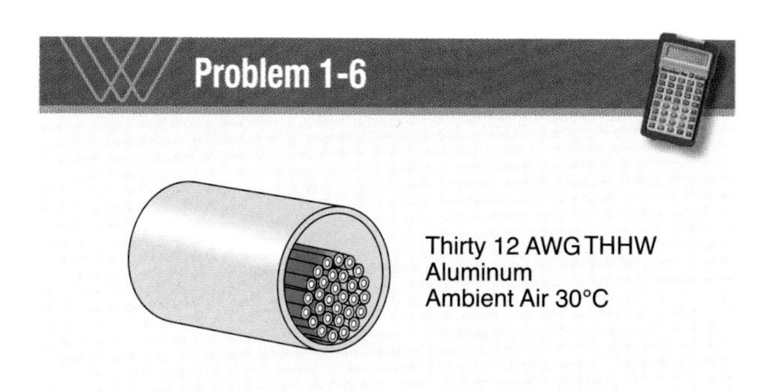

Thirty 12 AWG THHW aluminum current-carrying conductors are installed in the same conduit in a wet location area with a 30°C ambient temperature. What is the allowable ampacity of each conductor?

Solution
Table 310.104(A)
Type THHW insulation used in a wet location is limited to 75°C
Table 310.15(B)(16) Allowable Ampacity
12 AWG THHW aluminum at 75°C = 20 amps
Table 310.15(B)(3)(a) Adjustment Factors
30 conductors = 45%
20 amps × 0.45 = 9 amps
Answer: 9 amperes

Problem 1-7

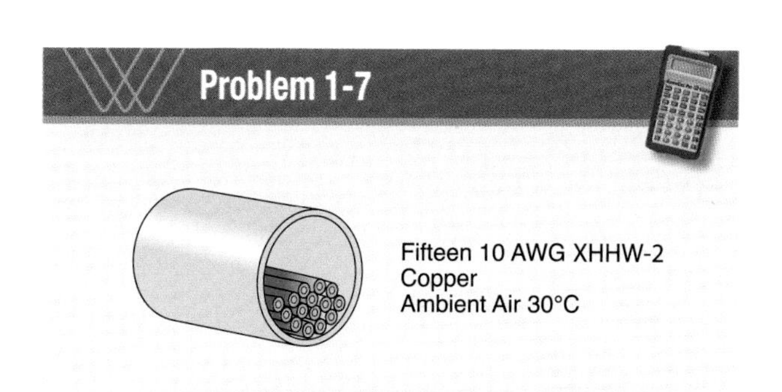

Fifteen 10 AWG XHHW-2 copper current-carrying conductors are installed in the same conduit in an area with an ambient temperature of 30°C. What is the allowable ampacity of each conductor?

Solution
Table 310.15(B)(16) Allowable Ampacity
10 AWG XHHW-2 copper = 40 amps
Table 310.15(B)(3)(a) Adjustment Factors
15 conductors = 50%
40 amps × 0.50 = 20 amps
Answer: 20 amperes

For the 2014 *NEC*, additional text was added to footnote No. 1 of **Table 310.15(B)(3)(a), Adjustment Factors for More Than Three Current-Carrying Conductors.** This revised text now requires "spare conductors" to be added to the count of current-carrying conductors, because at some time in the future, these conductors could become current-carrying conductors. In addition, the revised footnote excludes counting conductors that are not simultaneously energized. Finally, no other revisions were made to past methods of counting neutral conductors and of not counting grounding and bonding conductors.

Before continuing, refer back to **310.15 Ampacities for Conductors Rated 0-2000 Volts** reading again (A) and (B) toward an understanding that this text contains the practical Physics of electrical power wiring and are of the utmost importance to electrical safety.

1.7.1 Ampacity Adjustment Factors Using Different Sized Conductors

For an installation in which different sized conductors are installed in the same raceway, **Table 310.15(B)(3)(a) Adjustment Factors** is also applicable.

Problem 1-8

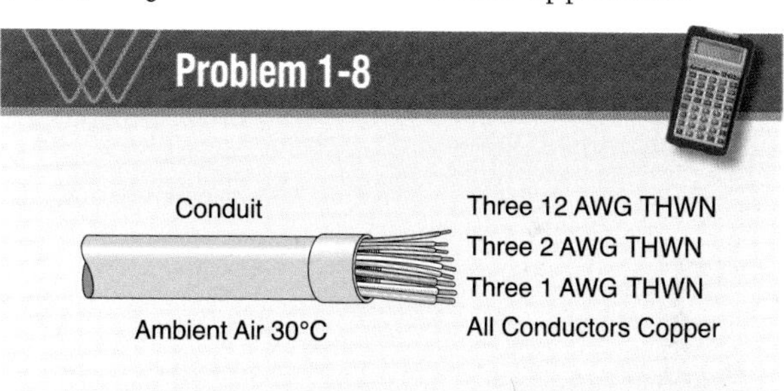

Three 2 AWG THWN, three 1 AWG THWN, and three 12 AWG THWN copper current-carrying conductors are installed in the same conduit in an area with an ambient temperature of 30°C. What is the allowable ampacity of the 2 AWG, the 1 AWG, and the 12 AWG conductors?

Solution
Number of conductors 3 + 3 + 3 = 9
Table 310.15(B)(3)(a) Adjustment Factors
9 conductors = 70%
Calculate conductors individually
Table 310.15(B)(16) Allowable Ampacity
2 AWG THWN copper; 115 amps × 0.70 = 80.5 amps
1 AWG THWN copper; 130 amps × 0.70 = 91.0 amps
12 AWG THWN copper; 25 amps × 0.70 = 17.5 amps
Answers: 2 THWN AWG = 80.5 amperes
1 THWN AWG = 91.0 amperes
12 THWN AWG = 17.5 amperes

1.7.2 Ampacity Adjustment Factors Using Different Insulation Temperatures and Different Sized Conductors

310.15(B)(1) General points to other sections of the *Code* which apply to conductors and the insulation placed on the conductors. In particular, **Section 310.15(A)(3)** sets the requirements for the temperature limits for conductors. This section stipulates that no conductor can be installed such that its operating temperature exceeds that designated for the type of insulation involved. As pointed out previously, where more than three current-carrying conductors are installed in the same raceway, **Table 310.15(B)(3)(a)** is also applicable.

The following problem points out the potential danger in placing conductors with different temperature ratings within the same raceway. Where mixed, the ampacity of the higher-temperature conductors may have to be lowered in the event that the higher-temperature conductor overheats (and presents the risk of damaging) the adjacent lower-temperature conductors.

IMC Thread Protector Caps

Color	Sizes	Examples
Orange	Inch sizes	1", 2", 3", 4"
Yellow	½" sizes	½", 1½", 2½", 3½"
Green	¼" sizes	¾", 1¼"

Understanding Insulation Abbreviations

T	Thermoplastic
R	Thermoset (previously rubber)
X	Thermoset (crossed linked)
S	Silicone
H	Heat resistant 75°C
HH	High Heat resistant 90°C
W	Water (moisture) resistant
N	Nylon jacket
MI	Mineral Insulated
MTW	Machine tool wiring
-2	90°C dry or wet location

Example: THHW translates to thermoplastic, high heat resistant 90°C, and water resistant

Rigid Thread Protector Caps

Color	Sizes	Examples
Blue	Inch sizes	1", 2", 3", 4", 5", 6"
Black	½" sizes	½", 1½", 2½", 3½"
Red	¼" sizes	¾", 1¼"

Problem 1-9

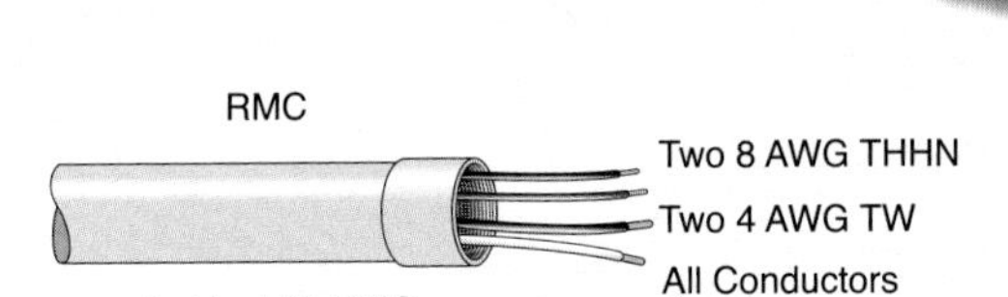

Two 8 AWG THHN and two 4 AWG TW copper current-carrying conductors are installed in the same rigid metal conduit, Type RMC, in an area with an ambient temperature of 30°C. Find the allowable ampacities of the 8 AWG THHN and the 4 AWG TW conductors.

Solution
Number of conductors 2 + 2 = 4
Table 310.15(B)(3)(a) Adjustment Factors
4 conductors = 80%
Table 310.15(B)(16) Allowable Ampacity
4 AWG TW (using the 60°C column) = 70 amps;
70 amps × 0.80 = 56 amps
8 AWG THHN (using the 90°C column) = 55 amps;
55 amps × 0.80 = 44 amps
Check 44 amps against 60°C column of Table 310.15(B)(16).
The 8 AWG (at 44 amps) exceeds the 60°C column maximum ampacity of 40 amps.
The allowable ampacity of 8 AWG must not exceed that of the 60°C column.
Therefore, the 8 AWG THHN is limited to a maximum of 40 amps.
Answers: 4 AWG TW = 56 amperes
8 AWG THHN = 40 amperes

Comment
Problem 1-9 contains two situations which require adjustment. First, there are more than three current-carrying conductors in a raceway, so each pair of example conductors must be adjusted to 80% of their table value. Second, since adjacent wires in a single conduit must not be subject to overheating, the lowest temperature insulation rating (60°C column ampacity from Table 310.15(B)(16)) must be applied to all conductors within a single conduit for Problem 1-9. Explaining it another way, since the most fragile insulation is the TW, the ampacity of THHN conductors must be reduced to the maximum operating temperature permitted for TW insulation, that is, the ampacity permitted in Table 310.15(B)(16), using the 60°C column for copper conductors.

1.7.3 Applying Both Correction Factors and Adjustment Factors

The phrase used at the top of percentage column of **Table 310.15(B)(3)(a)** "...as Adjusted for Ambient Temperature if Necessary" indicates that both the (temperature) correction factor and the (number of current-carrying conductors in a raceway or cable) adjustment factor must be applied whenever both are present. The following problems combine the ambient temperature correction factor and the adjustment factor.

Problem 1-10

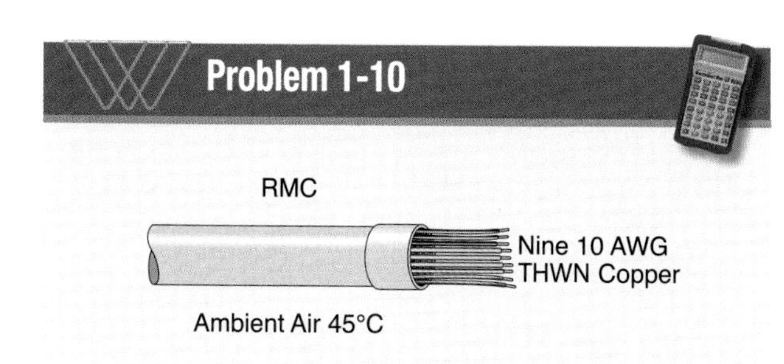

Nine 10 AWG THWN copper current-carrying conductors are installed in a 1-in. conduit, Type RMC, in an area with an ambient temperature of 45°C. What is the allowable ampacity of each conductor?

Solution
Table 310.15(B)(16) Allowable Ampacity
10 AWG THWN 75°C copper = 35 amps
Table 310.15(B)(2)(a) Correction Factors
75°C conductor in a 45°C ambient = 0.82
Table 310.15(B)(3)(a) Adjustment Factors
9 current-carrying conductors = 70%
35 amps x 0.82 x 0.70 = 20.09 amps
Answer: 20.09 amperes

Conduit installation within an electrical room.

Problem 1-11

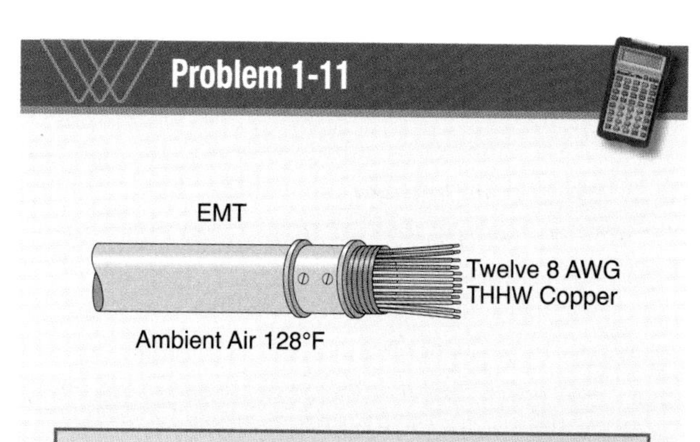

Twelve 8 AWG THHW copper current-carrying conductors are installed in a 2-in. conduit, Type EMT, located in a dry location with an ambient temperature area of 128°F. What is the allowable ampacity of each conductor?

Solution
Table 310.15(B)(16) Allowable Ampacity
8 AWG THHW 90°C copper = 55 amps
Table 310.15(B)(2)(a) Correction Factors
90°C conductor @ 128°F ambient = 0.76
Table 310.15(B)(3)(a) Adjustment Factors
12 conductors = 50%
55 amps x 0.76 x 0.50 = 20.9 amps
Answer: 20.9 amperes

Problem 1-12

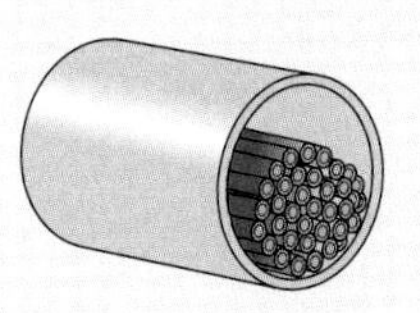

Thirty-six 12 AWG THHN copper current-carrying conductors are installed in the same raceway located in an ambient temperature of 50°C. What is the allowable ampacity of each conductor?

Solution
Table 310.15(B)(16) Allowable Ampacity
12 AWG THHN 90°C copper = 30 amps
Table 310.15(B)(2)(a) Correction Factors
90°C conductor @ 50°C ambient = 0.82
Table 310.15(B)(3)(a) Adjustment Factors
36 conductors = 40%
30 amps x 0.82 x 0.40 = 9.84 amps
Answer: 9.84 amperes

Comment
Installations with this many current-carrying conductors in a raceway often waste a significant amount of copper and should be avoided wherever possible.

1.7.4 Applying Adjustment and Temperature Correction Factors to Cable Assemblies

Up to now, only conductors installed in raceways have been considered. The allowable ampacity tables, the ambient temperature correction factors and the adjustment factors also apply to the installation of cable assemblies. The following are examples of cable assembly installations.

Problem 1-13

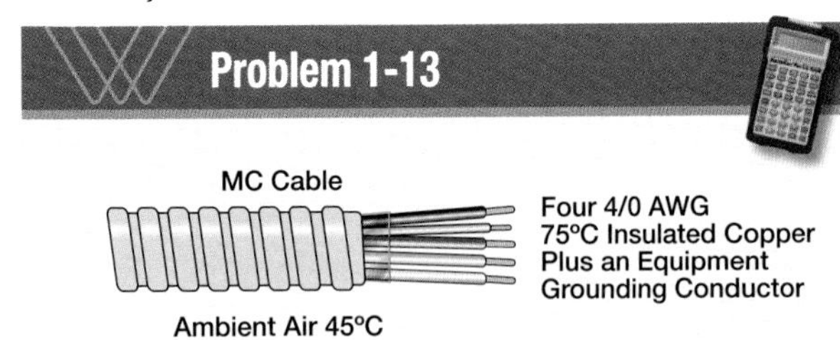

A 4-conductor, metal-clad cable is used as the feeder for a 120/208-volt panel supplying discharge lighting. The 4/0 AWG copper conductors have 75°C insulation and the cable is installed in an area with an ambient temperature of 45°C. What is the allowable ampacity of each 4/0 AWG conductor? (The discharge lighting requires all four conductors to be counted as current-carrying conductors since the neutral conductor is considered a current-carrying conductor in accordance with 310.15(B)(5)(c).

Solution
Table 310.15(B)(16) Allowable Ampacity
4/0 AWG @ 75°C copper = 230 amps
Table 310.15(B)(2)(a) Correction Factors
75°C conductor in 45°C ambient = 0.82
Table 310.15(B)(3)(a) Adjustment Factors
4 current-carrying conductors = 80%
230 amps × 0.82 × 0.80 = 150.88 amps
Answer: 150.88 amperes

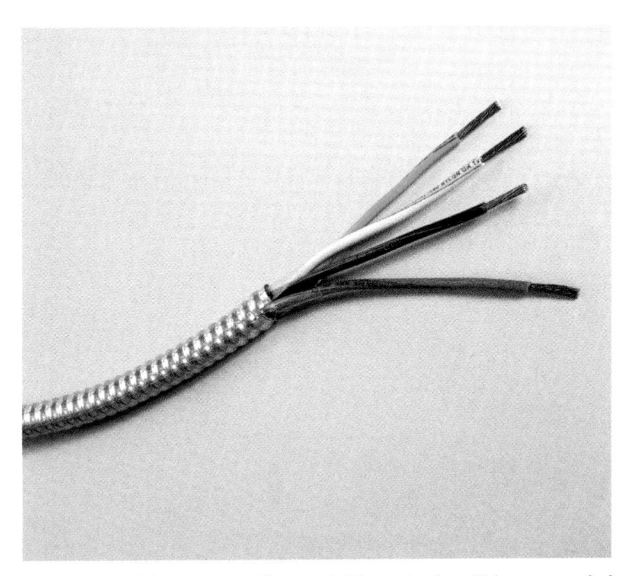

Type MC cable is generally available in both solid or stranded conductors in sizes 12 and 10 AWG.

Problem 1-14

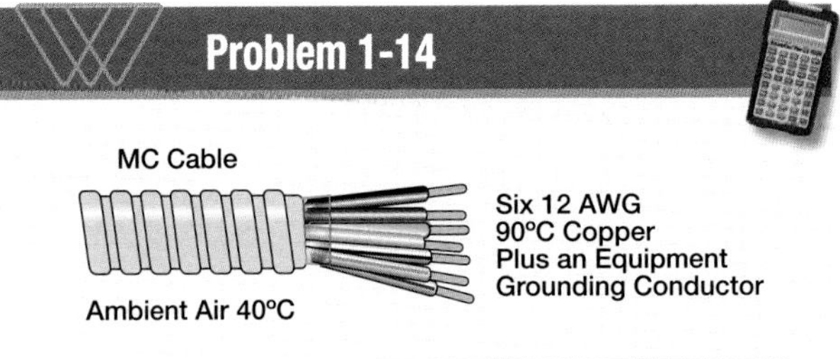

A 6-conductor (three circuit), Type MC, metal-clad cable is used to supply three 20-ampere fluorescent lighting circuits within an office environment. The cable is comprised of three ungrounded conductors, three grounded conductors, and a green equipment grounding conductor. The 12 AWG copper conductors have 90°C insulation, and the cable is installed in an area with an ambient temperature of 40°C. What is the allowable ampacity of each 12 AWG conductor?

Solution
Table 310.15(B)(16) Allowable Ampacity
12 AWG @ 90°C copper = 30 amps
Table 310.15(B)(2)(a) Correction Factors
90°C conductor in 40°C ambient = 0.91
Table 310.15(B)(2)(a) Adjustment Factors
6 current-carrying conductors = 80%
30 amps × 0.91 × 0.80 = 21.84 amps
Answer: 21.84 amperes

Comment
This is a practical example which will be used again in Chapter 2 to determine a circuit final ampacity.

Problem 1-15

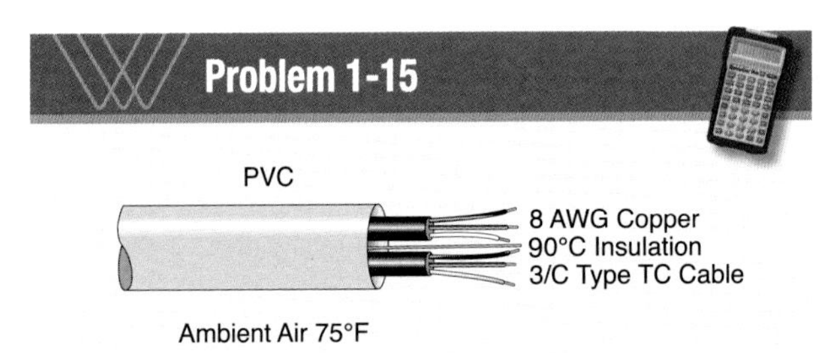

Two 3-conductor, Type TC cables are installed in a rigid PVC conduit in an area with an ambient temperature of 75°F. The conductors are 8 AWG copper with 90°C insulation. What is the allowable ampacity of each current-carrying conductor?

Solution
Table 310.15(B)(16) Allowable Ampacity
8 AWG copper @ 90°C = 55 amps
Table 310.15(B)(2)(a) Correction Factors
90°C copper in 75°F ambient = 1.04
2 cables × 3 = 6 conductors
Table 310.15(B)(3)(a) Adjustment Factors
6 conductors = 80%
55 amps × 1.04 × 0.80 = 45.76 amps
Answer: 45.76 amperes

1.7.5 Applying Adjustment Factors

There is significant text contained in **310.15(B)(3)(a) Adjustment Factors for More Than Three Current-Carrying Conductors.** The majority of these requirements are based on actual circuit properties and how heat is generated, dispersed and not permitted to be retained within the circuit conductors, raceways or cables or within surrounding circuits or raceways. Previously, fundamental methods of applying adjustment factors to correct circuit ampacity were presented. Also reviewed were spare conductors and neutral conductors. Within a raceway or cable, the actual conductor count may be adjusted or lowered by the number of neutral conductors which carry only the unbalanced current. Also the actual count may be reduced by the number of grounding and bonding conductors as well. Generally, the actual power and lighting conductors (and the spare if any) are considered current-carrying conductors. Other conductors could be control or signal conductors which carry small amounts of current and that may only last for short periods of time. Therefore, because they do not continuously carry current, they add little, if any, heat to the enclosure and therefore are not counted as current-carrying conductors where **Table 310.15(B)(3)(a)** is applied.

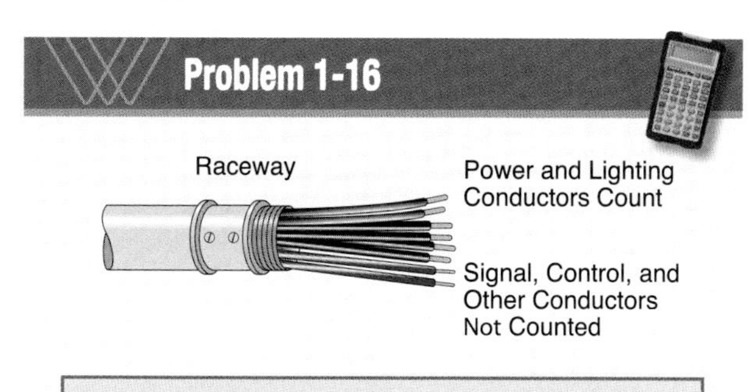

Two sets of 3-phase, 3-wire circuits, each comprised of 4 AWG THWN copper motor branch-circuit conductors, are installed in the same conduit with two 14 AWG THWN copper control circuit conductors in an ambient temperature of 30°C. What is the allowable ampacity of only the six 4 AWG THWN copper motor branch-circuit conductors?

Solution
Table 310.15(B)(16) Allowable Ampacity
4 AWG THWN copper = 85 amps
Number of conductors 6 + 2 = 8
Control circuit conductors do not count
Number of current-carrying conductors = 6
Table 310.15(B)(3)(a) Adjustment Factors
6 conductors = 80%
85 amps x 0.80 = 68 amps
Answer: 68 amperes

1.7.6 Applying Adjustment Factors to Cables in Cable Tray

310.15(B)(3)(a)(1) refers to **392.80(A)**, which covers the allowable ampacity of multiple conductor cables in cable tray. This section applies to multiconductor cables and requires the application of **Table 310.15(B)(3)(a)** to multiple conductor cables with more than three current-carrying conductors installed in cable trays. The adjustment factor applies to each individual cable as if it were in a separate raceway and not to all of the conductors collectively contained within the tray. This method of calculation applies only where cables are installed without maintained spacing of one cable diameter between cables. For cables installed with maintained spacing of one cable diameter between cables in a single layer in uncovered tray, the ampacity calculations for cables are much different. See **392.80(A)(1)(c)** for more details.

Photovoltaic (PV) systems are often installed on rooftops of dwelling and commercial structures.

Problem 1-17

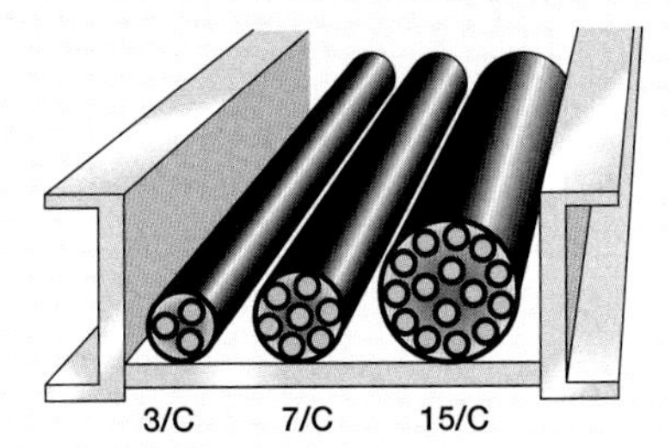

A 3-conductor, a 7-conductor, and a 15-conductor Type TC cable are all installed in a single uncovered cable tray without properly maintained cable diameter spacing. The installation area has an ambient temperature of 30°C. All of the individual conductors within the Type TC cables are 10 AWG 75°C insulated copper. What is the allowable ampacity of each current-carrying conductor?

Solution - Calculation 1
The 3/C cable using 75°C copper conductors
Table 310.15(B)(16) Allowable Ampacity
10 AWG 75°C insulated copper = 35 amps
392.80(A)(1)(a); adjustment factors do not apply
No adjustment necessary
3/Conductor 10 AWG 75°C copper = 35 amps
Answer: 3/C = 35 amperes

Solution - Calculation 2
The 7/C cable
Table 310.15(B)(16) Allowable Ampacity
10 AWG 75°C insulated copper = 35 amps
392.80(A)(1)(a) and Table 310.15(B)(3)(a)
Adjustment factor for 7 conductors = 70%
35 amps × 0.70 = 24.5 amps
Answer: 7/C = 24.5 amperes

Solution - Calculation 3
The 15/C cable
Table 310.15(B)(16) Allowable Ampacity
10 AWG 75°C insulated copper = 35 amps
392.80(A)(1)(a) and Table 310.15(B)(3)(a)
Adjustment factor for 15 conductors = 50%
35 amps × 0.50 = 17.5 amps
Answer: 15/C = 17.5 amperes

Cable tray is used to distribute multiconductor cables throughout a plant.

1.7.7 Adjustment Factors to Current-Carrying Conductors in Limited Length Raceways

310.15(B)(3)(a)(2) permits the installation of conductors in a raceway nipple, provided it is not over 24 inches in length, without applying any adjustment factor as shown in **Figure 1-2**.

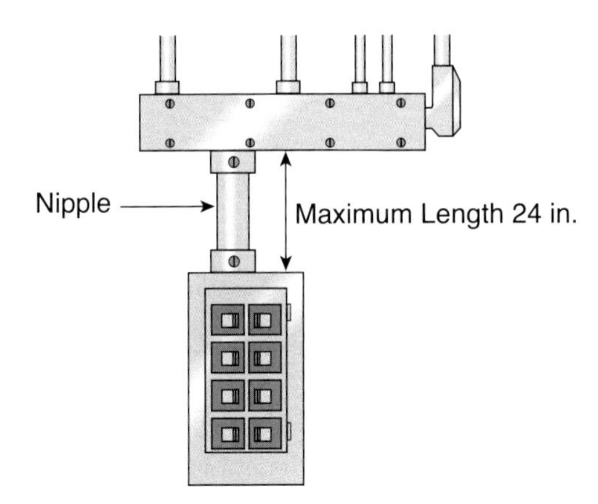

Figure 1-2. This 24 in. length of conduit or tubing is exempt from the adjustment factor requirements.

310.15(B)(3)(a)(3) permits the installation of conductors in a conduit coming out of the ground, without applying any adjustment factor, provided the rigid metal conduit, intermediate metal conduit, or rigid PVC conduit Schedule 80 is not over 10 feet long including above- and below-ground portions of the conduit. An application of this permission is shown in **Figure 1-3**.

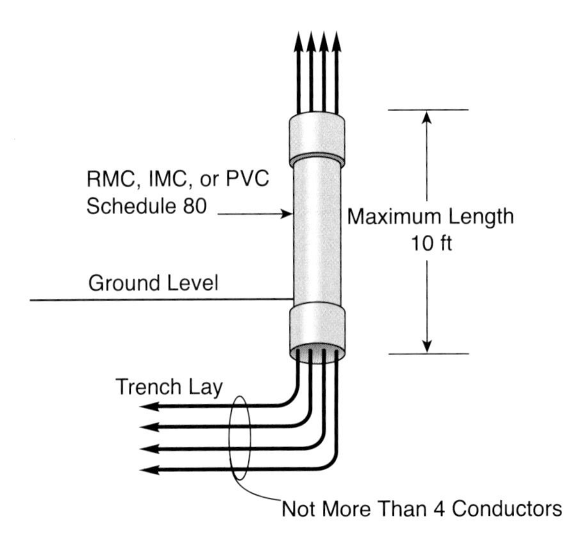

Figure 1-3. This 10 ft. length of conduit is exempt from the adjustment factor requirements.

Cable tray is used to gather communications cables inside an equipment room.

310.15(B)(3)(a)(4) is best understood as two separate parts of an exception.

First is the permission to bundle multiple MC or AC cables sized as 2-conductor 12 AWG (plus ground) or 3-conductor 12 AWG (plus ground) without applying any adjustment factor, provided the total number of current-carrying conductors in the bundle does not exceed 20.

Second, if the bundle of MC or AC cables exceeds 20 current-carrying conductors, a single adjustment factor of 60% must be applied to each current-carrying conductor within the bundle of cables.

Both the first part and the second part make the assumption that the neutral conductor is counted as a current-carrying conductor. For applications where the neutral conductor is not counted as a current-carrying conductor, in accordance with **310.15(B)(5)(a)**, 4-conductor 12 AWG (plus ground) Type MC or Type AC cables could be used in either of the two examples explaining **310.15(B)(3)(a)(4)**.

1.8 Allowable Ampacity of Bare Conductors

The main rule of **Section 230.41** requires service-entrance conductors to be insulated. The exception to this section permits the grounded conductor to be uninsulated or bare. For the purpose of determining ampacity, **310.15(B)(4)** requires that bare and covered conductors installed with insulated conductors have temperature ratings equal to the lowest temperature rating of the (adjacent) insulated conductors. The easiest way to determine the allowable ampacity of the bare or covered conductor where this mixture occurs is to consider the bare or covered conductor to have the same insulation as the ungrounded conductors and read the allowable ampacity directly from the table. This will ensure temperature compatibility of all of the conductors within the raceway and will prevent insulation damage to adjacent conductors.

Problem 1-18

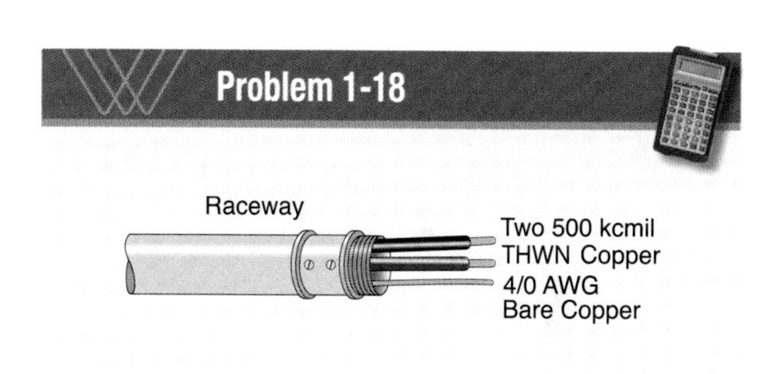

What is the allowable ampacity of a 4/0 AWG bare copper grounded conductor run in the same raceway with two 500-kcmil THWN copper service conductors?

Solution
Table 310.15(B)(16) Allowable Ampacity
4/0 AWG THWN copper = 230 amps
4/0 AWG bare copper = 230 amps
Answer: 230 amperes

Wire Facts:

The most common form of wire stranding is a concentric-stranded conductor. Each layer of this stranding (after the single initial core strand) has six more strands added. And each layer is applied in a direction opposite that of the layer under it. See the *NEC* Chapter 9, Table 10 for more information on wire stranding.

1.9 The Neutral Conductor and the Ampacity Adjustment Factor

310.15(B)(5) takes into consideration the three applications of a neutral conductor. Notice that when the neutral conductor is a current-carrying conductor in the following illustrations, it counts when considering the application of the adjustment factors of **Table 310.15(B)(3)(a)**. When the neutral conductor is not considered a current-carrying conductor, it is not counted.

310.15(B)(5)(a) covers the first application. The neutral conductor is treated as a noncurrent-carrying conductor and is not counted toward the adjustment factor. The neutral conductor of a 120/208-volt system supplying incandescent lighting will carry only the unbalanced current, and should the load be perfectly balanced, theoretically it will carry no current. Multiwire branch circuits, with a neutral conductor supplying only resistive loads (such as incandescent lighting and resistive heating), are circuits where the neutral conductor is not counted.

NOTE: The following illustrated problems are intended show the basic allowable ampacity only. No other factors, such as overcurrent protection, are taken into consideration at this time.

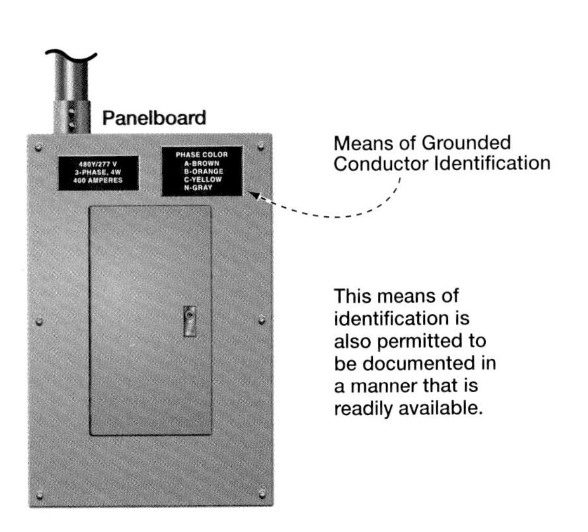

The means of identifying the grounded conductor must be documented in a manner that is readily available or permanently posted where conductors of different systems originate.

Problem 1-19

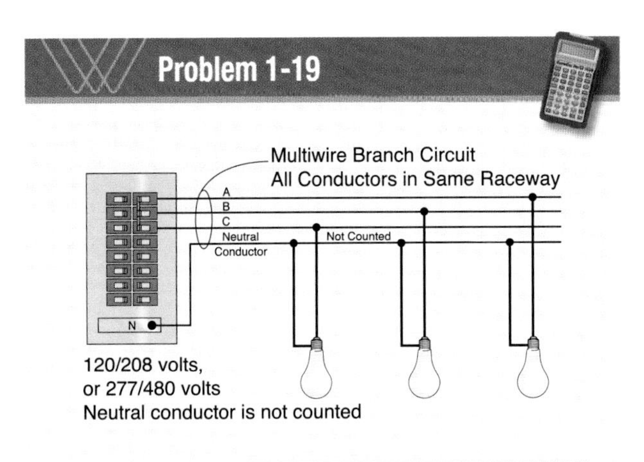

A 3-phase, 4-wire, multiwire branch circuit is installed using 12 AWG THWN copper conductors in Type EMT as the branch circuits for incandescent lighting in an area of 30°C ambient temperature. What is the allowable ampacity rating of each current-carrying conductor?

Solution
Table 310.16 Allowable Ampacity
12 AWG THWN = 25 amps
Total number of conductors = 4
310.15(B)(5)(a), neutral conductor is not counted
Number of current-carrying conductors = 3
Adjustment factor is not applicable
Answer: 25 amperes

Problem 1-20

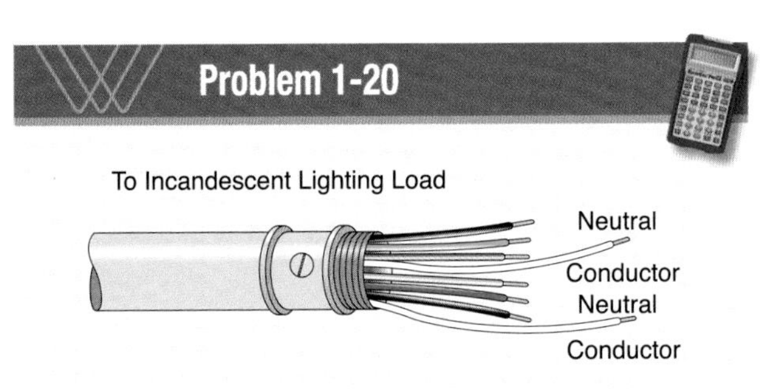

Two 3-phase, 4-wire, 120/208-volt multiwire branch circuits are installed using 14 AWG THWN copper conductors in Type EMT as the branch circuits for incandescent lighting in an area with an ambient temperature of 30°C. What is the allowable ampacity of the current-carrying conductors?

Solution
Table 310.15(B)(16) Allowable Ampacity
14 AWG THWN copper = 20 amps
Total number of conductors = 8
310.15(B)(5)(a), neutral is not counted
Number of current-carrying conductors = 6
Table 310.15(B)(3)(a) Adjustment Factors
6 conductors = 80%
20 amps × 0.80 = 16 amps
Answer: 16 amperes

Information

310.15(B)(5)(b) covers the second application. This takes into consideration the installation of a neutral conductor with only two of the phase conductors of a 3-phase, 4-wire system. An example may be a 1-phase, 3-wire feeder to an apartment or condominium within a high-rise multifamily complex. In this situation, the neutral conductor will carry approximately the same current as the phase conductors and is counted as a current-carrying conductor.

Problem 1-21

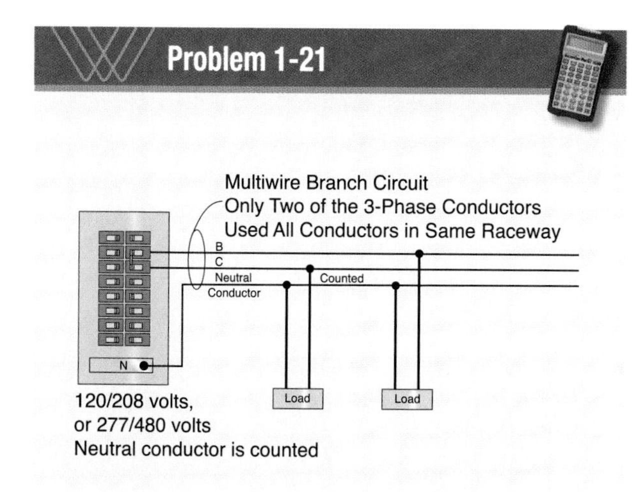

Two phase "B" conductors, two phase "C" conductors, and two neutral conductors of a 120/208-volt system are installed in the same rigid metal conduit in an area with an ambient temperature of 30°C. The conductors are 10 AWG THHN aluminum. What is the allowable ampacity of each current-carrying conductor?

(For clarity, only one 3-wire multiwire branch circuit is shown in the graphic.)

Solution
Table 310.15(B)(16) Allowable Ampacity
10 AWG THHN Aluminum = 35 amps
310.15(B)(5)(b)
All 6 conductors are current-carrying conductors
Table 310.15(B)(3)(a) Adjustment Factors
6 conductors = 80%
35 amps × 0.80 = 28 amps
Answer: 28 amperes

Information

310.15(B)(5)(c) covers the third application. This section requires the neutral conductor to be counted where the major portion of the load consists of nonlinear loads. Nonlinear loads cause harmonic currents in the neutral conductor. Electric discharge lighting and data processing equipment are examples of nonlinear loads which may cause harmonic currents. Due to harmonic currents, the neutral conductor will often carry as much or sometimes even more current than the line conductors. Therefore, the neutral conductors are counted as current-carrying conductors even where the load is balanced.

Problem 1-22

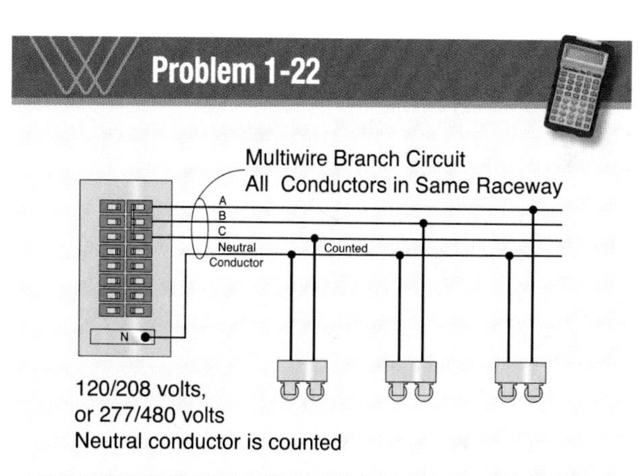

A 277/480-volt, 3-phase, 4-wire multiwire branch circuit supplies a fluorescent lighting load. The branch circuit is installed in Type EMT using four 12 AWG THHN copper conductors in an area with an ambient temperature of 90°F. What is the allowable ampacity of each of the current-carrying conductors?

Solution
Table 310.15(B)(16) Allowable Ampacity
12 AWG THHN copper = 30 amps
Table 310.15(B)(2)(a) Correction Factors
90°C insulation in 90°F ambient = 0.96
Table 310.15(B)(2)(a) Adjustment Factors
All conductors count
4 conductors = 80%
30 amps × 0.96 × 0.80 = 23.04 amps
Answer: 23.04 amperes

Harmonics and their related distortion are considered by many to be the most significant power quality problem today. However, the lighting industry in general has, over the past 20 years, made substantial progress in reducing the harmonics present on the fluorescent lighting circuits within buildings. Recently installed installations measure less than 5% total harmonic distortion (THD) most of the time and a great many measure less than 3% THD.

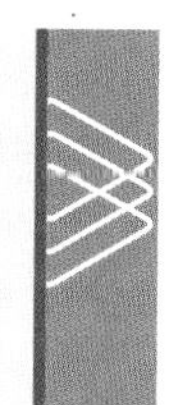

1.10 Counting Equipment Grounding and Bonding Conductors

310.15(B)(6) covers the installation of the equipment grounding and bonding conductors. The equipment grounding and bonding conductors do not carry current during normal operation, but rather carry current only during fault conditions. Therefore, they are not counted during the application of adjustment factors according to Footnote 1 of **Table 310.15(B)(3)(a)**.

Problem 1-23

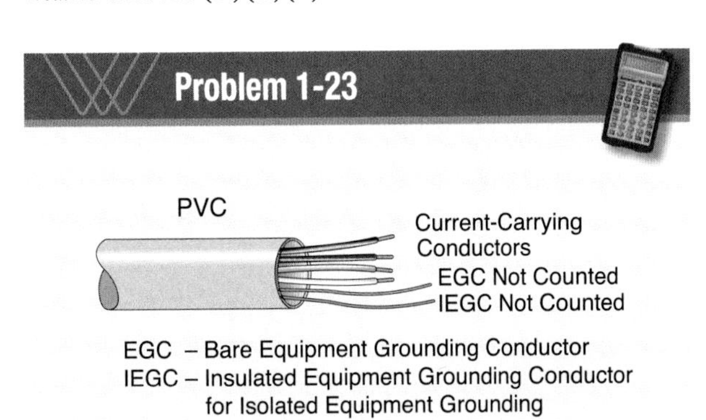

A 3-phase, 120/208-volt, 4-wire multiwire branch circuit supplies data processing equipment using isolated ground, Type IG, 120-volt receptacles in an ambient area of 30°C. The rigid conduit contains four 12 AWG THHN copper branch-circuit conductors, one bare equipment grounding conductor, and one insulated equipment grounding conductor. What is the ampacity of the branch-circuit conductors?

Solution

Table 310.15(B)(16) Allowable Ampacity
12 AWG THHN copper = 30 amps
Total number of conductors = 6
All three ungrounded conductors count
The load is data processing equipment; one neutral conductor counts
Neither of the two equipment grounding conductors counts
Number of current-carrying conductors = 4
Table 310.15(B)(3)(a) Adjustment Factors
4 conductors = 80%
30 amps x 0.80 = 24 amps
Answer: 24 amperes

1.11 Using 310.15(B)(7) for Service and Feeder Conductor Ampacity.

Although not stated as such, the newly revised section **310.15(B)(7)** is still sort of an exception to the ampacity table **310.15(B)(16)**. Under specific conditions, this section permits an ampacity of 83 percent of the ampacity requirements of **Table 310.15(B)(16)**. For the 2014 *NEC*, previous (2011 *NEC*) **Table 310.15(B)(7)** has been removed and in its place are more descriptive requirements and permissions to continue to use ampacity values of the past by simply using a multiplier not less than 0.83 for sizing 120/240-volt, single-phase dwelling services and feeders for one-family dwellings and the individual dwelling units of two-family and multifamily dwellings.

First, for a service rated 100 through 400 A, the service conductors, supplying the entire load associated with a one-family dwelling, or supplying the entire load associated with an individual dwelling unit in a two-family or multifamily dwelling, are permitted to have an ampacity not less than 83 percent of the service rating.

Second, for a feeder rated 100 through 400 A, the feeder conductors, supplying the entire load associated with a one family dwelling, or supplying the entire load associated with an individual dwelling, unit in a two-family or multifamily dwelling, are permitted to have an ampacity not less than 83 percent of the feeder rating.

Additionally, in no case could a feeder for an individual dwelling unit be required to have an ampacity greater than that specified in **310.15(B)(7)(1)** or **(2)**. Also, the permission to size grounded conductors smaller than the ungrounded conductors remains, provided that the requirements of **220.61** and **230.42** for service conductors or the

For the 2014 *NEC*, CMP-4 addressed the many proposals to change **310.15(B)(7)**, by analyzing the existing text. It was determined that the conductor sizes from the 2011 NEC **Table 310.15(B)(7)** are equivalent in all respects to those that would be used if a 0.83 multiplier were applied. Using this multiplier now results in the exact same wire size for a given ampacity as if **Table 310.15(B)(7)** were being applied.

requirements of **215.2** and **220.61** for feeder conductors are met. None of these permissive statements remove the requirements to apply ampacity correction or adjustment factors applicable to conductor installation(s) if necessary. Finally after all this discussion, it is important to point out that the grounded conductor is never required to be reduced, the NEC simply permits it in certain cases..

In addition, and where specifically permitted elsewhere in the *Code*, the grounded conductor is permitted to be smaller than the line conductors (ungrounded conductors). For these few examples using **310.15(B)(5)**, the grounded conductor could always be selected to equal the size of the ungrounded conductor. As shown in the next few examples, sometimes the calculated load of the grounded conductor may be less than the calculated load of the ungrounded conductor.

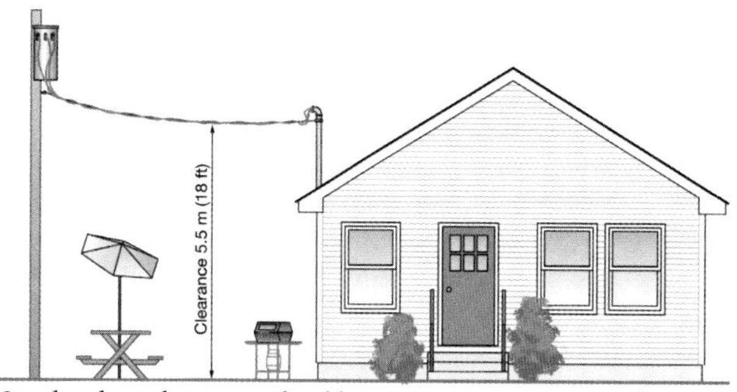

Overhead conductors and cables must meet minimum clearance distances.

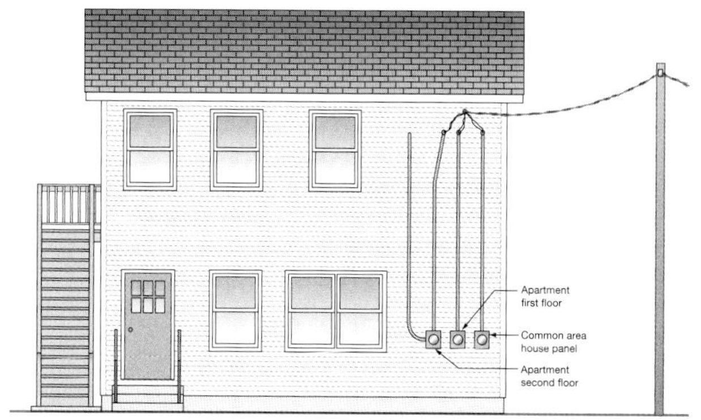

This multiple-occupancy building has multiple sets of service-entrance conductors.

Problem 1-24

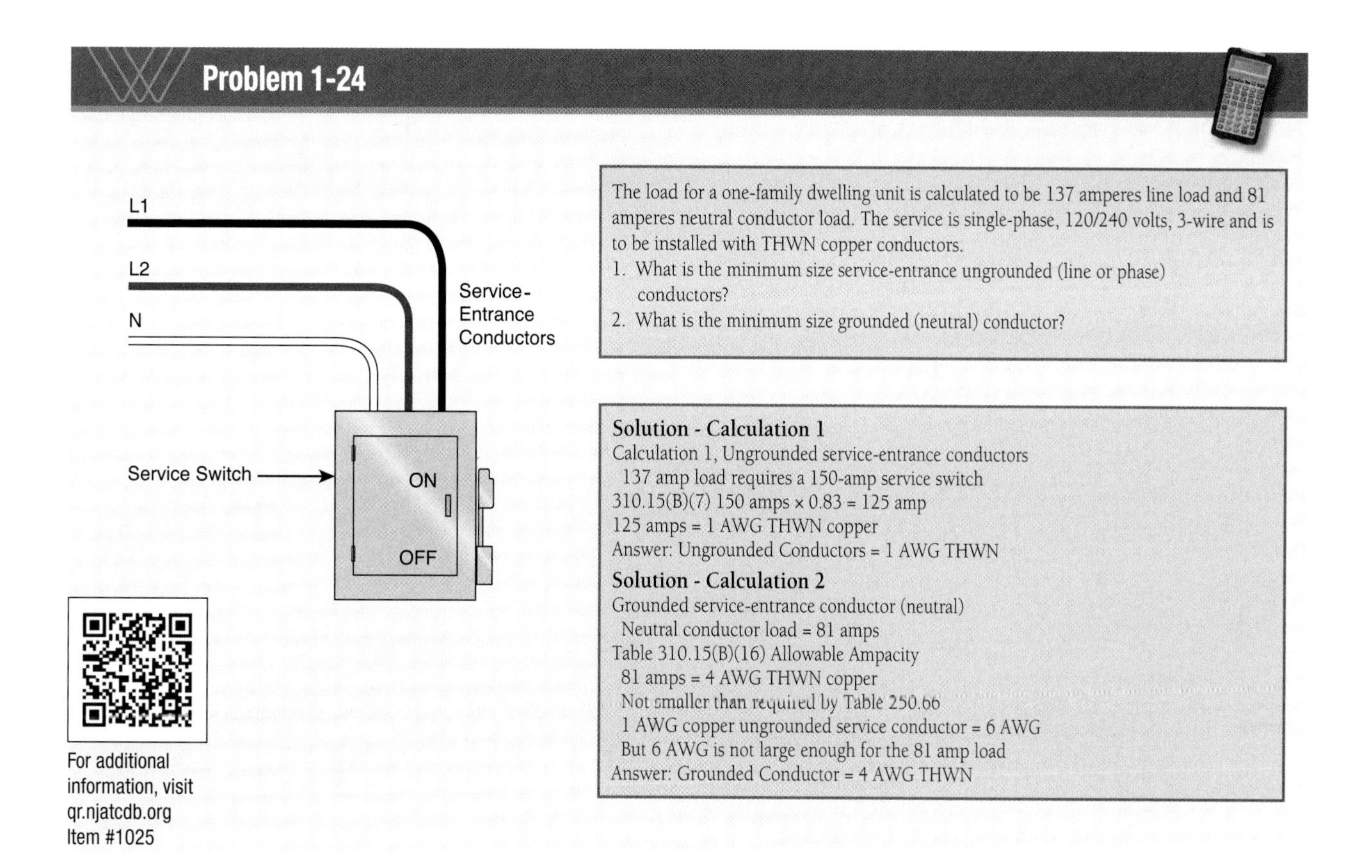

The load for a one-family dwelling unit is calculated to be 137 amperes line load and 81 amperes neutral conductor load. The service is single-phase, 120/240 volts, 3-wire and is to be installed with THWN copper conductors.

1. What is the minimum size service-entrance ungrounded (line or phase) conductors?
2. What is the minimum size grounded (neutral) conductor?

Solution - Calculation 1

Calculation 1, Ungrounded service-entrance conductors
137 amp load requires a 150-amp service switch
310.15(B)(7) 150 amps x 0.83 = 125 amp
125 amps = 1 AWG THWN copper
Answer: Ungrounded Conductors = 1 AWG THWN

Solution - Calculation 2

Grounded service-entrance conductor (neutral)
Neutral conductor load = 81 amps
Table 310.15(B)(16) Allowable Ampacity
81 amps = 4 AWG THWN copper
Not smaller than required by Table 250.66
1 AWG copper ungrounded service conductor = 6 AWG
But 6 AWG is not large enough for the 81 amp load
Answer: Grounded Conductor = 4 AWG THWN

For additional information, visit qr.njatcdb.org Item #1025

Problem 1-25

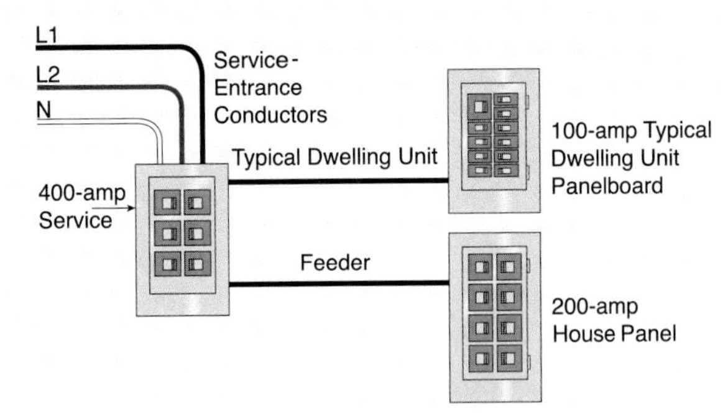

The single-phase, 120/240-volt, 3-wire, 400-ampere service for a multifamily dwelling is calculated to be 380 amperes on each ungrounded service conductor and 120-ampere load on the ungrounded service conductor with all the conductors being THWN copper.

1. What size of ungrounded (line or phase) service-entrance conductor is needed? (Hint: This service for this situation does not qualify under 310.15(B)(7).)
2. What is the minimum size grounded size service-entrance neutral conductor necessary?
3. What size of feeder conductors is needed for the 100-ampere panel located in each apartment all with 120-volt connected load? (Hint: Although the branch-circuit loads in each apartment are limited to 120 volts, the maximum unbalanced load on the ungrounded conductor cannot be accurately determined from this given information.)
4. What size feeder conductor is needed to the 200-ampere, 240-volt, single-phase power panel?

Solution - Calculation 1
Multifamily dwelling ungrounded service-entrance conductors
Table 310.15(B)(6) does not apply
Table 310.15(B)(7) does not apply
380 Amps THWN copper = 500 kcmil
Answer: 220.23(A) Load 380 amps THWN copper = 500 kcmil

Solution - Calculation 2
Multifamily dwelling grounded service-entrance conductor
Table 310.15(B)(16) Allowable Ampacity
120 amps THWN = 1 AWG
250.24(C)(1) Minimum Size
Table 250.66, Grounding Electrode Conductor
Service conductor of 500 kcmil = 1/0 AWG
Grounded service conductor cannot be smaller than 1/0 AWG
Answer: 1/0 AWG

Solution - Calculation 3
Typical dwelling unit ungrounded feeder conductors
310.15(B)(7) 100 amps × 0.83 = 83 amps
Using Table 310.15(B)(16)
83 amps @ 75° C = 4 AWG THWN
Grounded (neutral) conductor current for this feeder is not reduced in problem
Since Table 250.66 does not apply, match the full-size ungrounded conductor = 4 AWG
Answer: Line and Neutral 4 AWG

Solution - Calculation 4
House panel feeder
Table 310.15(B)(7) does not apply
Table 310.15(B)(16) Allowable Ampacity
200 amps THWN = 3/0 AWG copper
Answer: 3/0 AWG

1.12 Special Ampacity Information

The following is specific ampacity information for particular wiring methods and/or conditions.

1.12.1 Nonmetallic Sheathed Cable Type NM

Power conductors within Type NM (including NMC and NMS) cable are constructed with insulation rated at 90°C. The 90°C conductor rating may be used for temperature correction and ampacity adjustment calculations, but the final derated ampacity of these 90°C conductors must not exceed that of a 60°C rated conductor, according to **Section 334.80**.

Ampacity adjustments of NM cables may also be necessary where two or more cables are installed (1) through fire-stopped opening in wood framing members or (2) installed in direct contact with thermal insulation without maintaining spacing.

Problem 1-26

NM Cable

Bare

A 3-conductor, 12 AWG (with ground) copper Type NM cable is installed in an area with a 30°C ambient temperature. What is the allowable ampacity of the circuit conductors?

Solution
Section 334.80 Ampacity
Table 310.15(B)(16) Allowable Ampacity
12 AWG copper @ 90°C = 30 amps
Number of conductors = 4
Equipment grounding conductor not counted
Number of current-carrying conductors = 3
Table 310.15(B)(3)(a) Adjustment Factor does not apply
Table 310.15(B)(2)(a) Correction Factor does not apply
Table 310.15(B)(16) Allowable Ampacity
Final derated ampacity not to exceed 60°C
12 AWG copper @ 60°C = 20 amps
Answer: 20 amperes

Problem 1-27

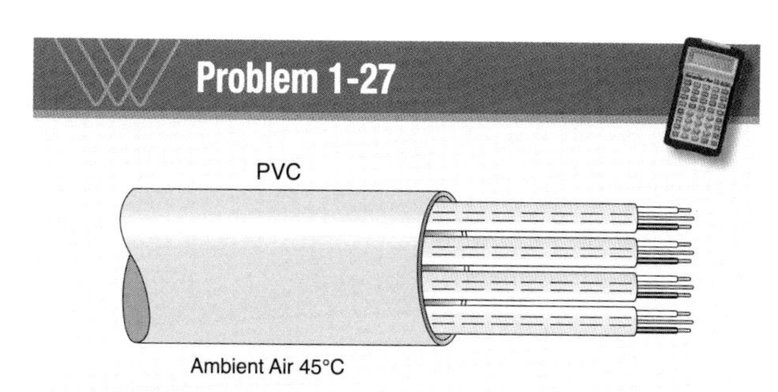

Four 2-conductor, Type NM cables are installed in a rigid PVC conduit in an area with an ambient temperature of 45°C. Each current-carrying conductor in the NM cables is 10 AWG aluminum with 90°C insulation. What is the allowable ampacity of each 10 AWG conductor?

Solution
Section 334.80 Ampacity
Table 310.15(B)(16) Allowable Ampacity
10 AWG aluminum NM @ 90°C = 35 amps
Table 310.15(B)(2)(a) Correction Factor
90°C insulation in 45°C ambient = 0.87
Table 310.15(B)(3)(a) Adjustment Factors
4 cables x 2 current-carrying conductors = 8
8 current-carrying conductors = 70%
35 amps x 0.87 x 0.70 = 21.31 amps
10 AWG aluminum @ 60°C = 25 amps
Calculated ampacity is less than @ 60°C
21.3 amps maximum permitted
Answer: 21.3 amperes

Problem 1-28

EMT

Two 2/C 8 AWG with Ground, Copper

Two 2-conductor, 8 AWG copper Type NM cables are installed in electrical metallic tubing, Type EMT, in an area with an ambient temperature of 35°C. What is the ampacity of each conductor?

Solution
Section 334.80 Ampacity
Table 310.15(B)(16), Allowable Ampacity
8 AWG copper @ 90°C = 55 amps
Table 310.15(B)(2)(a) Correction Factors
90°C insulated conductors in 35°C ambient = 0.96
Table 310.15(B)(3)(a) Adjustment Factors
2 cables x 2 conductors = 4 conductors
4 current-carrying conductors = 80%
55 amps x 0.96 x 0.80 = 42.24 amps
8 AWG copper at 60°C = 40 amps
42.24 amps exceeds 60°C ampacity
42.24 amps is not permitted
Answer: 40 amperes

1.12.2 Armored Cable Type AC

320.80(A) indicates the use of the allowable ampacity tables for the conductors of Type AC cable. Where the armored cable is installed in thermal-type insulation, only 90°C conductors are permitted to be used and the allowable ampacity of the 90°C conductor is limited to that of a 60°C rated conductor.

Problem 1-29

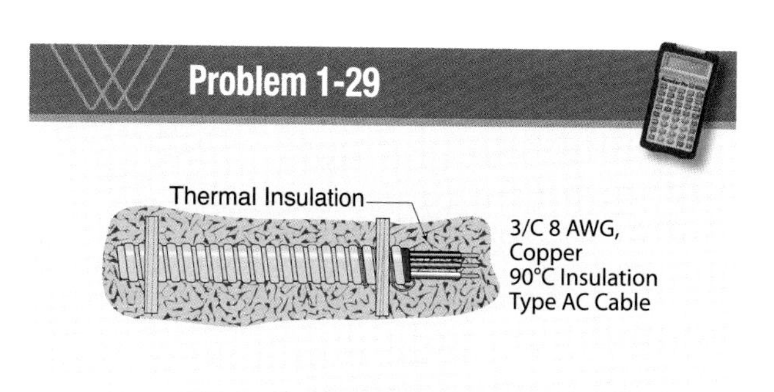

What is the allowable ampacity of 3-conductor, Type AC cable comprised of three 8 AWG, 90°C insulated copper conductors installed in thermal insulation?

Solution
Table 310.15(B)(16) Allowable Ampacity
60°C Column 8 AWG = 40 amps
Answer: 40 amperes

1.12.3 Metal Wireways and Sheet Metal Auxiliary Gutters

376.22(A) for metal wireways and **366.22(A)** for sheet metal auxiliary gutters directly relate to the permitted area of conductor fill allowed for these raceways. The maximum permitted fill of all contained conductors in both wiring methods is limited to (20%). There are no exceptions.

In addition, **376.22(B)** for metal wireways and **366.22(A)** for sheet metal auxiliary gutters require the application of adjustment factors according to **310.15(B)(3)(a)** only if the number of current-carrying conductors exceeds 30. Non-current-carrying conductors such as control and starting duty only conductors are not counted as current-carrying conductors according to **376.22(B)** for metal wireways and **366.22(A)** for sheet metal auxiliary gutters. We will now use two examples to explore different situations.

Installation Situation No. 1

Problem 1-30 limits the metal wireway cross-sectional area to 18% fill, thus complying with **376.22(A)**. No adjustment factors are necessary, because the problem is limited to 24 current-carrying conductors.

Problem 1-30

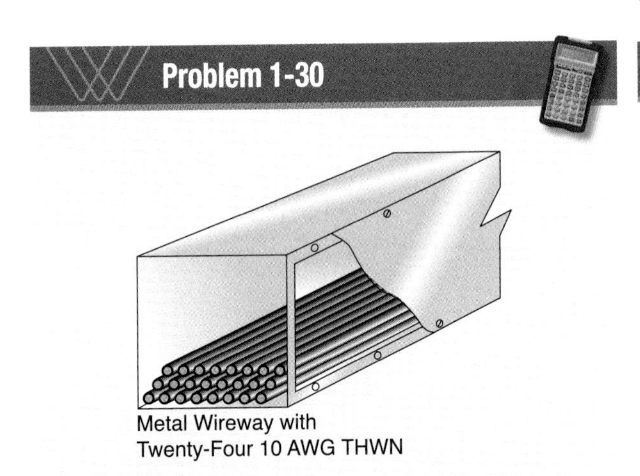

Metal Wireway with Twenty-Four 10 AWG THWN

What is the allowable ampacity of each of twenty-four 10 AWG THWN copper current-carrying conductors installed in a metal wireway, occupying 18% of the interior cross-sectional area of the wireway in an area with an ambient temperature of 30°C?

Solution
Table 310.15(B)(16) Allowable Ampacity
10 AWG THWN copper = 35 amps
Metal wireway contains less than 30 conductors
Metal wireway is less than 20% full
Table 310.15(B)(3)(a) Adjustment Factors does not apply
Answer: 35 amperes

Installation Situation No. 2

Problem 1-31 limits the metal wireway cross-sectional area to 19% fill, thus complying with **376.22(A)**. But, according to **376.22(B)**, an adjustment factor is necessary because the number of current-carrying conductors exceeds thirty.

Problem 1-31

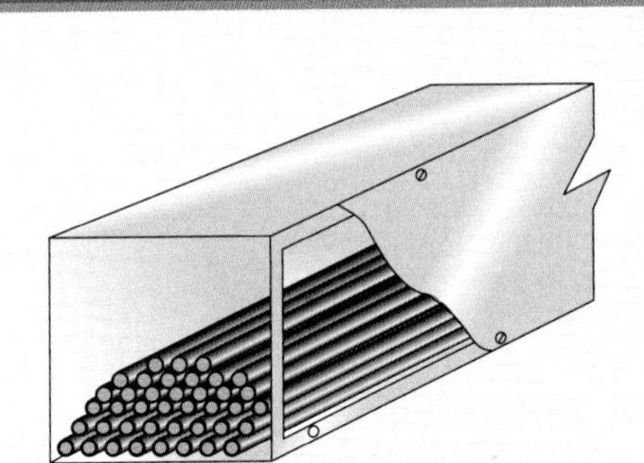

Metal Wireway with Forty 12 AWG THWN

What is the allowable ampacity of each of forty 12 AWG THWN, current-carrying copper conductors installed in a metal wireway, occupying 19% of the interior cross-sectional area, installed in an area with an ambient temperature of 30°C?

Solution
Table 310.15(B)(16) Allowable Ampacity
12 AWG THWN copper = 25 amps
Table 310.15(B)(3)(a) Adjustment Factors
40 conductor = 40%
25 amps × 0.40 = 10 amps
Answer: 10 amperes

Multiple sealed conduits are entering a wireway from below ground.

1.12.4 Nonmetallic Wireways and Nonmetallic Auxiliary Gutters

Section 378.22 addresses the ampacity and the number of conductors permitted in nonmetallic wireways.

1. The number of all contained conductors is limited to not more than 20% of the interior cross-sectional area of a nonmetallic wireway. There are no exceptions.
2. The adjustment factors of **310.15(B)(3)(a)** apply to all the current-carrying conductors in a nonmetallic wireway.
3. The adjustment factors of **310.15(B)(3)(a)** do not apply to conductors for signaling circuits or to controller conductors between a motor and its starter and used only for starting duty.

366.22(B) specifies the number of all contained conductors in a nonmetallic auxiliary gutter be limited to not more than 20% of the interior cross-sectional area of the gutter. There are no exceptions.

366.23(B) requires the adjustment factors of **310.15(B)(3)(a)** to be applied to the current-carrying conductors but not to signaling or control conductors installed in a nonmetallic auxiliary gutter.

It is important to always remember that the temperature correction factors of **Table 310.15(B)(2)(a)** apply to conductors installed in all (metal and nonmetallic) wireways or auxiliary gutters, regardless of the number of conductors.

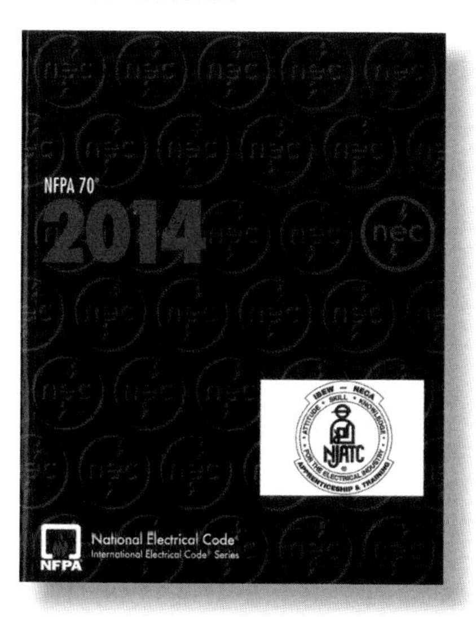

This is the 53rd *edition of the National Electrical Code®.*

These DC batteries are part of a large UPS system.

Problem 1-32

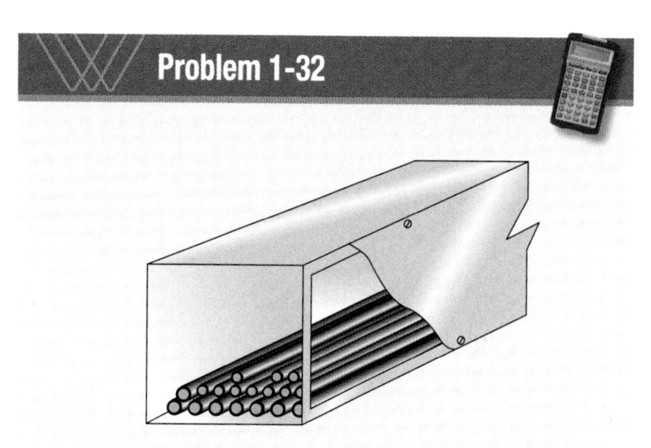

Nonmetallic Wireway with Ten 6 AWG THWN Copper and Eight 4 AWG THWN Copper

Ten 6 AWG THWN and eight 4 AWG THWN copper current-carrying conductors occupy 16% of the cross-sectional area of a nonmetallic wireway in an area with an ambient temperature of 30°C. What is the allowable ampacity of each 6 AWG and each 4 AWG conductor?

Solution

Table 310.15(B)(16) Allowable Ampacity
6 AWG THWN = 65 amps
4 AWG THWN = 85 amps
Total number of conductors
10 + 8 = 18 conductors
Table 310.15(B)(3)(a) Adjustment Factors
18 current-carrying conductors = 50%
65 amps × 0.50 = 32.5 amps
85 amps × 0.50 = 42.5 amps
Answers: 6 AWG = 32.5 amperes
4 AWG = 42.5 amperes

Definitions and Terms

Ampacity - The maximum current, in amperes, that a conductor can carry continuously under conditions of use without exceeding its temperature rating.

Allowable Ampacity - The maximum ampacity permitted by the values presented in the Tables 310.15(B)(16) through 310.15(B)(21).

Ambient temperature - The temperature of the air surrounding electrical circuit conductors.

Conditions of Use - A set of conditions specifically expressed in the title (or heading) of Tables 310.15(B)(16) through 310.15(B)(21).

Conductor, Bare - A conductor having no covering or electrical insulation whatsoever.

Conductor, Covered - A conductor encased within material of composition or thickness that is not recognized by this *Code* as electrical insulation. An example of a covered conductor is the green covered grounding conductor within a Type NM (Nonmetallic Sheathed) Cable.

Conductor, Insulated - A conductor encased within material of composition and thickness that is recognized by the *Code* as electrical insulation. Examples of insulation used with conductors are found in Tables 310.104(A) through (C).

Conductor Material - Basic electrical conductors are constructed of materials including copper, aluminum, and copper-clad aluminum. Other conductor materials reserved for high temperature applications include nickel and nickel-coated copper.

Equipment Grounding Conductor (EGC) - The conductive paths that provide a ground-fault current path and connects normally non–current-carrying metal parts of equipment together and to the system grounded conductor or to the grounding electrode conductor, or both. The equipment grounding conductor also performs bonding. Section 250.118 contains a list of acceptable equipment grounding conductors.

Grounded Conductor - A system or circuit conductor that is intentionally grounded.

Neutral Conductor - The conductor connected to the neutral point of a system that is intended to carry current under normal conditions.

Nonlinear Load - A load where the wave shape of the steady-state current does not follow the wave shape of the applied voltage. Examples of nonlinear loads may include electronic equipment, electronic/electric-discharge lighting, adjustable speed drive systems, and similar equipment.

Temperature Rating - The highest temperature that an insulated conductor can withstand for a long period of time before damage to the insulation occurs.

Summary

- The term *ampacity* means current-carrying capacity and is a coined word built from two common words, ampere and capacity.
- Insulated current-carrying conductors have two basic parts: the inner conducting metal and the outer covering of electrical insulation. Both parts are very important to ampacity.
- Ampacity varies according to conductor material, conductor insulation temperature, ambient temperature, heat generated internally in the conductor due to both fundamental and harmonic load current, the rate at which the internal heat is dissipated, and the number of adjacent load-carrying conductors present. The temperature of a conductor, therefore, is reached by adding these four heat "sources."
- The allowable ampacity of current-carrying conductors given in the *Code* tables is not the true ampacity of the conductor as defined in **Article 100**. Rather, it is the consideration of such factors as:
  - » Temperature compatibility with connected equipment
  - » Coordination with circuit and system overcurrent protection
  - » Compliance with the requirements of product listing
  - » Preservation of the safety benefits of established industry practices and standardized procedure
- Each ampacity table of **Article 310** uses the table heading as its "conditions of use."
- Correction factors and adjustment factors may need to be applied next.
- Based on specific load, the grounded and or neutral conductor may need to be sized as well.
- Lightly-loaded conductors within a raceway may not add heat to the raceway.
- Other factors will also negate or add the requirement for applying correction or adjustment factors such as:
  - » Neutral conductors in a single-phase tap of a three-phase wye system.
  - » Harmonic currents within a neutral conductor
  - » Different insulation types occupying the same raceway
  - » Control circuits within a power circuit raceway
  - » Circuit conductors in a conduit less than 24 inches in length
  - » Specific limitations of wiring method's ampacity calculations

Remember that the adequacy of the *NEC* is essentially free from hazard. Precisely following the calculation methods outlined in this chapter will provide an installation what is essentially free from hazard, but not necessarily efficient, convenient, or adequate for good service or future expansion of electrical use.

Conductor Allowable Ampacity II Calculations

Conductor allowable ampacity calculations include conductor terminations, the length of time a load is on, and the effects of "load time" or nearly constant heat on specific electrical equipment. Each one of these issues is related to the transfer of heat and the lack of heat dissipation. In addition, these issues are equipment-related issues which ultimately influence the selection of the "allowable ampacity" of a circuit conductor.

The *NEC* contains requirements for insulated conductor terminations on equipment. These **Article 110** provisions must be accurately applied for each and every power and lighting circuit. The *NEC* requirements concerning continuous and noncontinuous loads and the calculations involved will be studied in depth. Since the length of time a load is present affects the circuit parameters, numerous calculations will be performed to finalize the size of conductor and overcurrent device. Most branch circuits and feeder loads within commercial and industrial facilities are continuous loads. In addition, examples using 100% rated circuit breakers are addressed as well.

Objectives

- Understand and apply the termination temperature limitations.
- Understand and determine which loads are considered continuous loads.
- Apply continuous load factors to branch circuits, feeders, and services.
- Calculate conductor ampacities for continuous and noncontinuous loads.
- Calculate the proper size overcurrent protection for continuous and noncontinuous loads.
- Understand and properly calculate conductor size having both continuous and noncontinuous loads using 100% rated overcurrent devices.

Chapter 2

Table of Contents

2.1 Introduction

The study of ampacity in this chapter is more related to electrical equipment connected to conductors than it is to the specifics of various conductor current-carrying capacities. In the previous chapter, we dealt with the capacity of a conductor from a thermal conductor point of view. We looked at the physical properties of various types of conductors, the different causes of heat within and around electrical conductors, and the properties of an insulated conductor in a wiring method to manage and dissipate that heat into the surrounding atmosphere.

In this chapter, we will look at equipment-related issues which influence our selection of the proper *NEC* value for "allowable ampacity" of a circuit conductor. The transfer of heat and the lack of heat dissipation will again be studied as well. However, for this chapter, the focus will be on the conductor terminations, the length of time a load is on, and the effects of constant heat upon specific electrical equipment, such as overcurrent protective devices and their enclosures.

2.2 Temperature Limitations of Equipment

Section 110.3(B) requires that electrical equipment be installed and used according to any instructions included with the equipment's listing or labeling. Written manufacturer requirements and instructions as well as those found in the equipment listing and labeling information may duplicate or be more restrictive than the *Code* requirements.

Electrical conductors, circuit breakers, fuses, connectors, panelboards, and other electrical equipment all have temperature ratings. These temperature ratings are most often the maximum operating temperature permitted or the temperature limitations of equipment. Electrical conductors are terminated using connectors or lugs on circuit breakers, fused switches, and other electrical equipment which include temperature ratings or temperature limitations. These temperature ratings, such as 75°C terminations, are maximum limits and may not be exceeded.

The safety requirements of **110.14(C) Temperature Limitations** are a set of rules limiting the temperature of equipment terminations within various panelboards, disconnect switches, circuit breakers and other related equipment.

The basic rule states: "The temperature rating associated with the ampacity of a conductor shall be selected and coordinated so as not to exceed the lowest temperature rating of any connected termination, conductor, or device."

This basic rule is divided into two main parts:

> Part 1 deals with circuits and equipment rated 100 amperes or less, or conductors 14 AWG through 1 AWG.
>
> Part 2 deals with circuits and equipment rated over 100 amperes or conductors larger than 1 AWG.

Part 1, circuits and equipment rated 100 amperes or less, is further divided into four subcategories of conductors and their terminations as follows:

1. 60°C conductors connected to 60°C terminations of equipment
2. 75°C or 90°C conductors connected to 60°C terminations of equipment
3. 75°C or 90°C conductors connected to 75°C terminations of equipment
4. Certain types of motors permitted to use 75°C or 90°C conductors connecting to 75°C terminations

Part 2, circuits and equipment rated over 100 amperes, is further divided into two subcategories of conductors and their terminations as follows:

1. 75°C conductors connected to 75°C terminations of equipment
2. 90°C conductors connected to 75°C terminations of equipment

In order to simplify these rules, it is necessary to apply the weakest link principle: a chain is only as strong as its weakest link. When the principle is applied to temperature limitations, the final ampacity of a circuit is based upon the lowest temperature rating of any one part of the circuit and its connections. The application of this principle to electrical circuits will simplify the problems and allow a more thorough understanding of the calculations.

2.2.1 Circuits and Equipment Rated at 100 Amperes or Less

2.2.1.1 Using 60°C Insulated Conductors - 110.14(C)(1)(a)(1) requires equipment for circuits rated 100 amperes or less or 14 AWG through 1 AWG conductors to be rated for at least 60°C. This is the first and easiest of the four cases to understand. The *Code* recognizes that a rating of 60°C should be the fundamental (and lowest) temperature rating applied to conductors and equipment. As shown in **Figure 2-1**, both the circuit breaker terminations and the conductor insulation are rated for 60°C.

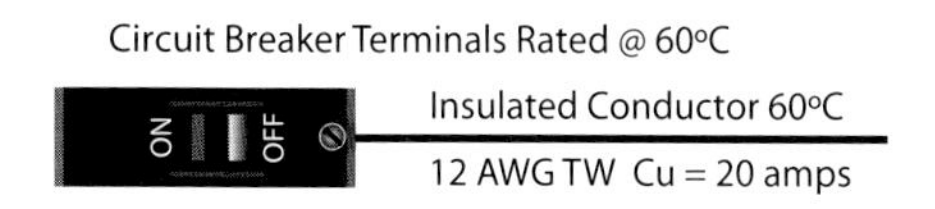

Figure 2-1. An example of an installation where the conductor insulation temperature and the circuit breaker conductor terminations are individually rated for 60°C according to 110.14(C)(1)(a)(1).

2.2.1.2 Using 75°C or 90°C Insulated Conductors - Moving to the second case, **110.14(C)(1)(a)(2)** permits a 75°C conductor to be used with devices marked (with temperature limitations of) 60°C, provided the allowable ampacity of the 75°C conductor is taken from the 60°C column of **Table 310.15(B)(16)**. This is done by using the 75°C copper (or aluminum) column from the table; go down to the row opposite the correct AWG size, then, before selecting the ampacity of the circuit, move horizontally to the left (in the same row) and select the ampacity within the 60°C column.

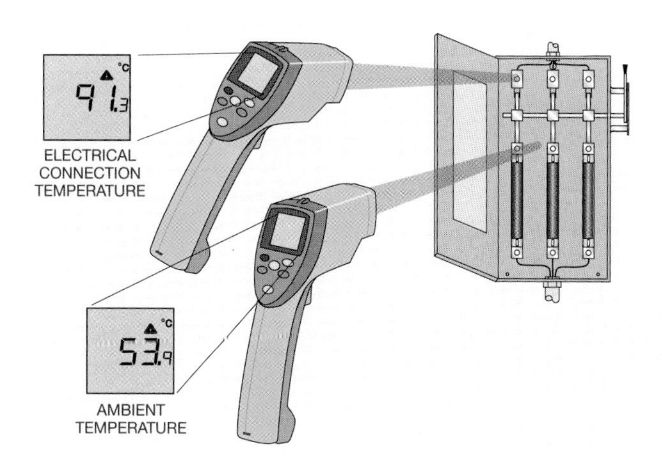

110.14(C)(1)(a)(2) also permits 90°C conductors to be used with devices marked with temperature limitations of 60°C, provided the allowable ampacity of the 90°C conductor again is taken from the 60°C column of **Table 310.15(B)(16)**. This is done by using the 90°C copper (or aluminum) column from the table; go down to the row opposite the correct AWG size, then, before selecting the ampacity of the circuit, move horizontally to the left two columns (in the same row) and select the ampacity within the 60°C column.

Problem 2-1

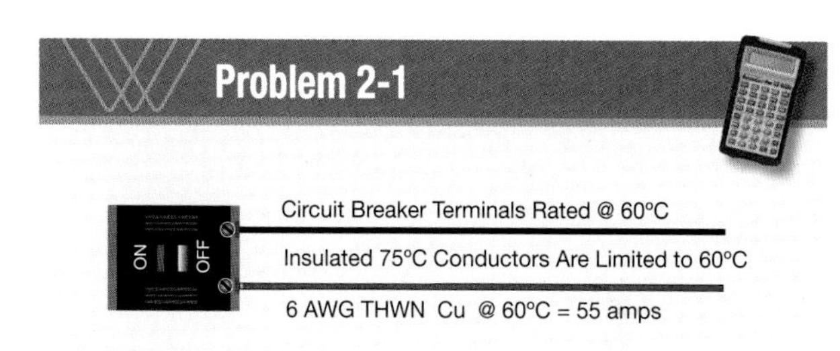

A 6 AWG THWN copper conductor is connected to a circuit breaker with termination temperature limitation marked (not to exceed) 60°C. What is the allowable ampacity of the 6 AWG THWN copper conductor now that it is connected to this circuit breaker?

Solution
110.14(C)(1)(a)(2) applies
CB terminations = 60°C
Table 310.15(B)(16) Allowable Ampacity
Limited by CB to 60°C
THWN ampacity at 75°C not permitted
Use ampacity of 6 AWG Cu at 60°C
6 AWG THWN copper limited to 60°C ampacity = 55 amps
Answer: 55 amperes

Applying the weakest link principle to **Problem 2-1**, both temperature ratings are 60°C, so the weakest link is 60°C. Therefore ampacity must be based upon the 60°C column of **Table 310.15(B)(16)**.

Infrared non-contact testers of the hand held variety are a very important part of an electrical equipment service or maintenance program. These testers can accurately measure the "in use" temperature of circuit conductors and terminals while the circuit is energized and operating. Any electrical worker performing "energized" tests must meet the definition of a "Qualified Person."

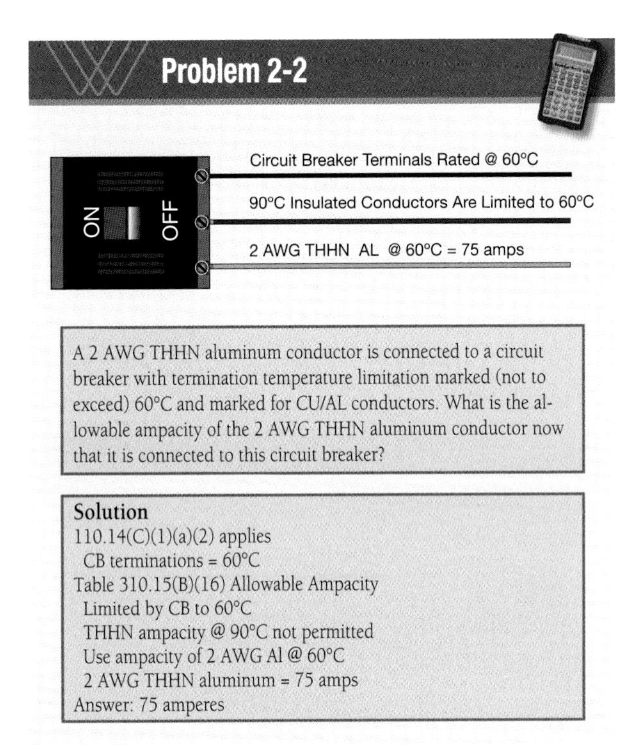

A 2 AWG THHN aluminum conductor is connected to a circuit breaker with termination temperature limitation marked (not to exceed) 60°C and marked for CU/AL conductors. What is the allowable ampacity of the 2 AWG THHN aluminum conductor now that it is connected to this circuit breaker?

Solution
110.14(C)(1)(a)(2) applies
CB terminations = 60°C
Table 310.15(B)(16) Allowable Ampacity
Limited by CB to 60°C
THHN ampacity @ 90°C not permitted
Use ampacity of 2 AWG Al @ 60°C
2 AWG THHN aluminum = 75 amps
Answer: 75 amperes

Applying the weakest link principle to **Problem 2-2**, the insulated conductor temperature rating is 90°C, the CB termination temperature rating is 60°C, so the weakest link is 60°C. Therefore, the ampacity of the circuit conductors must be based upon the 60°C column of **Table 310.15(B)(16)**.

2.2.1.3 For 100 Ampere Terminations or Less, at 75°C - Moving to the third case, **110.14(C)(1)(a)(3)** permits devices marked with a temperature limitation of 75°C to be used with 75°C conductors operating at their rated allowable ampacity. This applies to equipment for circuits rated 100 amperes or less, or marked for 14 AWG through 1 AWG. Use the 75°C column of **Table 310.15(B)(16)**. This case is common because new equipment is most frequently marked as acceptable to receive conductors rated 75°C. Also, new equipment may be dual rated and marked as 60/75°C, thereby acceptable for 75°C conductor terminations as well.

The example shown in **Figure 2-2** points out that both the circuit breaker conductor terminations and the conductor insulation are rated for 75°C.

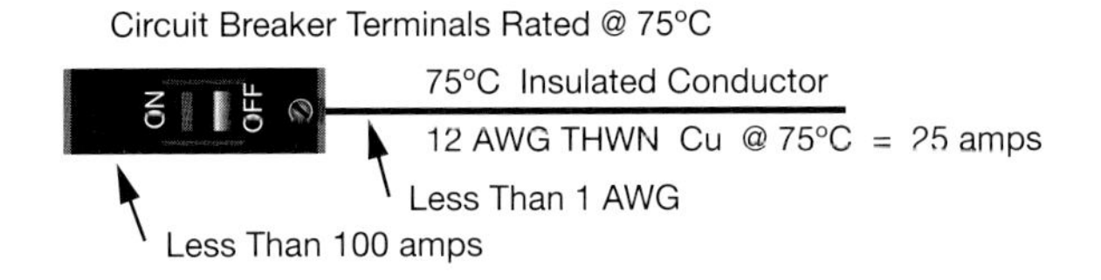

Figure 2-2. An example of an installation where the conductor insulation temperature and the circuit breaker conductor terminations both are rated and marked for 75°C and is acceptable according to 110.14(C)(1)(a)(3).

For additional information, visit qr.njatcdb.org Item #1026

2.2.2 Circuits and Equipment Rated Over 100 Amperes

110.14(C)(1)(b) deals with circuits and equipment rated over 100 amperes. This section is further divided into two subcategories:

1. Conductors rated 75°C connected to equipment rated 75°C
2. Conductors rated 90°C connected to equipment rated 75°C

2.2.2.1 Conductors Rated 75°C Connected to Equipment Rated 75°C - Moving to case four, **110.14(C)(1)(b)(1)** states that using 75°C insulated conductors requires equipment for circuits to be rated more than 100 amperes, or conductors larger than 1 AWG, to have a minimum temperature limitation marked (not to exceed) 75°C. **Figure 2-3** illustrates a circuit breaker sized greater than 100 amperes where both the conductor terminations and the conductor insulation are rated for 75°C.

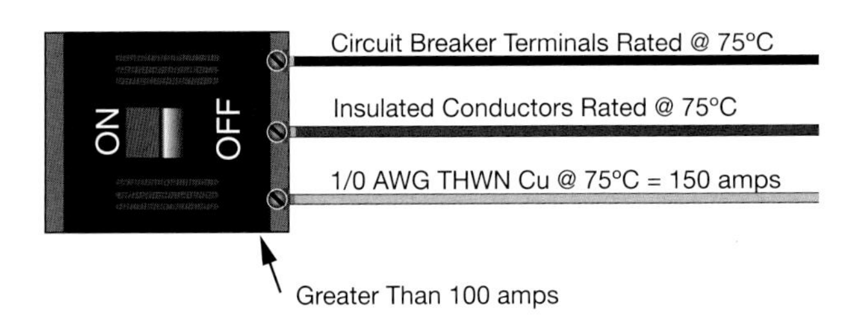

Figure 2-3. A circuit breaker with conductor terminations rated at 75°C and conductors larger than 1 AWG with THWN insulation rated at 75°C all in accordance with 110.14(C)(1)(b)(1).

2.2.2.2 Conductors Rated 90°C Connected to Equipment Rated 75°C - Finally, case five points out that **110.14(C)(1)(b)(2)** permits the use of 90°C conductors with terminations rated and marked at 75°C, provided the allowable ampacity of the 90°C conductors is taken from the 75°C column of **Table 310.15(B)(16)**.

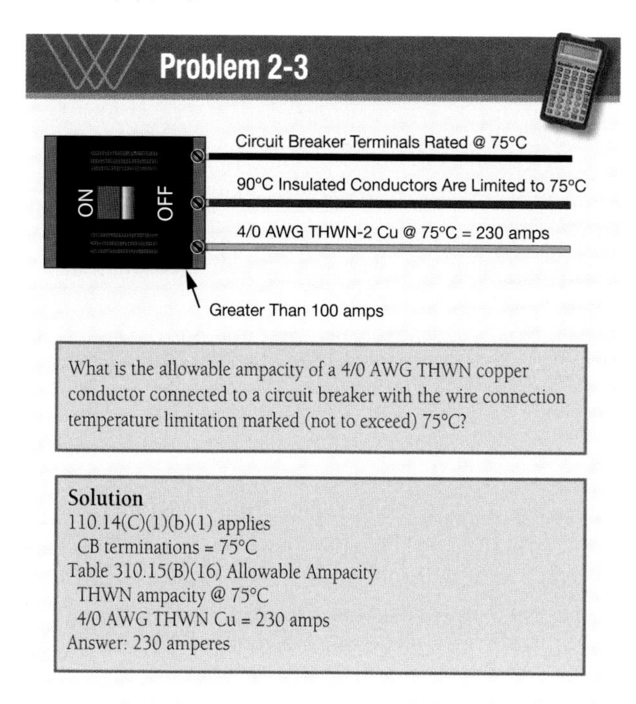

What is the allowable ampacity of a 4/0 AWG THWN copper conductor connected to a circuit breaker with the wire connection temperature limitation marked (not to exceed) 75°C?

Solution
110.14(C)(1)(b)(1) applies
CB terminations = 75°C
Table 310.15(B)(16) Allowable Ampacity
THWN ampacity @ 75°C
4/0 AWG THWN Cu = 230 amps
Answer: 230 amperes

2.2.3 Advantages of Using 90°C Conductors

Q. *So, what is the advantage of using 90°C conductors if they cannot be used at their full 90°C allowable ampacity from the tables?*

A. *There is an advantage to using 90°C insulated conductors when the conditions of use as shown in the heading of Table 310.15(B)(16) are changed. One example is where there are more than three current-carrying conductors in a raceway or cable. Another example is where the ambient temperature is above 86°F. Each of these conditions increases conductor temperature. Using 90°C insulated conductors allows the accommodation of some derating while not increasing the actual conductor size.*

Q. *Is there any equipment currently manufactured with terminations rated at 90°C?*

A. *If there is an installation where all the components are rated 90°C, the full ampacity of a 90°C conductor is permitted to be selected. However, for 600 volts or less, equipment with 90°C rated terminations is generally unavailable. However, large equipment rated over 600 volts with terminations rated 90°C is available and often used today.*

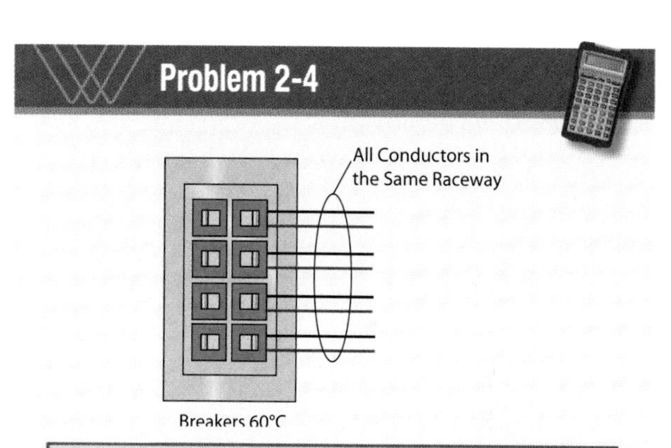

Eight 6 AWG THHN copper current-carrying conductors are installed to replace existing wiring within an existing single rigid metal conduit, Type RMC. The area of installation has an ambient temperature of 30°C. The new eight 6 AWG THHN conductors are connected to existing 50-ampere 2-pole circuit breakers with a marked terminal temperature rating of 60°C. What is the allowable ampacity of the conductors, and is this an acceptable installation?

Solution
Table 310.15(B)(16) Allowable Ampacity
6 AWG THHN @ 90°C = 75 amps
Table 310.15(B)(3)(a) Adjustment Factors
8 current-carrying conductors = 70%
75 amps x 0.70 = 52.5 amps
6 AWG in 60°C column = 55 amps
55 amps is not permitted
Allowable ampacity = 52.5 amps
Answer: 52.5 amperes

Comment
Although the heat developed by a 55-ampere load on 60°C equipment terminals is the maximum heating permitted on these circuit breaker terminals, the existing circuit breaker is rated for 50 amperes, which is below the maximum allowable ampacity of the circuit conductors. The current is limited by the 50-ampere circuit breaker to a value less than the maximum allowable circuit ampacity. Therefore, the installation is acceptable.

2.2.4 The Practical Solution Using 110.14(C)

In practice, the actual *NEC* rules are handled a little more easily and made more user-friendly by equipment manufacturers and listing agencies. Listed circuit breakers rated 125 amperes or less are marked with wire temperature ratings using one of three basic methods:

1. Wire temperature rating of 60°C.
2. Combination temperature rating of 60/75°C.
3. Wire temperature rating of 75°C only.

This means that the circuit breaker terminations are listed for electrical conductor connections using insulated conductors rated either 60°C or 75°C. Most 125 ampere or less circuit breakers today are rated and marked as 60/75°C. One advantage of this dual marking allows older 60°C wiring to be connected to more modern or replacement circuit breakers. Another advantage is that 90°C wiring can be used for derating, but still may qualify for the 75°C ampacity.

All listed circuit breakers rated over 125 amperes are suitable for 75°C conductor terminations. At the present time, there are no listed 600-volt circuit breakers with conductor terminations rated at 90°C.

Conductors rated for higher temperatures (such as 90°C THHN or XHHW) are most often installed today, but importantly, they must not be loaded to carry more current than that permitted by the 75°C or the 60°C ampacity of that size of conductor. This is because the temperature rating of conductor terminations on electrical equipment is most often the limiting factor (weakest link principle).

However, when applying adjustment and temperature correction factors, provided the actual load never exceeds the lowest temperature rating of the circuit conductor(s), equipment, or overcurrent protective devices, the circuit and all its connections are in compliance with **110.14(C)**.

2.3 Continuous Loads and Branch Circuits

Article 100 defines *continuous load* as "a load where the maximum current is expected to continue for 3 hours or more." Stores, offices, and show window lighting are normally considered continuous loads. A continuous load should not be confused with a continuous duty motor (load). The duty of a motor is determined by the application of the motor as defined in **Article 100** under the definition of *duty* and is not related to continuous loads.

The general *Code* requirements for all branch circuits connected to continuous loads are found in **210.19(A)** and **210.20(A)**. Some of the specific *Code* requirements for equipment connected to branch circuits include **422.10(A)** for appliance branch circuits, **Section 422.13** for storage-type water heaters, **424.3(B)** for fixed electric space-heating equipment, **Section 426.4** for fixed outdoor electric deicing and snow-melting equipment, and **Section 427.4** for fixed electric heating equipment for pipelines and vessels.

Other *Code* requirements for continuous loads are included in **215.2(A)** for feeders and **230.42(A)** for services.

All of these *Code* sections are based upon the electrical safety requirement that conductors and equipment should not be subjected to overheating. As equipment and conductors are analyzed for full load operation, the consistent trouble areas of a circuit are the overcurrent current devices and the wire terminations at overcurrent protective devices. Testing laboratories confirm that circuits loaded at 100% in a continuous load application can safely carry the load for three hours or more. But, this is true only if the test is conducted on overcurrent protective devices and conductors placed in "open air." Once the test is performed on circuit parts within enclosures, these same circuits clearly overheat and cannot safely handle the load.

The historical solution to prevent overheating in circuits which contain continuous loads has been to upsize both the conductors and the overcurrent protective devices.

The exact *NEC* language for the branch-circuit solution is as follows:

> **210.20(A)** - Where a branch circuit supplies continuous loads or any combination of continuous and noncontinuous loads, the rating of the overcurrent device shall not be less than the noncontinuous load plus 125% of the continuous load.

And...

210.19(A) Branch Circuits Not More Than 600 Volts.

(1) **General.** Branch-circuit conductors must have an ampacity not less than the maximum load to be served. Conductors must be sized to carry not less than the larger of **210.19(A)(1)(a)** or **(b)**.

(a) Where a branch circuit supplies continuous loads or any combination of continuous and noncontinuous loads, the minimum branch-circuit conductor size must have an allowable ampacity not less than the noncontinuous load plus 125 percent of the continuous load.

(b) The minimum branch-circuit conductor size must have an allowable ampacity not less than the maximum load to be served after the application of any adjustment or correction factors.

There is another solution in the *Code* (stated as an exception to both **210.19(A)(1)** and to **210.20(A)**) which permits certain overcurrent devices provided they are listed to carry their rated ampacity continuously (for three hours or more). These circuit breakers are referred to as 100% rated breakers. Although not widely used, they are available from many manufacturers. A few 100% rated calculations are provided in this chapter.

Different from an overcurrent protective device, an electrical conductor "can" carry its full ampacity rating continuously. After all, that is the definition of ampacity. So where conductors are applied to 100% rated circuit breakers, the conductors do not require "up sizing" since they are already considered 100% rated.

2.3.1 Branch Circuits Using Standard Circuit Breakers

210.19(A)(1) basically requires the rating of the branch circuit and the branch-circuit conductors to be not less than the noncontinuous load plus 125% of the continuous load. **210.20(A)** requires the branch-circuit overcurrent protection to be not less than the noncontinuous load plus 125% of the continuous load. Both requirements carry an exception allowing the use of listed 100% rated overcurrent protective devices.

By neglecting the exceptions, both the branch circuit and overcurrent protection requirements handle the "up sizing" requirements by using the same formula "100% of noncontinuous plus 125% of the continuous load."

For the overcurrent protection portion, there are more rules and permissions which need to be understood and applied. First and foremost is **Section 240.4** which states:

> Conductors, other than flexible cords, flexible cables, and fixture wires, shall be protected against overcurrent in accordance with their ampacities specified in **Section 310.15**, unless otherwise permitted or required in **240.4(A)** through **240.4(G)**.

Basically, this general rule of **Section 240.4** indicates that a conductor carrying 66 amperes must be protected by an overcurrent device rated at least 66 amperes. Since 66 amperes is not a standard size overcurrent device, **240.4(B)** permits using the next larger standard size according to **240.6(A)**. The choices in **240.6(A)** are either a 60- or a 70-ampere overcurrent device. The 70-ampere overcurrent device must be selected since the 60-ampere device would simply be too small for the load. In this case, being mindful of the fact that the 66 amperes is a minimum should make the selection process easy.

Problem 2-5 deals with continuous loads used to size a standard inverse-time circuit breaker whereas some later problems with continuous loads will apply 100% rated circuit breaker.

Problem 2-5

A 120-volt branch circuit supplies a continuous load (CL) of 21 amperes. The ambient temperature is 30°C, so temperature correction is unnecessary. What is the minimum standard size circuit breaker permitted for the branch-circuit overcurrent protection?

Solution
210.20(A) Overcurrent Protection
Min. OCPD = CL × 125%
= 21 × 125%
= 26.25 amps
240.6(A), next larger std rating
Next larger std rating = 30 amps
Answer: 30 ampere circuit breaker

2.3.2 Branch Circuit Load Using a 100% Rated Circuit Breaker

Applying 100% rated overcurrent devices to branch circuits is permitted by the exception to **210.20(A)**. However, in today's market place, 100% rated circuit breakers of the branch circuit variety are rare or nonexistent. Although understanding how to apply a 100% rated device products at the branch circuit level is valuable information, from a practical point of view, there seems to be few if any applications available at this time.

For additional information, visit qr.njatcdb.org Item #1027

Problem 2-6

A 480-volt, 2-wire branch circuit supplies a large electronic information roadway sign inside a tunnel. This lighting circuit has a continuous load of 47 amperes.

1. What is the minimum size circuit breaker using a standard inverse-time circuit breaker permitted for the branch-circuit overcurrent protection?
2. As an alternate solution, what is the minimum standard size 100% rated circuit breaker for this application?

Solution - Calculation 1

210.20(A)(1)

Min. CB = CL × 125%
= 47 amps × 1.25
= 58.75 amps

240.6(A)

Next larger standard size = 60 amps

Answer: 60 ampere inverse-time circuit breaker

Solution-Calculation 2

210.20(A)(1), Exception

Min. 100% CB = CL × 100%
= 47 amps × 1.00
= 47 amps

240.6(A)

Next larger std size = 50 amps

Answer: 50 ampere 100% rated circuit breaker

Estimators use manual or electronic rotometers to measure raceways and cable lengths from blueprints.

2.3.3 Continuous and Noncontinuous Load

Whether determining the ampacity of the conductors or the rating of the overcurrent device, the language is the same. The *Code* requires that each shall the capacity to handle "... not less than the noncontinuous load plus 125 percent of the continuous load."

From a practical point of view, after the continuous and noncontinuous loads are determined, the overcurrent device should be determined first, and then the size of the circuit conductors should be determined.

Problem 2-7

A 277-volt, single-phase circuit supplies a 6 kW continuous load and a 4 kW noncontinuous load.

1. What is the minimum standard size inverse-time circuit breaker (with terminations dual rated and marked at 60/75°C)?
2. Using XHHW-2 copper, determine the minimum size circuit conductors for this branch circuit.

Solution – Calculation 1

Continuous load (CL):

$$\text{CL amps} = \frac{\text{CL kW} \times 1{,}000}{277}$$

$$= \frac{6 \times 1{,}000}{277}$$

= 21.7amps

Answer: 21.7 amperes

Solution – Calculation 2

Noncontinuous load (NCL):

$$\text{NCL amps} = \frac{\text{NCL kW} \times 1{,}000}{277}$$

$$= \frac{4 \times 1{,}000}{277}$$

= 14.4 amps

Answer: 14.4 amperes

Solution – Calculation 3

Breaker rating:

210.20(A) Branch-circuit OCPD

Min. OCPD = (CL × 125%) + NCL
= (21.7 × 125%) + 14.4
= 41.5 amps

240.4(B) and 240.6(A), next larger std size

Next larger std size = 45 amps

Answer: 45 ampere circuit breaker (60°C/75°C rated)

Solution – Calculation 4

Conductor ampacity:

210.19(A)(1)

Min. ampacity = (CL × 125%) + NCL
= (21.7 × 1.25) + 14.4 amps
= 41.5 amps

110.14(C)(1)(a)(3)

90°C rated wire connected to 60°C/75°C rated CB

Table 310.15(B)(16) Ampacity using 75°C column

41.5 amps requires 8 AWG XHHW-2

Answer: 8 AWG XHHW-2 copper

2.3.4 Conductor Size and Continuous Load

210.19(A)(1) requires the minimum branch-circuit conductor size to have an allowable ampacity not less than the noncontinuous load plus 125% of the continuous load. In addition, this minimum conductor size must have an allowable ampacity not less than the maximum load to be served after the application of any adjustment or correction factors.

Problem 2-8

A 120-volt circuit supplies a continuous load of 7.5 kW.

1. What is the minimum standard size inverse-time circuit breaker permitted for branch- circuit overcurrent protection with conductor terminations rated at 75°C?
2. What size THWN copper conductors are needed so that the overcurrent device will protect the load and the conductors?

Solution – Calculation 1

210.20(A) Circuit breaker rating

Amps $= \dfrac{\text{kW} \times 1{,}000}{\text{E}}$

$= \dfrac{7.5 \times 1{,}000}{120}$

= 62.5 amps

Min. OCPD = CL x 125%
= 62.5 x 1.25
= 78.12 amps

240.4(B) and 240.6(A), next larger std size

Next larger std size = 80 amps

Answer: 80 ampere circuit breaker

Solution – Calculation 2

210.19(A)(1) Conductor Ampacity

Use 78.12 amps from Calculation 1

Table 310.15(B)(16) Ampacity, 75°C column

78.12 amps requires 4 AWG THWN rated at 85 amps

80 amp OCPD will protect conductors

Answer: 4 AWG THWN copper

Problem 2-9

A 120-volt circuit supplies a continuous load of 10.5 kW.

1. What is the minimum standard size 100% rated circuit breaker permitted for branch- circuit overcurrent protection with conductor terminations rated at 75°C?
2. What size THWN copper conductors are needed so that the overcurrent device will protect the load and the conductors?

Solution – Calculation 1

210.20(A) Circuit breaker rating

Min. OCPD = CL x 100%

Amps $= \dfrac{\text{kW} \times 1{,}000}{\text{E}}$

$= \dfrac{10.50 \times 1{,}000}{120}$

= 87.5 amps

Min. OCPD = 87.5 x 100% = 87.5 amps

240.4(B) and 240.6(A), next larger std size

Next larger std size = 90 amps

Answer: 90 ampere circuit breaker

Solution – Calculation 2

210.19(A)(1) Conductor Ampacity

Use 87.5 amps from Calculation 1

Table 310.15(B)(16) Ampacity, 75°C column

87.5 amps requires 3 AWG THWN Cu

4 AWG THWN Cu is too small

90 amp OCPD will protect 3 AWG THWN conductors

Answer: 3 AWG THWN copper

The most accurate way to convert temperature from one scale to another is by using the visual adaptations and math from this temperature conversion graphic. However, quick estimated conversions are possible by using NEC Table 310.15(B)(2)(a) and going between columns 1 and 5. For example, 11-15°C equals 51-59°F

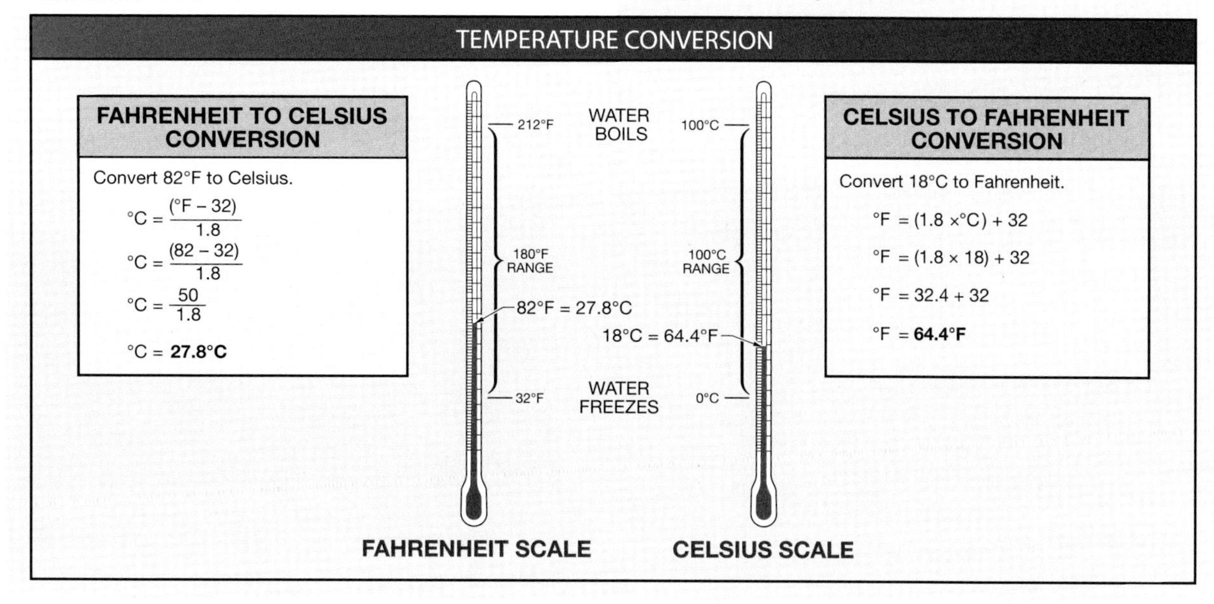

2.4 Continuous Loads and Feeders

Section 215.3 requires feeder overcurrent protective devices to be rated not less than the noncontinuous load plus 125% of the continuous load. **215.2(A)(1)** requires feeder conductors to have an ampacity of not less than 100% of the noncontinuous load plus 125% of the continuous load.

215.2(A)(1), Exception No. 1 permits the use of an assembly with an overcurrent device listed for operation at 100% of its rating. Where breakers of this type are used for continuous load, neither the overcurrent device nor the conductors are required to be increased by 125%. Either way, the selected overcurrent device must always protect the feeder conductors.

2.4.1 Continuous Loads Only

Sizing continuous loads only from a *Code* calculation point of view is simple and straight forward: 125% of the continuous load sizes the overcurrent device, and also sizes the conductors. But, a maintenance Electrical Worker knows intuitively that continuous loads operating 24 hours a day, seven days a week end up causing outages and finally damage to electrical equipment. Looking at it in a different way, the more hours a full load operates, the more the equipment is heated, without cooling or resting from heat. Other than motors, equipment heated for long periods of time generally becomes unreliable. Equipment which is operating at 100%, loaded continuously 24/7, is most likely vital equipment. This is the primary reason why many engineers design feeders and transformers with continuous loads much larger than *Code* minimums of 125%.

Problem 2-10

What size feeder overcurrent device is needed on a 240-volt, single-phase feeder supplying a 175-ampere continuous load?

Solution
Section 215.3 Feeder OCPD
Min. OCPD = load amps × 125%
= 175 × 1.25 = 218.75 amps
240.4(B) and 240.6(A), next larger std size
Next larger std size = 225 amps
Answer: 225 ampere fuse or circuit breaker

2.4.2 Continuous and Noncontinuous Loads

Whenever feeders for continuous and noncontinuous loads are calculated, there is an exception which permits the grounded conductor of the feeder to be sized at only 100% of the continuous plus noncontinuous loads. Since the grounded conductor is not connected to the overcurrent device protecting the feeder, sizing the conductors 25% larger than the load is not required for sizing the grounded conductor. **Problem 2-11** introduces and demonstrates this application of **Exception No. 2** of **215.2(A)(1)**.

Problem 2-11

A feeder supplies a continuous load of 100 amperes and a noncontinuous load of 35 amperes.
1. What is the minimum standard rating of time-delay fuses used for the feeder overcurrent protection?
2. What size THWN copper conductors are needed?
3. What size THWN grounded copper conductor is needed?

Solution – Calculation 1
Section 215.3 OCPD selection
Not less than (CL × 125%) + NCL
Min. OCPD = (100 × 1.25) + 35
= 160 amps
240.4(B) and 240.6(A), next larger std size
Next larger std size = 175 amps
Answer: 175 ampere time-delay fuses

Solution – Calculation 2
215.2(A)(1) Conductor size selection
Not less than (CL × 125%) + NCL
Min. OCPD = (100 x 1.25) +35
Min. amps = 160 amps
Table 310.15(B)(16) Ampacity
160 amps = 2/0 AWG THWN copper
Answer: 2/0 AWG THWN copper

Solution – Calculation 3
Grounded conductor (alternate solution)
Reduced size, 215.2(A)(1), Exception No. 2
Min. amps = Not less than 100% of CL + NCL
= 100 + 35 = 135 amps
Table 310.15(B)(16) Ampacity
135 amps = 1/0 AWG THWN Cu
Answer: 1/0 AWG THWN Cu

Temperature limitations of the conductor terminations are marked on the inside of equipment, such as this listed fused switch.

2.4.3 Continuous Loads in a Three-Phase Feeder

It is important to understand that a grounded conductor, with or without harmonic currents present, is permitted to be calculated according to **215.2(A)(1) Exception No. 2.**

Grounded conductors with and without harmonic currents are covered further in **220.61(A)**, which will be covered in Chapter 8.

Problem 2-12

A 120/208-volt, 3-phase feeder supplies a continuous lighting load of 42,400 VA, with no harmonic currents considered.

1. What is the minimum overcurrent protection for this feeder?
2. What is the minimum size THWN Cu conductors required for this feeder?
3. What is the minimum size THWN Cu grounded conductor required?

Solution – Calculation 1
Section 215.3 Overcurrent protection

$$CL = \frac{VA}{E \times 1.73} = \frac{42{,}400}{208 \times 1.73}$$
$= 117.8$ amps

Min. OCPD= CL × 125%
= 117.8 × 125%
= 147.25 amps

240.4(B) and 240.6(A), next larger std size
Next larger std size = 150 amps
Answer: 150 ampere breaker or fuse

Solution – Calculation 2
Ungrounded Conductors Size
215.2(A)(1)
The 147.25 amps in Calculation 1 is also for the conductors.
110.14(C)(1)(b)(2) CB terminals
Use Table 310.15(B)(16) 75°C column
Table 310.15(B)(16) Ampacity
147.25 amps = 1/0 AWG THWN copper
Answer: 1/0 AWG THWN copper

Solution – Calculation 3
Grounded Conductor Size (minimum)
215.2(A)(1), Exception No. 2
Not less than 100% of the CL + NCL
Min. amps = 117.8 amps
Table 310.15(B)(16) Ampacity
118 amps = 1 AWG THWN Cu
Answer: 1 AWG THWN Cu

2.5 Services with Continuous and Noncontinuous Loads

Service calculations according to **Section 230.42** are very similar to previous requirements for calculating sizes of overcurrent protective devices and conductors where continuous and noncontinuous loads are present. Service calculations are usually based upon anticipated loads prepared according to **Article 220**. In addition, **Part III** of **Article 220** permits the use of demand factors which will be dealt with in later chapters. So, for these calculations, we will simply show how to perform calculations using given loads postponing the application of demand factors of **Article 220** to later work.

Problem 2-13

A 277/480-volt, 3-phase, 4-wire service supplies a small retail commercial building. The calculated loads given are a continuous load of 75 kVA and a noncontinuous load of 60 kVA .

1. What is the minimum size THWN-2 copper conductors required for this service?
2. What is the minimum overcurrent protection for the service?

Solution– Calculation 1
Service size using THWN-2 copper
230.42(A)(1) Min. size

$$CL\ amps = \frac{kVA \times 1000}{E \times 1.73} = \frac{75 \times 1000}{480 \times 1.73}$$
$= 90.3$ amps

$$NCL\ amps = \frac{VA}{E \times 1.73} = \frac{60{,}000}{480 \times 1.73}$$
$= 72.3$ amps

Service amps = (CL × 125%) + NCL
= (90.3 × 1.25) + 72.3
= 185 amps

110.14(C)(1)(b)(2) CB Terminals
Use Table 310.15(B)(16) 75°C column
Table 310.15(B)(16) Ampacity
185 amps = 3/0 AWG
Answer: 3/0 THWN-2 Cu

Solution – Calculation 2
Service overcurrent protective device, minimum
Min. OCPD size = 185 amps
240.4(B) and 240.6(A), next larger std size
Next larger std size = 200 amps
Answer: 200 ampere breaker or fuse

Definitions and Terms

Branch Circuit - The circuit conductors between the final overcurrent device protecting the circuit and the outlet(s).

Branch-Circuit Overcurrent Protective Device. - A device capable of providing protection for service, feeder, and branch circuits and equipment over the full range of overcurrents between its rated current and its interrupting rating. Such devices are provided with interrupting ratings appropriate for the intended use, but no less than 5,000 amperes.

Continuous Load - A load where the maximum current is expected to continue for 3 hours or more.

Device - A unit of an electrical system, other than a conductor, that carries or controls electric energy as its principal function.

Equipment - A general term, including fittings, devices, appliances, luminaires, apparatus, machinery, and the like used as a part of, or in connection with, an electrical installation.

Feeder - All circuit conductors between the service equipment, the source of a separately derived system, or other power supply source and the final branch-circuit overcurrent device.

Interrupting Rating - The highest current at rated voltage that a device is identified to interrupt under standard test conditions.

Listed - Equipment, materials, or services included in a list published by an organization that is acceptable to the authority having jurisdiction and concerned with evaluation of products or services, that maintains periodic inspection of production of listed equipment or materials or periodic evaluation of services, and whose listing states that either the equipment, material, or service meets appropriate designated standards or has been tested and found suitable for a specified purpose.

Overcurrent - Any current in excess of the rated current of equipment or the ampacity of a conductor. It may result from overload, short circuit, or ground fault.

Utilization Equipment - Equipment that utilizes electric energy for electronic, electromechanical, chemical, heating, lighting, or similar purposes.

Abbreviations			
60°C/75°C	Dual-Rated Terminal Temperature Rating	OCPD	Overcurrent Protective Device
Al	Aluminum	Max	Maximum
CB	Circuit Breaker	Min	Minimum
CL	Continuous Load	NCL	Noncontinuous Load
Cu	Copper	Std	Standard

Summary

- **110.3(B)** requires that listed or labeled equipment be installed and used in accordance with any instructions included in the listing or labeling.
- Overcurrent devices and their enclosures are most often marked with temperature ratings or temperature limitations.
- The lowest temperature rating or limitation on electrical equipment must not be exceeded.
- The lowest temperature rating may be viewed as the weakest link in a chain. A chain will break upon stress if the strength of the weakest link is exceeded.
- Circuit breakers and equipment rated 100 amperes or less are governed by four separate rules according to **110.14(C)(1)(a)**:
  - » Conductors rated at 60°C connected to equipment terminations rated at 60°C
  - » Conductors rated at a higher temperature connected to equipment terminations rated at 60°C provided the 60°C ampacity of conductor size used
  - » Conductors rated at a higher temperature connected to equipment terminations listed and identified for such high temperature conductors
  - » For many common motors, conductors having an insulation rated at 75°C or higher can be used, provided the conductors do not exceed their 75°C ampacity.
- Circuit breakers and equipment rated over 100 amperes are governed by two separate rules according to **110.14(C)(1)(b).**
  - » Equipment terminal temperature ratings of 75°C
  - » Conductors rated at a higher temperature, provided the ampacity of such conductors does not exceed the 75°C ampacity of the conductor size used, or up to their ampacity if the equipment is listed and identified for use with such conductors.
- The advantage of higher temperature rated conductors is that correction factors and the adjustment factor may be applied to them, and the resulting allowable ampacity is often greater than the 60°C or 75°C ampacity rating of the same size conductor.
- A practical solution to temperature rated equipment is simplified by using a dual temperature rating (60°C/75°C) on terminations of equipment
- Overcurrent devices for branch circuits and feeders are sized based upon the sum of the noncontinuous load plus 125% of the continuous load. That amount is used to select a matching overcurrent device size or the next larger size according to **Section 240.6.**
- Circuit conductors for branch circuits and feeders are sized based upon the sum of the noncontinuous load plus 125% of the continuous load. The conductor sizes may need to be adjusted according to applicable adjustment and correction factors of **Table 310.15(B)(16).**
- If 100% rated circuit breakers are to be used, feeders are sized based upon the sum of the noncontinuous load plus the continuous load. That amount is used to select a matching overcurrent device size or the next larger size according to **Section 240.6.**
- Circuit conductors for feeders attached to 100% rated circuit breakers are sized based upon the sum of the noncontinuous load plus the continuous load. The conductor sizes may need to be adjusted according to applicable adjustment and correction factors of **Table 310.15(B)(16).**
- Services with continuous and noncontinuous loads are sized according to **Section 230.42** with their loads calculated according to **Part III** of **Article 220.**

Boxes

Electrical boxes are often used to begin, splice, and terminate various electrical wiring methods used for branch circuits and feeders. Boxes also play the same important role in low voltage systems and communication circuits. Electrical boxes permit accessibility to the actual circuit conductors, thereby allowing easy access for measurement, testing, and where necessary, additions and replacement of the circuit conductors. Electrical boxes are constructed of metal or nonmetallic (composite) material. Sometimes boxes and (accessible) fittings are simply used to permit raceways and cables to change direction or to bend around tight corners. Whatever their purpose and whatever their construction, the Electrical Worker must clearly understand the safety aspects of selecting a properly-sized box for the required conditions of use. This involves the physical aspects of layout and placement, understanding the environment the box is placed in, and understanding how the conductors and other various devices or equipment within the enclosure will be protected from a harsh environment outside of the enclosure. This chapter uses calculations driven by **Article 314** to determine the safe amount of physical space or volume within the box required for a particular set of installation conditions.

The title of **Article 314** is actually a description of the various boxes and associated items covered within the article. Included are outlet and device boxes, pull and junction boxes, conduit bodies, fittings, and handhole enclosures. Each box or enclosure installed is judged (and often inspected) for its ability to permit a quantity of conductors, splices, devices, and equipment to operate safely, thereby preventing shock or fire hazard to persons or property using these enclosures and the devices placed within them.

Objectives

- Identify and distinguish between a conduit fitting and a device.
- Recognize and describe standard outlet boxes.
- Determine and apply proper volume allowances for each type of box fill.
- Calculate minimum outlet box sizes using both **Table 314.16(A)** and **Table 314.16(B)**.
- Compare and contrast the various methods of determining conductor fill for standard boxes.
- Calculate the minimum dimensions for pull and junction boxes for straight and angle pulls of conductors 1000 volts, nominal, or less.
- Calculate the minimum dimensions for pull and junction boxes for straight and angle pulls of conductors over 1000 volts.
- Calculate the minimum dimensions for handhole enclosures for conductors 1000 volts, nominal, or less.

Chapter 3

Table of Contents

3.1 Determining the Number of Conductors in a Box

According to **Section 314.16**, boxes and conduit bodies must be an approved size to provide free space for all enclosed conductors. However, it does not apply to motor terminal housings. These requirements apply only to wire sizes 18 AWG through 6 AWG. Conductor sizes 4 AWG and larger will be addressed later in this chapter.

To determine the maximum number of conductors which may be placed within a box, we need to determine the total space within an empty box, the space taken up by the conductors within the box, and the space taken up by various other devices, equipment, and fittings within the box. The fundamental equation used to determine the maximum number of conductors permitted inside a box is equal to the volume of space within the box minus the space required for the volume of conductors. Where the maximum number of conductors permitted inside a box is determined in accordance with **Section 314.16**, there will be sufficient free space for the enclosed conductors.

3.2 Box Fill for Standard Boxes

Table 314.16(A) lists the maximum number of conductors permitted in a box, providing all of the following are true:

- The box is a standard size
- There is nothing in the box except conductors
- All the conductors are the same AWG size
- No conductor is larger than 6 AWG

When equipment other than conductors (such as fittings and devices) are installed in a standard box, adjustments have to be made for the number of conductors permitted. **314.16(B)(1)** through **314.16(B)(5)** list the deductions from **Table 314.16(A)** which need to be made. **Section 3.2.1** through **Section 3.2.5** summarize the volume allowance which will be required. Remember, where extension boxes, extensions rings, and plaster frames are stacked atop of standard boxes, the total available box volume becomes the sum of the volumes of the items assembled together.

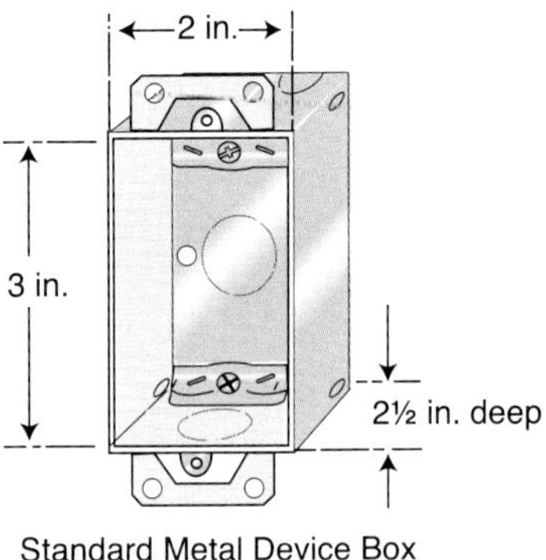

Figure 3-1. An example of a standard metal device box found in Table 314.16(A) with a minimum volume of 12.5 in.^{3}

3.2.1 Conductors

1. Each conductor which originates outside the box and terminates within the box is counted as one conductor.
2. Each conductor which originates outside the box and is spliced within the box counts as one conductor. A conductor entering a box is spliced to a conductor leaving the box, it counts as two conductors.
3. Each conductor which passes through the box without splice or termination counts as one conductor.
4. A conductor, no part of which leaves the box, shall not be counted. If the unbroken conductor is as long or longer than twice the minimum length, the long conductor counts as two conductors. The minimum conductor length required within boxes for conductor terminations and splices is explained in **Section 300.14**. The minimum required length of free conductor is 6 inches and is measured from where it emerges from the raceway or cable sheath. For boxes with small openings (an opening dimension less than 8 in.), the minimum required length for each conductor is 6 inches, with at least 3 inches of the conductor outside the opening.

Using the 4 inch metal box shown in **Figure 3-8** along with **Section 300.14**, the minimum length of a free conductor would be 6 inches. So, the unspliced length of free conductor could not exceed 12 inches (6 in. × 2) before the conductor would have to be counted as two conductors.

3.2.2 Equipment Grounding Conductors

1. One or more bare, insulated, or covered system equipment grounding conductors, all the same size, count as one conductor.
2. One or more bare, insulated, or covered system equipment grounding conductors of various sizes count as one conductor based upon the largest equipment grounding conductor.
3. One or more separate isolated equipment grounding conductors count as one conductor in addition to the system equipment grounding conductor(s), based upon the size of the largest isolated equipment grounding conductor in the box.

3.2.3 Clamps

Clamps installed inside the box by the factory or installed in the field are counted as one conductor. In both cases, the conductor size is based upon largest conductor present in the box. No allowance is required for a cable connector if the clamping mechanism is outside the box. Clamp assemblies with cable terminations are not covered in this chapter.

3.2.4 Support Fittings

One or more fixture studs and fixture hickeys count as one conductor. In both cases, the conductor size is based upon the largest conductor present in the box.

3.2.5 Device(s) or Equipment in a Box

1. Yoke or strap with switch, count as two conductors.
2. Yoke or strap with duplex receptacle, count as two conductors.
3. Yoke or strap with triplex receptacle, count as two conductors.
4. Yoke or strap with switch and light on the same yoke or strap, count as two conductors.
5. Yoke with any combination of three devices or equipment, on the same yoke or strap, count as two conductors.
6. Where a device or utilization equipment is wider than that required for a single gang device box (2 inches), each gang of width required for mounting must count as two conductors.

In all cases, the conductor size used here is based upon the conductor size connected to the device or equipment on the yoke or strap.

3.3 Examples of Individual Volume Allowances

The following examples illustrate the individual volume allowances for various installations.

3.3.1 Counting Clamps and Fittings

Determining the volume allowances for clamps and fittings is covered in **314.16(B)(2)** and **314.16(B)(3)**. See **Figure 3-2** for an example of counting fixture studs, hickeys and clamps within a standard metal box.

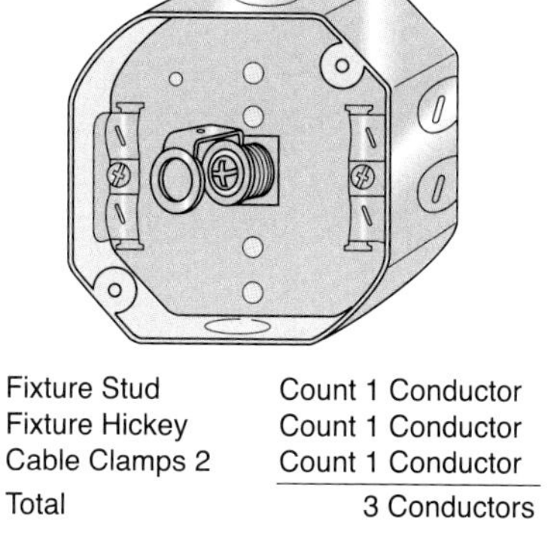

Fixture Stud	Count 1 Conductor
Fixture Hickey	Count 1 Conductor
Cable Clamps 2	Count 1 Conductor
Total	3 Conductors

Figure 3-2. An example of counting fixture studs, hickeys and clamps within a 4 × 1 1/2 in. round/octagonal standard metal box used in Table 310.16(A).

3.3.2 Counting Devices

Counting devices inside a box is not the same as it used to be. **314.16(B)(4)** Device or Equipment Fill covers these requirements. The definition of utilization equipment, found in **Article 100**, includes modern devices which use or consume small amounts of energy. And clearly, box-mounted equipment is included as well.

For each yoke, mounting assembly, or strap containing one or more devices, a double volume (conductor) allowance is made in accordance with **Table 314.16(A)**, based upon the largest conductor connected to the device. Where a device or utilization equipment is wider than a standard device width box (2 inches), a double volume allowance must be made for each standard device width. The graphic example of this section clearly shows the number of conductors of volume necessary for each single-gang device represented. Not shown, however, is extra wide equipment such as fire alarm horn/light (flush and semirecessed) assemblies requiring standard two-gang boxes. Also not shown are standard two-gang electrical devices, such as a 4-wire, 30-ampere electric dryer receptacle. For these types of devices or equipment, a double volume (conductor) allowance must be made for each gang or each 2 inches of width the device requires, in addition to the other fill requirements of **Section 314.16**.

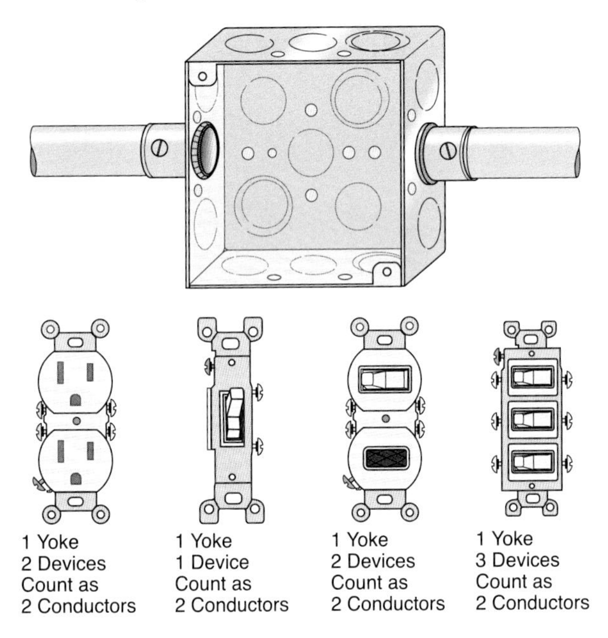

Figure 3-3. An example of counting devices which may be located within the standard metal box according to Table 310.16(A).

3.3.3 Counting Equipment Grounding Conductors

Two types of grounding conductors are:

1. The equipment grounding conductor (EGC) used for general grounding of metal equipment and raceways, which can be insulated, covered, or bare
2. The isolated equipment grounding conductor (IEGC) used specifically for isolated equipment grounding as permitted by **250.146(D)**.

For additional information, visit qr.njatcdb.org Item #1028

This 4-wire 30-amp household dryer receptacles is an example of using a double volume allowance for each gang requiring mounting in a 2-gang outlet.

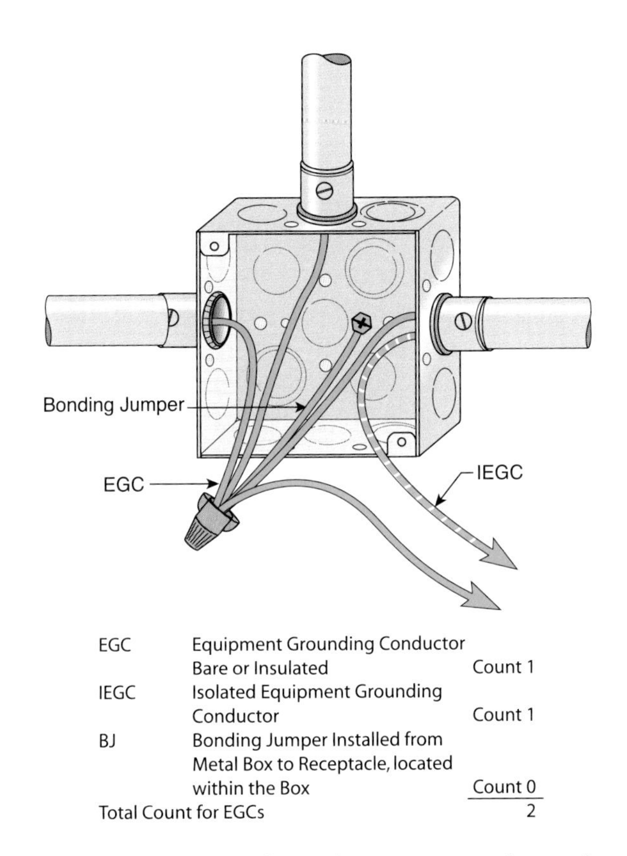

Bonding Jumper
EGC
IEGC
EGC Equipment Grounding Conductor
Bare or Insulated Count 1
IEGC Isolated Equipment Grounding
Conductor Count 1
BJ Bonding Jumper Installed from
Metal Box to Receptacle, located
within the Box Count 0
Total Count for EGCs 2

Figure 3-4. An example of several equipment grounding conductors (EGC) plus an isolated equipment grounding conductor (IEGC) in the same metal box.

3.3.4 Counting Terminated Conductors

Each conductor which terminates in the box is counted as a single conductor.

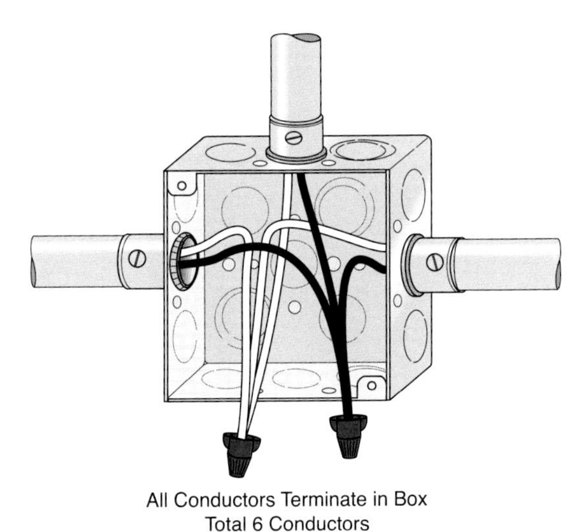

Figure 3-5. An example of counting conductors terminated within a standard metal box.

3.3.5 Counting Conductors Pulled Straight Through

When an unspliced conductor is pulled straight through the box, it counts as one conductor. **Figure 3-6** illustrates an example of a metal box with conductors pulled straight through without a splice.

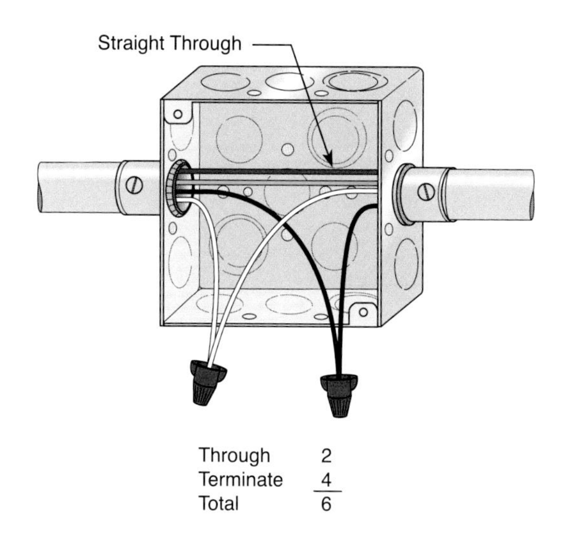

Figure 3-6. An example of counting "straight through" conductors within a standard metal box.

3.3.6 Counting Fixture Wires

When fixture wires are installed to connect to a fixture outside the box, they count as conductors inside the box.

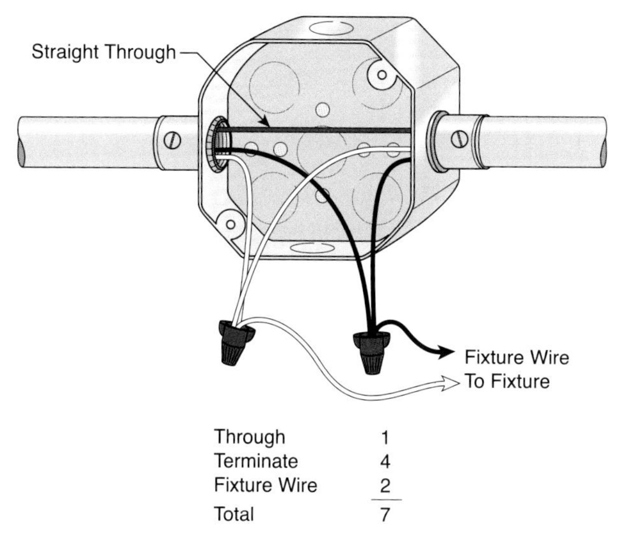

Figure 3-7. An example of counting fixture wires when they leave the metal box.

3.3.7 Counting Looped and Unbroken Conductors

When conductors are looped through the box in unbroken lengths but terminate on a device, they count as two conductors. An unbroken length is defined as "not less than twice the minimum length of free conductor as required by **Section 300.14**..." (usually 6 in. x 2 or a 12 in. min. length). See **Figure 3-8** for an example of unbroken looped conductors which will be connected to a device mounted within the metal box.

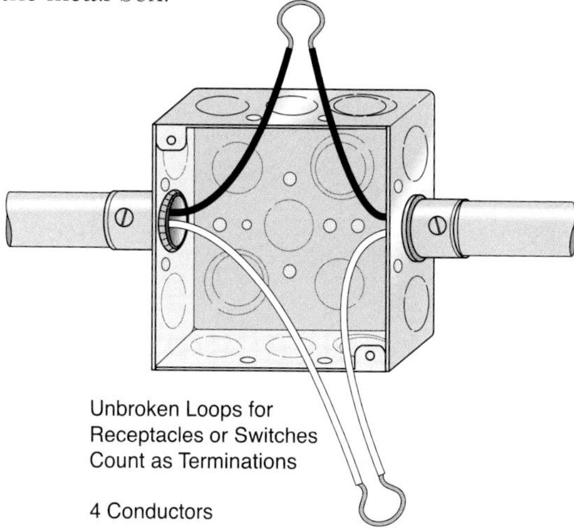

Figure 3-8. An example of unbroken looped conductors which will be connected to a device mounted within the metal box.

3.3.8 Counting Jumper Conductors

Jumpers, sometimes called pigtails, which do not leave the box are not counted. Jumpers are usually connected to devices in the box, such as receptacles and switches.

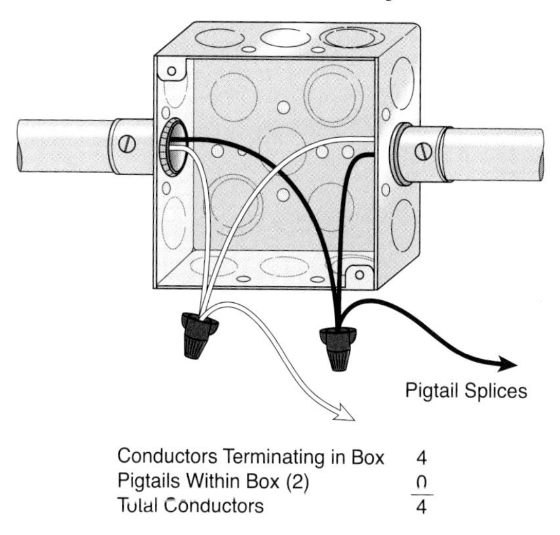

Figure 3-9. An illustration of the exception where some of the conductors within a domed cover are not required to be counted as fill within a box.

3.3.9 Exceptions to Counting Conductors

An exception is made for a dome-type canopy, such as one used with combination fan/light fixture installation. Where either or both of the following terminate in the box, they are not counted in the box calculation:

For additional information, visit qr.njatcdb.org Item #1029

1. Equipment grounding conductor.
2. Not more than four 16 AWG or smaller fixture wires.
3. Or both of the above.
4. Where there are more than four fixture wires, each fixture wire over four counts as 1.

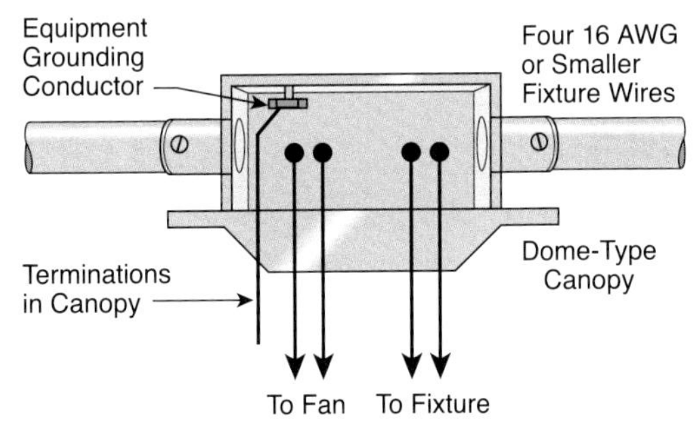

Figure 3-10. An illustration of the exception where some of the conductors within a domed cover are not required to be counted as fill within a box.

3.3.10 Determining Clamp Allowances

According to **314.16(B)(2)**, where there are one or more factory-installed or field-installed internal clamps mounted within a box, all of the clamps count only as one single conductor. Where the cable clamping mechanism is outside of the box, there is no conductor allowance necessary for the cable clamps. Also, there is no allowance necessary for locknuts or bushings installed within a box. **Figure 3-11** shows a standard metal box using Type NM cable clamps.

314.16(B)(4) covers switching and dimming devices and requires a double volume allowance for each of these single gang devices.

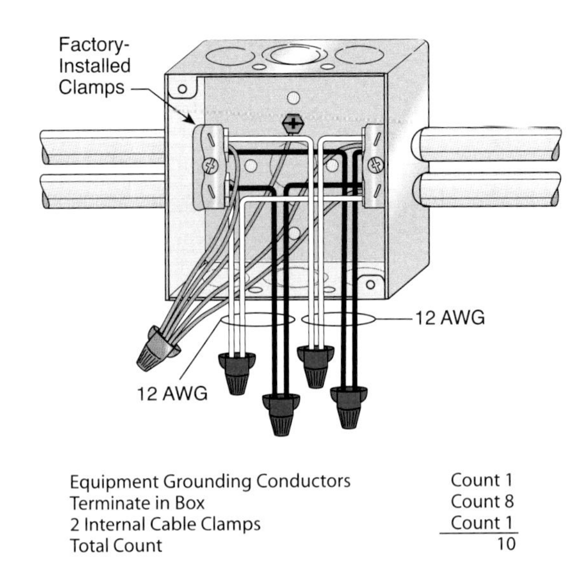

Figure 3-11. An example of Type NM cable clamps used in a standard metal box.

3.3.11 Determining Fitting Allowances

According to **314.16(B)(2)** and **314.16(B)(3)**, the actual conductor allowance varies where different sized conductors occupy the same box. Specifically, when calculating box size for mixed conductor sizes, the cubic inch area (or volume allowance) used for a fixture stud, a cable clamp, or a support hickey is the same as the cubic inch area of the largest conductor entering the box. An example of such a situation is shown in **Figure 3-12**.

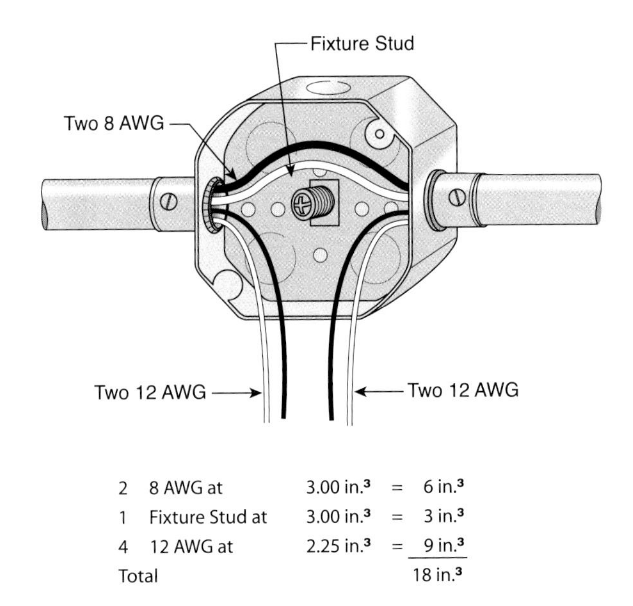

Figure 3-12. An example of different sized (AWG) conductors and a fixture stud support inside a standard metal box.

3.3.12 Determining Yoke or Strap Allowances

The yoke or strap is counted as two conductors according to 314.16(B)(4). When calculating the box size for mixed sizes of conductors, the cubic inch area for the yoke or strap is the cubic inch area of the largest conductor connected to the yoke or strap.

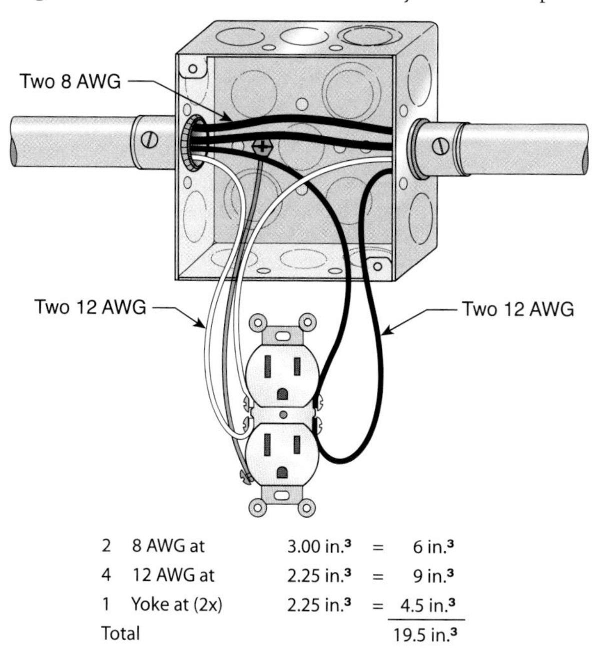

2	8 AWG at	3.00 in.³	=	6 in.³
4	12 AWG at	2.25 in.³	=	9 in.³
1	Yoke at (2x)	2.25 in.³	=	4.5 in.³
Total				19.5 in.³

Figure 3-13. An example of counting a simple device mounted in a box.

3.3.13 Determining EGC Allowances

When more than one equipment grounding conductor (EGC) enters a box, according to 314.16(B)(5), they are counted as one conductor. When calculating the box size for mixed sizes of conductors and there is more than one size equipment grounding conductor, the cubic inch area of the equipment grounding conductor is that of the largest equipment grounding conductor entering the box. The determination of this size conductor is illustrated in **Figure 3-14**.

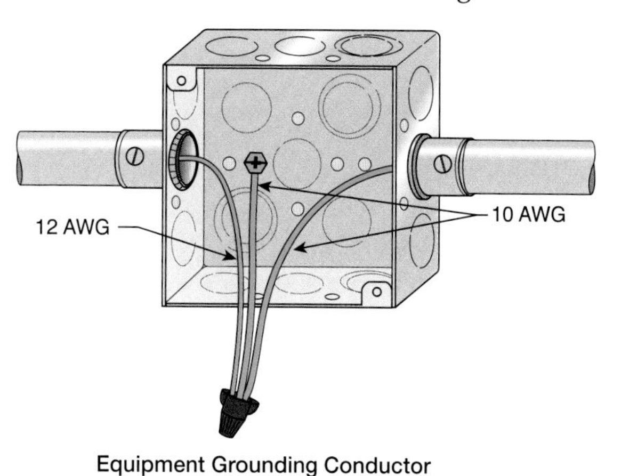

Equipment Grounding Conductor
Use one 10 AWG at 2.5 in.³

Figure 3-14. An example of using the largest Equipment Grounding Conductor (EGC) entering the box.

Enclosure Considerations for Equipment in Ordinary Locations (AALZ)

Enclosure Type Number	Provides a Degree of Protection Against the Following Environmental Conditions*
1	Indoor use
2	Indoor use, limited amounts of falling water
3R	Outdoor use, undamaged by the formation of ice on the enclosure**
3	Same as 3R plus windblown dust
3S	Same as 3R plus windblown dust; external mechanisms remain operable while ice laden
4	Outdoor use, splashing water, windblown dust, hose-directed water, undamaged by the formation of ice on the enclosure**
4X	Same as 4 plus resists corrosion
5	Indoor use to provide a degree of protection against settling airborne dust, falling dirt, and dripping noncorrosive liquids
6	Same as 3R plus entry of water during temporary submersion at a limited depth
6P	Same as 3R plus entry of water during prolonged submersion at a limited depth
12, 12K	Indoor use, dust, dripping noncorrosive liquids
13	Indoor use, dust, spraying water, oil, and noncorrosive coolants

*All enclosure types provide a degree of protection against ordinary corrosion and against accidental contact with the enclosed equipment when doors or covers are closed and in place. All types of enclosures provide protection against a limited amount of falling dirt.

**All outdoor-type enclosures provide a degree of protection against rain, snow, and sleet. Outdoor enclosures are also suitable for use indoors if they meet the environmental conditions present. *Source: Underwriters Laboratories, General Information for Electrical Equipment, White Book, 2010 edition.*

3.4 Using Table 314.16(A) and Table 314.16(B)

3.4.1 Conductors of Different Sizes

When all the conductors in a box are not the same size, the maximum number of conductors is based upon the volume fill of the box. Volume is measured in cubic inches. **Table 314.16(B)** shown in **Figure 3-16** lists the volume fill that must be allowed for each conductor when a combination of sizes is installed.

Table 314.16(A) has a column headed "Minimum Volume (in.^{3})" listing the cubic inch volume of each box. The type of insulation or conductor material is not considered when calculating box fill. **Table 314.16(A)** and **Table 314.16(B)** apply for both copper and aluminum conductors.

Table 314.16(B) Volume Allowance Required per Conductor

Size of Conductor (AWG)	Free Space Within Box for Each Conductor	
	cm^3	in.^{3}
18	24.6	1.50
16	28.7	1.75
14	32.8	2.00
12	36.9	2.25
10	41.0	2.50
8	49.2	3.00
6	81.9	5.00

Figure 3-16. NEC Table 314.16 (B).

Table 314.16(A) Metal Boxes

Box Trade Size			Minimum Volume		Maximum Number of Conductors* (arranged by AWG size)						
mm	in.		cm^3	in.^{3}	18	16	14	12	10	8	6
100 × 32	(4 × 1¼)	round/octagonal	205	12.5	8	7	6	5	5	5	2
100 × 38	(4 × 1½)	round/octagonal	254	15.5	10	8	7	6	6	5	3
100 × 54	(4 × 2⅛)	round/octagonal	353	21.5	14	12	10	9	8	7	4
100 × 32	(4 × 1¼)	square	295	18.0	12	10	9	8	7	6	3
100 × 38	(4 × 1½)	square	344	21.0	14	12	10	9	8	7	4
100 × 54	(4 × 2⅛)	square	497	30.3	20	17	15	13	12	10	6
120 × 32	(4 11/16 × 1¼)	square	418	25.5	17	14	12	11	10	8	5
120 × 38	(4 11/16 × 1½)	square	484	29.5	19	16	14	13	11	9	5
120 × 54	(4 11/16 × 2⅛)	square	689	42.0	28	24	21	18	16	14	8
75 × 50 × 38	(3 × 2 × 1½)	device	123	7.5	5	4	3	3	3	2	1
75 × 50 × 50	(3 × 2 × 2)	device	164	10.0	6	5	5	4	4	3	2
75× 50 × 57	(3× 2 × 2¼)	device	172	10.5	7	6	5	4	4	3	2
75 × 50 × 65	(3× 2 × 2½)	device	205	12.5	8	7	6	5	5	4	2
75 × 50 × 70	(3× 2 × 2¾)	device	230	14.0	9	8	7	6	5	4	2
75 × 50 × 90	(3 × 2 × 3½)	device	295	18.0	12	10	9	8	7	6	3
100 × 54 × 38	(4 × 2⅛ × 1½)	device	169	10.3	6	5	5	4	4	3	2
100 × 54 × 48	(4 × 2⅛ × 1⅞)	device	213	13.0	8	7	6	5	5	4	2
100 × 54 × 54	(4 × 2⅛ × 2⅛)	device	238	14.5	9	8	7	6	5	4	2
95 × 50 × 65	(3¾ × 2 × 2½)	masonry box/gang	230	14.0	9	8	7	6	5	4	2
95 × 50 × 90	(3¾ × 2 × 3½)	masonry box/gang	344	21.0	14	12	10	9	8	7	4
min. 44.5 depth	FS — single cover/gang (1¾)		221	13.5	9	7	6	6	5	4	2
min. 60.3 depth	FD — single cover/gang (2⅜)		295	18.0	12	10	9	8	7	6	3
min. 44.5 depth	FS — multiple cover/gang (1¾)		295	18.0	12	10	9	8	7	6	3
min. 60.3 depth	FD — multiple cover/gang (2⅜)		395	24.0	16	13	12	10	9	8	4

*Where no volume allowances are required by 314.16(B)(2) through (B)(5).

Figure 3-15. NEC Table 314.16 (A).

Problem 3-1

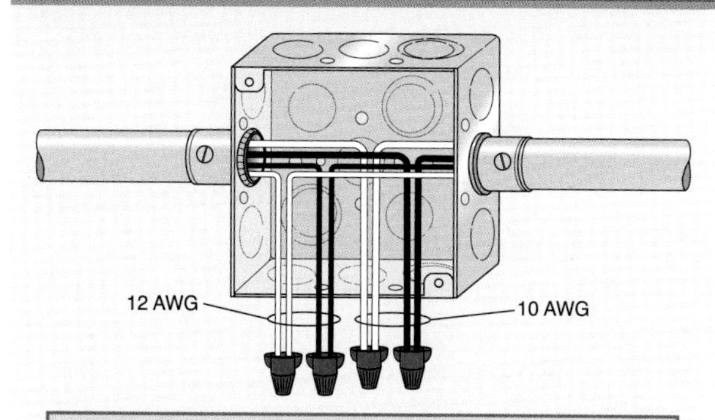

What is the maximum box volume needed for the installation of four 12 AWG THHN copper conductors and four 10 AWG THW aluminum conductors?

Solution

Table 314.16(B)
12 AWG = 2.25 in.^{3}
10 AWG = 2.50 in.^{3}
Total volume

12 AWG = 2.25 in.^{3} × 4	=	9 in.^{3}
10 AWG = 2.50 in.^{3} × 4	=	10 in.^{3}
Total volume required		19 in.^{3}

Answer: 19 in.^{3} min. volume

Problem 3-2

Which of the following four boxes could be used for the installation shown in Problem 3-1?

$4 \times 1\frac{1}{2}$ octagonal	$4 \times 2\frac{1}{8}$ octagonal
$4 \times 1\frac{1}{2}$ square	$4\frac{11}{16} \times 1\frac{1}{4}$ square

Solution

Minimum cubic inch (in.^{3}) volume needed = 19 in.^{3}
Table 314.16(A): in.^{3} volume of boxes

$4 \times 1\frac{1}{2}$ octagonal	=	15.5 in.^{3}	**Not acceptable**
$4 \times 2\frac{1}{8}$ octagonal	=	21.5 in.^{3}	**Acceptable**
$4 \times 1\frac{1}{2}$ square	=	21.0 in.^{3}	**Acceptable**
$4\frac{11}{16} \times 1\frac{1}{4}$ square	=	25.5 in.^{3}	**Acceptable**

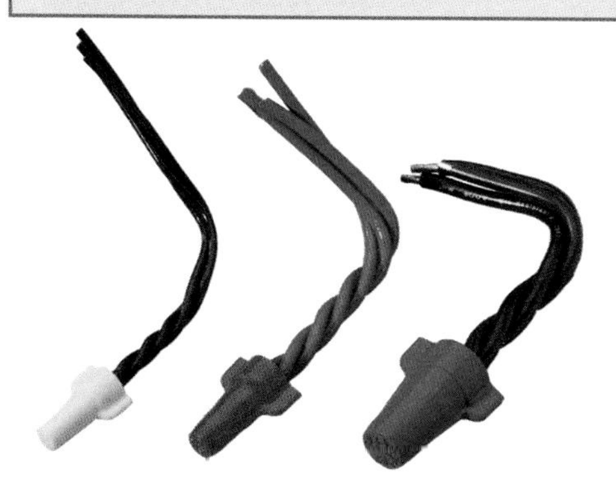

Wire connectors and splicing devices do not count in volume allowance calculations.

Problem 3-3

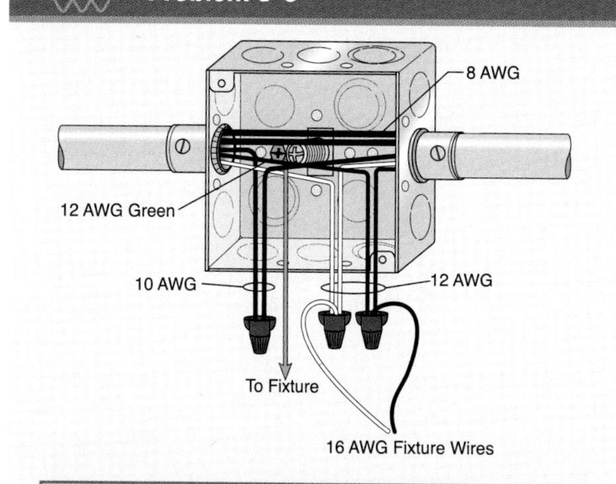

What is the minimum box volume required for the illustration?

Solution

EGCs all count as one 12 AWG
Fixture stud counts the same as the largest conductor
Fixture wires count as two conductors
Table 314.16(A) and Table 314.16(B)
Cubic inch displacement from Table 314.16(B)

16 AWG = 1.75 in.^{3} × 2	=	3.50 in.^{3}
12 AWG = 2.25 in.^{3} × 5	=	11.25 in.^{3}
10 AWG = 2.50 in.^{3} × 2	=	5.00 in.^{3}
8 AWG = 3.00 in.^{3} × 2	=	6.00 in.^{3}
Fixture stud		
8 AWG = 3.00 in.^{3} × 1	=	3.00 in.^{3}
Total box volume		28.75 in.^{3}

Answer: 28.75 in.^{3} min. volume

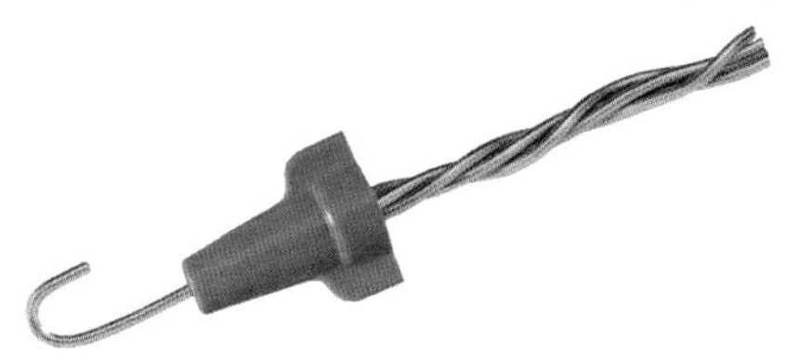

All of the equipment grounding conductors count as a single volume allowance.

Table 314.16(A) is limited to only standard metal boxes.

3.4.2 Nonstandard Boxes

"Nonstandard boxes" are boxes which do not appear in **Table 314.16(A)**. One significant group of non-standard boxes are nonmetallic boxes including round/ octagonal boxes, square boxes, and device boxes. **314.16(B)** covers nonstandard boxes found in **Table 314.16(A)** and boxes 100 cubic inches or less.

The volume of a box is determined by reading the volume stamped on the box or maybe in some cases, reviewing manufacturer's data. Under no circumstance can the volume stamped on the box during manufacture (plus the volume(s) of any assembled section(s)) be less than the fill calculation in accordance with **314.16(B)**.

Problem 3-4

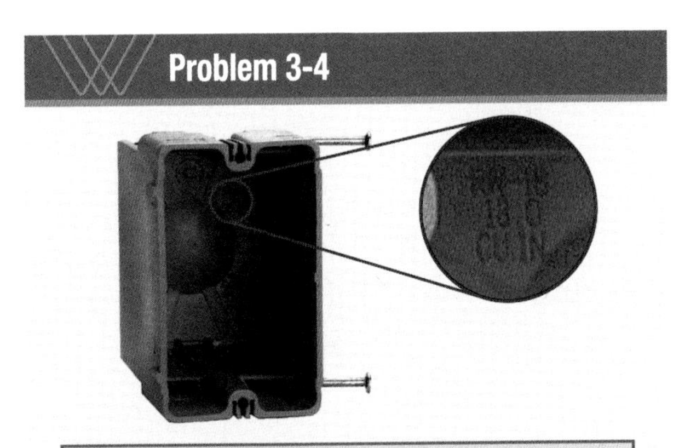

What is the maximum number of 12 AWG conductors which may be installed in a nonmetallic switch box with a marked volume of 18 in.^{3}?

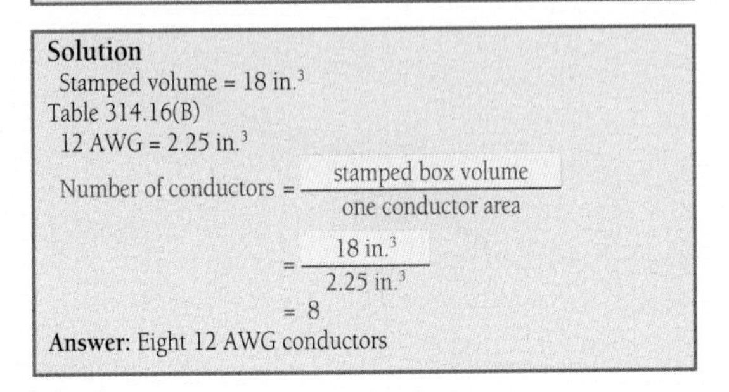

Solution

Stamped volume = 18 in.^{3}

Table 314.16(B)

12 AWG = 2.25 in.^{3}

$$\text{Number of conductors} = \frac{\text{stamped box volume}}{\text{one conductor area}} = \frac{18 \text{ in.}^3}{2.25 \text{ in.}^3} = 8$$

Answer: Eight 12 AWG conductors

Volume calculations of listed nonmetallic boxes with marked clamp assemblies incorporating a cable termination for cable conductors do not require a cable-clamp allowance for the cable clamp(s) within the box, since the marked volume on the box has been adjusted by the manufacturer in accordance with 314.16(B)(2).

3.4.3 Adding to an Existing Box

When conductors are added to existing boxes, the same calculations apply.

Problem 3-5

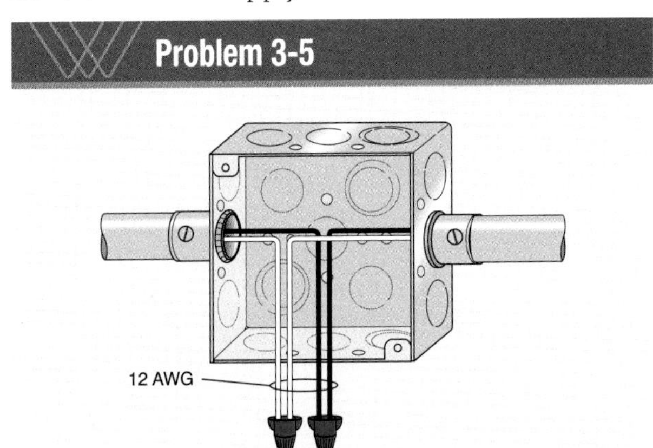

A 4 in. x $1\frac{1}{2}$ in. square box contains four 12 AWG conductors. Assuming that the EMT is large enough, how many additional 14 AWG conductors can be added and pulled directly through the box?

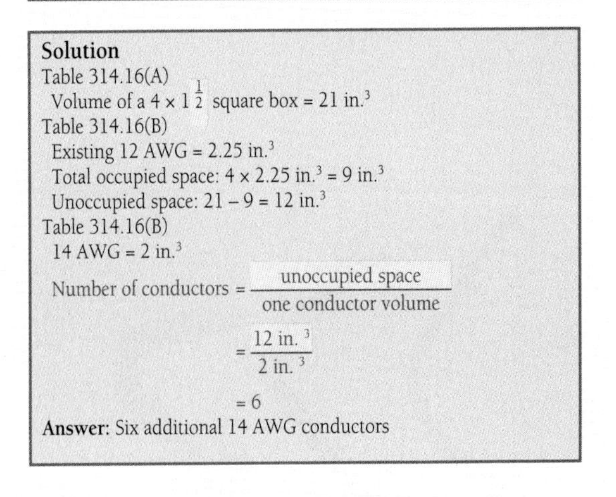

Solution

Table 314.16(A)

Volume of a 4 x $1\frac{1}{2}$ square box = 21 in.^{3}

Table 314.16(B)

Existing 12 AWG = 2.25 in.^{3}

Total occupied space: 4 x 2.25 in.^{3} = 9 in.^{3}

Unoccupied space: 21 – 9 = 12 in.^{3}

Table 314.16(B)

14 AWG = 2 in.^{3}

$$\text{Number of conductors} = \frac{\text{unoccupied space}}{\text{one conductor volume}} = \frac{12 \text{ in.}^3}{2 \text{ in.}^3} = 6$$

Answer: Six additional 14 AWG conductors

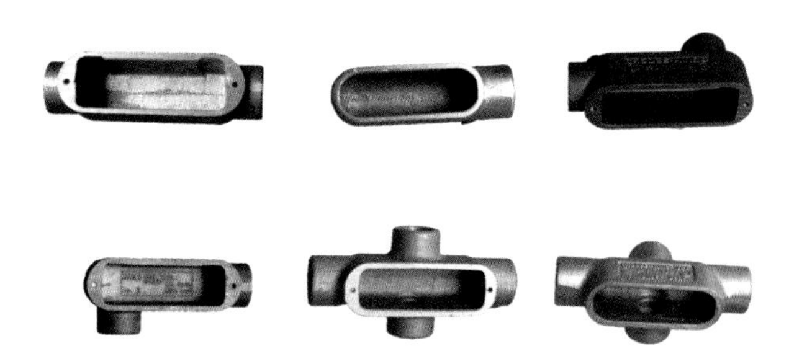

Various conduit bodies

3.4.4 Conduit Body

314.16(C) covers the number of conductors permitted in a conduit body. The maximum number of conductors permitted in a conduit body must be the maximum number permitted by **Chapter 9, Table 1** for the conduit to which the conduit body is attached.

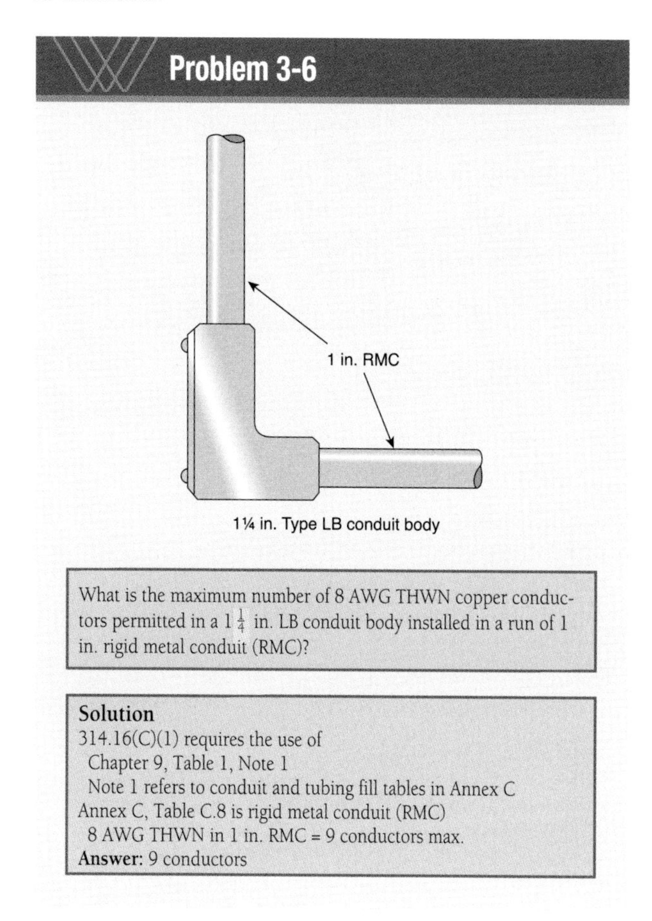

What is the maximum number of 8 AWG THWN copper conductors permitted in a $1\frac{1}{4}$ in. LB conduit body installed in a run of 1 in. rigid metal conduit (RMC)?

Solution
314.16(C)(1) requires the use of
Chapter 9, Table 1, Note 1
Note 1 refers to conduit and tubing fill tables in Annex C
Annex C, Table C.8 is rigid metal conduit (RMC)
8 AWG THWN in 1 in. RMC = 9 conductors max.
Answer: 9 conductors

3.5 Sizing Pull and Junction Boxes for Conductors 1000 Volts or Less

3.5.1 Straight-Through Pull

The largest conductor listed in **Table 314.16(A)** is 6 AWG. For conductors 4 AWG or larger, **Section 314.28** is used for calculating the minimum size box needed. The size of the box is based upon the trade size (standard trade diameter) of the largest raceway or cable assembly entering the box. According to 314.28(A)(1), for straight-through pulls where conductors are installed without splices, the minimum dimension for the length of the box must be at least eight times the raceway trade size. However, as stated in 314.28(A)(2) for straight-through pulls where conductors are spliced, the minimum dimension is allowed to be reduced from eight times to six times the raceway trade size.

Where cables are used, the cable size should match the trade size raceway required for the same conductors.

For junction boxes for conductors over 1000 volts, nominal, a different set of measurements is used. Although the *Code* does not specify the depth of the box, it must be large enough to accommodate the largest raceway or cable fittings, such as locknuts and bushings, which will be used within the box.

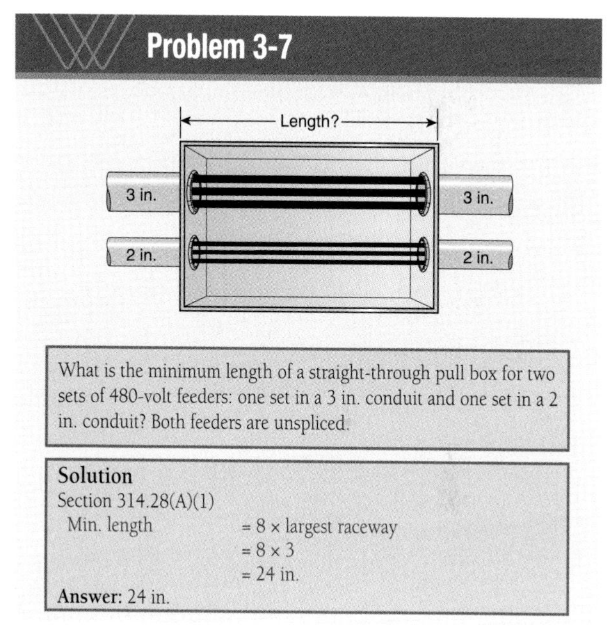

What is the minimum length of a straight-through pull box for two sets of 480-volt feeders: one set in a 3 in. conduit and one set in a 2 in. conduit? Both feeders are unspliced.

Solution
Section 314.28(A)(1)
Min. length = 8 × largest raceway
= 8 × 3
= 24 in.
Answer: 24 in.

These complex back to back bends consist of three 90° bends fabricated in the same length of conduit.

3.5.2 Angle Pull

314.28(A)(2) covers angle pulls or raceway entries which are located on adjacent walls of a junction box enclosure. The distance between raceways entries must not be less than six times the raceway trade size. This distance must be increased by the sum of the trade size diameters of raceways in same row on the same side of the enclosure.

When there is an angle pull box, both rows of raceways need to be calculated and the largest measurement used. For the angle pull, the box can be considered square (length and width measurements the same). As a general rule, the size of the box with the largest diameter raceway usually results in the largest measurement, but not always. Therefore, both sides need to be calculated and the largest calculation used.

Problem 3-8

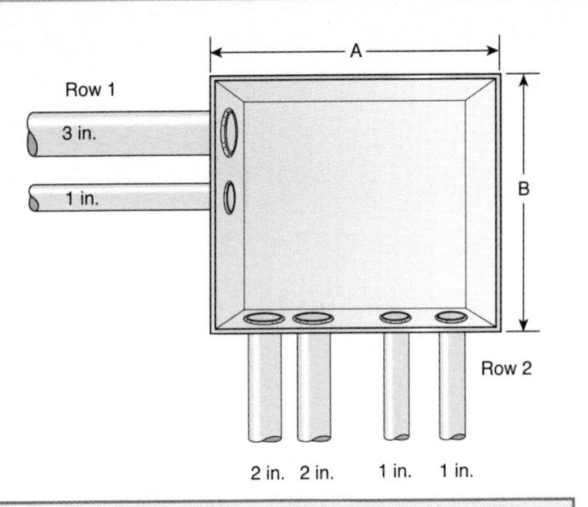

What is the length and width for the following illustrated angle pull box?

Solution
Table 314.28(A)(2)
For any one row = (largest raceway in row × 6) + (other raceway(s) in same row × 1)
Repeat calculation for each row and use the greatest distance calculated for any one row
Row 1 (Dimension A)
= (3 in. × 6) + 1 in.
= 19 in.
Row 2 (Dimension B)
= (2 × 6) + (2 + 1 + 1)
= 16 in.
Using only the minimum dimension = 19 in. × 16 in.
Answer: 19 in. × 16 in.

3.5.3 Dimensions Between Conduit Entries

In addition to sizing the box according to the raceway or cable entries, the entrance and exit of a cable are taken into consideration so the cable is not required to be bent too short. Both measurements must be taken into consideration.

Problem 3-9

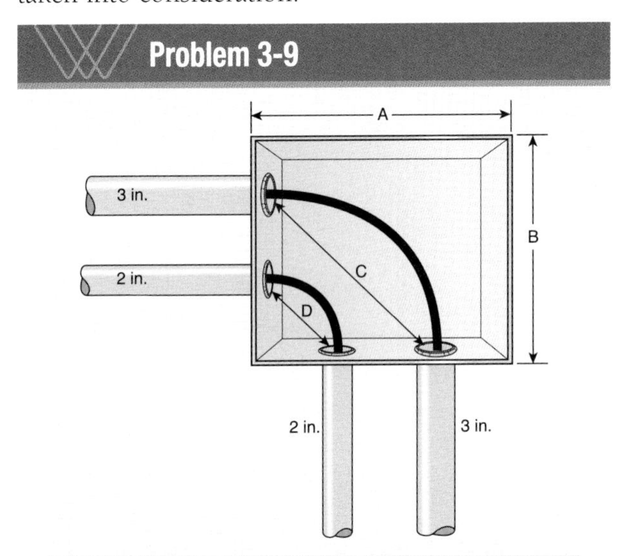

What is the length and width of the box needed and what is the minimum distance between the entry and exit point for the conductors in the following angle pull box?

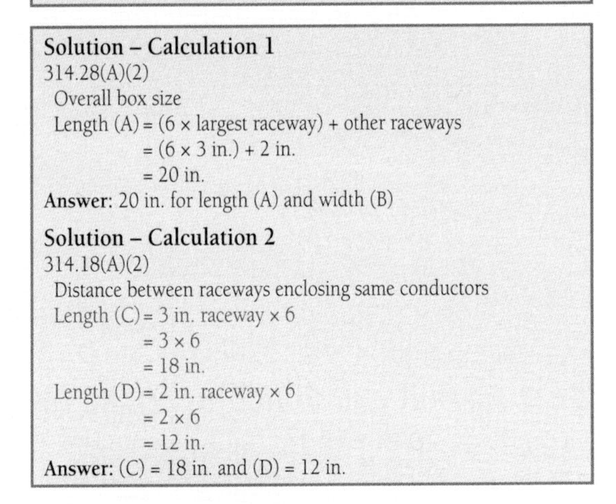
Solution – Calculation 1
314.28(A)(2)
Overall box size
Length (A) = (6 × largest raceway) + other raceways
= (6 × 3 in.) + 2 in.
= 20 in.
Answer: 20 in. for length (A) and width (B)

Solution – Calculation 2
314.18(A)(2)
Distance between raceways enclosing same conductors
Length (C) = 3 in. raceway × 6
= 3 × 6
= 18 in.
Length (D) = 2 in. raceway × 6
= 2 × 6
= 12 in.
Answer: (C) = 18 in. and (D) = 12 in.

Pull and junction boxes located in hazardous locations must be identified as suitable according to 500.8(A).

3.5.4 Raceways Entering Opposite a Removable Cover

The exception for angle pulls permits a long narrow box, very similar to a conduit body. Where a box has a removable cover opposite a raceway entering the box, the depth of the box is sized according to **Table 312.6(A)**. **Table 312.6(A)** is a minimum bending space table for conductors and makes the depth of the box dependent upon the size of the conductors entering the box. The length of the box is not calculated the same as an angle pull. Rather, the distance from the bottom raceway entry to the furthest side is six times the trade size of the largest raceway. The minimum final height of the box is a practical calculation of the sum of 'B' distance plus the outside diameter of the 2 in. locknut plus a fraction for locknut clearance.

Problem 3-10

A
2 in.
Three 4/0 AWG conductors
B
Removable cover
2 in.

Three 4/0 AWG conductors in a 480-volt feeder circuit are pulled through an angle pull box with a removable cover directly opposite the raceway entry. Calculate the length and width of the box.

Solution – Calculation 1
314.28(A)(2) Exception and Table 312.6(A)
'A' dimension is based upon the conductor size entering through the raceway opposite the removable cover and it is based on one conductor termination from Table 312.6(A).
One 4/0 AWG termination = 4 in.
Answer: 4 in.

Solution – Calculation 2
314.28(A)(2)
Distance between raceway entries
'B' dimension = 6 × largest raceway
= 6 × 2
= 12 in.
Answer: 12 in.

3.5.5 Combination Straight and Angle Pulls

When there is a combination of a straight-through pull and an angle pull in the same pull box, both straight-through and angle calculations apply.

Problem 3-11

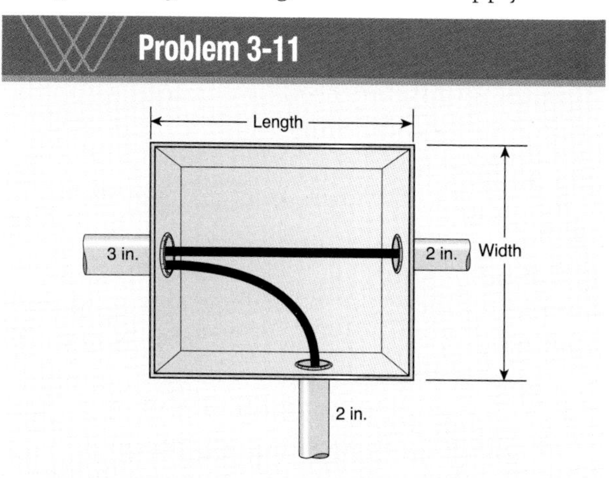

A pull box has a 3 in. conduit entering the box and a 2 in. conduit leaving the box for a straight-through pull and a 2 in. conduit leaving the box for an angle pull. Calculate the length and width for the pull box.

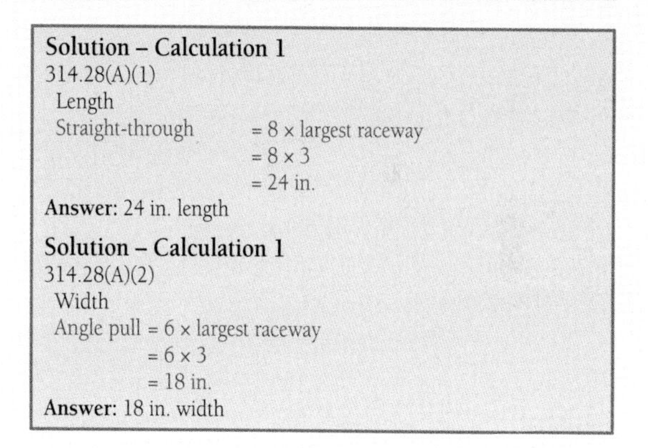

Solution – Calculation 1
314.28(A)(1)
Length
Straight-through = 8 × largest raceway
= 8 × 3
= 24 in.
Answer: 24 in. length

Solution – Calculation 1
314.28(A)(2)
Width
Angle pull = 6 × largest raceway
= 6 × 3
= 18 in.
Answer: 18 in. width

Pull boxes are often installed in larger sizes than required for a more efficient and convenient installation.

3.6 Sizing Pull and Junction Boxes for Conductors Over 1000 Volts

Part IV of Article 314 deals with pull and junction boxes for use on systems over 1000 volts. **314.70(A)** sets the ground rules for all the requirements within Part IV. **314.70(A)** specifically omits any mention of **314.28(A)**, thereby omitting conduit size from the basic calculation of determining the final pull box size where conductors over 1000 volts are installed.

Section 314.71 covers the sizing of pull and junction boxes, conduit bodies, and handhole enclosures containing conductors or cables over 1000 volts nominal (i.e. a 4,160-volt feeder with conductor insulation rated 5,000 volts). The size of the pull or junction box depends upon the type of covering the higher voltage conductors or cables have. Each type of covering requires a different bending radius and a different size pull or junction box. Three types of conductor coverings are used for the higher voltage cables.

1. Shielded
2. Nonshielded
3. Lead covered

As pointed out in the second sentence of **310.10(E)**, most cables above 2 kV will be of the shielded variety. Shielded cables are the most often used. Nonshielded cables are not as common and are generally used as jumpers or connecting cables within switchgear and equipment. Lead-covered or lead-sheathed cables are most often found in older underground distribution installations or in some utility underground installations. The installation of new lead-covered cables is rapidly diminishing in the electrical industry.

3.6.1 Straight-Through Pull Using Single Conductor Cables Over 1000 Volts

The following problems for sizing pull and junction boxes, conduit bodies and handhole enclosures for conductors or cables over 1000 volts, nominal, will illustrate the calculations for each type of conductor or cable covering. Use the largest outside diameter (OD) of any one conductor or cable.

314.71(A) requires the following minimum measurements for straight-through pull or junction boxes:

MINIMUM LENGTH (L) FOR SHIELDED CABLES
= LARGEST CABLE OD × 48

MINIMUM LENGTH (L) FOR NONSHIELDED CABLES
= LARGEST CABLE OD × 32

MINIMUM LENGTH (L) FOR LEAD-COVERED CABLES
= LARGEST CABLE OD × 48

Problem 3-12

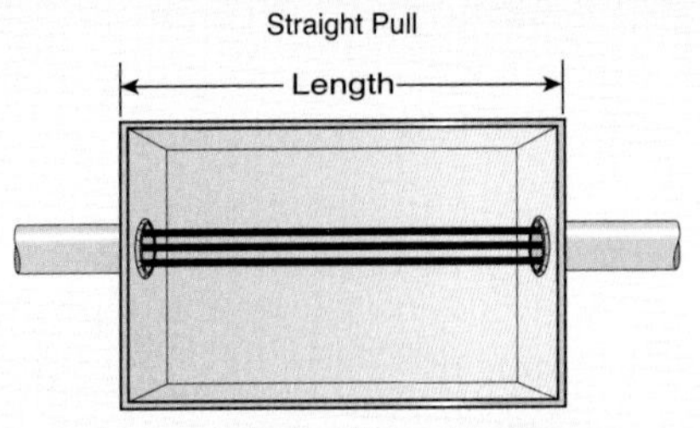

Perform three separate calculations to find the length and width of a straight-through pull box for three 5 kV conductors with an outside diameter of 1 in. when all the conductors are either:
1. Shielded Conductors
2. Nonshielded Conductors
3. Lead covered Conductors

Solution – Calculation 1
314.71(A) Shielded conductors
Length = OD of conductor × 48
= 1 in. × 48
= 48 in.
Answer: 48 in. length

Solution – Calculation 2
314.71(A) Nonshielded conductors
Length = OD of conductor × 32
= 1 in. × 32
= 32 in.
Answer: 32 in. length

Solution – Calculation 3
314.71(A): Lead-covered conductors
Length = OD of conductor × 48
= 1 in. × 48
= 48 in.
Answer: 48 in. length

Comment
The problems associated with conductors over 600 volts omit all references to conduit sizes to ensure the student understands that these pull boxes are not related to conduit sizes. Rather, the *Code* stipulates that only the diameter and type of cable determine the minimum pull or junction box size. The maximum size box or enclosure is not considered by the *NEC*.
Most often, many practical considerations force the increase of the box size above any minimum mandated by the *NEC*. These considerations are based on several factors, such as quantity, size, type of cable splices (if any), and whether cable racks or other supports are installed for conductor(s) supports within the box. For multiconductor cables, these factors require even more consideration. Consultation with the cable manufacturer is also critical for a satisfactory installation.

3.6.2 Straight-Through Pull Using Multiconductor Cables Rated Over 1000 Volts

Problem 3-13

Straight Pull

Length

Cable OD 1.25 in.

Perform two separate calculations to determine the length dimension of the junction box illustrated for a 1.25 in. diameter multiconductor cable when the single cable is either:

1. Shielded
2. Nonshielded

Solution – Calculation 1
314.71(A)
Shielded cable
Length = OD of cable × 48
= 1.25 × 48
= 60 in.
Answer: 60 in.

Solution – Calculation 2
314.71(A)
Nonshielded cable
Length = OD of cable × 32
= 1.25 × 32
= 40 in.
Answer: 40 in.

Segmented bends all have the same center point, but each has a different radius.

3.6.3 Angle Pull and Conduit Entry Dimension Using Cables Rated Over 1000 Volts

Just as with the lower-voltage cables, when an angle pull is made, both the size of the box and the point of entry of the cable or conductors and the exit angle are considered. The following problems will take both into consideration.

314.71(B)(1) requires the following as minimum measurements for angle pull or junction boxes:

Minimum distance between entry and exit = OD of largest cable × 36

Shielded single conductors = (OD of largest cable × 36) + ODs of other cables

Shielded multiple-conductor cables = (OD of largest cable × 36) + ODs of other cables

314.71(B)(1) Exception No. 2 covers the installation of nonshielded conductors and cables and requires the following as the minimum measurements for angle pull or junction boxes:

Minimum dimension for distance between entry and exit = OD of largest cable × 24

Minimum distance from cable entry to the opposite wall of the box for nonshielded single cables = (OD of largest cable × 24) + ODs of other cables

Minimum distance from cable entry to the opposite wall of the box for nonshielded multiple conductor cables = (OD of largest cable × 24) + ODs of other cables

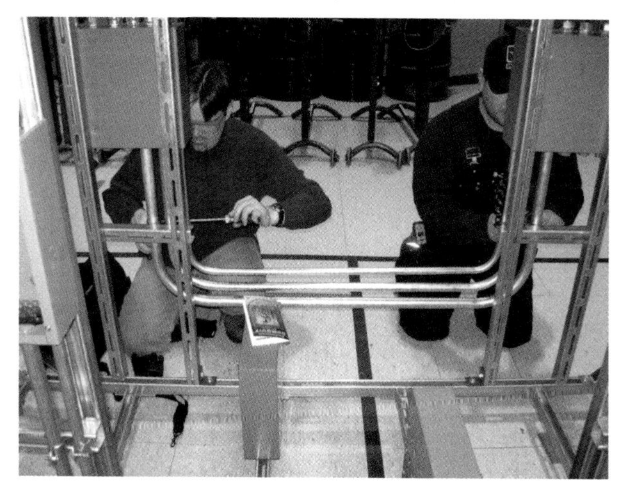

Back to back 90° bends consist of two 90° bends fabricated in the same length of conduit.

Problem 3-14

1. Calculate the minimum size pull box measurements for dimensions A, B, C and D using the following pull box containing only shielded single-conductor high-voltage cables.
2. Calculate the minimum size pull box measurements for dimensions A, B, C and D using the following pull box containing only nonshielded single-conductor high-voltage cables.

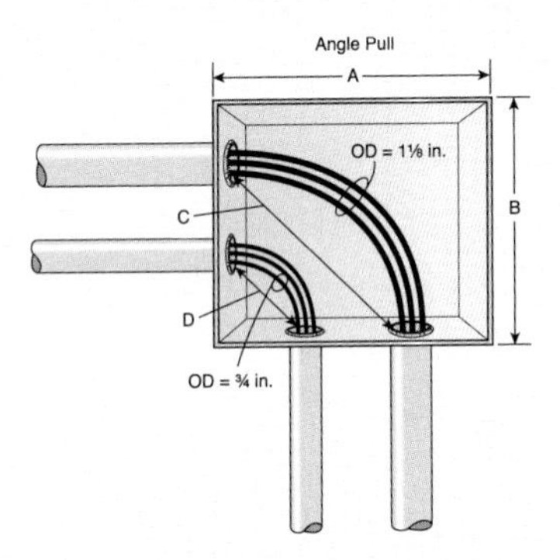

Dimension (A) = Length
Dimension (B) = Width
Dimension (C) = Diagonal for large conductors
Dimension (D) = Diagonal for small conductors

Solution 1 Shielded Conductors - 314.71(B)(1)
Find: Min. dimensions for (A) length and (B) width
Min. length of side = (OD of largest cable × 36) + OD of other cables

OD of largest cable	= $1\frac{1}{8}$ in.	= 1.125 in.	
OD of smallest	= $\frac{3}{4}$ in.	= 0.75 in.	
OD of largest × 36	= 1.125 in. × 36 =		40.50 in.
OD of others	= 2 @ 1.125 in. =		2.25 in.
	= 3 @ 0.75 in. =		2.25 in.
Total			45.00 in.

$L = (OD_L \times 36) + (n \times OD_1) + (n \times OD_2) + (n \times OD_3)$
$= (1.125 \times 36) + (2 \times 1.125) + (3 \times 0.75)$
$= 45.00$

Answer: Both (A) length and (B) height = 45 in. min.

Shielded Conductors - 314.71(B)(1)
Find: Min. length of diagonal lines (C) and (D)
Min. length of diagonal = OD of cable × 36

OD of largest (C)	= $1\frac{1}{8}$ in	= 1.125 in.
OD of smallest (D)	= 3/4 in.	= 0.75 in.
(C) diagonal length	= 1.125 in. × 36	= 40.5 in
(D) diagonal length	= 0.75 in. × 36	= 27.0 in.

Answers: (C) diagonal = 40.5 in.
(D) diagonal = 27 in. min.

Solution 2 Nonshielded Conductors - 314.71(B)(1) Exc. No. 2
Find: Min. dimensions for (A) length and (B) width
Min length of side = (OD of largest cable × 24) + OD of other cables

OD of largest cable	= $1\frac{1}{8}$ in.	= 1.125 in.	
OD of smallest	= $\frac{3}{4}$ in.	= 0.75 in.	
OD of largest × 24	= 1.125 in × 24		= 27.00 in.
OD of others	= 2 @ 1.125 in.		= 2.25 in.
	= 3 @ 0.75 in.		= 2.25 in
Total			31.50 in.

$L = (OD_L \times 36) + (n \times OD_1) + (n \times OD_3)$
$= (1.125 \times 36) + (2 \times 0.125) + (3 \times 0.75)$
$= 31.50$

Answer: 31.50 in. or 32 in. minimum length or width

Nonshielded Conductors
Find: Min. length of diagonal lines (C) and (D)
Min. length of diagonal = OD of largest × 24

OD of largest (C)	= $1\frac{1}{8}$ in	= 1.125 in.
OD of smallest (D)	= 3/4 in.	= 0.75 in.
(C) diagonal length	= 1.125 in. × 24	= 27 in
(D) diagonal length	= 0.75 in. × 24	= 18 in.

Answers: (C) diagonal = 27 in. min.
(D) diagonal = 18 in. min.

Power-driven threaders, often called "power ponies" are very similar to hand-driven threaders. However, power-driven threaders use a reversible electric motor to turn the die head. Power ponies require the use of a hand oiler and can thread 1/2" to 2" conduit. They are often used as the power supply for die heads directly mounted on large conduit.

The equipment shown is a chain vice (background) holding the rigid steel conduit being threaded. Also a bench yoke type safety bracket is clamped to the conduit being threaded. The "power pony" is slid over the conduit end. The Electrical Worker then operates the power switch for the power pony and applies oil as necessary to cut a thread on the the conduit.

Problem 3-15

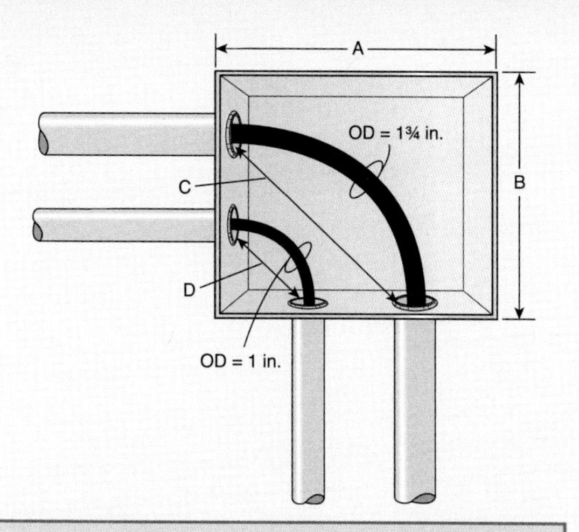

Calculate the dimensions A and B and the diagonal entry and exit measurements (C and D) for the following junction box using multiple single-conductor neoprene jacketed cables for both a shielded and a nonshielded cable installation. Note that since both the entry side and exiting side cables are the same, the box will be square, and therefore dimension A equals dimension B.

Solution 1– Shielded Cable - 314.71(B)
Find: Min. dimensions for (A) length and (B) width
Length (A) = (OD of largest cable × 36) + OD of other cables
Length (A) = (1.75 in. × 36) + 1 in.
= 64 in.
Answer: (A)= 64 in. and (B) = 64 in.

Find: Min length of diagonal lines (C) and (D)
Min. length of diagonal = OD of cable × 36

(C) diagonal length	= 1.75 in. × 36	= 63 in
(D) diagonal length	= 1.00 in. × 36	= 36 in.

Answers: (C) diagonal = 63 in. min.
(D) diagonal = 36 in. min.

Answer: 43 in. dimensions A and B

Solution 2– Nonshielded Conductors - 314.71(B), Exc. No. 2
Find: Min. dimensions for (A) length and (B) width
Length (A = (1.75 in. × 24) + 1 in.
= 42 in. + 1 in.
= 43 in.
Answer: (A)= 43 in. and (B) = 43 in.

Find: Min length of diagonal lines (C) and (D)
Min. length of diagonal = OD of cable × 36

(C) diagonal length	= 1.75 in. × 24	= 42 in
(D) diagonal length	= 1.00 in. × 24	= 24 in.

Answers: (C) diagonal = 42 in. min.
(D) diagonal = 24 in. min.

3.6.4 Cable Entry Opposite Removable Cover

314.71(B)(1), Exception No. 1, gives permission to reduce the box dimension where a conductor or cable rated over 1000 Volts enters from the wall of a box opposite a removable cover. Where this type of cableenters a box in this manner, the distance from the wall of the box to the cover is based on the diameter of the conductor or cable according to **Section 300.34**. **Section 300.34** requires the following minimum bending radius:

Single Conductor	
Shielded	12 × overall diameter
Nonshielded	8 × overall diameter
Lead-covered	12 × overall diameter

Multiconductor or multiplexed single-conductors, having individually shielded conductors
The greater of the two:

Individually shielded conductors	12 × diameter
Overall diameter of cable	7 × overall diameter

Problem 3-16

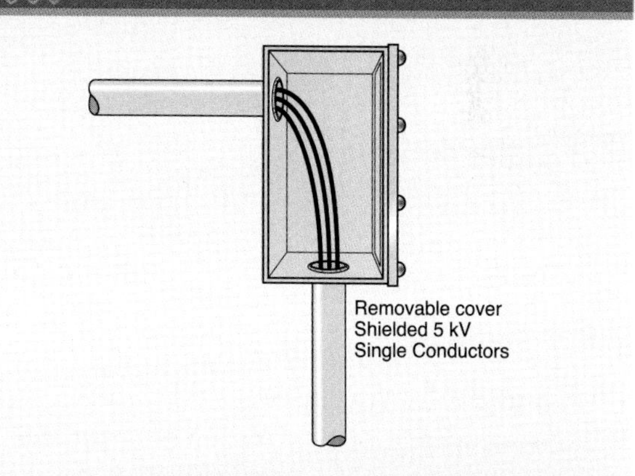

Three single-conductor 4/0 AWG shielded 5 kV feeder conductors are pulled through an angle box with a removable cover directly opposite the raceway entry. Calculate the minimum depth of the box. (The OD for a single 4/0 AWG shielded 5 kV cable is 0.825 in.)

Solution
314.71(B) Exception No.1 and 300.34
Depth = OD of cable × 12
= 0.825 × 12
= 9.9 or 10 in.
Answer: 9.9 or 10 in. minimum

Problem 3-17

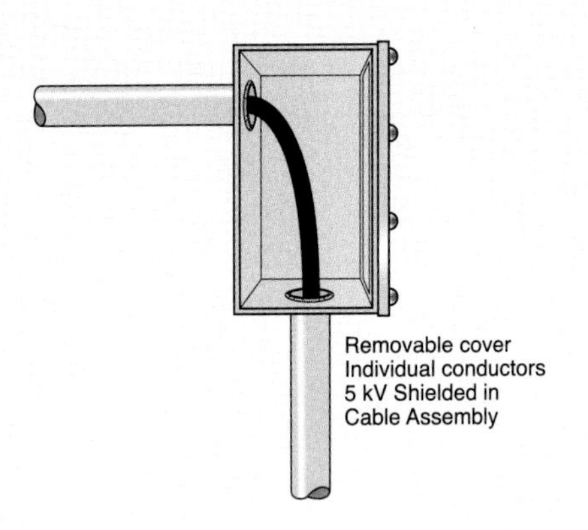

A multiple conductor 5 kV cable, consisting of four 2 AWG shielded conductors, is pulled through an angle box with a removable cover directly opposite the raceway entry. Calculate the minimum depth of the box. (The OD of each 2 AWG shielded 5 kV conductor in the cable assembly is 0.572 in. The OD of the overall cable assembly is 1.28 in.)

Solution
314.71(B) Exception No. 1 and 300.34
Depth = OD of cable × 12
= 0.572 × 12
= 6.864 in.
Depth = OD of cable assembly × 7
= 1.28 × 7
= 8.96 in.
Use the greater of the two calculations
Answer: 8.96 or 9 in. minimum

Definitions and Terms

Conduit Body - A separate portion of a conduit or tubing system that provides access through a removable cover(s) to the interior of the system at a junction of two or more sections of the system or at a terminal point of the system. Boxes such as FS and FD or larger cast or sheet metal boxes are not classified as conduit bodies.

Connector, Pressure (Solderless) - A device that establishes a connection between two or more conductors or between one or more conductors and a terminal by means of mechanical pressure and without the use of solder.

Device - A unit of an electrical system that carries or controls electric energy as its principal function.

Equipment - A general term, including fittings, devices, appliances, luminaires, apparatus, machinery, and the like used as a part of, or in connection with, an electrical installation.

Fitting - An accessory such as a locknut, bushing, or other part of a wiring system that is intended primarily to perform a mechanical rather than an electrical function.

Handhole Enclosure - An enclosure for use in underground systems, provided with an open or closed bottom, and sized to allow personnel to reach into, but not enter, for the purpose of installing, operating, or maintaining equipment or wiring or both.

Lighting Outlet - An outlet intended for the direct connection of a lampholder or luminaire.

Outlet - A point on the wiring system at which current is taken to supply utilization equipment.

Receptacle - A receptacle is a contact device installed at the outlet for the connection of an attachment plug. A single receptacle is a single contact device with no other contact device on the same yoke. A multiple receptacle is two or more contact devices on the same yoke.

Receptacle Outlet - An outlet where one or more receptacles are installed.

Summary

- Boxes and conduit bodies containing conductors 6 AWG and smaller must be of an approved size to provide sufficient free space for all enclosed conductors.
- Box volume is the total volume of the assembled boxes together with other parts (with additional volume) such as plaster frames, extension boxes, dome type covers, etc.
- Box fill is a total volume of all the conductors, clamps, supports, devices, equipment, and equipment grounding conductors placed within the box which take up a specified amount of space.
- **Table 314.16(A)** is used to look up the dimensions and volumes for twenty-four different standard size metal boxes.
- Nonstandard metal boxes 100 $in.^3$ or less and nonmetallic boxes are required to be marked by the manufacturer with their volume.
- **Table 314.16(B)** is used to look up the volume allowances (or space required) for each conductor size from 18 AWG through 6 AWG.
- Various items (other than conductors) which take up space within a box are assigned a volume allowance equal to a certain quantity of conductors.
- Conduit body fill is generally determined by using the maximum fill permitted for the same size and type of conduit or tubing attached.
- Conduit body fill with conductor splices is determined by **Table 314.16(B)** provided the conduit body is clearly marked with its volume.
- Pull and junction boxes with insulated conductors 4 AWG and larger are sized based upon the trade size diameter of entering raceways.
- Pull and junction boxes installed in straight pulls (and not containing splices) are sized at eight times the trade size diameter of the largest raceway.
- Where splices, or angle pulls, or U pulls are made, the distance between each raceway entry inside the box and the opposite wall must be sized at least six times the trade size diameter of the largest raceway.
- Where multiple raceways enter from one wall of the angle or U pull box (or a box with splices) an additional dimension equal to the sum of the trade size diameters must be added to the 6 times distance to the opposite wall.
- For pull and junction boxes used in systems over 1000 volts, nominal, the calculations are based upon the largest outside diameter of shielded and nonshielded conductors or cables within each raceway.

Raceway Fill

Determining the number of conductors permitted to occupy a particular raceway can range from a simple process to a daunting task. Generally, the Electrical Worker reviews the electrical plans and determines the number of conductors and the size of raceway required. However, where the Electrical Worker is required to determine the proper size of conductors and/or conduits without the assistance of prepared plans or specifications, the process can sometimes seem to be overwhelming. Breaking down the raceway fill calculations into simple and organized step-by-step processes eases the pain, and actually proves that the process is not so hard after all. This chapter contains a comprehensive review of the raceway fill calculation process using the Chapter 3 wiring methods of the *National Electrical Code* (*NEC*). First, conduit and tubing wire fill calculations including single and multiconductor cables are introduced. This process demonstrates step-by-step detailed solutions to the varied types of circular raceway fill calculations. Next, round and rectangular underfloor raceways are considered. Again, the process is step-by-step, providing detailed solutions along the way. The process is again repeated for square and rectangular raceways manufactured as metallic and nonmetallic wireways and auxiliary gutters. Calculating fill for these wiring methods uses a different process, which is explained in detail. Finally, this chapter approaches alternate raceways such as those for surface metal and nonmetallic raceways and uses manufacturer-specific look-up tables to determine specific conductor (and device) fill parameters.

Objectives

- Correctly apply the conductor fill requirements for raceways with one or more conductors installed.
- Calculate the maximum number of conductors permitted in specified raceways where all of the conductors are of the same size.
- Calculate the maximum number of conductors permitted in specified raceways where the conductors are of different sizes.
- Apply the Notes to **Table 1, Chapter 9**, as they relate to raceway fill.
- Determine, understand, and apply the different raceway-specific requirements as they apply to conductor fill calculations.
- Understand and apply manufacturer-specific conductor fill tables to specific surface raceways.

Chapter 4

Table of Contents

4.1 Conduit and Tubing Fills

The *NEC* contains strict requirements to limit number and size of conductors drawn into a conduit or tubing, but the *NEC* does not regulate the overall length of a conduit in most cases.

4.1.1 Wiring Methods Covered

The conduit and tubing wiring methods recognized in **Chapter 3** of the *NEC* refer to **Chapter 9, Table 1** for the maximum number of conductors permitted in the conduit or tubing. The following is a list of sections and wiring methods which refer to **Chapter 9, Table 1**:

342.22 Intermediate Metal Conduit: Type IMC
344.22 Rigid Metal Conduit: Type RMC
348.22* Flexible Metal Conduit: Type FMC
350.22(A)* Liquidtight Flexible Metal Conduit: Type LFMC
352.22 Rigid Polyvinyl Chloride Conduit: Type PVC
353.22 High Density Polyethylene Conduit: Type HDPE Conduit
354.22 Nonmetallic Underground Conduit with Conductors: Type NUCC
355.22 Reinforced Thermosetting Resin Conduit: Type RTRC
356.22 Liquidtight Flexible Nonmetallic Conduit: Type LFNC
358.22 Electrical Metallic Tubing: Type EMT
360.22(A)* Flexible Metallic Tubing: Type FMT
362.22 Electric Nonmetallic Tubing: Type ENT

*In addition, these wiring methods are available in the trade size of 3/8 in. and refer specifically to **Table 348.22** for insulated conductor fill.

For a more complete list, see **Section 300.17**.

In addition to conduit and tubing, other wiring methods such as underfloor raceways, auxiliary gutters, and wireways are covered in this chapter.

4.1.2 Chapter 9, Table 1 and Accompanying Notes

Chapter 9, Table 1 gives the maximum conduit and tubing fills permitted for any standard size conduit or tubing wiring method. It applies when all the conductors are the same size with the same insulation, and it applies when the conductor sizes and/or insulations are mixed.

When there are more than two conductors in the conduit or tubing, **Table 1** limits the raceway or conduit fill to 40% of the internal cross-sectional area of all the standard sizes of conduit or tubing. A simpler way to say it is: if there are three conductors or more, the raceway fill is limited to 40%.

When all the conductors and/or fixture wires have the same insulation and are the same size, **Table 1, Note (1)** refers the user to **Tables** in **Annex C** of the *Code*. The *Code* uses the annexes to present additional information. Because the tables in **Annex C** fulfill the correct fill requirements of **Chapter 9, Table 1** the tables are often usable in the field. The tables are also based on conductors meeting all of the following requirements:

1. All the conductors are the same size.
2. All the conductors have the same insulation type.

The **Annex C** tables are entitled **Conduit and Tubing Fill Tables for Conductors and Fixture Wires of the Same Size**. These tables are used for solid or stranded copper conductors. Some tables are reserved for "compact conductors" only. These are usable for all compact conductors, including both copper and aluminum. Compact stranding is

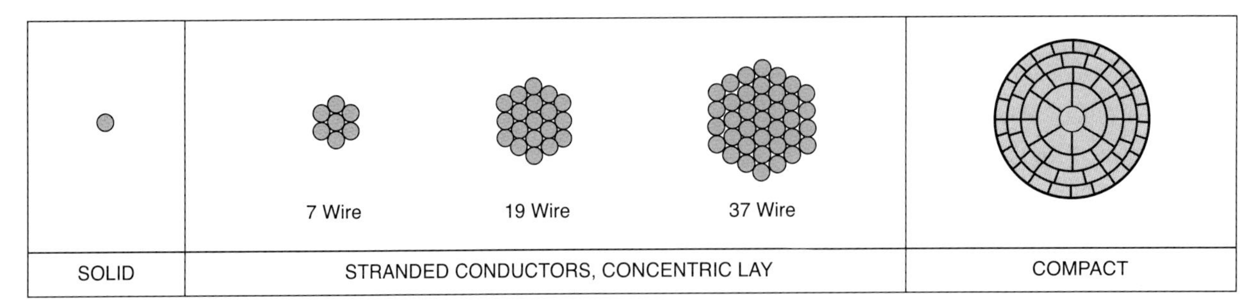

Figure 4-1. Types of Standard and Compact Wire Stranding

the result of a manufacturing process where the stranded conductor is compressed to the extent that the interstices (voids between strand wires) are virtually eliminated. **Figure 4-1** illustrates five different types of wire assemblies.

For additional information, visit qr.njatcdb.org Item #1030

4.1.3 Annex C Tables

There are twenty-four separate fill tables in **Annex C**. The fills listed in these tables are based on the fill limitation required in **Chapter 9, Table 1**. The calculations for the fill are already made in the tables for the various standard trade sizes. There are twelve tables that serve solid and stranded conductors and twelve separate tables which serve compact stranded conductors.

Because the inside cross-sectional areas of the various conduits and tubings are different, the actual fill will be different. Therefore, there is a separate table for almost every raceway containing conductors, and a separate table for almost every raceway containing compact conductors. The compact conductor tables are close to one page in length, whereas the solid and concentric (most common) stranded conductor tables are roughly four pages in length.

- Tables identified as "C" are for solid and concentric stranded conductors.
- Tables identified as "C(A)" are only for compact stranded conductors.
- Reinforced Thermosetting Resin Conduit (RTRC) continues to be omitted from **Table 4** and **Annex C**. The manufacturer should be contacted for appropriate technical dimensions to determine proper raceway fill.
- There are two tables for each of the following raceways: (Notice that these tables are in alphabetical order)

C.1 Electrical Metallic Tubing (EMT)
C.2 Electrical Nonmetallic Tubing (ENT)
C.3 Flexible Metal Conduit (FMC)
C.4 Intermediate Metal Conduit (IMC)
C.5 Liquidtight Flexible Nonmetallic Conduit (Type LFNC-B)
C.6 Liquidtight Flexible Nonmetallic Conduit (Type LFNC-A)
C.7 Liquidtight Flexible Metal Conduit (LFMC)
C.8 Rigid Metal Conduit (RMC)
C.9 Rigid PVC Conduit, Schedule 80
C.10 Rigid PVC Conduit, Schedule 40 and HDPE Conduit
C.11 Type A, Rigid PVC Conduit
C.12 Type EB, PVC Conduit

Add an "A" to the above table numbers [example: **C.1(A)**] and another complete set of tables is presented for compact conductors. This results in more compacted conductors of a corresponding size and insulation being permitted to be installed in the same standard size conduit than ordinary stranded copper conductors. For example, in a 2 in. rigid metal conduit using 1 AWG THHN conductor, the maximum number of conductors permitted is:

C.8 1 AWG THHN stranded = 8 conductors
C.8(A) 1 AWG THHN compact = 10 conductors

Compact conductor cables are available in either copper or aluminum construction, and the dimensions found in **Chapter 9, Table 5A** reflect both compact aluminum and compact copper building wire.

Raceway Fill Comparison
Maximum 8 AWG XHHW-2 within a 1½ in. Circular Raceway

Type of Raceway	*Table No.*	*No. of Conductors*	*Compact Table No.*	*No. of Conductors*
EMT	C.1	18	C.1(A)	20
ENT	C.2	18	C.2(A)	20
FMC	C.3	17	C.3(A)	19
IMC	C.4	20	C.4(A)	22
LFNC-B	C.5	18	C.5(A)	20
LFMC	C.7	18	C.7(A)	20
RMC	C.8	19	C.8(A)	21
PVC Schedule 80	C.9	15	C.9(A)	17
Note: XHHW and not XHHW-2. Often XHHW and XHHW-2 use the same dimensions, as shown in Chapter 9, Table 5.				

Table 4-1. Comparison of Raceway Fills Using the Maximum Number of 8 AWG XHHW-2 in a 1-1/2 in. Circular Raceway

There are ten notes and two informational notes to **Chapter 9, Table 1**. Many of these notes cover specific installations which will be addressed individually. **Note 2**: When a conduit or tubing is used for physical protection of the conductors, the fill is not limited by **Table 1**. This includes such installations as cable sleeves and supplemental protection of direct burial conductor emerging from the ground. However, the fundamental requirements of **Section 300.17** must be implemented to prevent cable or conductor damage.

A new note 10 was added for the 2014 *NEC*. This note applies to **Tables 5** and **5A**, and indicates that where the actual values of conductor diameter and area are known, they are permitted to be used.

4.1.4 Using Annex C Tables for Conduit and Tubing Fill

To avoid errors working with look-up tables in Annex C, use the following steps:

1. Read the table header to make sure you are using the correct table.
2. Determine which two columns you will be using.
3. Next, select the proper row.
4. Using a straight edge just below the selected row, follow the edge to the correct column and select the proper number.
5. Double check your answer.

Problem 4-1

What is the maximum number of 4 AWG THHN copper conductors permitted in a 2 in. rigid metal conduit (RMC)?

Solution
Annex C Table of Contents points to Table C.8 for Rigid Metal Conduit (RMC)
Table C.8, Rigid Metal Conduit (RMC)
Max. 4 AWG THHN in 2 in. = 16 conductors
Answer: 16 conductors

Problem 4-2

What is the maximum number of 4 AWG THHN compact aluminum conductors permitted in a 2 in. intermediate metal conduit (IMC)?

Solution
Annex C Table of Contents points to Table C.4(A) for Intermediate Metal Conduit (IMC)
Table C.4(A) Intermediate Metal Conduit (IMC)
Max. 4 AWG THHN in 2 in. = 20 conductors
Answer: 20 conductors

4.1.5 Insulation Thickness

When determining the number of conductors permitted in a conduit or tubing, the thickness of the insulation and the jacket or outer covering, when used, is also taken into consideration. When an outer covering is added to the insulation of a conductor during manufacturing, it increases the diameter of the conductor and the cross-sectional area of the conductor. Therefore, it in turn will reduce the number of conductors permitted to be installed in a conduit or tubing without exceeding the 40% fill for three or more insulated conductors. Some Type RHW, RHW-2, and RHH insulations do not require an outer covering and are marked with an asterisk in **Chapter 9, Table 5** as well as in the **Annex C** tables. Other outer coverings are found in **Table 310.104(A)** and the accompanying footnotes. Include the outer covering dimension only where an outer covering is specifically mentioned in the problem.

Problem 4-3

1. What is the maximum number of 12 AWG RHW conductors without outer covering permitted in a 1 in. rigid metal conduit (RMC)?
2. What is the maximum number of 12 AWG RHW conductors with outer covering permitted in a 1 in. rigid metal conduit (RMC)?

Solution – Calculation 1
RHW without outer covering (RHW*)
Table C.8 Rigid Metal Conduit
Max. 12 AWG RHW in 1 in. RMC = 13 conductors
Answer: 13 conductors
Solution – Calculation 2
RHW with outer covering (RHW)
Table C.8 Rigid Metal Conduit
Max. 12 AWG RHW in 1 in. RMC = 10 conductors
Answer: 10 conductors

4.1.6 Fixture Wires

Although traditionally used for luminaire wiring, fixture wire is also permitted to be used for Class 1 circuits in **Section 725.49** and Class 2 and 3 circuits in **725.130(B)**, provided they are installed in and protected by a Chapter 3 wiring method. For fire alarm circuit wiring, **Article 760** has similar permissions found in **760.49**. Another traditional use of fixture wire is for motor control circuits in compliance with **Section 430.72**.

Section 402.7 refers to **Chapter 9, Table 1** for determining the maximum number of fixture wires which are permitted in a conduit or tubing. **Chapter 9, Table 1** refers to the fill tables in **Annex C**. Fixture wires are found in all of the **Annex C** tables, but fixture wires are not listed in the compact tables.

Problem 4-4

What is the maximum number of 18 AWG, Type SFF-2 fixture wires permitted in a ¾ in. electrical metallic tubing (EMT)?

Solution
Table C.1 Electrical Metallic Tubing
Max. 18 AWG SFF-2 in ¾ in. EMT = 18 conductors
Answer: 18 conductors

4.2 Using Tables 4 and 5 of Chapter 9

According to **Chapter 9, Table 1, Note 6**, where there is a combination of conductors with different sizes and/or insulations, **Tables 4, 5,** and **5A** of **Chapter 9** are used for calculating the minimum size conduit or tubing. For calculating fills of combinations of conductors, the square inch area of a cross section of a raceway is used. The **Table 4** charts contain the calculated square inch areas of a cross section of standard size conduits and the permitted fill percentage. **Table 5** and **Table 5A** contain the square inch area of a cross section of conductors including insulation and covering (if any).

Because each type of conduit and tubing has a different interior cross-sectional area, **Table 4** consists of 12 separate charts. There is one chart for each type of conduit or tubing. The fills listed in each chart of **Table 4** are based on **Table 1** fill percentages. Illustrated in **Figure 4-2** is a portion of **Table 4** limited to electrical metallic tubing (EMT). **Table 4** contains many other charts specific to other wiring methods. Each chart is laid out identically to the EMT chart. Note, each chart gives 100% of the cross-sectional area, in square inches, of each of the standard size raceways in both metric and English units. The table also has a calculated 40% fill column, which is 40% of the 100% column. These two values will come in very handy when calculating conduit and tubing fills. The column "Over 2 Wires, 40%" will be the most used column and is now close to trade size column to reduce "lookup" errors.

Table 4 Dimensions and Percent Area of Conduit and Tubing
(Areas of Conduit or Tubing for the Combinations of Wires Permitted in Table 1, Chapter 9)

Article 358 — Electrical Metallic Tubing (EMT)

Metric Designator	Trade Size	Over 2 Wires 40%		60%		1 Wire 53%		2 Wires 31%		Nominal Internal Diameter		Total Area 100%	
		mm²	in.²	mm²	in.²	mm²	in.²	mm²	in.²	mm	in.	mm²	in.²
16	½	78	0.122	118	0.182	104	0.161	61	0.094	15.8	0.622	196	0.304
21	¾	137	0.213	206	0.320	182	0.283	106	0.165	20.9	0.824	343	0.533
27	1	222	0.346	333	0.519	295	0.458	172	0.268	26.6	1.049	556	0.864
35	1¼	387	0.598	581	0.897	513	0.793	300	0.464	35.1	1.380	968	1.496
41	1½	526	0.814	788	1.221	696	1.079	407	0.631	40.9	1.610	1314	2.036
53	2	866	1.342	1299	2.013	1147	1.778	671	1.040	52.5	2.067	2165	3.356
63	2½	1513	2.343	2270	3.515	2005	3.105	1173	1.816	69.4	2.731	3783	5.858
78	3	2280	3.538	3421	5.307	3022	4.688	1767	2.742	85.2	3.356	5701	8.846
91	3½	2980	4.618	4471	6.927	3949	6.119	2310	3.579	97.4	3.834	7451	11.545
103	4	3808	5.901	5712	8.852	5046	7.819	2951	4.573	110.1	4.334	9521	14.753

Figure 4-2. An excerpt from Chapter 9, Table 4 dealing with Electrical Metallic Tubing (EMT)

Chapter 9, Table 5, shown here as **Figure 4-3**, is dedicated to conductors other than compact conductors. **Table 5** is applicable to solid and stranded copper conductors. Each table lists a multitude of dimensions for insulated conductors and fixture wire. One column of the chart lists the square inch area of the conductor and is now closer to the size column. A new note 10 was added for the 2014 *NEC*. This note applies to **Tables 5** and **5A**, and indicates that where the actual values of conductor diameter and area are known, they are permitted to be used.

Table 5 ***Continued***

Type	Size (AWG or kcmil)	Approximate Area		Approximate Diameter	
		mm²	in.²	mm	in.
TW, THHW, THW, THW-2	12	11.68	0.0181	3.861	0.152
	10	15.68	0.0243	4.470	0.176
	8	28.19	0.0437	5.994	0.236
RHH*, RHW*, RHW-2*	14	13.48	0.0209	4.140	0.163
RHH*, RHW*, RHW-2*, XF, XFF	12	16.77	0.0260	4.623	0.182
Type: RHH*, RHW*, RHW-2*, THHN, THHW, THW, THW-2, TFN, TFFN, THWN, THWN-2, XF, XFF					
RHH,* RHW,* RHW-2,* XF, XFF	10	21.48	0.0333	5.232	0.206
RHH*, RHW*, RHW-2*	8	35.87	0.0556	6.756	0.266
TW, THW, THHW, THW-2, RHH*, RHW*, RHW-2*	6	46.84	0.0726	7.722	0.304
	4	62.77	0.0973	8.941	0.352
	3	73.16	0.1134	9.652	0.380
	2	86.00	0.1333	10.46	0.412
	1	122.6	0.1901	12.50	0.492
	1/0	143.4	0.2223	13.51	0.532
	2/0	169.3	0.2624	14.68	0.578
	3/0	201.1	0.3117	16.00	0.630
	4/0	239.9	0.3718	17.48	0.688
	250	296.5	0.4596	19.43	0.765
	300	340.7	0.5281	20.83	0.820
	350	384.4	0.5958	22.12	0.871
	400	427.0	0.6619	23.32	0.918
	500	509.7	0.7901	25.48	1.003
	600	627.7	0.9729	28.27	1.113
	700	710.3	1.1010	30.07	1.184
	750	751.7	1.1652	30.94	1.218
	800	791.7	1.2272	31.75	1.250
	900	874.9	1.3561	33.38	1.314
	1000	953.8	1.4784	34.85	1.372
	1250	1200	1.8602	39.09	1.539
	1500	1400	2.1695	42.21	1.662
	1750	1598	2.4773	45.11	1.776
	2000	1795	2.7818	47.80	1.882
TFN, TFFN	18	3.548	0.0055	2.134	0.084
	16	4.645	0.0072	2.438	0.096

(Continues)

Figure 4-3. An excerpt from Table 5 of Chapter 9 showing the cross-sectional areas of conductors

Problem 4-5

What size electrical metallic tubing is needed for three 2/0 AWG XHHW-2 copper conductors and four 4/0 AWG XHHW-2 copper conductors?

Solution
Table 5
Cross-sectional area of individual conductors
2/0 AWG XHHW-2 = 0.2190 in.^{2}
0.2190 in.^{2} × 3 = 0.6570
4/0 AWG XHHW-2 = 0.3197 in.^{2}
0.3197 in.^{2} × 4 = 1.2788
Total square inch area: 1.9358
Table 4
EMT chart, 40% fill column
40% fill of $2\frac{1}{2}$ in. EMT = 2.343 in.^{2}
1.9358 in.^{2} requires a $2\frac{1}{2}$ in. EMT
Answer: $2\frac{1}{2}$ in. EMT

For additional information, visit qr.njatcdb.org Item #1031

4.2.1 Insulation Outer Covering

It takes five charts in **Chapter 9, Table 5** to list all the various insulations and conductor sizes. The asterisk (*) note to one of the tables refers to Type RHH, RHW, and RHW-2 insulation, without outer covering. There are two types of these conductors: one with outer covering and one without outer covering.

Problem 4-6

What is the minimum size intermediate metal conduit (IMC) needed for the installation of four 14 AWG RHH copper conductors with outer coverings and six 12 AWG RHH copper conductors without outer coverings?

Solution
Table 5
Cross-sectional area of individual conductors; first chart
14 RHH AWG with outer covering = 0.0293 in.^{2}
0.0293 in.^{2} × 4 = 0.1172
Table 5
Second chart
12 RHH AWG without outer covering = 0.0260 in.^{2}
0.0260 in.^{2} × 6 = 0.1560
Total square inch area: 0.2732
Table 4
IMC chart; 40% fill column
40% fill for 1 in. IMC = 0.384 in.^{2}
0.2732 in.^{2} requires a 1 in. IMC
Answer: 1 in. IMC

4.2.2 Counting All Conductors

When calculating conduit and tubing fills, all conductors installed in the raceway contribute to the conductor occupied space or conductor fill. This includes current-carrying conductors, noncurrent-carrying conductors, part-time current-carrying conductors, equipment bonding jumpers, and equipment grounding conductors.

Problem 4-7

What is the minimum size intermediate metal conduit (IMC) required for the installation of three 2/0 AWG THWN copper conductors feeding a motor and three 12 AWG THHN copper conductors for the motor control circuit?

Solution
Table 5
Cross-sectional area of individual conductors
2/0 AWG THWN = 0.2223 in.^{2}
0.2223 in.^{2} × 3 = 0.6669
12 AWG THHN = 0.0133 in.^{2}
0.0133 in.^{2} × 3 = 0.0399
Total square inch area: 0.7068
Table 4
IMC table; 40% fill column over 2 wires
40% fill for $1\frac{1}{2}$ in. IMC = 0.890 in.^{2}
0.7068 in.^{2} requires a $1\frac{1}{2}$ in. IMC
Answer: $1\frac{1}{2}$ in. IMC

The NFPA Codes and Standards Development Process

The NFPA process encourages public participation in the development of its codes and standards. All NFPA codes and standards (also referred to here as NFPA "Standards") are revised and updated every three to five years in revision cycles that begin twice each year and that normally take approximately two years to complete. Each revision cycle proceeds according to a published schedule that includes final dates for all major events in the process.

The NFPA Codes and Standards process contains four basic steps:

1. Input Stage
2. Comment Stage
3. Association Technical Meeting
4. Council Appeals and Issuance of Standard

http://www.nfpa.org/codes-and-standards/standards-development-process/how-codes-and-standards-are-developed/standards-development-process

4.2.3 Insulated and Bare Grounding and Bonding Conductors

Chapter 9, Table 1, Note 3 requires that equipment grounding or bonding conductors, where installed, must be included in the conduit or tubing fill calculation. The actual dimensions of the equipment grounding or bonding conductor (insulated or bare) have to be used in this calculation.

Where a bare conductor is selected as the equipment grounding or bonding conductor, **Table 8 of Chapter 9** is used to determine the conductor cross-section area. The column marked "Conductors, Overall Area, in.^{2}" is used to determine the appropriate conductor area. This area is then added to the calculated area used by the other conductors to determine the total square inch area. See **Figure 4-5**.

Problem 4-8

Four 3/0 AWG THHN copper conductors with a 6 AWG THHN copper equipment grounding conductor are to be installed in the same rigid PVC schedule 40 conduit. What is the minimum size PVC schedule 40 conduit required?

Solution
Table 5
Cross-sectional area in square inches
3/0 AWG THHN = 0.2679 in.^{2}
0.2679 in.^{2} × 4 = 1.0716
6 AWG THHN = 0.0507 in.^{2}
0.0507 in.^{2} × 1 = 0.0507
Total square inch area 1.1223
Table 4
Rigid PVC Conduit, Schedule 40 chart
40% fill column over 2 wires
40% fill for 2 in. PVC = 1.316 in.^{2}
1.1223 in.^{2} fill requires a 2 in. PVC, Schedule 40
Answer: 2 in. Rigid PVC Conduit, Schedule 40

4.3 Using Tables 4 and 5A of Chapter 9

Table 5A, as illustrated in **Figure 4-4**, lists the cross-sectional area in square inches for individual compact copper and aluminum conductors. Industry sources have included compact building wire Types RHH, RHW, USE, THW, THHW, THHN, XHHW, and Bare in sizes 8 AWG through 1,000 kcmil in **Table 5A**. The following is a comparison of the square inch area of a concentric stranded conductor and a compact conductor for a 1 AWG size:

Table 5 - Size 1 AWG THHN concentric stranded conductor = 0.1562 in.^{2} area.
Table 5A - Size 1 AWG THHN compact stranded conductor = 0.1352 in.^{2} area.

Problem 4-9

What size Schedule 80 Rigid PVC Conduit is needed for the installation of six 1 AWG XHHW compact aluminum conductors and four 6 AWG XHHW compact aluminum conductors?

Solution
Table 5A
Cross-sectional area in square inches
1 AWG XHHW = 0.1352 in.^{2}
0.1352 in.^{2} × 6 = 0.8112
6 AWG XHHW = 0.0530 in.^{2}
0.0530 in.^{2} × 4 = 0.2120
Total square inches 1.0232
Table 4
Rigid PVC Conduit, Schedule 80 chart
40% fill column for over 2 wires
40% fill for 2 in. PVC = 1.150 in.^{2}
1.0232 in.^{2} requires a 2 in. PVC
Answer: 2 in. Rigid PVC Conduit, Schedule 80

NJATC Textbook covers

Table 5A Compact Copper and Aluminum Building Wire Nominal Dimensions* and Areas

Size (AWG or kcmil)	Bare Conductor Diameter		Types RHH**, RHW**, or USE Approximate Diameter		Types RHH**, RHW**, or USE Approximate Area		Types THW and THHW Approximate Diameter		Types THW and THHW Approximate Area		Type THHN Approximate Diameter		Type THHN Approximate Area		Type XHHW Approximate Diameter		Type XHHW Approximate Area		Size (AWG or kcmil)
	mm	in.	mm	in.	mm²	in.²	mm	in.	mm²	in.²	mm	in.	mm²	in.²	mm	in.	mm²	in.²	
8	3.404	0.134	6.604	0.260	34.25	0.0531	6.477	0.255	32.90	0.0510	—	—	—	—	5.690	0.224	25.42	0.0394	8
6	4.293	0.169	7.493	0.295	44.10	0.0683	7.366	0.290	42.58	0.0660	6.096	0.240	29.16	0.0452	6.604	0.260	34.19	0.0530	6
4	5.410	0.213	8.509	0.335	56.84	0.0881	8.509	0.335	56.84	0.0881	7.747	0.305	47.10	0.0730	7.747	0.305	47.10	0.0730	4
2	6.807	0.268	9.906	0.390	77.03	0.1194	9.906	0.390	77.03	0.1194	9.144	0.360	65.61	0.1017	9.144	0.360	65.61	0.1017	2
1	7.595	0.299	11.81	0.465	109.5	0.1698	11.81	0.465	109.5	0.1698	10.54	0.415	87.23	0.1352	10.54	0.415	87.23	0.1352	1
1/0	8.534	0.336	12.70	0.500	126.6	0.1963	12.70	0.500	126.6	0.1963	11.43	0.450	102.6	0.1590	11.43	0.450	102.6	0.1590	1/0
2/0	9.550	0.376	13.72	0.540	147.8	0.2290	13.84	0.545	150.5	0.2332	12.57	0.495	124.1	0.1924	12.45	0.490	121.6	0.1885	2/0
3/0	10.74	0.423	14.99	0.590	176.3	0.2733	14.99	0.590	176.3	0.2733	13.72	0.540	147.7	0.2290	13.72	0.540	147.7	0.2290	3/0
4/0	12.07	0.475	16.26	0.640	207.6	0.3217	16.38	0.645	210.8	0.3267	15.11	0.595	179.4	0.2780	14.99	0.590	176.3	0.2733	4/0
250	13.21	0.520	18.16	0.715	259.0	0.4015	18.42	0.725	266.3	0.4128	17.02	0.670	227.4	0.3525	16.76	0.660	220.7	0.3421	250
300	14.48	0.570	19.43	0.765	296.5	0.4596	19.69	0.775	304.3	0.4717	18.29	0.720	262.6	0.4071	18.16	0.715	259.0	0.4015	300
350	15.65	0.616	20.57	0.810	332.3	0.5153	20.83	0.820	340.7	0.5281	19.56	0.770	300.4	0.4656	19.30	0.760	292.6	0.4536	350
400	16.74	0.659	21.72	0.855	370.5	0.5741	21.97	0.865	379.1	0.5876	20.70	0.815	336.5	0.5216	20.32	0.800	324.3	0.5026	400
500	18.69	0.736	23.62	0.930	438.2	0.6793	23.88	0.940	447.7	0.6939	22.48	0.885	396.8	0.6151	22.35	0.880	392.4	0.6082	500
600	20.65	0.813	26.29	1.035	542.8	0.8413	26.67	1.050	558.6	0.8659	25.02	0.985	491.6	0.7620	24.89	0.980	486.6	0.7542	600
700	22.28	0.877	27.94	1.100	613.1	0.9503	28.19	1.110	624.3	0.9676	26.67	1.050	558.6	0.8659	26.67	1.050	558.6	0.8659	700
750	23.06	0.908	28.83	1.135	652.8	1.0118	29.21	1.150	670.1	1.0386	27.31	1.075	585.5	0.9076	27.69	1.090	602.0	0.9331	750
900	25.37	0.999	31.50	1.240	779.3	1.2076	31.09	1.224	759.1	1.1766	30.33	1.194	722.5	1.1196	29.69	1.169	692.3	1.0733	900
1000	26.92	1.060	32.64	1.285	836.6	1.2968	32.64	1.285	836.6	1.2968	31.88	1.255	798.1	1.2370	31.24	1.230	766.6	1.1882	1000

*Dimensions are from industry sources.

**Types RHH and RHW without outer coverings.

Figure 4-4. Table 5A Dimensions for Compact Conductors.

Excerpt from the NFPA Directory - Establishing the Consensus Body

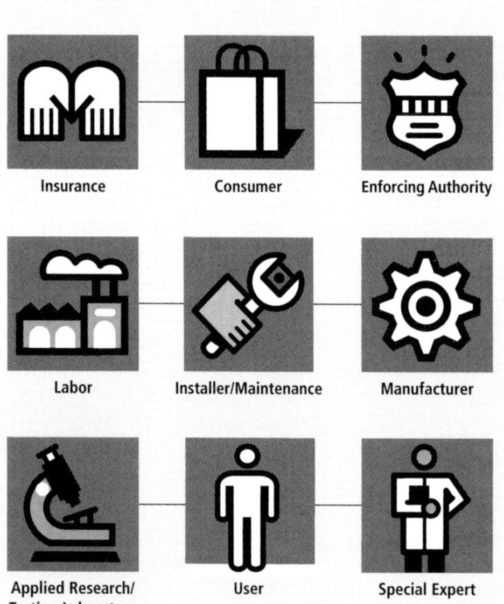

In the NFPA codes and standards development process, NFPA Technical Committees and Code Making Panels serve as the principal consensus bodies responsible for developing and regularly updating all NFPA codes and standards. Committees and Panels are appointed by the Standards Council and typically consist of no more than thirty voting members representing a balance of interests. NFPA membership is not required in order to participate on an NFPA Technical Committee, and appointment is based on such factors as technical expertise, professional standing, commitment to public safety and the ability to bring to the table the point of view of a category of interested people or groups. Each Technical Committee is constituted so as to contain a balance of affected interests, with no more than one-third of the Committee from the same interest category. The categories generally used by the Standards Council to classify Committee members are summarized here. For the National Electrical Code Committee, an additional category of Utility (UT) is added. The Committee must reach a consensus in order to take action on an item.

4.4 Using Table 8 of Chapter 9 for Bare Conductors

A bare conductor is permitted:

250.118(1) permits the use of a bare equipment grounding conductor.
230.41 Exception permits the use of a bare grounded conductor as a service-entrance conductor.

Where bare conductors are installed, **Chapter 9, Table 1, Note 8** permits the use of **Chapter 9, Table 8** for bare conductor dimensions. **Table 8, Conductor Properties** gives various dimensions of uninsulated or bare conductors, including cross-section square inch area. **Chapter 9, Table 8** is illustrated in **Figure 4-5**.

Table 8 makes no distinction between the square inch area of a bare copper or a bare aluminum conductor. Therefore, the table is applicable to solid as well as concentric stranded copper and aluminum conductors. The insulated conductors are calculated according to **Table 5** or 5A. See **Table 8** in **Figure 4-5**. For additional footnotes, refer to the actual **Chapter 9, Table 8** in the *NEC*.

Problem 4-10

Four 4/0 AWG THHN copper conductors and a 6 AWG bare copper equipment grounding conductor are to be installed in the same Rigid PVC Conduit, Schedule 80. What size conduit is needed?

Solution
Table 5
Cross-sectional area of insulated conductors
Table 8
Square in. area column for bare conductors
4/0 AWG THHN = 0.3237 in.^{2}
0.3237 in.^{2} x 4 = 1.2948
6 AWG bare = 0.0270 in.^{2}
0.0270 in.^{2} x 1 = 0.0270
Total square inches 1.3218
Table 4
Rigid PVC Conduit, Schedule 80 chart
40% fill column for over 2 wires
40% fill for 2½ in. PVC = 1.647 in.^{2}
1.3218 in.^{2} requires a 2½ in. PVC, Schedule 80
Answer: 2½ in. Rigid PVC Conduit, Schedule 80

However, where bare conductors (aluminum or copper) of compact construction are used, the dimensions of **Table 5A** may be used in lieu of **Table 8**.

Problem 4-11

A 3-phase, 4-wire, high-leg delta service is calculated to need one 4/0 AWG and two 250 kcmil phase THHW copper conductors and a bare 3/0 AWG grounded copper conductor as service-entrance conductors. What is the minimum size rigid metal conduit required?

Solution
Table 5
Cross-sectional insulated conductors
Table 8
Cross-sectional bare conductor
250 kcmil THHW = 0.4596 in.^{2}
0.4596 in.^{2} x 2 = 0.9192
4/0 AWG THHW = 0.3718 in.^{2}
0.3718 in.^{2} x 1 = 0.3718
3/0 AWG bare = 0.1730 in.^{2}
0.1730 in.^{2} x 1 = 0.1730
Total square inches 1.4640
Table 4
Rigid Metal Conduit (RMC) chart
40% fill for over two conductors
40% fill for 2½ in. RMC = 1.946 in.^{2}
1.4640 in.^{2} requires a 2½ in. RMC
Answer: 2½ in. RMC

Problem 4-12

A 3-phase, 4-wire, 120/208 volt service uses three 250 kcmil XHHW compact copper phase conductors and one bare concentric copper 4/0 AWG grounded conductor. What is the minimum conduit size required when using Schedule 40 PVC conduit?

Solution
Table 5A
Cross-sectional area of compact conductors
Table 8
Bare conductor cross-sectional area
250 kcmil XHHW = 0.3421 in.^{2}
0.3421 in.^{2} x 3 = 1.0263
4/0 AWG bare = 0.2190 in.^{2}
0.2190 in.^{2} x 1 = 0.2190
Total square inches 1.2453
Table 4
Rigid PVC Conduit, Schedule 40 chart
40% fill for over 2 conductors
40% fill for ^{2} in. PVC = 1.316 in.^{2}
1.2453 in.^{2} requires a 2 in. PVC
Answer: 2 in. Rigid PVC Conduit, Schedule 40

Table 8 Conductor Properties

Size	Area		Conductors: Stranding			Conductors: Overall				Direct Current Resistance at 75°C (167°F): Copper				Aluminum	
			Quantity	Diameter		Diameter		Area		Uncoated		Coated			
(AWG or kc.mil)	mm²	Circular mils	Quantity	mm	in.	mm	in.	mm²	in²	ohm/km	ohm/kFT	ohm/km	ohm/kFT	ohm/km	ohm/kFT
18	0.823	1620	1	—	—	1.02	0.040	0.823	0.001	25.5	7.77	26.5	8.08	42.0	12.8
18	0.823	1620	7	0.39	0.015	1.16	0.046	1.06	0.002	26.1	7.95	27.7	8.45	42.8	13.1
16	1.31	2580	1	—	—	1.29	0.051	1.31	0.002	16.0	4.89	16.7	5.08	26.4	8.05
16	1.31	2580	7	0.49	0.019	1.46	0.058	1.68	0.003	16.4	4.99	17.3	5.29	26.9	8.21
14	2.08	4110	1	—	—	1.63	0.064	2.08	0.003	10.1	3.07	10.4	3.10	16.6	5.06
14	2.08	4110	7	0.62	0.024	1.85	0.073	2.68	0.004	10.3	3.14	10.7	3.26	16.9	5.17
12	3.31	6530	1	—	—	2.05	0.081	3.31	0.005	6.34	1.93	6.57	2.01	10.45	3.18
12	3.31	6530	7	0.78	0.030	2.32	0.092	4.25	0.006	6.50	1.98	6.73	2.05	10.69	3.25
10	5.261	10380	1	—	—	2.588	0.102	5.26	0.008	3.984	1.21	4.148	1.26	6.561	2.00
10	5.261	10380	7	0.98	0.038	2.95	0.116	6.76	0.011	4.070	1.24	4.226	1.29	6.679	2.04
8	8.367	16510	1	—	—	3.264	0.128	8.37	0.013	2.506	0.764	2.579	0.786	4.125	1.26
8	8.367	16510	7	1.23	0.049	3.71	0.146	10.76	0.017	2.551	0.778	2.653	0.809	4.204	1.28
6	13.30	26240	7	1.56	0.061	4.67	0.184	17.09	0.027	1.608	0.491	1.671	0.510	2.652	0.808
4	21.15	41740	7	1.96	0.077	5.89	0.232	27.19	0.042	1.010	0.308	1.053	0.321	1.666	0.508
3	26.67	52620	7	2.20	0.087	6.60	0.260	34.28	0.053	0.802	0.245	0.833	0.254	1.320	0.403
2	33.62	66360	7	2.47	0.097	7.42	0.292	43.23	0.067	0.634	0.194	0.661	0.201	1.045	0.319
1	42.41	83690	19	1.69	0.066	8.43	0.332	55.80	0.087	0.505	0.154	0.524	0.160	0.829	0.253
1/0	53.49	105600	19	1.89	0.074	9.45	0.372	70.41	0.109	0.399	0.122	0.415	0.127	0.660	0.201
2/0	67.43	133100	19	2.13	0.084	10.62	0.418	88.74	0.137	0.3170	0.0967	0.329	0.101	0.523	0.159
3/0	85.01	167800	19	2.39	0.094	11.94	0.470	111.9	0.173	0.2512	0.0766	0.2610	0.0797	0.413	0.126
4/0	107.2	211600	19	2.68	0.106	13.41	0.528	141.1	0.219	0.1996	0.0608	0.2050	0.0626	0.328	0.100
250	127	—	37	2.09	0.082	14.61	0.575	168	0.260	0.1687	0.0515	0.1753	0.0535	0.2778	0.0847
300	152	—	37	2.29	0.090	16.00	0.630	201	0.312	0.1409	0.0429.	0.1463	0.0446	0.2318	0.0707
350	177	—	37	2.47	0.097	17.30	0.681	235	0.364	0.1205	0.0367	0.1252	0.0382	0.1984	0.0605
400	203	—	37	2.64	0.104	18.49	0.728	268	0.416	0.1053	0.0321	0.1084	0.0331	0.1737	0.0529
500	253	—	31	2.95	0.116	20.65	0.813	336	0.519	0.0845	0.0258	0.0869	0.0265	0.1391	0.0424
600	304	—	61	2.52	0.099	22.68	0.893	404	0.626	0.0704	0.0214	0.0732	0.0223	0.1159	0.0353
700	355	—	61	2.72	0.107	24.49	0.964	471	0.730	0.0603	0.0184	0.0622	0.0189	0.0994	0.0303
750	380	—	61	2.82	0.111	25.35	0.998	505	0.782	0.0563	0.0171	0.0579	0.0176	0.0927	0.0282
800	405	—	61	2.91	0.114	26.16	1.030	538	0.834	0.0528	0.0161	0.0544	0.0166	0.0868	0.0265
900	456	—	61	3.09	0.122	27.79	1.094	606	0.940	0.0470	0.0143	0.0481	0.0147	0.0770	0.0235
1000	507	—	61	3.25	0.128	29.26	1.152	673	1.042	0.0423	0.0129	0.0434	0.0132	0.0695	0.0212
1250	633	—	91	2.98	0.117	32.74	1.289	842	1.305	0.0338	0.0103	0.0347	0.0106	0.0554	0.0169
1500	760	—	91	3.26	0.128	35.86	1.412	1011	1.566	0.02814	0.00858	0.02814	0.00883	0.0464	0.0141
1750	887	—	127	2.98	0.117	38.76	1.526	1180	1.829	0.02410	0.00735	0.02410	0.00756	0.0397	0.0121
2000	1013	—	127	3.19	0.126	41.45	1.632	1349	2.092	0.02109	0.00643	0.02109	0.00662	0.0348	0.0106

Notes:

1. These resistance values are valid **only** for the parameters as given. Using conductors having coated strands, different stranding type, and, especially, other temperatures changes the resistance.
2. Equation for temperature change: $R_2 = R_1, [1 + \alpha (T_2 - 75)]$ where $\alpha_{cu} = 0.00323$, $\alpha_{AL} = 0.00330$ at 75°C.
3. Conductors with compact and compressed stranding have about 9 percent and 3 percent, respectively, smaller bare conductor diameters than those shown. See Table 5A for actual compact cable dimensions.
4. The IACS conductivities used: bare copper = 100%, aluminum = 61%.
5. Class B stranding is listed as well as solid for some sizes. Its overall diameter and area are those of its circumscribing circle.

Informational Note: The construction information is in accordance with NEMA WC/70-2009 or ANSI/UL 1581-2011. The resistance is calculated in accordance with National Bureau of Standards Handbook 100, dated 1966, and Handbook 109. dated 1972.

Figure 4-5. Table 8 Conductor Properties

Bare conductors can be used for services, feeders, or branch circuits. Grounded service-entrance conductors are permitted to be bare if they used in a raceway or in an auxiliary gutter according to **Section 230.41**. As an equipment grounding conductor used in every feeder and branch circuit, they may be installed as a separate conductor where used in a metal raceway system and must be installed if used in nonmetallic raceway systems. Equipment grounding conductors are generally green or bare, in accordance with **250.119**.

Problem 4-13

A 3/0 AWG bare copper equipment bonding jumper is required to be installed in liquidtight flexible metal conduit (LFMC) with six 500 kcmil THHN copper conductors. What is the minimum size liquidtight flexible metal conduit needed between a current transformer (CT) cabinet and the service disconnecting means?

Solution
Table 5
Cross-sectional area of insulated conductors
Table 8
Cross-sectional area of bare conductors
500 kcmil THHN = 0.7073 in.^{2}
0.7073 in.^{2} × 6 = 4.2438
3/0 bare = 0.1730 in.^{2}
0.1730 in.^{2} × 1 = 0.1730
Total square inches 4.4168
Table 4
Liquidtight Flexible Metal Conduit chart
40% fill for over 2 conductors
40% fill for 4 in. LFMC = 5.077 in.^{2}
4.4168 in.^{2} requires a 4 in. LFMC
Answer: 4 in. LFMC

4.5 Short Nipple Fill Using Note 4 to Table 1 of Chapter 9

Chapter 9, Table 1, Note 4 permits nipples between boxes, cabinets, or the like, when not over 24 inches in length, to be filled to 60% of the cross-sectional area of the conduit or tubing. Each of the raceway tables within **Table 4** contains a 60% column. Also, the adjustment factors of **310.15(B)(3)(a)** need not apply where more than three current-carrying conductors are installed in a raceway.

Problem 4-14

What is the maximum number of 1/0 AWG THWN-2 copper conductors permitted to be installed in a 3 in. intermediate metal conduit nipple, 18 inches in length, between an auxiliary gutter and a switchboard?

Solution
Nipple is less than 24 inches in length.
Table 1 Note 4
Max. fill of 60% permitted
Table 4
IMC chart; 60% fill column
60% fill for a 3 in. IMC = 4.753 in.^{2}
Table 5
Cross-sectional area of conductors
1/0 AWG THWN-2 = 0.1855 in.^{2}

$$\text{Number of conductors} = \frac{60\%\ \text{fill area}}{\text{single conductor in.}^2} = \frac{4.753}{0.1855} = 25.62 \text{ or } 25$$

Answer: 25 conductors

Each of the three conduit nipples shown here may be filled to a maximum of 60% fill according to Chapter 9, Table 1, Note 4.

4.6 Using Note 7 to Table 1 of Chapter 9

When calculating the fill of a single conductor or a number of conductors or cables in a conduit or tubing, with all the conductors the same size, including the insulation, and the decimal is equal to or greater than 0.8, **Note 7** to **Table 1** of **Chapter 9** permits the next higher number of conductors.

Problem 4-15

What is the maximum number of 4 AWG XHHW copper conductors permitted in a $\frac{3}{4}$ in. electrical metallic tubing nipple, 20 inches in length?

Solution
Nipple is less than 24 inches in length
Table 1 Note 4
Max. fill of 60% permitted
Table 4
EMT chart; 60% fill column
60% fill for a $\frac{3}{4}$ in. EMT = 0.320 in.^{2}
Table 5
Cross-sectional area conductor
4 AWG XHHW copper = 0.0814 in.^{2}

$$\text{Number of conductors} = \frac{60\%\ \text{fill area}}{\text{single conductor in.}^2} = \frac{0.320}{0.0814} = 3.93$$

Table 1 Note 7
Next larger whole number permitted if the answer results in a decimal 0.8 or greater
0.93 is greater than 0.8; therefore round up
Answer: 4 conductors

4.7 Multiconductor Cables in Conduit or Tubing

Chapter 9, Table 1, Notes 5 and 9 set the method for calculating conduit or tubing size for multiconductor cables, optical fiber cables, and, where specifically permitted, flexible cords. The following factors are taken into consideration:

1. The actual diameter of the electrical cable, optical fiber cable, or flexible cord needs to be known or measured. From this equation, the area of a cable whose diameter is known is calculated as follows:
   r = radius of cable or flexible cord
   D = diameter of cable or flexible cord
   A = area of cable or flexible cord
   $$r = \frac{D}{2}$$
   $$A = \pi \times r^2 = 3.1416 \times r^2$$
2. An elliptical shaped cable or cord is treated as a round cable and the largest diameter is used.
3. Each multiconductor electrical cable, optical fiber cable, or cord is treated as a single conductor.

Problem 4-16

What size electrical metallic tubing is needed for two 4/C 12 AWG with ground, Type NM cables measuring 0.72 inches in diameter?

Solution

$$r = \frac{D}{2} = \frac{0.72}{2} = 0.36\ \text{in.}$$

$$A = 3.1416 \times r^2 \times \text{number of cables} = 3.1416 \times (0.36 \times 0.36) \times 2 = 0.814\ \text{in.}^2$$

Two cables = 2 conductors
Table 4
EMT chart; 2 wires, 31% fill column
31% fill for 2 in. EMT = 1.040 in.^{2}
0.814 in.^{2} requires a 2 in. EMT
Answer: 2 in. EMT

Problem 4-17

A 6 ft length of 2 in. electrical metallic tubing is installed between a ceiling pull box and a lighting panel. How many runs of 3-conductor 12 AWG, Type NM cable with ground can be installed in the 2 in. EMT without exceeding permitted fill? The diameter of the Type NM cable is 0.65 in.

Solution

$$r = \frac{D}{2} = \frac{0.65}{2} = 0.325\ \text{in.}$$

$$A = 3.1416 \times r^2 = 3.1416 \times (0.325 \times 0.325) = 0.3318\ \text{in.}^2$$

Table 4
EMT chart
40% fill column for (assumed) over 2 wires
40% fill for 2 in. EMT = 1.342 in.^{2}

$$\text{Number of cables} = \frac{40\%\ \text{fill area}}{\text{one cable area}} = \frac{1.342}{0.3318} = 4.04\ \text{or 4 cables}$$

Answer: 4 cables in a 2 in. EMT

4.8 Adding Conductors to Existing Conduit or Tubing

Tables 4, 5, and 5A may also be used for calculating the number of conductors which are permitted to be added to an existing conduit without exceeding the fill limit.

Problem 4-18

An existing 1 in. electrical metallic tubing contains four 8 AWG THWN copper conductors. How many 10 AWG THWN copper conductors can be added without exceeding the 40% fill?

Solution
Table 4
EMT chart; 40% fill column for over 2 wires
40% fill for 1 in. EMT = 0.3460 in.^{2}
Table 5
Cross-sectional area for insulated conductors
Space occupied by existing conductors
8 AWG THWN = 0.0366 in.^{2}
0.0366 x 4 = 0.1464 in.^{2}
Space available = permitted space – occupied space
= 0.3460 – 0.1464
= 0.1996 in.^{2}
10 AWG THWN = 0.0211 in.^{2}

$$\text{Number permitted} = \frac{\text{space available}}{\text{conductor area}} = \frac{0.1996}{0.0211} = 9.46 \text{ or } 9$$

Answer: Permitted to add an additional nine 10 AWG THWN conductors

4.9 Tables for 3/8 in. Flexible Metal Conduit

Table 348.22 is a special table for the raceway fill where a 3/8 in. flexible wiring method is used. The table is located in **Article 348 Flexible Metal Conduit: Type FMC**, but is referred to in other sections of the *Code* dealing with different types of raceways.

For example:

350.22(B) - for 3/8 in. liquidtight flexible metal conduit
360.22(B) - for 3/8 in. flexible metallic tubing

When a termination fitting, such as a box connector, is installed inside flexible metal conduit, it will decrease the internal cross-sectional area. Therefore, fewer conductors will be permitted in a flexible metal conduit with the fittings installed inside the conduit than if the fittings are installed outside the conduit. **Table 348.22** lists two columns for the two types of installations for each of four styles of conductors.

Table 348.22 lists the conductors permitted to be installed where the fitting is installed inside the conduit and where the fitting is installed outside the conduit.

The footnote to the table indicates an equipment grounding conductor of the same size is also permitted to be installed in the conduit with the other conductors.

Problem 4-19

What is the maximum number of 16 AWG TFN conductors permitted in a 3/8 in. run of flexible metal conduit where fittings are installed inside the conduit? What is the maximum number conductors permitted where the fittings are installed outside the conduit?

Solution – Calculation 1
Table 348.22
16 AWG TFN with fittings inside = 4
Answer: 4 conductors + one Equipment Grounding Conductor

Solution – Calculation 2
Table 348.22
16 AWG TFN with fittings outside = 6
Answer: 6 conductors + one Equipment Grounding Conductor

Problem 4-20

What is the maximum number of 18 AWG FEP conductors permitted in a 3/8 in. run of liquidtight flexible metal conduit with fittings installed inside the conduit with an 18 AWG bare equipment grounding conductor?

Solution
350.22(B) refers to Table 348.22
18 AWG FEP fittings inside = 5
Table footnote permits one equipment grounding conductor to be added
Answer: 5 conductors and 1 equipment grounding conductor

4.10 Underfloor Type Raceways

***NEC* Chapter 3 Wiring Methods and Materials** contains three article devoted to cellular and underfloor raceways. These raceways provide an economical method to deliver electricity, data and communication to office furniture while maintaining the "open space" design so many architects and designers specify.

4.10.1 General

The *Code* lists three types of raceways which are manufactured in a variety of cross-sectional areas and installed under the floor or as a part of the floor. The maximum number of conductors permitted in these raceways is limited to 40% fill, just as conduits and tubing are limited to 40% fill for three or more conductors. Other than **Table 1**, there is no *Code* table set up for these various raceways. The manufacturer's data should be consulted in all cases. Raceway fill is calculated using **Tables 5** and **5A** for the conductor areas.

372.11 Cellular Concrete Floor Raceway 40% fill
374.5 Cellular Metal Floor Raceway 40% fill
390.6 Underfloor Raceway 40% fill

These raceways are not standard sizes and they do not have standard cross-sectional areas. In fact, some of these raceways are of an irregular shape. In most, if not all cases, the manufacturer's literature must be consulted to determine the actual cross-sectional area available for conductors or to determine maximum number of conductors permitted within these specialty raceways. Raceways are also limited to a maximum size conductor.

372.10 Cellular Concrete Floor Raceway, No Conductor Larger Than 1/0 AWG
374.4 Cellular Metal Floor Raceway, No Conductor Larger than 1/0 AWG
390.5 Underfloor Raceway limited by manufacturer

Also important, the ampacity adjustment factors of **Table 310.15(B)(3)(a)** must be applied to conductors installed in each of these raceways according to **Section 372.17**, **Section 374.17**, and **Section 390.17**.

4.10.2 Cross-Sectional Area Calculation for a Rectangular Underfloor Raceway

A typical view of a rectangular underfloor raceway is shown in **Figure 4-6**. Listed, labeled, or identified underfloor raceways often have installation instructions for these types of raceways. Wire capacity charts and tables also accompany the installation instruction. If the raceways have wire capacity charts, those charts must be used as instructed. The following examples are meant to supplement and add to the learning process. But only manufacturer-stated internal cross-section area or the given wire charts supplied by the manufacturer are permitted to be used as the final determination of wire capacity for a given underfloor raceway.

To determine the 100% cross-sectional area available for conductors for this style of raceway, we calculate the area of the inside of the rectangle.

Area (internal) = *Length* × *Width*

$A = L \times W$

$= 3 \text{ in.} \times 2 \text{ in.}$

$= 6 \text{ in.}^2$ (100% of the internal raceway area)

Since manufacturers comply with 40% fill for more than 2 conductors, the usable area for this raceway will be:

$A = 6 \text{ in.}^2 \times 0.4$

$= 2.4 \text{ in.}^2$ (40% of internal usable area)

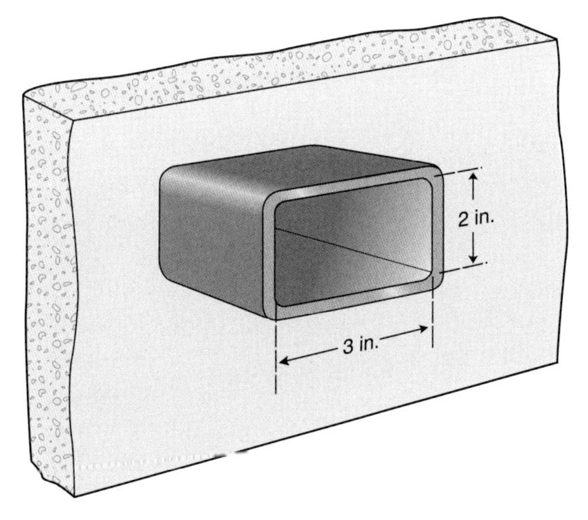

Figure 4-6. An Isometric Cross-Sectional View of a Common Rectangular Underfloor Raceway.

4.10.3 Cross-Sectional Area Calculation for a Circular Underfloor Raceway

A circular underfloor raceway may have an internal diameter of 3 inches. To determine the cross-sectional area available for conductors for this style of raceway, we simply calculate the area of the inside of the circle.

Area (internal) $= \pi \times r^2$

$$r = \frac{D}{2}$$

$$\pi = 3.1416$$

$$A = \pi \times r^2$$

$$= 3.1416 \times r^2$$

$$= 3.1416 \times (1.5 \times 1.5)$$

$= 7.0686$ in.^{2} (100% of the internal raceway area)

Courtesy of Pass and Seymour Legrand®/Wiremold

Problem 4-21

What is the maximum number of 12 AWG THHW copper conductors permitted to be installed in a rectangular underfloor raceway with internal dimensions of 1¼ in. high and 3 in. wide where manufacturer data is not available?

Solution

$A = H \times W$
= 1.25 in. × 3 in.
= 3.75 in.^{2}

Section 390.6
Max. fill = area × 40%
= 3.75 × 0.40
= 1.5 in.^{2}

Table 5
12 AWG THHW = 0.0181 in.^{2}

$$\text{Number of conductors} = \frac{\text{max. fill}}{\text{one conductor area}}$$

$$= \frac{1.5}{0.0181}$$

= 82.87 or 82

Answer: 82 Conductors

Note: In most cases, it is prudent to reduce this number because **Table 310.15(B)(3)** may need to be applied.

Problem 4-22

What is the maximum number of 8 AWG THHW compact aluminum conductors permitted to be installed in a cell of a cellular metal underfloor raceway with a stated internal area of 2 in.^{2}?

Solution

$A = 2.0$ in.^{2}

Section 374.5
Area max. fill = A × 40%
= 2.0 × 0.40
= 0.8 in.^{2}

Table 5A
8 AWG THHW = 0.0510 in.^{2}

$$\text{Number of conductors} = \frac{\text{max. fill}}{\text{one conductor area}}$$

$$= \frac{0.8}{0.0510}$$

= 15.6 or 15

Answer: 15 conductors

Problem 4-23

What is the maximum number of 10 AWG THWN-2, solid copper conductors permitted to be installed in a cellular metal floor raceway with internal manufacturer's dimensions of 2 in. x 2 in.?

Solution

$A = L \times W$
= 2 × 2
= 4 in.^{2}

Section 374.5
Area max. fill = A × 40%
= 4 × 0.40
= 1.6 in.^{2}

Table 5
10 AWG THWN-2 = 0.0211 in.^{2}

$$\text{Number of conductors} = \frac{\text{max. fill}}{\text{one conductor area}}$$

$$= \frac{1.6}{0.0211}$$

= 75.8 or 75

Answer: 75 conductors

Cellular metal floor raceway is covered by Article 374. Section 374.17 requires that the ampacity adjustments in 310.15(B)(3)(a) be applied to all current-carrying conductors installed in cellular metal floor raceways. For Problem 4-23, the current-carrying ampacity of the seventy-five 10 AWG THWN-2 conductors would be reduced to 35% of the original table ampacity.

Problem 4-24

What is the maximum number of 10 AWG THWN copper conductors permitted to be installed in a cell of a cellular concrete underfloor raceway with an internal diameter of 2 in.? The manufacturer's data indicates that the 2 in. raceway has a usable total cross-sectional area of 2.057 in.².

Solution
Section 372.11
Area max. fill = A × 40%
= 2.057 × 0.40
= 0.8228 in.²
Table 5
10 AWG THWN = 0.0211 in.²

$$\text{Number of conductors} = \frac{\text{balance}}{\text{one conductor}} = \frac{0.8228}{0.0211} = 38.99$$

Chapter 9, Table 1, Note 7 does not apply to this raceway
Answer: 38 conductors

Problem 4-25

An existing underfloor raceway with an internal cross-sectional area of 3 in.² contains sixteen 8 AWG THW copper conductors. How many more 8 AWG THWN copper conductors are permitted to be installed in the same raceway?

Solution
Section 390.6
Area max. fill = A × 40%
= 3 × 0.40
= 1.2 in.²
Table 5
8 AWG THW = 0.0437 in.²
8 AWG THWN = 0.0366 in.²
Percent fill = 16 × 0.0437
= 0.6992 in.²
Balance fill = max. fill – present fill
= 1.2 – 0.6992
= 0.5008 in.²

$$\text{Number of conductors} = \frac{\text{balance}}{\text{one conductor}} = \frac{0.5008}{0.0366} = 13.68$$

Answer: 13 conductors size 8 AWG THWN

4.11 Metal Wireways, Nonmetallic Wireways, and Auxiliary Gutters

One of the basic differences between a metal wireway and an auxiliary gutter is that an auxiliary gutter is limited to 30 ft in length while a wireway is not limited in length. Another basic difference between wireways and auxiliary gutters is that wireways are limited to a maximum size of a single conductor according to both **Section 376.21** and **Section 378.21**. Auxiliary gutters have no such limitation. The basic rules for the fill of a metal wireway or an auxiliary gutter are very much the same when it comes to the installation of electrical conductors, other than busbars. There is also a commonality between nonmetallic wireways and nonmetallic auxiliary gutters.

Section 376.21 indicates that "no conductor larger than that for which the wireway is designed shall be installed in any wireway." This statement leaves the user without an actual maximum wire size for a given size wireway. However, the largest wire permitted within a wireway or auxiliary gutter can be determined by using the wire size column and the 1 Wire per Terminal column of **Table 312.6(A)**.

For example, a 500 kcmil conductor (one wire per terminal) requires at least a 6 in. x 6 in. wireway or auxiliary gutter. Another example, as an alternate, for an 4 in. x 4 in. size wireway, the maximum size conductor permitted is 4/0 AWG. These maximum wire sizes may also be determined by using *ANSI/UL 870, Standard for Safety for Wireways, Auxiliary Gutters, and Associated Fittings.*

4.11.1 Article 376 Metal Wireways

376.22(A) permits a maximum of 20% fill at any interior cross-sectional area. **376.22(B)** permits a maximum of 30 current-carrying conductors without applying adjustment factors. Signaling and motor conductors used only for starting are not counted as current-carrying conductors.

Splices and taps are permitted in wireways, provided they are accessible. Specifically, **376.56(A)** requires that splices and taps made in metal wireways must not exceed 75% of the cross-sectional area at splice or tap location.

Problem 4-26

How many 1 AWG THWN copper conductors are permitted to be installed in a metal wireway with internal dimensions of 3 in. by 4 in.?

Solution

A = L × W

= 4 in. × 3 in.

= 12 in.²

376.22(A)

Area max. fill = A × 20%

= 12 × 0.20

= 2.4 in.²

Table 5

1 AWG THWN = 0.1562 in.²

$$\text{Number of conductors} = \frac{\text{max. fill}}{\text{one conductor}} = \frac{2.4}{0.1562} = 15.3$$

Answer: 15 conductors

Problem 4-27

A metal wireway with inside dimensions of 3 in. by 3 in. contains twenty 12 AWG THWN and ten 10 AWG THWN current-carrying conductors of copper. Could 12 more 10 AWG THWN current-carrying conductors be added and stay within raceway fill limitations without applying the ampacity adjustment factors to the conductors?

Solution

Section 376.22

Without applying adjustment factors, the maximum number of conductors permitted is 30

Twenty (12 AWG) + ten (10 AWG) + twelve (10 AWG) = 42 conductors

Max. number of conductors will be exceeded

The cross-sectional area of 20% may not be exceeded, but the 30 conductor limitation would be exceeded.

Answer: No. Ampacity adjustment factors would need to be applied

Requirements	Metal Wireways	Nonmetallic Wireways
Listing	No	Yes
Fill	20%	20%
Supports	5 ft.	3 or 4 ft.*
Ampacity Adjustment	Only over 30 conductors	Current-carrying conductors

See 378.30(A) and (B)

4.11.2 Article 378 Nonmetallic Wireways

According to **Section 378.22** for nonmetallic wireways:

- A maximum of 20% fill at any interior cross-sectional area cannot be exceeded. There are no exceptions.
- The number of conductors is not limited.
- Adjustment factors apply to all current-carrying conductors.
- Signaling and motor conductors used for starting are not counted as current-carrying conductors.

Splices and taps are permitted in nonmetallic wireways, provided they are accessible. **Section 378.56** requires that splices and taps must not exceed 75% of the cross-sectional area at splice or tap location.

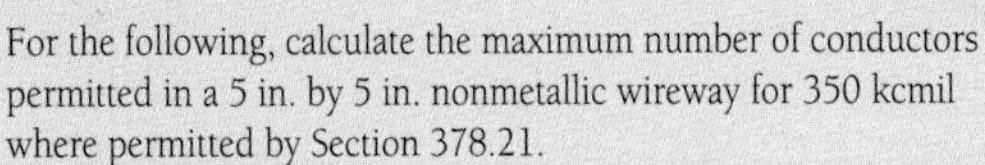

For the following, calculate the maximum number of conductors permitted in a 5 in. by 5 in. nonmetallic wireway for 350 kcmil where permitted by Section 378.21.

1. 350 kcmil THHW concentric stranded copper
2. 350 kcmil THHW compact stranded aluminum

Solution - Calculation 1

Wireway area = 5 in. × 5 in.

= 25 in.²

Section 378.22

Max. fill = A × 20%

= 25 × 0.20

= 5.0 in.²

Table 5

350 kcmil THHW = 0.5958 in.²

$$\text{Number of conductors} = \frac{\text{max. fill}}{\text{one conductor}} = \frac{5}{0.5958} = 8.39 \text{ or } 8$$

Answer: Eight 350 kcmil THHW concentric copper conductors

Solution – Calculation 2

Table 5A

350 kcmil THHW compact aluminum = 0.5281 in.²

$$\text{Number of conductors} = \frac{\text{max. fill}}{\text{one conductor}} = \frac{5}{0.5281} = 9.47 \text{ or } 9$$

Answer: Nine 350 kcmil THHW compact aluminum conductors

4.11.3 Article 366 Auxiliary Gutters

According to **366.22(A)** for sheet metallic auxiliary gutters:

- A maximum of 20% fill at any interior cross-sectional area is permitted.
- A maximum of 30 current-carrying conductors without de-rating is permitted.
- Signaling and motor conductors used for starting are not counted.

Problem 4-29

A square sheet metal auxiliary gutter with internal dimensions of 3 in. by 3 in. contains six 1/0 AWG THW copper conductors. How many 4 AWG THHW copper conductors are permitted to be added without exceeding the maximum fill of the auxiliary gutter?

Solution

A = L x W
= 3 in. x 3 in.
= 9 in.^{2}

366.22(A)

Max. fill = A x 20%
= 9 x 0.20
= 1.8 in.^{2}

Table 5

1/0 AWG THW copper = 0.2223 in.^{2}

Present fill = 0.2223 x 6 conductors
= 1.3338 in.^{2}

Balance = max. fill – present fill
= 1.8 – 1.3338
= 0.4662 in.^{2}

Table 5

4 AWG THHW = 0.0973 in.^{2}

$$\text{Number of conductors} = \frac{\text{balance area}}{\text{one conductor}}$$

$$\text{Number of conductors} = \frac{0.4662}{0.0973}$$

= 4.79 or 4

Answer: 4 conductors

- Where used in accordance with **Article 620** for elevators, escalators, and the like, **Section 366.6** does not apply.

Splices and taps are permitted in auxiliary gutters, provided they are accessible. The regulations are very much the same for wireways and auxiliary gutters. According to **366.56(A)**, splices and taps made in auxiliary gutters must not exceed 75% of the interior cross-sectional area at tap or splice location.

Problem 4-30

What is the maximum number of 10 AWG THHN current-carrying copper conductors permitted to be installed in a sheet metalallic auxiliary gutter with a 4 in. by 4 in. internal area?

Solution

A = L x W
= 4 in. x 4 in.
= 16 in.^{2}

366.22(A)

Max. fill = A x 20%
= 16 x 0.20
= 3.2 in.^{2}

Table 5

10 AWG THHN = 0.0211 in.^{2}

$$\text{Number of conductors} = \frac{\text{max. fill}}{\text{one conductor}}$$

$$= \frac{3.2}{0.0211}$$

= 151.6 or 151

Answer: 151 conductors

Comment

The 20% fill has not been exceeded, therefore 151 conductors are permitted. However, when the 30-conductor limitation is exceeded, the ampacity adjustment factors of Table 310.15(B)(2)(a) will apply and the conductors must be adjusted accordingly.

The following definition from the NFPA Regulations governing the Development of NFPA Standards, Section 3.3.6.1, represents an official definition for all NFPA documents:

Approved - Acceptable to the authority having jurisdiction.

NOTE: The *National Fire Protection Association* does not approve, inspect, or certify any installations, procedures, equipment, or materials nor does it approve or evaluate testing laboratories. In determining the acceptability of installations or procedures, equipment, or materials, the "authority having jurisdiction" may base acceptance on compliance with NFPA or other appropriate standards. In the absence of such standards, said authority may require evidence of proper installation, procedure, or use. The "authority having jurisdiction" may also refer to the listings or labeling practices of an organization that is concerned with product evaluations and is thus in a position to determine compliance with appropriate standards for the current production of listed items.

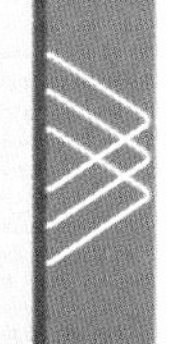

Problem 4-31

Three 4/0 AWG THHN copper conductors are to be spliced in a sheet metallic auxiliary gutter with the splices staggered. Will an auxiliary gutter with inside dimensions of 2 ½ in. deep by 3 in. wide be large enough for the installation? (For this example only, assume the spliced cross-sectional area of any one conductor is three times the normal cross-sectional area of the conductor.)

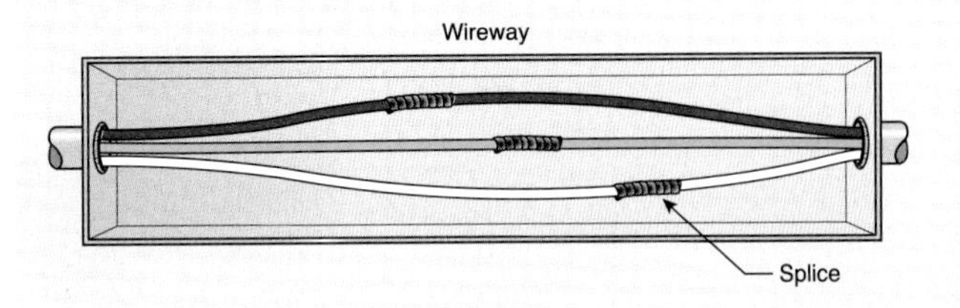

Solution

$A = H \times W$
$= 2.5 \text{ in.} \times 3 \text{ in.}$
$= 7.5 \text{ in.}^2$

366.56(A)

Max. fill = $A \times 75\%$
$= 7.5 \times 0.75$
$= 5.625 \text{ in.}^2$

Table 5

Cross-sectional area of conductors

4/0 AWG THHN = 0.3237 in.^2

Spliced conductor tripled = $0.3237 \text{ in.}^2 \times 3$ = 0.9711

Remaining 2 conductors = $0.3237 \text{ in.}^2 \times 2$ = 0.6474

Total 1.6185

Since the splice plus the other two conductors occupy less than 5.625 in^2, then answer is yes, staggered splices are permitted.

Answer: Yes, the size given is sufficient and complies with 366.56(A).

Definitions and Terms

Compact Stranding - The result of a manufacturing process where the standard conductor is compressed to the extent that the interstices (voids between strand wires) are virtually eliminated.

Conductor, Bare - A conductor having no covering or electrical insulation whatsoever.

Conductor, Covered - A conductor encased within material of composition or thickness that is not recognized by the *NEC* as electrical insulation.

Conductor, Insulated - A conductor encased within material of composition and thickness that is recognized by the *NEC* as electrical insulation (recognized insulation is found in **Table 310.104(A)**).

Informative Annex - All Informative Annexes are not a part of the requirements of the *NEC*, but are included in the *NEC* for informational purposes only.

Nipple - A short piece of raceway usually 24 in. or less in length.

Raceway - An enclosed channel of metallic or nonmetallic materials designed expressly for holding wires, cables, or busbars, with additional functions as permitted in this *Code*. **Informational Note:** A Raceway is identified within specific article definitions.

Summary

- Adhering to the conductor fill table requirements of **Chapter 9** permits conductors to safely dissipate heat, permits ease in conductor installation and removal, and prevents premature insulation failure due to shorts and grounds.
- Conduit and tubing are the most common raceways requiring the use of conduit fill look-up tables or individual calculations.
- The performance requirements for conduit fill are found in **Section 300.17**. Each specific *NEC* section containing fill requirements is found in the Informational Note following **Section 300.17**.
- **Chapter 9, Table 1** gives the maximum conduit and tubing fills permitted for all standard size conduits and tubing wiring methods.
- The maximum permitted percentage of the cross section of conduit and tubing for conductors is 53% for a one conductor, 31% for two conductors, and 40% for more than two conductors.
- The cross-sectional areas of each and every conductor (current carrying or not) which occupies space within a conduit must be summed and used to calculate the filled area within the interior cross section of the conduit or tubing.
- Where all the conductors are the same size and the same insulation, the look-up tables of **Annex C** are permitted to be used to determine the number of conductors permitted in a raceway.
- Where all the conductors are the same size, the same insulation, and of the compact-type stranding, the look-up tables of **Annex C** are also permitted to be used.
- Where the conductors are of mixed sizes or insulation types, the values found in **Chapter 9, Tables 4** and **5** must be used to calculate the proper conductor fill.
- Where bare conductors are used in conduit or tubing, the values expressed in **Chapter 9, Table 8** are permitted to be used for fill calculations.
- For short lengths of conduit or tubing not exceeding 24 inches in length, the conductors are permitted to fill the raceway up to 60% of the raceway interior cross-sectional area.
- The cross-sectional area of multiconductor cables or flexible cords is permitted to be calculated to determine overall raceway fill.
- Adding conductors to an existing raceway requires a calculation to ensure *Code* compliance before the installation is made.
- Trade size 3/8 in. flexible metal conduit, flexible metal tubing, and liquidtight flexible metal conduit all use **Table 348.22** for conduit and tubing fill calculations.
- For specialized raceways, it is important to thoroughly review manufacturer's literature to determine the maximum conductor size and the permitted conductor fill (tables) for their product. See **110.3(B)**.
- The allowed maximum conductor size of 1/0 AWG and the permitted conductor fill of 40% are requirements for the cells and headers of both cellular concrete floor raceways and cellular metal floor raceways.

Motor Calculations

According to **Section 90.3**, **Article 430** contains the general rules for motors. These general requirements may be altered by the requirements of Chapters 5, 6, or 7 for special occupancies, special equipment, or special conditions.

Article 430 is a large and somewhat complex group of electrical safety requirements for motors. But, these safety requirements are dealt with in an extremely organized fashion. A thorough treatment of safety to each and every portion of a motor circuit is provided. The *Code* requirements are easily applied to both small and large systems. They can be applied from the simplest circuits to the most complex equipment. The method of organization of **Article 430** is first shown in the *NEC* **Table of Contents** and contains fourteen separate parts. Also, an electrical single-line drawing is provided in **Figure 430.1** as a visual outline which shows the reader graphically how the various parts of **Article 430** are organized and where in the motor circuit they apply.

The successful completion of this chapter on motor calculations and the demonstration of the calculation skills shown within this chapter will equate to a major step into the world of a competent commercial and industrial Electrical Worker.

Objectives

- Calculate the ampacities of motor circuits and hermetically sealed air-conditioning motor branch circuits.
- Calculate and select the motor circuit copper conductor size using **Table 310.15(B)(16)**.
- Calculate and select the proper size branch-circuit short-circuit and ground-fault protection device for motor circuits using **430.52** and **240.6(A)**.
- Calculate the size of the motor overload protection for motors in general.
- Using **430.7(B)**, **Table 430.51(A)** and **Table 430.51(B)**, calculate the locked-rotor current for various motors.
- Calculate the minimum ampacity and horsepower rating necessary for a motor disconnecting means.

Chapter 5

Table of Contents

5.1 Motor Branch-Circuit Conductors

5.1.1 Introduction

A branch circuit is that portion of the circuit which extends beyond the last overcurrent protection device. A motor branch circuit includes all conductors between the branch-circuit protection device and the motor as shown in **Figure 5-1**. The motor overload relays are for the running protection of the motor and are not considered the last or final overcurrent protection device.

430.6(A)(1) gives some basic information for motor calculations. It particularly identifies when the nameplate full-load current rating must be used and when the ampacity values listed in full-load current **Table 430.248**, **Table 430.249**, and **Table 430.250** must be used.

5.1.2 General Application

1. Conductors must be selected from the allowable ampacity tables (primarily **Table 310.15(B)(16)** in accordance with **310.15(B)**).
2. **Section 240.4(D)** is a general overcurrent protection requirement for small conductors referenced at the bottom of **Table 310.15(B)(16)**. This general requirement is amended for motor and motor-control circuits through the reference to specific overcurrent protection requirements identified in **Table 240.4(G)**. This table points directly to **Article 430, Parts III.** and **IV.** dealing with motor branch-circuit overload protection, and motor branch-circuit short-circuit and ground-fault protection. Therefore, the overcurrent protection requirements of **240.4(D)** do not apply to motor-circuit overcurrent protection. Rather, the user is directed to the **Article 430** for this protection.
3. The ampacity values given in the motor full-load current tables must be used to calculate the ampacity of motor branch-circuit conductors whether or not the full-load current is given on the nameplate.
4. The values given in **Table 430.248**, **Table 430.249**, and **Table 430.250** are the full-load currents (FLCs) for most motors of normal torque values and common speeds.

For additional information, visit qr.njatcdb.org Item #1033

5.1.2.1 Low-Speed and Multispeed Motors - Low-speed and high-torque motors usually have higher full-load currents than those found in the tables, so the nameplate currents must be used for these types of motors. For multispeed motors, the full-load current will vary with the speed and again, the nameplate values must be used.

5.1.2.2 Listed Appliances and Equipment of a Specific Type - According to **430.6(A)(1)**, **Exception No. 2** and **No. 3**, listed appliances and specific types of blower motors are allowed to use their equipment nameplate current rating for calculating the ampacity of these branch-circuit conductors instead of the **Article 430** table values.

5.1.2.3 Torque Motors - According to **430.6(B)**, the motor rated current for torque motors is the locked-rotor current marked on the nameplate and must be used to calculate the ampacity of the branch-circuit conductors.

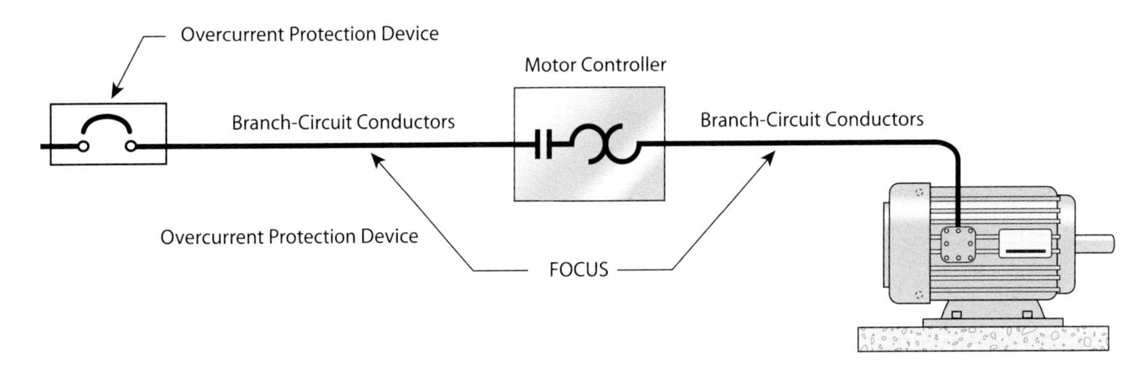

Figure 5-1. Shown here is a complete branch circuit, beginning at the final overcurrent protective device and ending at the motor terminal box.

5.1.2.4 Motors Used in AC Adjustable Voltage, Variable Torque Drive Systems - AC adjustable voltage motors and variable torque drive system conductors are covered in 430.6(C).

1. The maximum operating current given on the nameplate of the equipment controller or motor must be used to calculate the ampacity of the branch-circuit conductors.
2. Where the maximum operating current is not marked on the nameplate, 150% of the values given in **Table 430.249** and **Table 430.250** must be used.

The branch-circuit conductors supplying the conversion equipment for an adjustable-speed drive system are required to be 125% of the rated input current of the power conversion equipment according to **430.122(A)**.

5.1.3 Sizing Motor Branch-Circuit Conductors

5.1.3.1 Continuous-Duty Motors - Section 430.22 requires the ampacity of motor branch-circuit conductors supplying a single motor used in a continuous-duty application to be 125% of the motor full-load current (FLC).

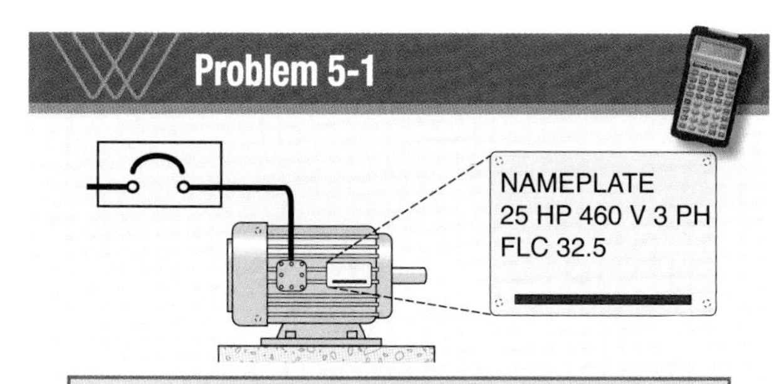

What is the minimum ampacity of the branch-circuit conductors for a 25-hp, 460-volt, 3-phase squirrel-cage motor with a nameplate full-load current rating of 32.5 amperes?

Solution
Nameplate FLC is not used
430.6(A)(1)
Use table value for FLC
Table 430.250
25 hp at 460 volts
Table value FLC = 34 amps
Section 430.22
Ampacity = FLC × 125%
= 34 × 1.25
= 42.5 amps
Answer: 42.5 amperes

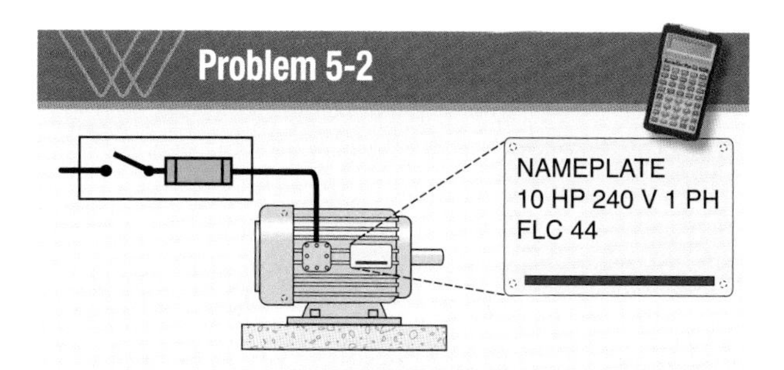

What is the minimum size THHN copper conductors required for a 10-hp, 240-volt, single-phase motor used in a continuous-duty application with a nameplate rating of 44 amperes?

Solution
Nameplate FLC is not used
430.6(A)
Use table value for FLC
Table 430.248
10 hp at 240 volts
Table value FLC = 50 amps
Section 430.22
Ampacity = FLC × 125%
= 50 × 1.25
= 62.5 amps
Table 310.15(B)(16)
75°C column
62.5 amps requires 6 AWG THHN
Answer: 6 AWG THHN

Parallel conductors are covered in 310.10(H).

Construction dirt and dust often makes new equipment look unsightly.

5.1.3.2 Multispeed Motors - A multispeed motor will have more than one set of windings. How the windings are connected determines the speed of the motor. Or, the motor may have separate windings designated for a particular speed, and only one set of windings is used at any one time. The installation of multispeed motors is covered in **430.6(A)(1), Exception No. 1** and **430.22(B)**. The ampacity of the branch-circuit conductors between the controller and the motor must not be less than 125% of the current rating of the winding(s) that the conductors energize.

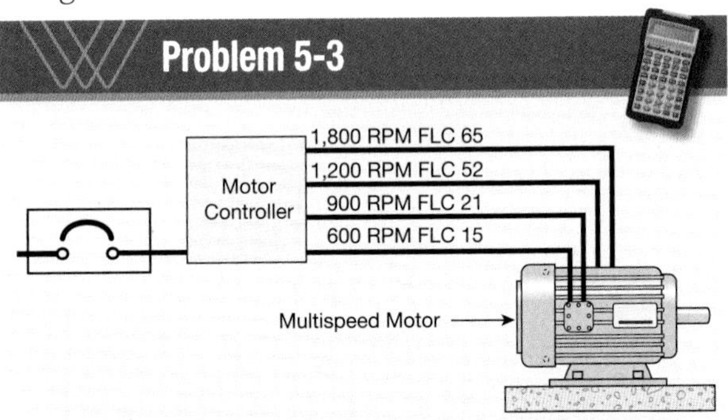

What is the minimum ampacity of the branch-circuit conductors between the controller and the 3-phase, 460-volt multispeed motor for each of the following speeds and nameplate full-load currents?

50 HP	1,800 RPM	FLC = 65 amps
40 HP	1,200 RPM	FLC = 52 amps
15 HP	900 RPM	FLC = 21 amps
10 HP	600 RPM	FLC = 15 amps

Solution
430.22(B)
Use nameplate current ratings
Ampacity = FLC nameplate × 125%

1,800 RPM	= 65 × 1.25	**= 81.25 amps**
1,200 RPM	= 52 × 1.25	**= 65 amps**
900 RPM	= 21 × 1.25	**= 26.25 amps**
600 RPM	= 15 × 1.25	**= 18.75 amps**

Electric motors of the industrial variety are often built for harsh environments.

Baldor Electric Co.

Information

430.22(B) requires the ampacity of the branch-circuit conductors on the line side of a multispeed motor controller to be not less than 125% of the highest current rating of any one speed.

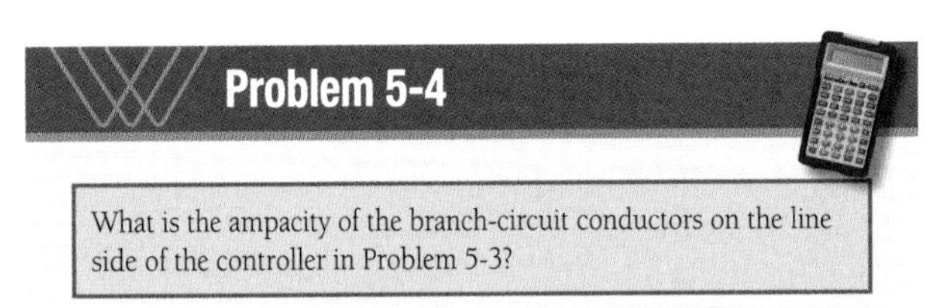
Problem 5-4

What is the ampacity of the branch-circuit conductors on the line side of the controller in Problem 5-3?

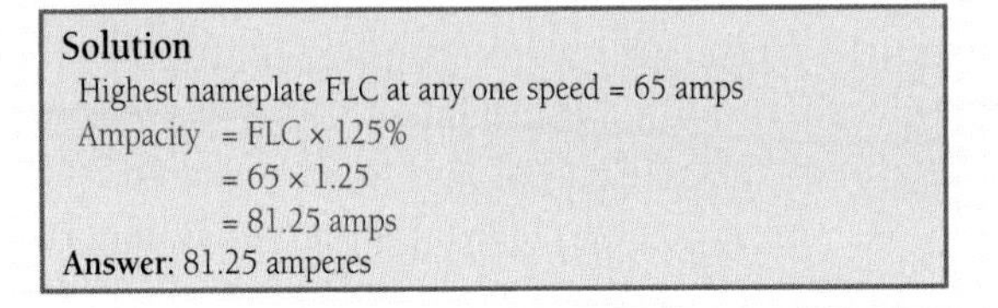
Solution
Highest nameplate FLC at any one speed = 65 amps
Ampacity = FLC × 125%
= 65 × 1.25
= 81.25 amps
Answer: 81.25 amperes

Note:
The branch-circuit conductors on the line side of the controller are the same ampacity as the conductors with the highest ampacity rating on the motor side of the controller.

5.1.3.3 Periodic and Varying Duty Motors The ampacity of conductors for motors used for short-time, intermittent, periodic, or varying duty is permitted to have the branch-circuit conductor ampacity calculated using a special table, **Table 430.22(E) Duty-Cycle Service**.

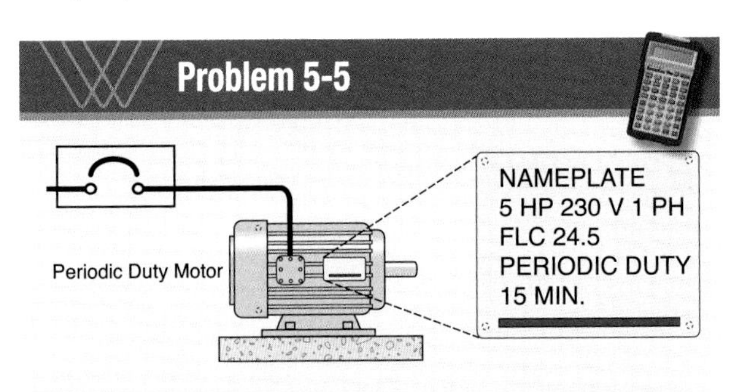

What is the minimum ampacity of the branch-circuit conductors for a 5-hp, single-phase, 230-volt motor with a nameplate current of 24.5 amperes, operating on a periodic duty cycle of 15 minutes?

Solution
Nameplate current rating is used
Table 430.22(E)
FLC 15 min. periodic duty = 90%
Ampacity = FLC (nameplate) × 90%
= 24.5 × 0.90
= 22.05 amps
Answer: 22.05 amperes

Problem 5-6

What is the minimum ampacity of the branch-circuit conductors for a continuous rated 15-hp, 230-volt, 3-phase varying duty type motor with a nameplate full-load current rating of 16.5 amperes?

Solution
Nameplate FLC is used
Table 430.22(E)
Continuous rated motor, varying duty = 200%
Ampacity = FLC (nameplate) × 200%
= 16.5 × 2.00
= 33 amps
Answer: 33 amperes

Problem 5-7

What is the minimum size of THWN copper conductors needed for the branch-circuit conductors of Problem 5-6?

Solution
Table 310.15(B)(16)
THWN, 75°C copper column
33 amp THWN requires 10 AWG copper
Answer: 10 AWG THWN copper

5.1.3.4 Torque Motor - A torque motor is designed primarily to exert torque while stalled or rotating slowly. The term locked-rotor current (LRC) is used in conjunction with the torque motor. Locked-rotor current is the current a motor will draw with the shaft of the motor held stationary. A torque motor will drive the load until the shaft of the motor is stationary. Think of a torque motor as a motor used to shut a valve. When the torque motor has driven the valve completely shut, the shaft of the motor can move no further, and locked-rotor current will flow. The circuit may be opened by means of a torque or limit switch if need be.

According to **430.6(B)**, when calculating the ampacity of the branch-circuit conductors for a torque motor, the nameplate locked-rotor current will be used.

Valve Actuated Motor

Section **430.6(D)** clarifies that valve actuator motor assemblies are not torque motors and that the nameplate full-load current is the rated motor current to be used for calculations.

Problem 5-8

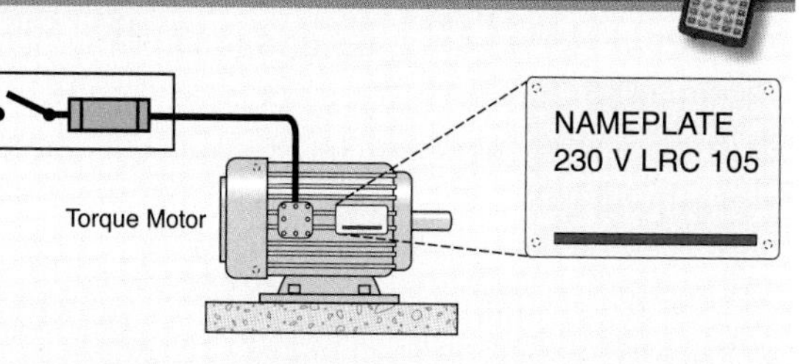

What is the minimum ampacity of the branch-circuit conductors for a 230-volt, single-phase torque motor with a nameplate locked-rotor current rating of 105 amperes?

Solution
430.6(B)
Use nameplate locked-rotor current
430.22
Ampacity = LRC (nameplate) × 125%
= 105 × 1.25
= 131.25 amps
Answer: 131.25 amperes

Problem 5-9

What is the minimum size XHHW branch-circuit copper conductors needed for the motor in Problem 5-8?

Solution
Table 310.15(B)(16)
XHHW, 90°C column
110.14(C)(1)(b)(2)
Limited to 75°C column
131.25 amps requires 1/0 AWG XHHW copper
Answer: 1/0 AWG XHHW copper

Two motors are controlled from a common control system located in a single cabinet.

5.1.3.5 Wye-Start, Delta-Run Motor - The wye-start, delta-run motor is one technique used for reduced-voltage starting, resulting in a lower starting current. **430.22(C)** requires the line conductors to the controller to be calculated at 125% of the motor full-load current. This section also requires that the six (load) conductors between the controller and the motor to be calculated at 72% of the motor full-load current. The informational note following **430.22(C)** explains that the 72% is actually the product of the 58% multiplied by 125%.

Problem 5-10

Determine the minimum ampacity of the following branch-circuit conductors for a 20-hp, 3-phase, 440-volt wye-start, delta-run motor.

1. Line conductors to controller
2. Conductors between controller and motor

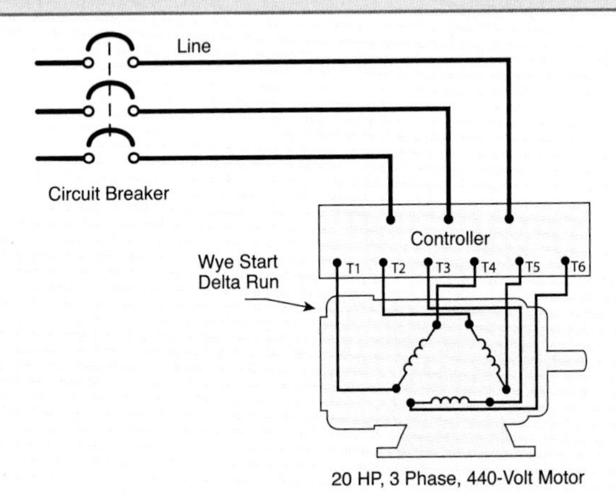

Solution - Calculation 1
Line conductors to controller
Table 430.250
20 hp at 440 volts
FLC = 27 amps
430.22(C)
Ampacity = FLC × 125%
= 27 × 1.25
= 33.75 amps
Answer: 33.75 amperes

Solution - Calculation 2
All conductors between controller and motor
430.22(C)
Ampacity = FLC × 72%
= 27 × 0.72
= 19.44 amps
Answer: 19.44 amperes

Problem 5-11

Determine the minimum size of the following THWN copper branch-circuit conductors for the installation of a 20-hp, 3-phase, 460-volt adjustable speed motor and a conversion unit with an input rating of 30 amperes.

1. Input conductors to conversion unit
2. Conductors between conversion unit and motor

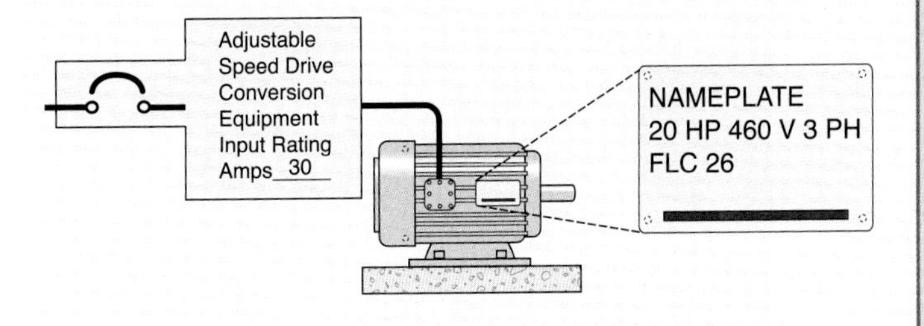

Solution - Calculation 1
Circuit conductors to drive unit
430.6(C) and 430.122(A)
Ampacity = input current × 125%
= 30 × 1.25
= 37.5 amps
Table 310.15(B)(16) Ampacity
THWN, 75°C copper column
37.5 amps requires 8 AWG THWN
Answer: 8 AWG THWN

Solution - Calculation 2
Conductors from drive unit to motor
Section 430.22
Use table value
Table 430.250
20 hp at 460 volts
FLC = 27 amps
Ampacity = FLC × 125%
= 27 × 1.25
= 33.75 amps
Table 310.15(B)(16) Ampacity
THWN, 75°C copper column
33.75 amps requires 10 AWG THWN
Answer: 10 AWG THWN

5.1.3.6 Adjustable Speed Drive Motor - 430.122(A) Circuit conductors supplying power conversion equipment included as part of an adjustable-speed drive system require the input conductors to be 125% of the input rating of the conversion unit. This compensates for the current used in the conversion unit.

5.1.3.7 Wound-Rotor Induction Motor - The wound-rotor induction motor is like a transformer with a primary and a secondary. The primary is the stationary or stator winding, and the secondary winding is wound on the rotor and allowed to turn.

There are no line electrical connections to the secondary or rotor of a squirrel-cage induction motor. However, there are electrical control connections to the secondary (rotor) of the wound-rotor motor. These connections are used to connect resistance into the secondary circuit to control the starting and the speed of the motor. Therefore, there are two sets of conductors to calculate for the branch-circuit conductors of a wound-rotor motor. The branch-circuit conductors going to the motor, often referred to as the primary conductors, are connected to the stator. The leads used to connect the rotor, often referred to as the secondary conductors, are connected to a remote bank of resistors via slip rings. The resistance can be built into the controller, as for a drum controller, or the resistance can be a bank of resistors entirely separate or apart from the controller. **See Figure 5-2**.

Three different installation situations can affect the sizing of the secondary conductors:

1. Resistors within a controller rated as continuous duty
2. Resistors within a controller rated other than continuous duty (short-time, intermittent, periodic, or varying)
3. Resistors located apart from the controller

430.23(A) covers **Problem 5-12**. For a continuous-duty motor with the resistors within the controller, the conductors connecting the secondary (rotor) to the controller must have an ampacity of not less than 125% of the nameplate full-load current of the motor secondary.

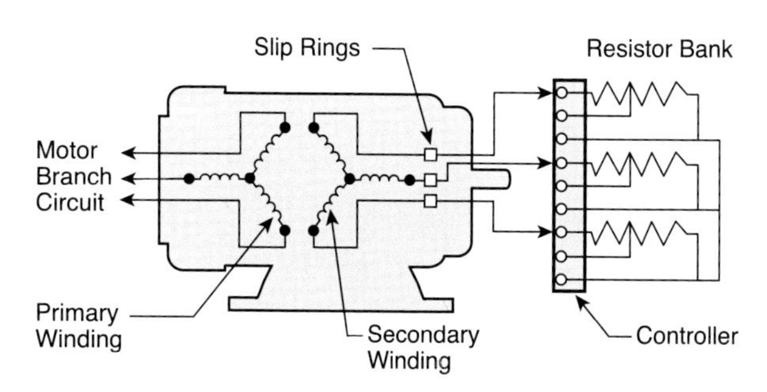

Figure 5-2. A wiring diagram of wound-rotor induction motor including the secondary winding and resistor bank.

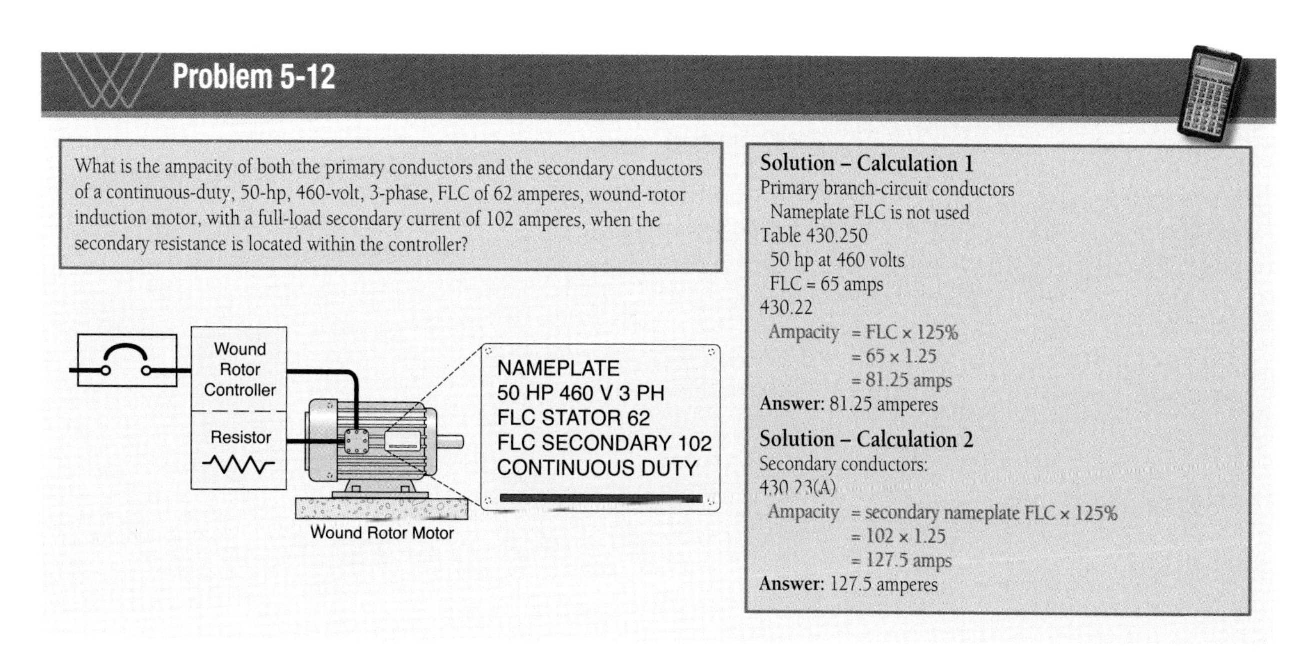

Problem 5-12

What is the ampacity of both the primary conductors and the secondary conductors of a continuous-duty, 50-hp, 460-volt, 3-phase, FLC of 62 amperes, wound-rotor induction motor, with a full-load secondary current of 102 amperes, when the secondary resistance is located within the controller?

Solution – Calculation 1
Primary branch-circuit conductors
Nameplate FLC is not used
Table 430.250
50 hp at 460 volts
FLC = 65 amps
430.22
Ampacity = FLC × 125%
= 65 × 1.25
= 81.25 amps
Answer: 81.25 amperes

Solution – Calculation 2
Secondary conductors:
430.23(A)
Ampacity = secondary nameplate FLC × 125%
= 102 × 1.25
= 127.5 amps
Answer: 127.5 amperes

Information

430.23(B) covers the wound-rotor secondary conductors of **Problem 5-13**. For a wound-rotor motor with the resistor within the controller and a short-time, intermittent, periodic, or varying duty rating, the ampacity of the secondary conductors must not be less than the nameplate secondary current multiplied by the percentage given in **Table 430.22(E)**. Wound-rotor motors are used less today due to modern adjustable speed drive AC motors.

Information

430.23(C) covers **Problem 5-14**. When the secondary resistor is separate from the controller, the ampacity of the conductors between the controller and resistor must not be less than that given in **Table 430.23(C)**.

5.1.3.8 Circuit Conductors Supplying More Than One Motor - Section 430.24 covers the installation of two or more motors of the same or different sizes installed on the same branch circuit.

Problem 5-13

What is the minimum ampacity of the primary branch-circuit conductors and secondary conductors for a continuous-duty, wound-rotor induction motor, with an intermittent duty rated secondary resistors that are installed within the controller and the following information on the nameplate: 75 hp, 460 volts, 3-phase, primary FLC 88 amperes and secondary FLC 150 amperes?

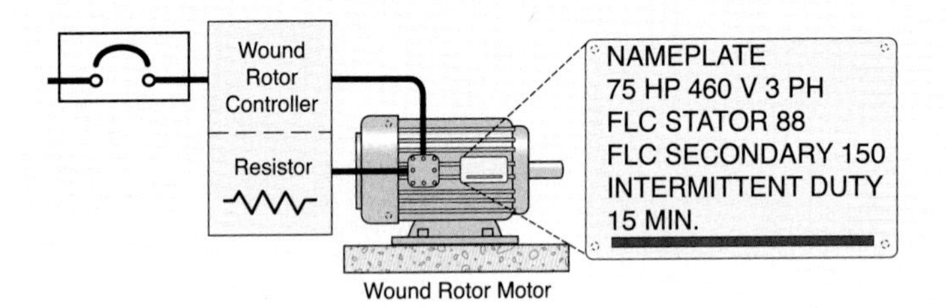

Wound Rotor Motor

Solution – Calculation 1
Primary branch-circuit conductors
Nameplate FLC not used
Table 430.250
75 hp at 460 volts
FLC = 96 amps
430.22(A)
Use table value
Ampacity = FLC × 125%
= 96 × 1.25
= 120 amps
Answer: 120 amperes

Solution – Calculation 2
Secondary conductors
430.23(B)
Use Table 430.22(E)
Intermittent duty 15 min. = 85%
Ampacity = secondary nameplate FLC × 85%
= 150 × 0.85
= 127.5 amps
Answer: 127.5 amperes

Problem 5-14

What is the ampacity of the primary and secondary conductors for a continuous-duty, wound-rotor induction motor with the heavy intermittent-duty secondary resistors mounted in a separate enclosure from the controller with the following motor information on the nameplate: 25 hp, 3-phase, 208 volts, primary FLC 75 amperes and secondary FLC 142 amperes?

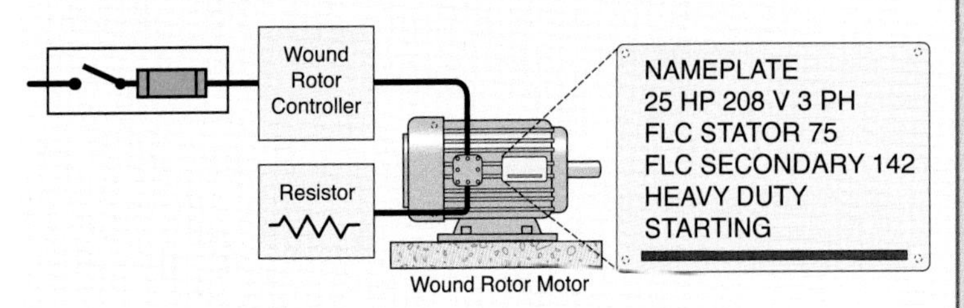

Wound Rotor Motor

Solution – Calculation 1
Primary branch-circuit conductors
Nameplate not used
Table 430.250
25 hp at 208 volts
FLC = 74.8 amps
430.22(A)
Ampacity = FLC × 125%
= 74.8 × 1.25
= 93.5 amps
Answer: 93.5 amperes

Solution – Calculation 2
Secondary conductors
Table 430.23(C)
Heavy intermittent duty = 85%
Ampacity = secondary nameplate FLC × 85%
= 142 amps × 0.85
= 120.7 amps
Answer: 120.7 amperes

Three basic steps are used:

Step 1: Decide which motor has the largest FLC rating. (Note: largest hp may not be largest FLC)
Step 2: Multiply the largest FLC rating of one motor by 125%.
Step 3: Add FLC rating of all other motors.

When two motors of equal rating are installed on the same branch circuit, the ampacity of the branch-circuit conductors must be 125% of the full-load current of one motor plus the full-load current of the second motor.

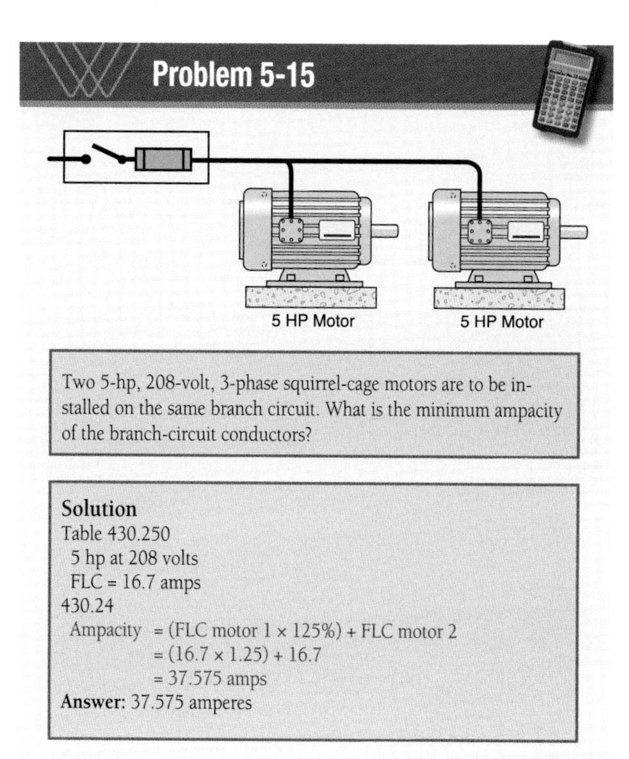

Two 5-hp, 208-volt, 3-phase squirrel-cage motors are to be installed on the same branch circuit. What is the minimum ampacity of the branch-circuit conductors?

Solution
Table 430.250
5 hp at 208 volts
FLC = 16.7 amps
430.24
Ampacity = (FLC motor 1 × 125%) + FLC motor 2
= (16.7 × 1.25) + 16.7
= 37.575 amps
Answer: 37.575 amperes

Information

When two motors of unequal rating are installed on the same branch circuit, the ampacity of the branch-circuit conductors must be 125% of the full-load current of the largest motor plus the full-load current of the second motor.

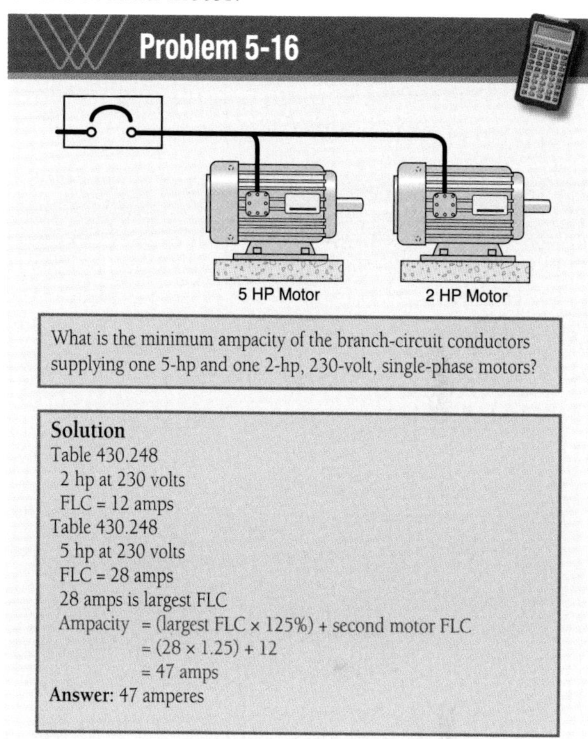

What is the minimum ampacity of the branch-circuit conductors supplying one 5-hp and one 2-hp, 230-volt, single-phase motors?

Solution
Table 430.248
2 hp at 230 volts
FLC = 12 amps
Table 430.248
5 hp at 230 volts
FLC = 28 amps
28 amps is largest FLC
Ampacity = (largest FLC × 125%) + second motor FLC
= (28 × 1.25) + 12
= 47 amps
Answer: 47 amperes

Information

When several motors are installed on the same branch circuit, the ampacity of the branch-circuit conductors must be 125% of the largest motor full-load current plus the full-load currents of the other motors.

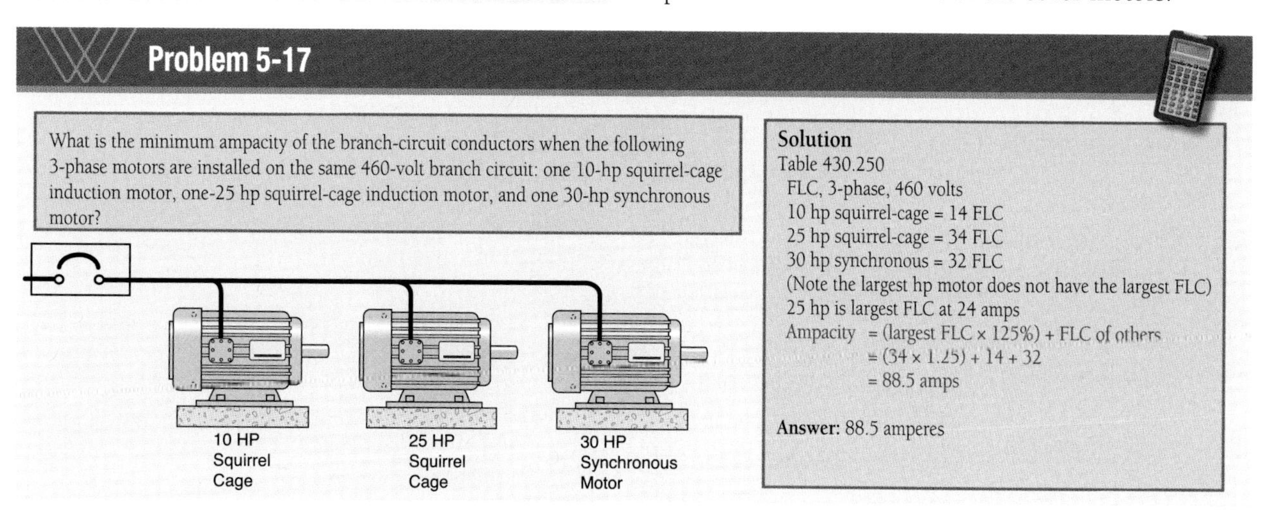

What is the minimum ampacity of the branch-circuit conductors when the following 3-phase motors are installed on the same 460-volt branch circuit: one 10-hp squirrel-cage induction motor, one-25 hp squirrel-cage induction motor, and one 30-hp synchronous motor?

Solution
Table 430.250
FLC, 3-phase, 460 volts
10 hp squirrel-cage = 14 FLC
25 hp squirrel-cage = 34 FLC
30 hp synchronous = 32 FLC
(Note the largest hp motor does not have the largest FLC)
25 hp is largest FLC at 24 amps
Ampacity = (largest FLC × 125%) + FLC of others
= (34 × 1.25) + 14 + 32
= 88.5 amps

Answer: 88.5 amperes

5.1.3.9 Circuit Conductors Combining Motors with Other Loads - Section 430.24 covers motor circuit conductors supplying both motor- and non-motor-operated equipment, such as lighting, heating, or appliances. For these circuits, the minimum ampacity of the branch-circuit conductors must not be less than 125% of the FLC of the largest motor, plus the full-load currents of the other motors, plus 100% of the noncontinuous load of the other equipment and 125% of the continuous load of the other equipment on the circuit.

Problem 5-18

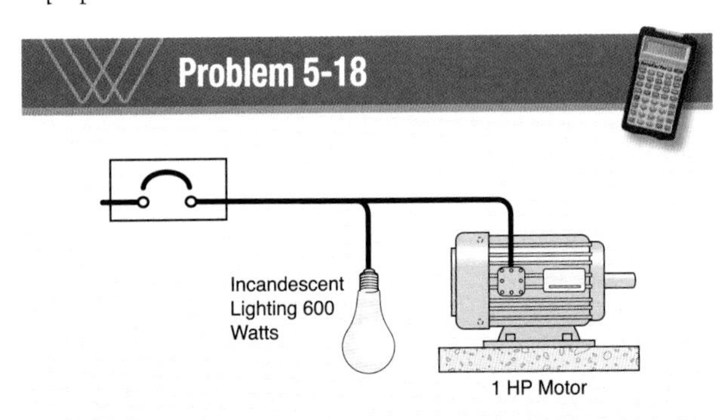

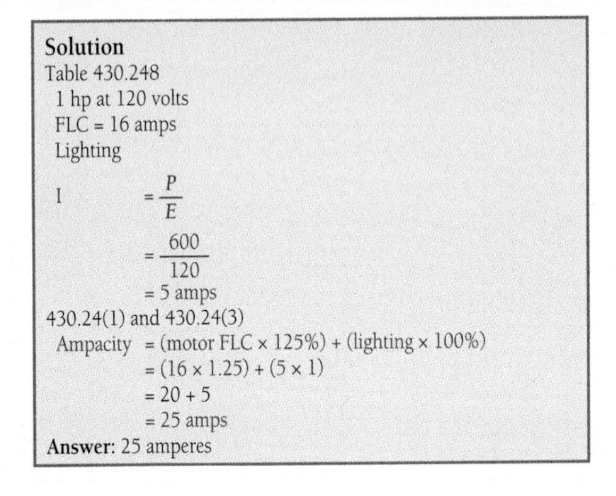

What is the minimum ampacity of the branch-circuit conductors supplying power to a 1-hp, 120-volt, single-phase, squirrel-cage motor and 600 watts of noncontinuous incandescent lighting load?

Solution

Table 430.248

1 hp at 120 volts

FLC = 16 amps

Lighting

$$I = \frac{P}{E} = \frac{600}{120} = 5 \text{ amps}$$

430.24(1) and 430.24(3)

Ampacity = (motor FLC × 125%) + (lighting × 100%)
= (16 × 1.25) + (5 × 1)
= 20 + 5
= 25 amps

Answer: 25 amperes

An individual disconnecting means is generally required to be provided for each motor controller, to disconnect the controller, and be located within sight from the controller.

Information

When a motor and fixed electrical space heating are installed on the same branch circuit, **Section 430.24 Exception No. 2** indicates that **424.3(B)** will take precedence. **424.3(B) Branch-Circuit Sizing**, points out that these heater conductors are continuous duty and require the branch-circuit conductors for the heater portion of the circuit to be increased to 125% of the heater ampacity. The result is that both the motor and heater require the common branch circuit to be sized at 125% of their combined full load current.

Problem 5-19

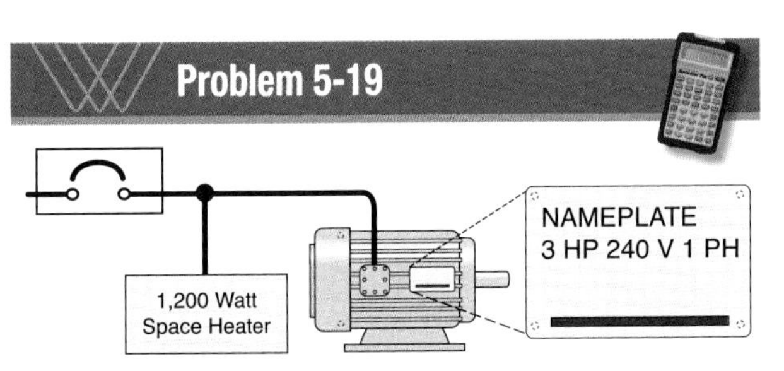

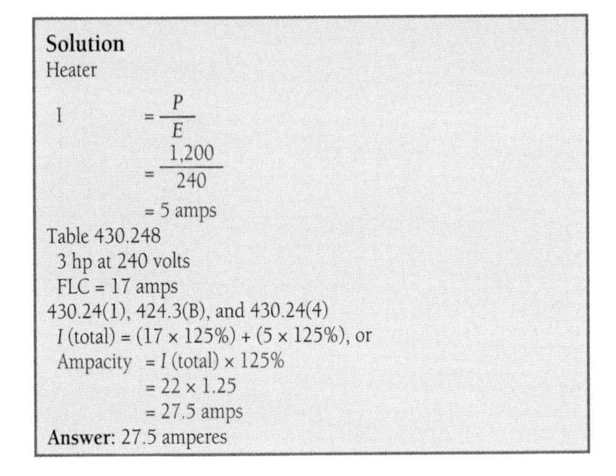

A fixed electric space heater has 1,200 watts of resistance heating and an associated 3-hp, 240-volt, single-phase fan motor. The fixed space heater is to be installed on a single branch circuit. Calculate the minimum ampacity of the branch-circuit conductors.

Solution

Heater

$$I = \frac{P}{E} = \frac{1{,}200}{240} = 5 \text{ amps}$$

Table 430.248

3 hp at 240 volts

FLC = 17 amps

430.24(1), 424.3(B), and 430.24(4)

I (total) = (17 × 125%) + (5 × 125%), or

Ampacity = I (total) × 125%
= 22 × 1.25
= 27.5 amps

Answer: 27.5 amperes

Information

When a group of motors is installed on the same circuit and one of the motors is short-time, intermittent, periodic, or varying duty, that motor's branch circuit current is permitted to be calculated according to the table for duty-cycle service. **Section 430.24, Exception No. 1**, references **430.22(E)** for these calculations.

Problem 5-20

A single-phase, 240-volt branch circuit consists of two 10-hp continuous-duty motors and one 5-hp, 15-minute periodic duty motor. The nameplate FLC for the 5-hp motor is 27 amperes. What is the minimum ampacity of the branch-circuit conductors?

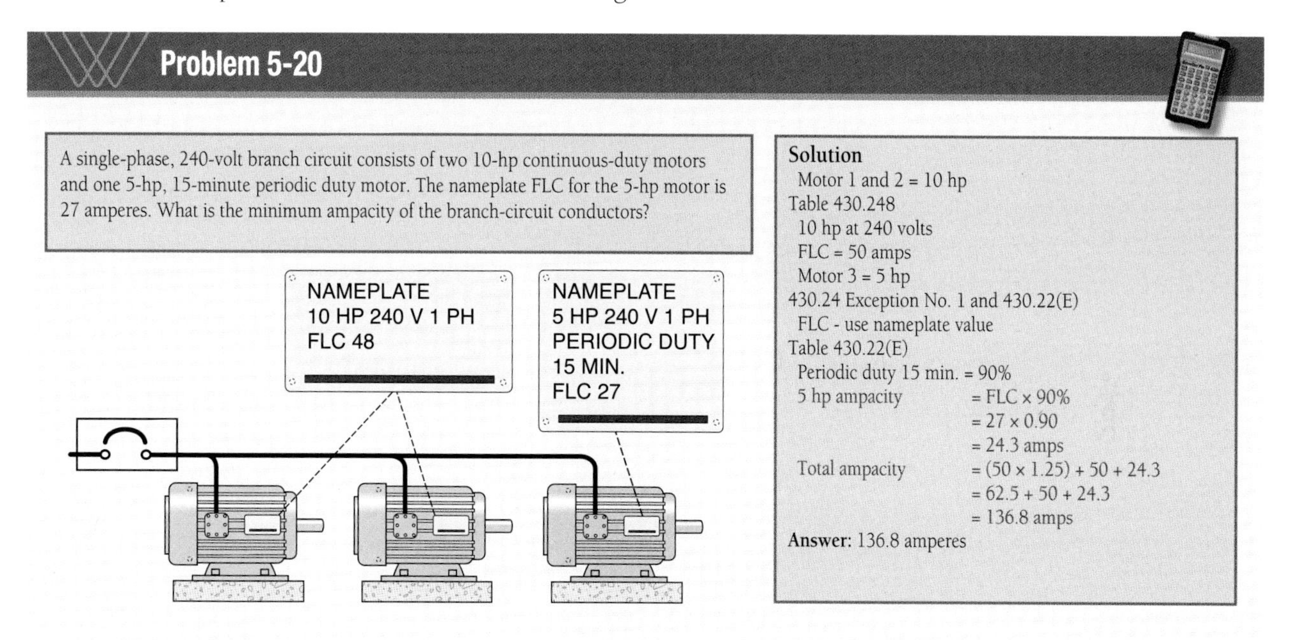

Solution

Motor 1 and 2 = 10 hp
Table 430.248
10 hp at 240 volts
FLC = 50 amps
Motor 3 = 5 hp
430.24 Exception No. 1 and 430.22(E)
FLC - use nameplate value
Table 430.22(E)
Periodic duty 15 min. = 90%
5 hp ampacity = FLC × 90%
= 27 × 0.90
= 24.3 amps
Total ampacity = (50 × 1.25) + 50 + 24.3
= 62.5 + 50 + 24.3
= 136.8 amps

Answer: 136.8 amperes

Information

When two motors, or one motor and other loads, are on the same branch circuit and they are interlocked to prevent simultaneous operation, the branch-circuit conductors are permitted to be sized at 125% of the largest load (or total simultaneous load) according to **Section 430.24 Exception No. 3**.

Problem 5-21

A 20-hp and a 40-hp 3-phase, 460-volt motor are on the same branch circuit and are interlocked so that only one of the motors can operate at any given time. What is the minimum ampacity of the branch circuit conductors?

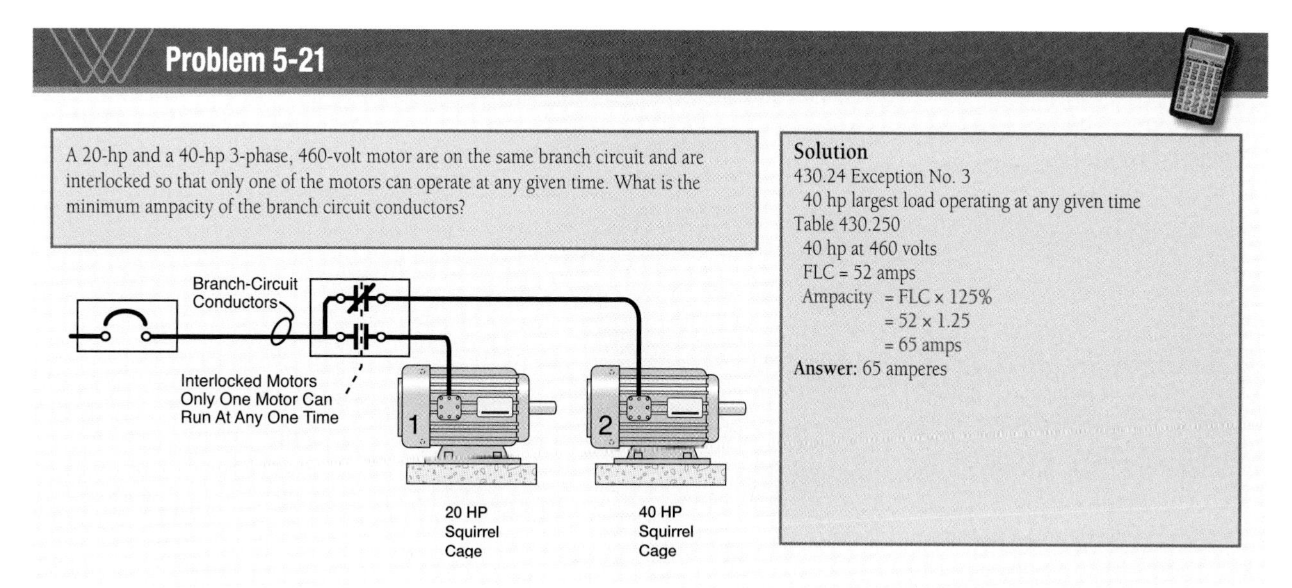

Solution

430.24 Exception No. 3
40 hp largest load operating at any given time
Table 430.250
40 hp at 460 volts
FLC = 52 amps
Ampacity = FLC × 125%
= 52 × 1.25
= 65 amps

Answer: 65 amperes

5.2 Motor Branch-Circuit Short-Circuit and Ground-Fault Protection

5.2.1 General

The purpose of motor branch-circuit short-circuit and ground-fault protection is to protect the circuit conductors, motors, and motor-controller equipment against overcurrent due to short circuits or ground faults. **See Figure 5-3.**

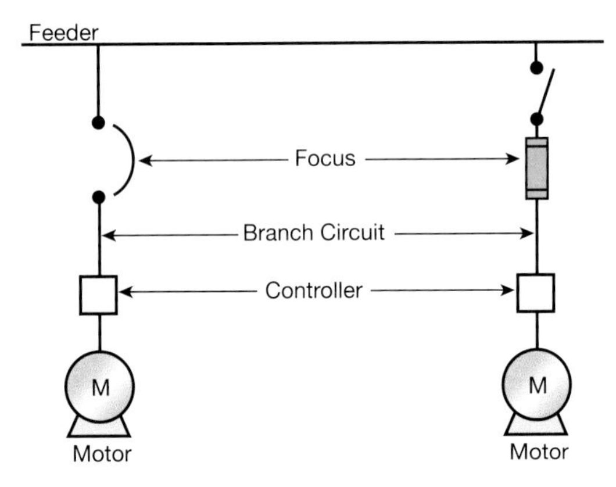

Figure 5-3. The focus of this section is on the circuit breakers or fuses which serve as the motor branch-circuit short-circuit and ground-fault protection of a motor circuit as set forth in Part IV of Article 430.

Circuit breakers, fuses, and thermal protectors are permitted to be used as the overcurrent protection device for motor branch-circuit short-circuit and ground-fault protection.

Three types of fuses are used:

1. Nontime-delay fuses
2. Time-delay fuses
3. Class CC fuses

Three types of circuit breakers are used:

1. Fixed or inverse-time circuit breakers
2. Adjustable-trip circuit breakers
3. Instantaneous-trip circuit breakers. Also known as motor circuit protectors (MCPs)

Calculations for inverse-time and adjustable-trip circuit breakers use the same values. A more common term for inverse-time circuit breaker is thermal-magnetic circuit breaker.

5.2.1.1 Standard Size Ratings for Overcurrent Devices - The standard ampere ratings for fuses and fixed circuit breakers are listed in **240.6(A)**. The two types of fixed circuit breaker is the inverse-time circuit breaker and the nonadjustable circuit breaker.

The three parts to be looked at in **240.6(A)** are:

1. The standard ampere ratings for fuses and inverse-time circuit breakers are listed for 15 through 6,000 amperes. (Note: 15 amperes is the smallest standard rating listed for a circuit breaker.)
2. Special ampere ratings for fuses are listed at 1, 3, 6, 10, and 601.
3. The use of nonstandard rated fuses or circuit breakers is permitted. Fuses are available in almost any size to fit the installation at hand.

430.6(A) requires the use of the full-load current **Table 430.247, Table 430.248, Table 430.249**, and **Table 430.250**, as given in the *Code*, for calculating the size of the motor branch-circuit short-circuit and ground-fault protection. The motor nameplate full-load current rating is generally NOT used. The values given in the tables are the full-load currents (FLCs), which are suitable to be used for most motors running at usual speeds and with normal torque characteristics.

The calculations for the motor branch-circuit short-circuit and ground-fault protection for all squirrel-cage motors are the same when using a nontime-delay fuse, a time-delay fuse, an adjustable-trip circuit breaker, or an inverse-time circuit breaker. The percentages given in **Table 430.250** are the same for both types of alternating-current polyphase motors. The calculations differ only when an instantaneous-trip circuit breaker is used, because the percentage increases are different. The instantaneous-trip circuit breaker calculations will be looked at separately.

The starting current is also the locked-rotor current of a motor. The locked-rotor current of a motor depends on the type and construction of the motor. When a motor is starting, the motor inrush current or motor starting current is much higher than the motor full-load current. A general rule of thumb: calculate the locked-rotor current for other than a Design B energy-efficient motor at six times the motor full-load current. The starting current for a Design B energy-efficient motor will run eight or more times the full-load current of the motor.

5.2.1.2 Basic Protection - 430.52(A) sets the stage for providing a motor branch-circuit short-circuit and ground-fault protection overcurrent device.

430.52(B) requires the motor branch-circuit short-circuit and ground-fault protective device to be able to carry the starting current.

430.52(C) lists each of the various methods for calculating the rating or setting of the various overcurrent protective devices used for motor branch-circuit short-circuit ground-fault protection.

The fundamental rule, **430.52(C)(1)**, requires that the maximum rating or setting of the protective device not exceed the values calculated according to **Table 430.52** of the *Code*. **See Figure 5-4**. From a practical point of view, **Exception No. 1** permits the next higher standard size overcurrent device to be used in many cases. The exact language of **Exception No. 1** is important. This allowance is very often referred to as "rounding up."

430.52(C)(2) gives the manufacturer's ratings preference over any calculations made. Where maximum branch-circuit and ground-fault protection ratings are shown in the manufacturer's overload relay table for use with a motor controller or otherwise marked on the equipment, the calculated standard size must not exceed the listed manufacturer's rating. This applies to all calculations when made with the basic rules or with the exceptions for starting current. This section also follows the second sentence of Section 110.9 such that the overload devices must be rated for not less than the current that must be interrupted.

The symbol OCPD will be used to indicate overcurrent protective device.

The symbol CB will be used for circuit breaker.

Table 430.52 Maximum Rating or Setting of Motor Branch-Circuit Short-Circuit and Ground-Fault Protective Devices

	Percentage of Full-Load Current			
Type of Motor	**Nontime Delay Fuse[1]**	**Dual Element (Time-Delay) Fuse[1]**	**Instantaneous Trip Breaker**	**Inverse Time Breaker[2]**
Single-phase motors	300	175	800	250
AC polyphase motors other than wound-rotor	300	175	800	250
Squirrel cage — other than Design B energy-efficient	300	175	800	250
Design B energy-efficient	300	175	1100	250
Synchronous[3]	300	175	800	250
Wound rotor	150	150	800	150
DC (constant voltage)	150	150	250	150

Note: For certain exceptions to the values specified, see 430.54.

[1]The values in the Nontime Delay Fuse column apply to Time-Delay Class CC fuses.

[2]The values given in the last column also cover the ratings of nonadjustable inverse time types of circuit breakers that may be modified as in 430.52(C)(1), Exception No. 1 and No. 2.

[3]Synchronous motors of the low-torque, low-speed type (usually 450 rpm or lower), such as are used to drive reciprocating compressors, pumps, and so forth, that start unloaded, do not require a fuse rating or circuit-breaker setting in excess of 200 percent of full-load current.

Figure 5-4. Table 430.52 A wiring diagram of wound-rotor induction motor including the secondary winding and resistor bank.

(C) Rating or Setting.

(1) In Accordance with Table 430.52. A protective device that has a rating or setting not exceeding the value calculated according to the values given in Table 430.52 shall be used.

Exception No. 1: Where the values for branch-circuit short-circuit and ground-fault protective devices determined by Table 430.52 do not correspond to the standard sizes or ratings of fuses, nonadjustable circuit breakers, thermal protective devices, or possible settings of adjustable circuit breakers, a higher size, rating, or possible setting that does not exceed the next higher standard ampere rating shall be permitted.

430.52(C)(1), Exception No. 1 is most often used to determine the actual setting of the motor protective device.

5.2.1.3 Standard Example - From **Figure 5-5**, determine the maximum overcurrent protection permitted according to **Table 430.52** for a typical 20 hp, 3-phase, 460-volt, Design B, squirrel-cage or synchronous motor. The overcurrent protection devices selected for this example include time-delay fuses, nontime-delay fuses, and inverse-time circuit breakers. Since the example does not deal with instantaneous-trip circuit breakers, a close reading of **Table 430.52** will show that this example also applies to a Design B energy-efficient squirrel-cage motor.

For additional information, visit qr.njatcdb.org Item #1034

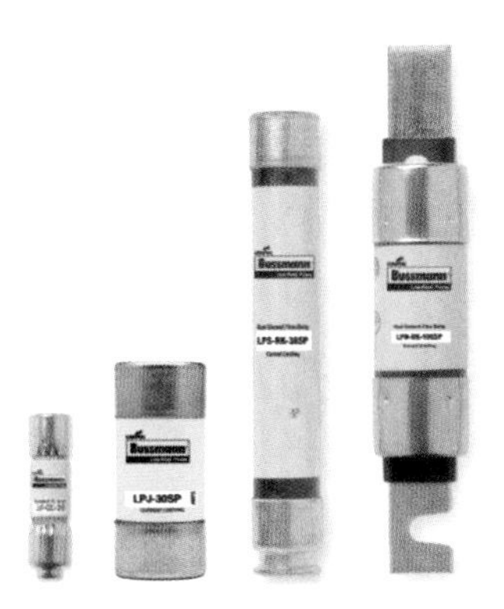

Dual element fuses and inverse time circuit breakers are the most common form of motor branch-circuit and ground-fault protective devices used today.

Example

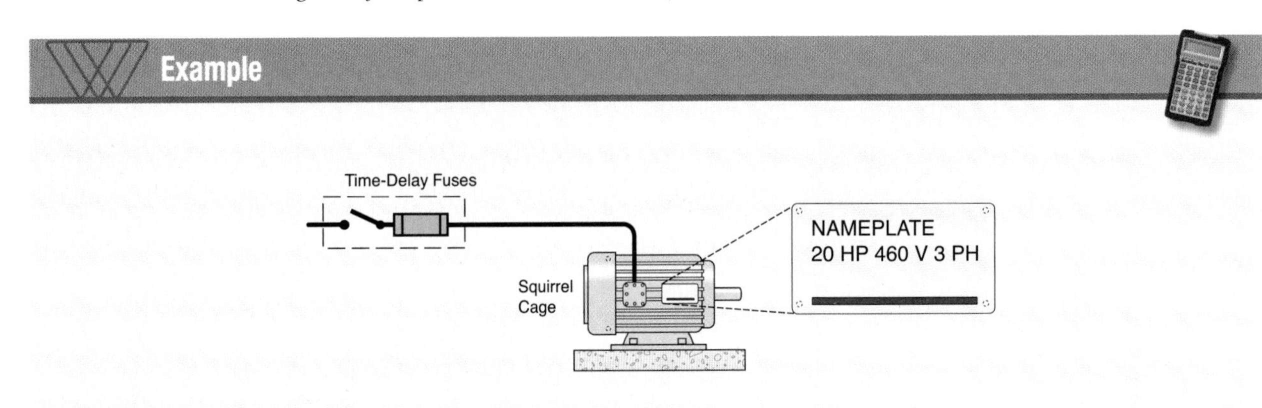

Solution - Calculation 1	Solution - Calculation 2	Solution - Calculation 3
Time-delay fuses (as shown)	Nontime-delay fuses (option 1)	Inverse-time circuit breaker (option 2)
Table 430.250	Table 430.250	Table 430.250
20 hp at 460 volts	20 hp at 460 volts	20 hp at 460 volts
FLC = 27 amps	FLC = 27 amps	FLC = 27 amps
Table 430.52	Table 430.52	Table 430.52
Time-delay fuse = 175%	Nontime-delay fuse = 300%	Inverse time circuit breaker = 250%
OCPD = FLC x 175%	OCPD = FLC x 300%	OCPD = FLC x 250%
= 27 x 1.75	= 27 x 3.00	= 27 x 2.50
= 47.25 amps	= 81 amps	= 67.5 amps
430.52(C)(1) Exception No. 1	430.52(C)(1) Exception No. 1	430.52(C)(1) Exception No. 1
Next larger standard size permitted	Next larger standard size permitted	Next larger standard size permitted
240.6(A)	240.6(A)	240.6(A)
Next larger size = 50 amps	Next larger size = 90 amps	Next larger size = 70 amps
50 amps is the maximum rating of the time-delay fuse for the 20 hp motor	90 amps is the maximum rating of the non time-delay fuses for the 20 hp motor.	70 amps is the maximum rating of the inverse-time circuit breaker for the 20 hp motor.
Answer: 50 ampere time-delay fuses	**Answer:** 90 ampere nontime-delay fuses	**Answer:** 70 ampere inverse time circuit breaker

Figure 5-5. This single-line drawing shows a time-delay fuse serving as the motor branch-circuit short-circuit and ground-fault protection for a 20-hp, 460-volt, 3-phase motor. Parallel calculations provide three different methods of protection.

5.2.2 Standard Sizes for Overcurrent Device Ratings (Basic Protection)

The next two problems demonstrate the necessary steps to determine the size of maximum overcurrent protection device using time-delay fuses and nontime-delay fuses. Reviewing **Table 430.52(C)**, each type of overcurrent device is given a maximum rating or setting depending on the type of motor. For example, except for wound rotor motors and direct current motors, the maximum rating of non time delay fuse values are 300% of the table value of the full-load current. For time-delay fuses, the rating is maximum value of 175% of the full-load current. These examples also show how **Exception No. 1** is applied using the standard overcurrent device values of **240.6(A)**.

In practice, the maximum value stated in the *NEC* is not always desirable. But, neither is the minimum value. Rather, the middle ground for motor protection most often is to avoid nuisance interruption of the motor while affording maximum protection for the motor.

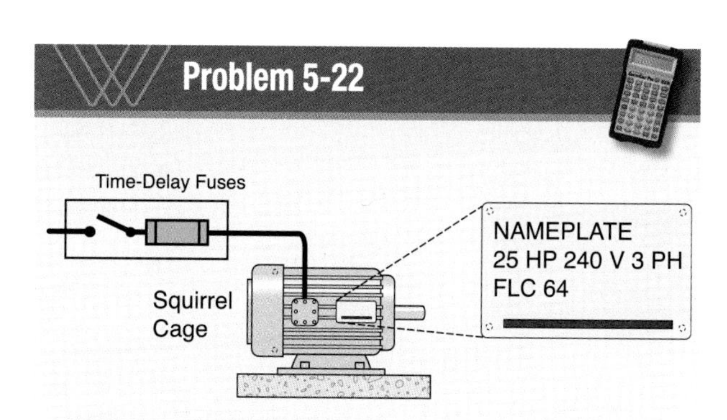

Determine the maximum standard size time-delay fuses permitted for the motor branch-circuit short-circuit and ground-fault protection for a 25-hp, 240-volt, 3-phase, squirrel-cage motor with a nameplate full-load current of 64 amperes.

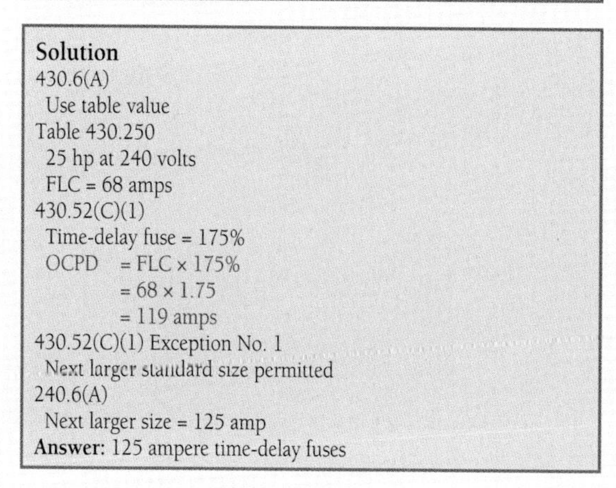

Solution
430.6(A)
Use table value
Table 430.250
25 hp at 240 volts
FLC = 68 amps
430.52(C)(1)
Time-delay fuse = 175%
OCPD = FLC x 175%
= 68 x 1.75
= 119 amps
430.52(C)(1) Exception No. 1
Next larger standard size permitted
240.6(A)
Next larger size = 125 amp
Answer: 125 ampere time-delay fuses

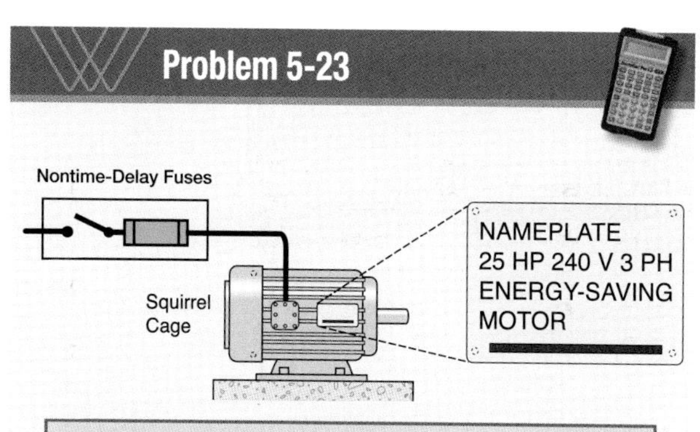

Determine the maximum standard size nontime-delay fuses permitted for the motor branch-circuit short-circuit and ground-fault protection for a 25-hp, 240-volt, 3-phase, squirrel-cage, Design B energy-efficient motor.

Solution
Table 430.250
25 hp at 240 volts
FLC = 68 amps
430.52(C)(1)
Nontime-delay fuses = 300%
OCPD = FLC x 300%
= 68 x 3.00
= 204 amps
430.52(C)(1) Exception No. 1
Next larger standard size permitted
240.6(A)
Next larger size = 225 amps
Answer: 225 ampere nontime-delay fuses

Comment
Take a practical look at Problem 5-23. The basic calculation of 204 amperes is permitted by the *Code* to go to the next higher size of 225 amperes. However, for some applications, the closer the overcurrent protection device is to protecting the equipment may be the best practice. In this case, the best practical solution may be to select a 200 ampere fuse.

5.2.3 Calculations With and Without a Starting Current Problem

5.2.3.1 The Basic Overcurrent Protection Rules - The following chart illustrates the use of the most common types of overcurrent protection devices which might be used for motor branch-circuit short-circuit and ground-fault protection when starting current is not a problem. The motor is a 40-hp, 460-volt, 3-phase, squirrel-cage motor. The calculations illustrated in the following chart are again the maximum permitted values which incorporate the permission afforded by **430.52(C), Exception No. 1** to further increase those values. The chart applies to both Design B motors and Design B energy-efficient motors.

Type of OCPD	460 Volt 3-Phase FLC	Percentage of Full Load Current Table 430.52	Next Higher Standard OCPD Rating
Nontime-delay Fuses	52	× 300% = 156 amps	175 amps
Time-Delay Fuses	52	× 175% = 91 amps	100 amps
Inverse-Time Breaker	52	× 250% = 130 amps	150 amps

5.2.3.2 Exceptions to the Basic Overcurrent Protection Rules - Now, if upon trying to start the motor and the previous calculated size will not carry the starting current, a different problem exists. The overcurrent device used must be able to carry the starting current. **430.52(C)(1) Exception No. 2** permits increasing the size of the overcurrent device when the basic calculations using **Table 430.52** and rounding up will not hold for the starting current. A separate part of the exception applies to each type of overcurrent device.

The following values apply to Design B motors and Design B energy-efficient motors.

Type of OCPD	OCPD Rating Limit	Maximum Percent of FLC	430.52(C)(1) Exception
Nontime-delay Fuses	600 amps or less	400%	No. 2(a)
Nontime-delay Fuses (Type CC only)	None	400%	No. 2(a)
Time-Delay Fuses	None	225%	No. 2(b)
Large-Size Fuses	601 through 6,000 amps	300%	No. 2(d)
Inverse-Time CB	100 amps or less	400%	No. 2(c)
Inverse-Time CB	Greater than 100 amps	300%	No. 2(c)

When referring to **Exception No. 2** of **430.52(C)(1)**, it is important to note the one thing that all of the lettered exceptions have in common: the phrase "shall in no case exceed..." This means that when the calculations are made using this exception, the next smaller standard size overcurrent protective device must be selected. This is sometimes referred to as "rounding down."

The following problem illustrates the calculation of overcurrent protection device ratings when the starting current is a problem. It will be necessary to apply **430.52(C)(1), Exception No. 2**.

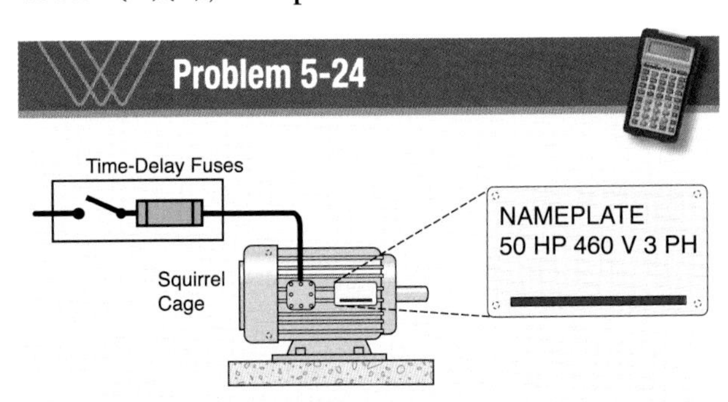

Determine the maximum standard size time-delay fuses permitted for a 50-hp, 460-volt, 3-phase, squirrel-cage, Design B motor when the starting current is a problem.

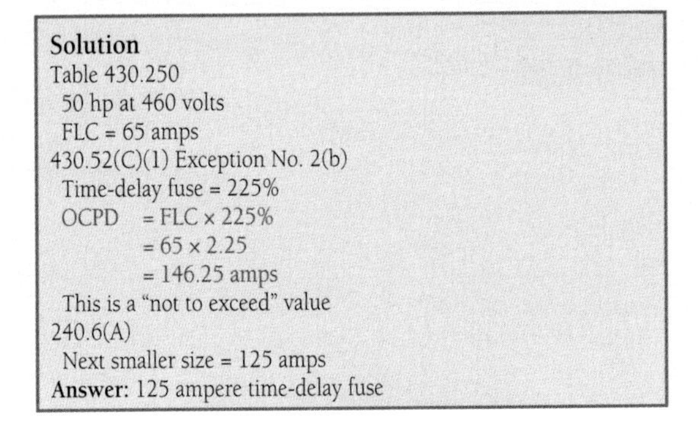

Solution
Table 430.250
50 hp at 460 volts
FLC = 65 amps
430.52(C)(1) Exception No. 2(b)
Time-delay fuse = 225%
OCPD = FLC × 225%
= 65 × 2.25
= 146.25 amps
This is a "not to exceed" value
240.6(A)
Next smaller size = 125 amps
Answer: 125 ampere time-delay fuse

Motor circuit overcurrent protective devices are typically located at the same location as the motor controller.

Information

The following chart illustrates the use of the various protection devices when starting current is a problem. The previously illustrated 40-hp, 460-volt, 3-phase squirrel-cage motor is used, and these values apply to Design B motors and Design B energy-efficient motors only when starting currents are a problem.

Type of OCPD	Section 430.52 (C)(1) Exception	FLC	Percentage Table 430.52	Next Lower Standard OCPD Rating
Nontime-delay Fuses	No. 2(a)	52	× 400% = 208	200 amps
Class CC Fuses	No. 2(a)	52	× 400% = 208	200 amps
Time-Delay Fuses	No. 2(b)	52	× 225% = 117	110 amps
Inverse-Time Breaker	No. 2(c)	52	× 400% = 208	200 amps

In all cases, when **Exception No. 2** is used, the next smaller standard size overcurrent protection device is required.

Type CC fuses are available in fractional sizes through 30 amperes in both time-delay and nontime-delay. The voltage rating is 600 volts AC. These fuses have an interrupting rating of 200,000 amperes rms symmetrical.

Courtesy of Eaton's Bussmann Business
Two examples of 600-volt rated Class CC fuses.

Information

Problem 5-25 gives a comparison of the basic calculations and the calculations for starting current using **430.52(C)(1), Exception No. 2(a)** for nontime-delay fuses.

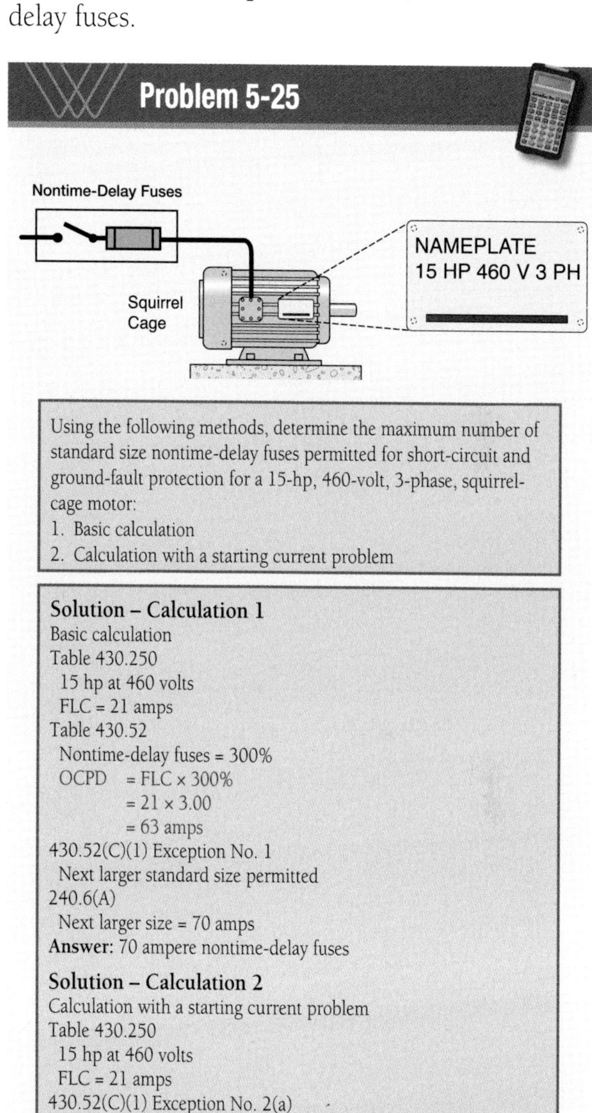

Using the following methods, determine the maximum number of standard size nontime-delay fuses permitted for short-circuit and ground-fault protection for a 15-hp, 460-volt, 3-phase, squirrel-cage motor:

1. Basic calculation
2. Calculation with a starting current problem

Solution – Calculation 1
Basic calculation
Table 430.250
15 hp at 460 volts
FLC = 21 amps
Table 430.52
Nontime-delay fuses = 300%
OCPD = FLC × 300%
= 21 × 3.00
= 63 amps
430.52(C)(1) Exception No. 1
Next larger standard size permitted
240.6(A)
Next larger size = 70 amps
Answer: 70 ampere nontime-delay fuses

Solution – Calculation 2
Calculation with a starting current problem
Table 430.250
15 hp at 460 volts
FLC = 21 amps
430.52(C)(1) Exception No. 2(a)
Nontime-delay fuses = 400%
OCPD = FLC × 400%
= 21 × 4.00
= 84 amps
240.6(A)
Next smaller size = 80 amps
Answer: 80 ampere nontime-delay fuses

Information

The next problem compares the methods of basic calculation and calculation for starting current using **430.52(C)(1), Exception No. 2(b)** for time-delay fuses. Calculations using **430.52(C)(1), Exception No. 2(b)**, with time-delay fuses are a reliable form of high torque motor protection.

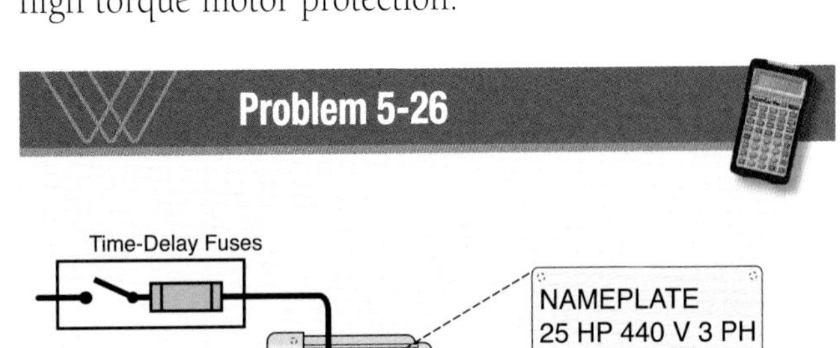

Problem 5-26

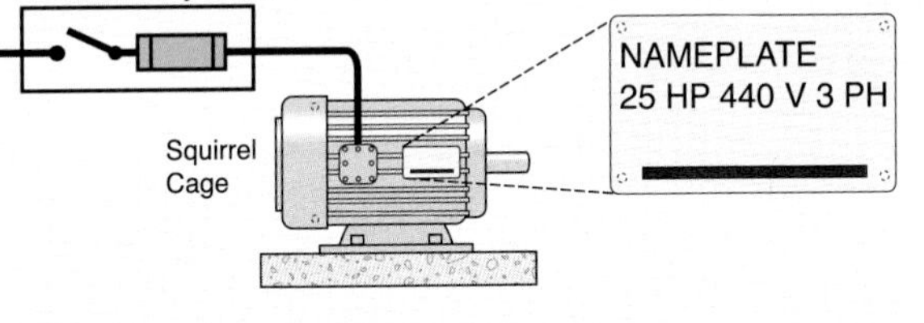

Using the following methods, determine the maximum standard size time-delay fuses permitted for branch-circuit short-circuit ground-fault protection for a 25-hp, 440-volt, 3-phase squirrel-cage motor:

1. Basic protection
2. Absolute maximum under any possible condition

Solution – Calculation 1

Basic protection
Table 430.250
25 hp at 440 volts
FLC = 34 amps
Table 430.52
Time delay fuses = 175%
OCPD = FLC × 175%
= 34 × 1.75
= 59.5 amps
430.52(C)(1) Exception No. 1
Next larger standard size permitted
240.6(A)
Next larger size = 60 amps
Answer: 60 ampere time-delay fuse

Solution – Calculation 2

Absolute maximum includes starting current
Table 430.250
25 hp at 440 volts
FLC = 34 amps
430.52(C)(1) Exception No. 2(a)
Permits time-delay fuses up to 225%
OCPD = FLC × 225%
= 34 × 2.25
= 76.5 amps
240.6(A)
Next smaller size = 70 amps
Answer: 70 ampere nontime-delay fuses

Information

The next problem compares the basic calculation and the calculation for starting current using **Exception No. 2(c)** of **430.52(C)(1)** for an inverse-time circuit breaker. Calculations permitted **by 430.52(C)(1), Exception No. 2(c)** using inverse time circuit breakers are permitted.

Problem 5-27

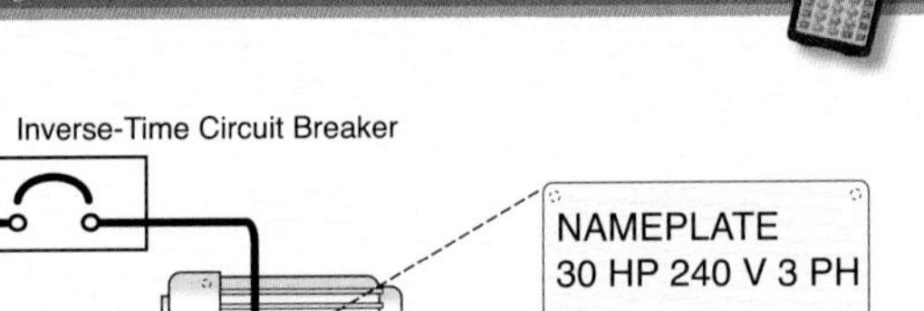

Using the following methods, determine the maximum standard size inverse-time circuit breaker permitted for branch-circuit short-circuit ground-fault protection for a 30-hp, 240-volt, 3-phase squirrel-cage motor:

1. Basic protection
2. Absolute maximum under any possible condition

Solution – Calculation 1

Basic protection
Table 430.250
30 hp at 240 volts
FLC = 80 amps
Table 430.52
Inverse-time circuit breaker = 250%
OCPD = FLC × 250%
= 80 × 2.50
= 200 amps
240.6(A)
200 amps is a standard rating size
Answer: 200 ampere inverse-time circuit breaker

Solution – Calculation 2

Absolute maximum includes starting current
Table 430.250
30 hp at 240 volts
FLC = 80 amps
430.52(C)(1) Exception No. 2(c):
Inverse-time circuit breakers (100 amps or less) = 400%
OCPD = FLC × 400%
= 80 × 4.00
= 320 amps
240.6(A)
Next smaller size = 300 amps
Answer: 300 ampere inverse-time circuit breaker

5.2.3.3 Single-Phase Motors - Table 430.52 provides a listing for single-phase motors. The calculations are also based upon **Section 430.52** and its exceptions.

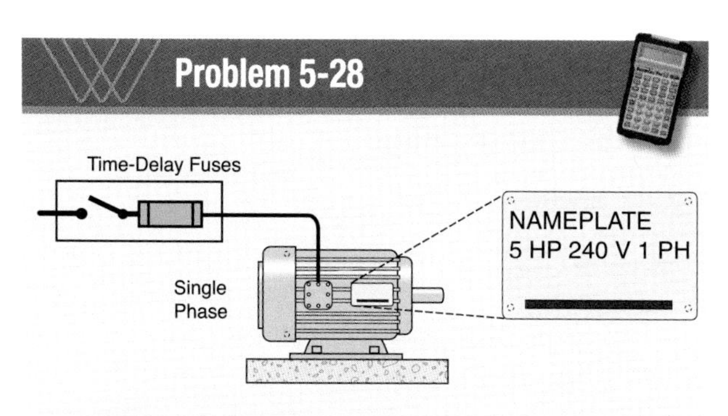

Using the following methods, determine the maximum standard size time-delay fuses permitted for branch-circuit short-circuit ground-fault protection for a 5-hp, 240-volt, 1-phase capacitor-start motor.

1. Basic protection
2. Absolute maximum under any possible condition

Solution – Calculation 1
Basic protection
Table 430.248
5 hp at 240 volts
FLC = 28 amps
Table 430.52
Time-delay fuse (single-phase) = 175%
OCPD = FLC x 175%
= 28 x 1.75
= 49 amps
430.52(C)(1) Exception No. 1
Next larger standard size permitted
240.6(A)
Next larger size = 50 amps
Answer: 50 ampere time-delay fuses

Solution – Calculation 2
Absolute maximum includes starting current
Table 430.248
5 hp at 240 volts
FLC = 28 amps
430.52(C)(1) Exception No. 2(b)
Time-delay fuses = 225%
OCPD = FLC x 225%
= 28 x 2.25
= 63 amps
240.6(A)
Next smaller size = 60 amps
Answer: 60 ampere time-delay fuses

Five horsepower, single-phase motors are hard to find and more costly as compared to 3-phase motors of the same size. Due to installation plus operating costs, alternative solutions should always be investigated.

5.2.3.4 Wound-Rotor Motors - Wound-rotor motors are also considered in **Table 430.52**.

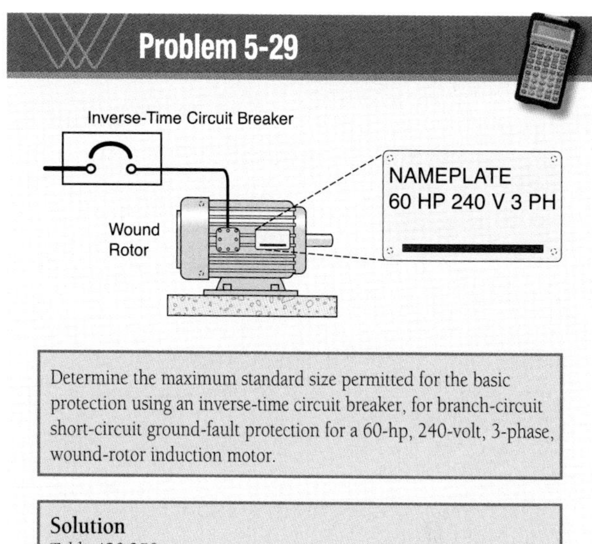

Determine the maximum standard size permitted for the basic protection using an inverse-time circuit breaker, for branch-circuit short-circuit ground-fault protection for a 60-hp, 240-volt, 3-phase, wound-rotor induction motor.

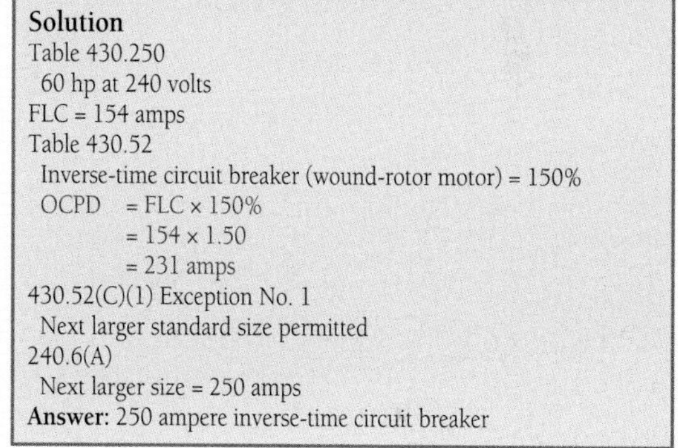

Solution
Table 430.250
60 hp at 240 volts
FLC = 154 amps
Table 430.52
Inverse-time circuit breaker (wound-rotor motor) = 150%
OCPD = FLC x 150%
= 154 x 1.50
= 231 amps
430.52(C)(1) Exception No. 1
Next larger standard size permitted
240.6(A)
Next larger size = 250 amps
Answer: 250 ampere inverse-time circuit breaker

5.2.3.5 Instantaneous-Trip Circuit Breakers - The instantaneous-trip circuit breaker [also known as motor-circuit protectors (MCPs)] is covered separately in **430.52(C)(3)**. The instantaneous-trip circuit breaker is adjustable and can basically be set to trip at any particular ampere rating within its range. Therefore, there is no rounding up or rounding down or using the next larger or smaller standard size. The ampere rating calculated is the ampere setting for the instantaneous-trip circuit breaker. **430.52(C)(3)** puts several limitations on an instantaneous-trip circuit breaker. For example:

1. It must be part of a listed combination motor controller (such as a combination breaker and controller in the same enclosure).
2. There must be coordinated motor overload protection in each conductor.
3. There must be coordinated short-circuit ground-fault protection in each conductor.
4. The instantaneous-trip circuit breaker must be adjustable.

Where instantaneous-trip circuit breakers are selected, there are two different basic values for the percent of motor full-load current. Single-phase, synchronous, and squirrel-cage motors (other than Design B energy-efficient motors) are permitted to be rated up to 800% of the motor full-load current. Because the Design B energy-efficient motors have a much higher starting current, they are permitted to be rated up to 1,100% of the motor full-load current. The Design B high-efficiency motor is handled in the same way as the obsolete Design E energy-efficient motor of the past.

Where instantaneous-trip circuit breakers are used and there is a problem with starting currents, **430.52(C)(3), Exception No. 1** applies. The beginning portion of this exception permits two different values for the percent of motor full-load current. Single-phase, synchronous, and squirrel-cage motors (other than Design B energy-efficient motors) are permitted to be rated up to 1,300% of the motor full-load current. Because Design B energy-efficient motors have a much higher starting current, they are permitted to be rated up to 1,700% of the motor full-load current.

The latter portion of **430.52(C)(3), Exception No.1** provides that, when an engineering analysis is performed and a need is demonstrated, any values above the previously discussed 800% and 1,100% are allowed.

Finally, **430.52(C)(3), Exception No. 1** also permits the electrical engineer's evaluations to be used without actually trying the instantaneous-trip circuit breaker at a lower rating.

Basic Calculation

Table 430.52
- Squirrel-cage motor
- Other than Design B energy-efficient motors at 800% max.
- Design B energy-efficient motors at 1,100% max.

Calculations for Starting Current Problem

430.52(C)(3), Exception No. 1
- Other than Design B energy-efficient motors at 1,300%
- Design B energy-efficient motors at 1,700%

5.2.3.6 Design B Energy-Efficient Motors - This next problem is a unique problem. Although most often applied by electrical engineers, Electrical Workers must be able to understand and apply these *Code* rules.

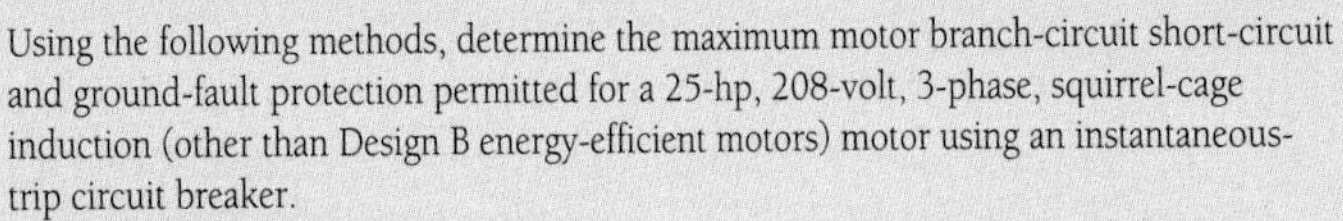
Using the following methods, determine the maximum motor branch-circuit short-circuit and ground-fault protection permitted for a 25-hp, 208-volt, 3-phase, squirrel-cage induction (other than Design B energy-efficient motors) motor using an instantaneous-trip circuit breaker.
1. Basic calculations
2. Calculations for starting current

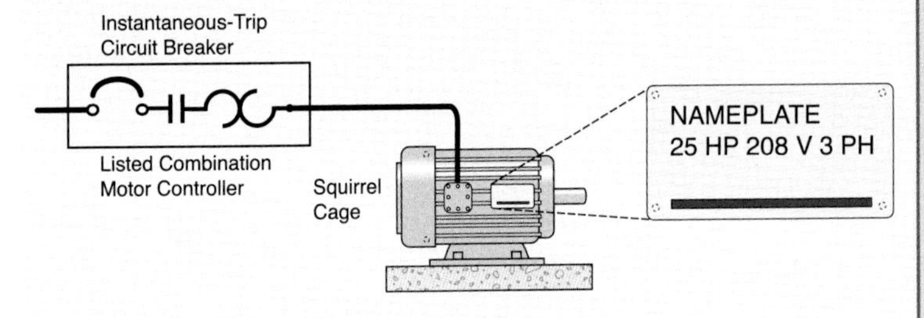

Solution – Calculation 1
Basic protection
Table 430.250
25 hp at 208 volts
FLC = 74.8 amps
Table 430.52
Other than Design B energy-efficient motors = 800%
OCPD = FLC × 800%
= 74.8 × 8.00
= 598.4 amps
Answer: 598.4 ampere instantaneous-trip circuit breaker

Solution – Calculation 2
Absolute maximum includes starting current
Table 430.250
25 hp at 208 volts
FLC = 74.8 amps
430.52(C)(3) Exception No. 1
Permits 1,300% of the FLC
OCPD = FLC × 1,300%
= 74.8 × 13.00
= 972.4 amps
Answer: 972.4 ampere instantaneous-trip circuit breaker

Problem 5-31

Using the following methods, determine the maximum motor branch-circuit short-circuit and ground-fault protection permitted for a 25-hp, 460-volt, 3-phase, squirrel-cage induction, Design B energy-efficient motor using an instantaneous-trip circuit breaker.
1. Basic calculations
2. Calculations for starting current

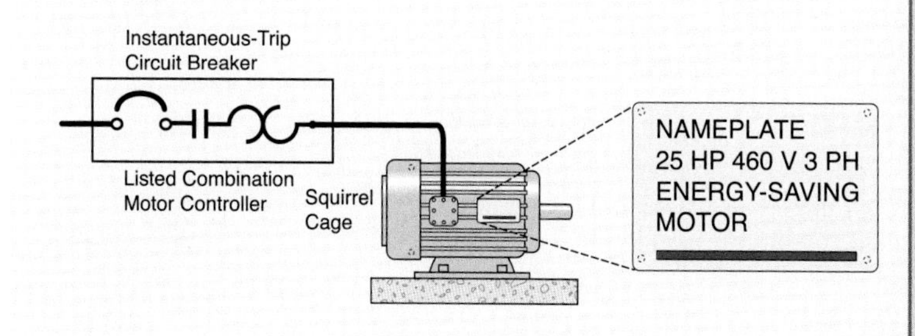

Solution – Calculation 1
Basic protection
Table 430.250
25 hp at 460 volts
FLC = 34 amps
Table 430.52
Design B energy-efficient motors = 1,100%
OCPD = FLC x 1,100%
= 34 x 11.00
= 374 amps
Answer: 374 ampere instantaneous-trip circuit breaker

Solution – Calculation 2
Absolute maximum includes starting current
Table 430.250
25 hp at 460 volts
FLC = 34 amps
430.52(C)(3) Exception No. 1
Permits 1,700% of the FLC
OCPD = FLC x 1,700%
= 34 x 17.00
= 578 amps
Answer: 578 ampere instantaneous-trip circuit breaker

5.2.3.7 Synchronous Motors - Two things are common to the synchronous motor:

1. The power factor is considered in the calculations.
2. The percentages in **Table 430.52** are different.

The full-load current values given in **Table 430.250** are for the unity power factor. When no power factor is indicated, it is assumed to be the unity power factor and the motor full-load current is read directly from **Table 430.250**. The footnote to **Table 430.250** calls attention to the fact that the motor full-load current must be increased using the power factor.

1. When the power factor is 90%, motor FLC = Table FLC x 1.10
2. When the power factor is 80%, motor FLC = Table FLC x 1.25

The synchronous motor will have less full-load current than a squirrel-cage or wound-rotor induction motor, and the percentages permitted in **Table 430.52** can be different. The following problems are based upon the calculations for a synchronous motor.

Problem 5-32

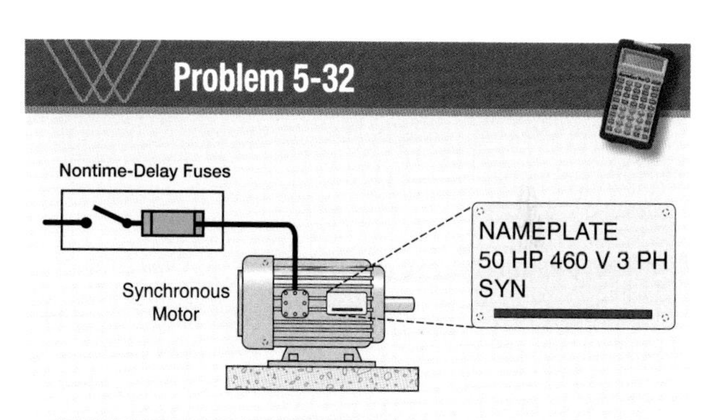

Determine the motor branch-circuit short-circuit and ground-fault protection for a 50-hp, 460-volt, 3-phase synchronous motor using nontime-delay fuses. The nameplate abbreviation for synchronous is syn.

Solution
Table 430.250
50 hp at 460 volts
FLC = 52 amps
Table 430.52
Synchronous motor, nontime-delay fuses = 300%
OCPD = FLC x 300%
= 52 x 3.00
= 156 amps
430.52(C)(1) Exception No. 1
Next larger standard size permitted
240.6(A)
Next larger size = 175 amps
Answer: 175 ampere nontime-delay fuses

Problem 5-33

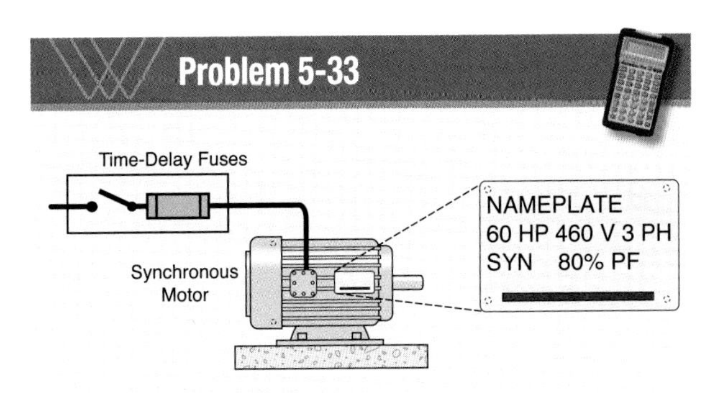

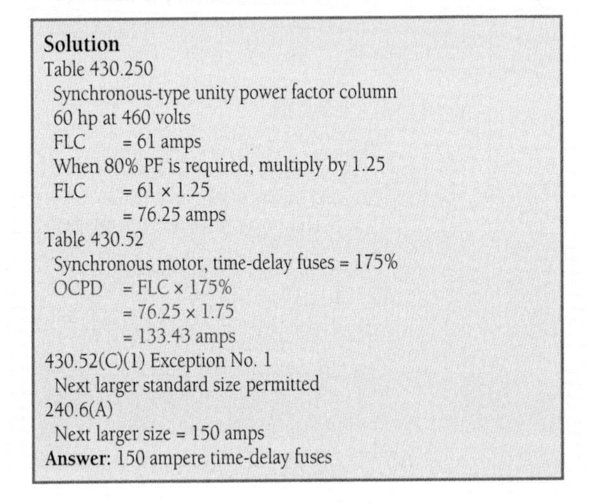

Determine the motor branch-circuit short-circuit and ground-fault protection for a 60-hp, 460-volt, 3-phase synchronous motor with an 80% power factor using time-delay fuses. The symbol for power factor is PF.

Solution

Table 430.250
Synchronous-type unity power factor column
60 hp at 460 volts
FLC = 61 amps
When 80% PF is required, multiply by 1.25
FLC = 61 × 1.25
= 76.25 amps
Table 430.52
Synchronous motor, time-delay fuses = 175%
OCPD = FLC × 175%
= 76.25 × 1.75
= 133.43 amps
430.52(C)(1) Exception No. 1
Next larger standard size permitted
240.6(A)
Next larger size = 150 amps
Answer: 150 ampere time-delay fuses

Problem 5-34

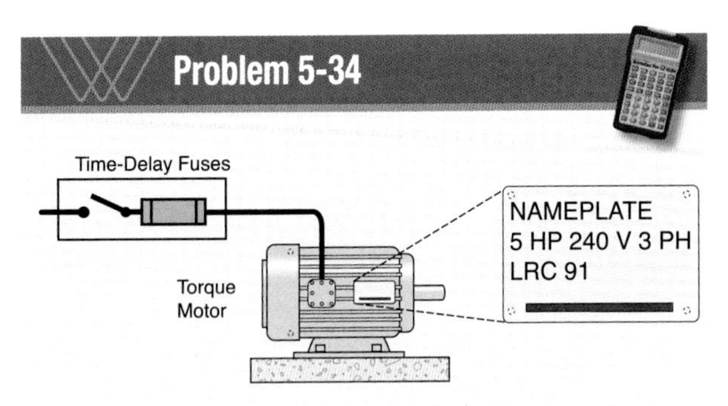

Determine the maximum motor branch-circuit short-circuit and ground-fault protection for a 5-hp, 240-volt, 3-phase torque motor with a nameplate full-load current of 91 amperes, installed with THW copper conductors and using time-delay fuses.

Solution

430.6(B)
Nameplate LRC = 91 amps
430.22
Ampacity = LRC × 125%
= 91 × 1.25
= 113.75 amps
Table 310.15(B)(16) Ampacity
113.75 amps = 2 AWG THW rated 115 amps
240.4(B)
Next larger standard size permitted
240.6(A)
Next larger size = 125 amps
Answer: 125 ampere time-delay fuses

5.2.3.8 Torque Motors - The motor branch-circuit short-circuit and ground-fault protection for a torque motor is based upon the amount of conductor use. The conductor size is based upon the nameplate full-load current of the motor, which is the locked-rotor current of the torque motor.

430.52(D) applies to the rating or setting for torque motors and refers to **240.4(B)**.

When the conductor's ampacity does not correspond to standard size overcurrent devices, **240.4(B)(3)** permits the next higher overcurrent device for a conductor up to 800 amperes.

430.6(B) requires that the nameplate full-load current rating be used for calculating the minimum size of branch-circuit conductors. For a torque motor, the nameplate full-load current is the locked-rotor current (LRC).

5.2.3.9 Basic Protection For Motor Circuits Containing Power Conversion Equipment - Adjustable–speed drives are covered by **Article 430, Parts I through IX** unless modified or supplemented by Part X. Within Part X of **Article 430**, branch-circuit short circuit and ground-fault protection for single and multiple motor circuits containing power conversion equipment, there are special requirements related to overcurrent protective devices. New **Sections 430.130** and **430.131** provide the user with the required overcurrent protection requirements for power conversion equipment such as adjustable speed drive controllers.

First the type of overcurrent protective device, its rating and setting is often marked on or provided within the installation instructions of the controller or equipment. Where this information is provided with the power conversion equipment, it must be followed exactly due to the requirements of **110.3(B)**.

Otherwise, where detailed protection requirements are not provided with the equipment, Sections **430.130** and **430.131** directs the user to follow specific protection requirements within **430.130**. Based on specific equipment, the equipment protection must comply with the standard ratings and setting as specified within either **430.52(C)(1)** using **Table 430.52(C)(3)** using instantaneous trip circuit breakers, **(C)(5)** using power electronic devices, or **(C)(6)** using self-protected combination controllers, all as determined by **430.6**.

In addition, where by-pass circuits are incorporated, branch-circuit short circuit and ground-fault protection must also be incorporated to protect these additional circuits and added equipment as well. Be aware that specific limitations on maximum permitted size of overcurrent protective devices used to protect power conversion are often specified or required. Finally, should a single adjustable-speed drive be used to control several motors, this circuit must be protected by an overcurrent protection device in accordance with **430.131** and **430.53**.

5.3 Motor Overload Protection

5.3.1 General Requirements

The purpose of motor overload (OL) protection is to protect the motor, motor control apparatus, and motor branch-circuit conductors against excessive heating due to overloads. The overload current is a current that, when it persists for a sufficient length of time, can damage the equipment and/or the conductors. **See Figure 5-5**.

Overloads are caused by the following:

1. Failure to start
2. Excessive load on motor
3. Worn motor bearings
4. Other mechanical problems

Motor overloads cause a motor to draw more current. The added current produces more heat within the motor. This addition heat is often more than the motor was designed for, which reduces the life of the motor.

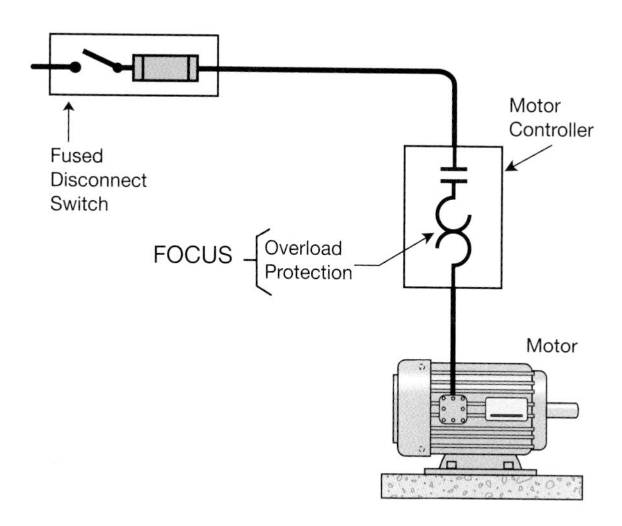

Figure 5-5. Motor branch-circuit overload protection as covered in Part III of Article 430 is the focus of this section.

A motor branch-circuit short-circuit and ground-fault protection device protects the motor from short circuits and line to ground faults; it does not protect the motor from overloads.

Overload protection is designed to sense the current of the motor and when the setting of the overload device is exceeded, the device causes automatic disconnection of the motor. Overload protection can be accomplished by directly disconnecting line current or causing the control circuit to stop the motor. Motor overload protection is often referred to as motor running protection.

Heaters installed in a magnetic motor starter are the most common form of overload protection. Fuses, circuit breakers, and thermal protectors may also be used to provide overload protection, provided they fulfill the specific requirements for motor overload protection.

430.6(A) and **430.32(A)(1)** both require the nameplate FLC (full-load current) to be used when calculating motor overload protection. The overload calculations are the same for Design B, C, and D motors.

430.32(A)(1) requires that overload protection devices for continuous-duty motors rated more than one horsepower be sized according to the motor nameplate FLC and the following motor nameplate information. These percentages are used for calculating the maximum overload protection.

Maximum OL protection (not to exceed)

1. Marked service factor 125%
2. Marked temperature rise 40°C or less 125%
3. All other motors 115%

The service factor of 1.15 means that a motor is built to operate at 115% of its nameplate rating and may sustain this small overload without motor damage.

The 40°C temperature rise limit is rather common for most motors. The motor with a service factor of 1.15 or larger and a temperature rise of 40°C indicates a well-built motor, and the overload protection is permitted to be increased in size accordingly. Should the service factor or the temperature rise not be marked on the motor, or a temperature rise be greater than 40°C, the size of the motor overload protection should be reduced to a service factor of 1.15.

Problem 5-35

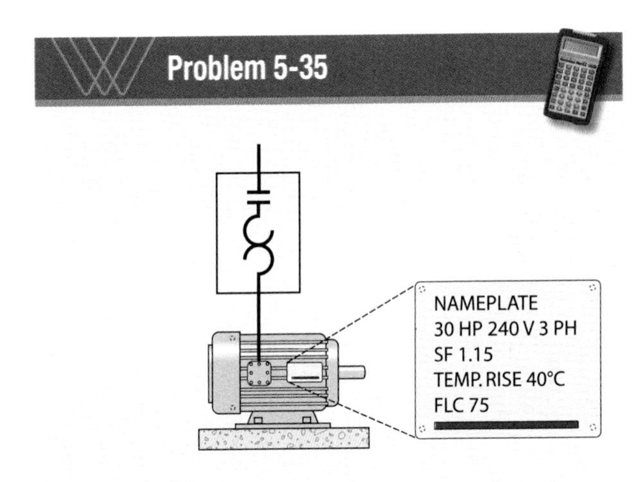

Determine the maximum overload protection (OL protection) for a 30-hp, 240-volt, 3-phase, squirrel-cage induction motor with a nameplate full-load current of 75 amperes, service factor (SF) of 1.15, and a temperature rise of 40°C.

Solution
430.32(A)(1)
Max. OL protection
Use nameplate FLC
Service factor 1.15, temperature rise 40°C, 125%
OL protection = FLC x 125%
= 75 x 1.25
= 93.75 amps

Answer: 93.75 amperes

5.3.2 Overload Protection With and Without Problem Starting Currents

Once again, consideration is given to the starting current. Two terms need to be considered: maximum overload protection, as calculated above, and maximum overload protection with starting current. Where the overload relays calculated by the percentages given in **430.32(A)(1)** do not permit the motor to start because of the starting current, **430.32(C)** permits the use of the percentages listed in the table following **Problem 5-36**. The values of **430.32(C)** allow about 15% more current before causing automatic disconnection of the motor.

One of the goals of overload protection is to protect a piece of electrical equipment as close to its rated full-load current as possible, while not having nuisance tripping during the starting current period.

430.32(C) Selection of Overload Relay is followed by an informational note. The informational note calls attention to the classification of motor over¬load relays. There are Class 10 and 10A, Class 20, and Class 30 overload relays. A Class 20 overload relay will provide a longer motor acceleration time than a Class 10 or 10A overload relay. A Class 30 overload relay will provide a longer motor acceleration time than a Class 20 overload relay. Therefore, selecting the higher class overload relay may preclude the need for selecting a higher trip cur¬rent. Once the acceleration period has passed, the optimal situation is to have the motor overload relay protecting the motor as close as practical.

Problem 5-36

Using the circuit diagram for Problem 5-35, determine the maximum overload protection, when starting current is a problem, for a 30-hp, 240-volt, 3-phase, squirrel-cage induction motor with a nameplate full-load current of 75 amperes, a service factor of 1.15, and a temperature rise of 40°C.

Solution
Maximum protection with problem starting current:
430.32(C)
Use nameplate FLC
Service factor 1.15, temperature rise 40°C, 140%
OL protection = FLC x 140%
= 75 x 1.40
= 105 amps

Answer: 105 amperes

Maximum OL protection where starting current is a problem:

1. Marked service factor of 1.15 or greater 140%
2. Marked temperature rise to 40°C or less 140%
3. All other motors 130%

Information

It is necessary to know when the starting current (or the motor acceleration time) is and is not, considered a problem in order to calculate the maximum rating permitted for the overload protection device. The use of time-delay overload relays in some cases, may preclude the need to select a higher overload trip setting of 130% or 140%.

Problem 5-37

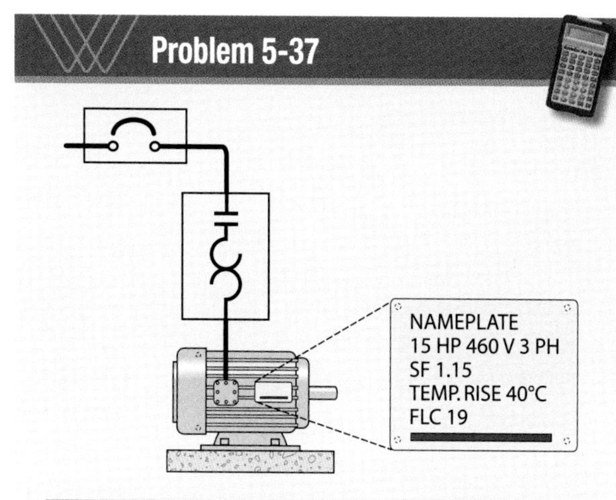

Determine overload rating of the overload relays for a 15-hp, 460-volt, 3-phase motor with a service factor (SF) of 1.15, a temperature rise of 40°C, and a nameplate current rating of 19 amperes.

1. What is the maximum overload protection?
2. What is the maximum overload protection with starting current?

Solution – Calculation 1
Max. OL protection
430.32(A)(1)
Use nameplate FLC
Service factor 1.15, temperature rise of 40°C, 125%
OL protection = FLC x 125%
= 19 x 1.25
= 23.75 amps
Answer: 23.75 amperes

Solution – Calculation 2
Max. OL protection with problem starting current
430.32(C)
Use nameplate FLC
Service factor 1.15, temperature rise of 40°C, 140%
OL protection = FLC x 140%
= 19 x 1.40
= 26.6 amps
Answer: 26.6 amperes

Problem 5-38

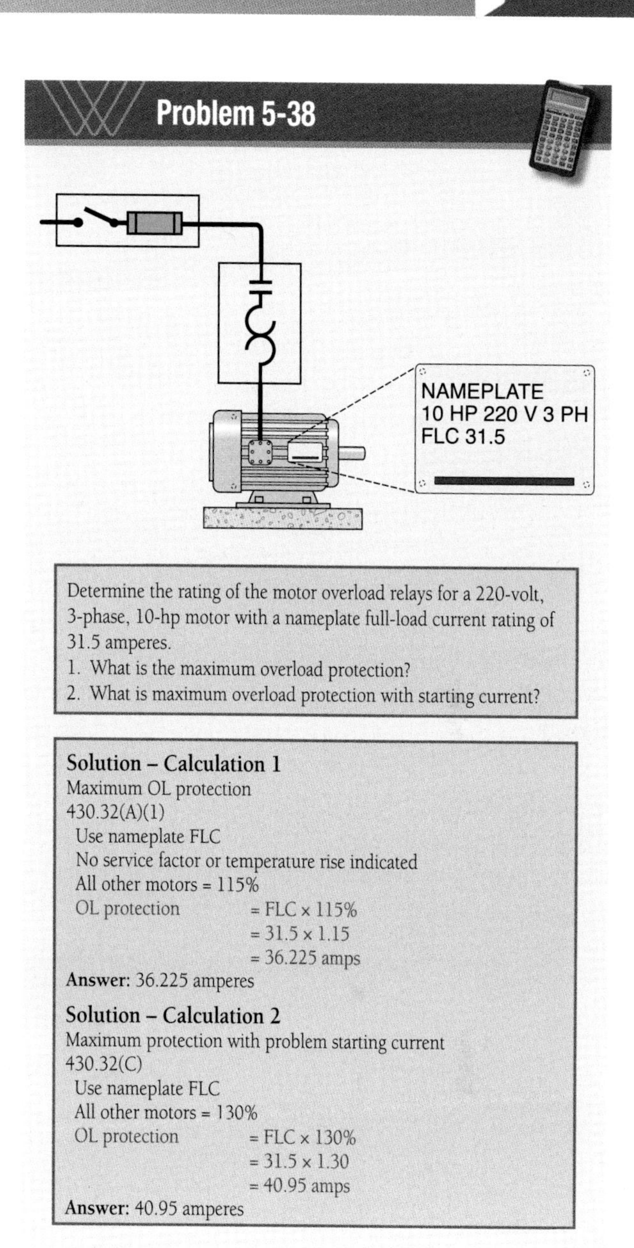

Determine the rating of the motor overload relays for a 220-volt, 3-phase, 10-hp motor with a nameplate full-load current rating of 31.5 amperes.

1. What is the maximum overload protection?
2. What is maximum overload protection with starting current?

Solution – Calculation 1
Maximum OL protection
430.32(A)(1)
Use nameplate FLC
No service factor or temperature rise indicated
All other motors = 115%
OL protection = FLC x 115%
= 31.5 x 1.15
= 36.225 amps
Answer: 36.225 amperes

Solution – Calculation 2
Maximum protection with problem starting current
430.32(C)
Use nameplate FLC
All other motors = 130%
OL protection = FLC x 130%
= 31.5 x 1.30
= 40.95 amps
Answer: 40.95 amperes

5.3.3 Using Thermal Protectors

When the overload protection is a thermal protector and is an integral part of the motor, **430.32(A)(2)** gives a special percentage for it. Note that for these calculations, the motor full-load current table values are to be used. The larger the motor full-load current, the less the percentage of increase is for sizing thermal protectors:

1. Motor FLC 9 amperes or less 170%
2. Motor FLC 9.1 through 20 amperes 156%
3. Motor FLC greater than 20 amperes 140%

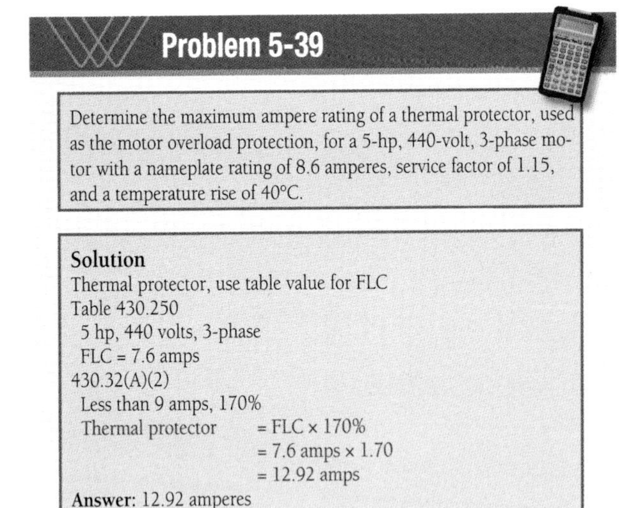

Problem 5-39

Determine the maximum ampere rating of a thermal protector, used as the motor overload protection, for a 5-hp, 440-volt, 3-phase motor with a nameplate rating of 8.6 amperes, service factor of 1.15, and a temperature rise of 40°C.

Solution

Thermal protector, use table value for FLC
Table 430.250
5 hp, 440 volts, 3-phase
FLC = 7.6 amps
430.32(A)(2)
Less than 9 amps, 170%
Thermal protector = FLC × 170%
= 7.6 amps × 1.70
= 12.92 amps

Answer: 12.92 amperes

5.3.4 Using Fuses and Circuit Breakers as Motor Overload Protection

430.32(A) requires a separate overload device responsive to motor current. This could be a fuse or circuit breaker. A fuse may be used as the motor overload protection and as the motor branch-circuit short-circuit and ground-fault protection, provided it is sized for the overload protection as well.

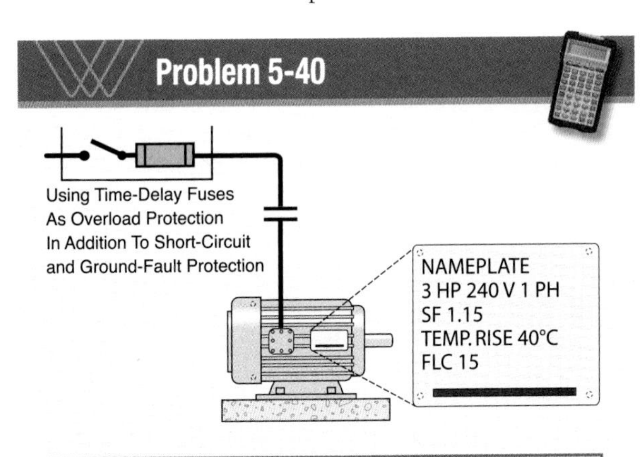

Problem 5-40

Determine the largest time-delay fuse permitted for the overload protection, when the starting current is a problem, for a 3-hp, single-phase, 240-volt motor, with a service factor of 1.15, temperature rise of 40°C, and a nameplate current rating of 15 amperes.

Solution

430.32(C)
Service factor 1.15, temperature rise 40°C, 140%
Maximum OL amps not to exceed nameplate
OL protection = FLC × 140%
= 15 × 1.40
= 21 amps
Not to exceed 21 amps
240.6(A)
Next smaller standard size = 20 amps

Answer: 20 ampere time-delay fuses

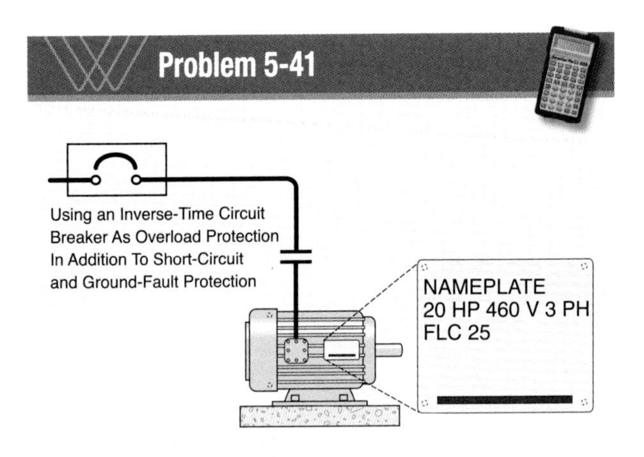

Problem 5-41

Determine the maximum size inverse-time circuit breaker that is permitted to be used as the motor overload protection for a 20-hp, 460-volt, 3-phase motor with a nameplate full-load current rating of 25 amperes when the starting current is a problem.

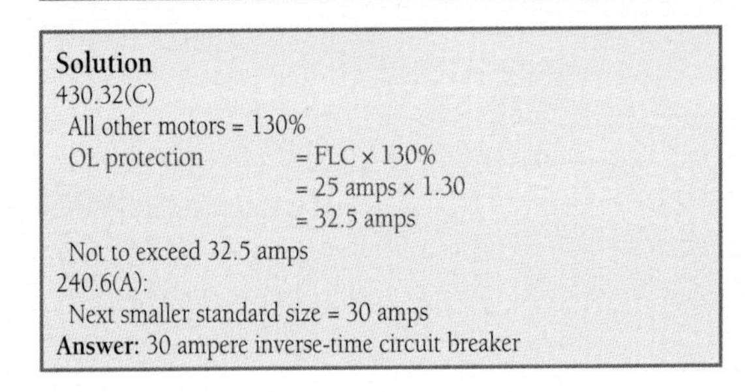

Solution

430.32(C)
All other motors = 130%
OL protection = FLC × 130%
= 25 amps × 1.30
= 32.5 amps
Not to exceed 32.5 amps
240.6(A):
Next smaller standard size = 30 amps

Answer: 30 ampere inverse-time circuit breaker

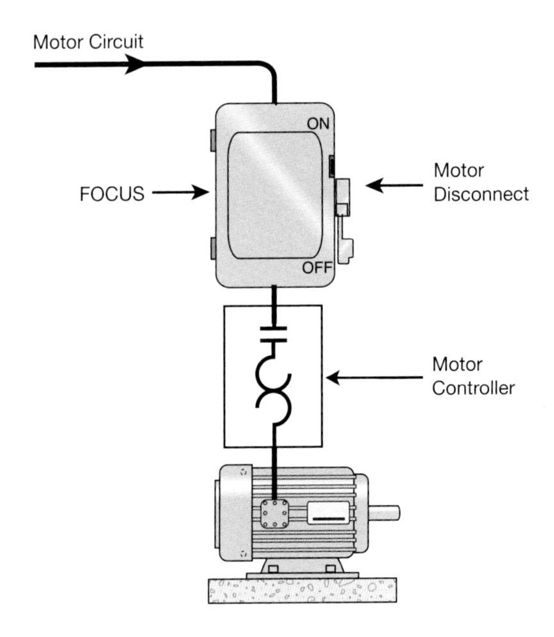

Figure 5-6. Motor disconnecting means as covered in Part IX of Article 430 is the focus of this section.

5.3.5 Overload Protection with Power Factor Corrected Motors

Where a motor installation includes a capacitor connected on the load side of the motor overload device, the rating or setting of the motor overload device is required to be based upon the improved power factor of the motor circuit according to **Section 460.9**. Where a motor is power factor corrected, the voltage of the circuit remains the same. But as the power factor is improved (moves toward 100%), the motor running current will decrease.

Problem 5-42

Determine the maximum size overload protection for a 10-hp, 240-volt, 3-phase motor, with a full-load current rating of 33 amperes at a power factor of 85% and a service factor of 1.25 when capacitors are connected on the load side of the overload relays for power factor correction and the circuit draws 29 amperes after power factor correction.

Solution
430.32(A)(1)
Service factor = 125%
Nameplate FLC = 33 amps
Power factor correction = FLC × 85%
= 33 × 0.85
= 29 amps
Power factor-corrected FLC = 29 amps
Section 460.9
Use power factor-corrected amps
OL protection = FLC (corrected) × 125%
= 29 × 1.25
= 36.25 amps
Answer: 36.25 amperes

5.4 Motor Disconnecting Means

5.4.1 General Requirements

A motor disconnecting means, commonly referred to as the motor disconnect switch, has three responsibilities:

1. To safely carry the motor full-load current under normal operation
2. To be capable of being opened under load
3. To be capable of being opened under locked-rotor conditions

See **Figure 5-6** for an example of where the motor disconnection means is placed within the motor circuit.

To fulfill the first responsibility, the *Code* requires the motor disconnecting means to have an ampere rating. **430.110(A)** requires the motor disconnecting means to have an ampere rating of at least 115% of the motor full-load current.

To fulfill the second and third responsibility, **Section 430.109** requires the motor-circuit switch to be horsepower rated for the locked-rotor current and to be listed. The circuit breaker and molded-case switch are not horsepower rated but are tested and identified for interrupting locked-rotor current at six times their rated ampacity.

Section 430.109 permits the motor disconnecting switch to be one of the following "listed" pieces of equipment:

1. Motor-Circuit Switch: A motor-circuit switch is a fused switch rated in horsepower and used as a disconnecting means.
2. Molded-Case Circuit Breaker: A molded-case circuit breaker contains overcurrent protection and is used as a disconnecting means.
3. Molded-Case Switch: A molded-case switch has no overcurrent protection and serves as a disconnecting means.
4. Instantaneous-Trip Circuit Breaker: An instantaneous trip circuit breaker is part of a listed combination motor controller.
5. Self-protected combination controller
6. Manual motor controller: A manual motor controller is marked "Suitable as Motor Disconnect." It is installed between the motor branch-circuit short-circuit and ground-fault protection and the motor. It could be used as an isolating switch.

A motor disconnecting means must be capable of interrupting locked-rotor currents (LRC). The horsepower rating of a disconnecting means indicates the amount of locked-rotor current it is capable of interrupting. The motor-circuit switch is required to be horsepower rated.

5.4.2 Locked-Rotor Current Calculations

Locked-rotor current (LRC) is the amount of current a motor will draw when the rotor of the motor is unable to turn. No counter-electromotive force is developed in the motor rotor, and the current is limited only by the low resistance of the rotor windings. Locked-rotor current can take place at the time of starting as the motor rotor and the load are at rest, or when a fault takes place and the rotor is unable to turn. A general rule of thumb indicates that the locked-rotor current of a motor is the starting current of the motor.

430.7(A)(8) requires motors to be marked with a code letter or the locked-rotor amperes of the motor to be marked on the nameplate of the motor. The code letter indicates the range of the locked-rotor current for a particular motor and is usually found on Design B, C, and D motors as well as Design B energy-efficient motors. Remember, Design B energy-efficient motors typically have a very high starting current. See **Figure 5-7**.

Table 430.7(B) Locked-Rotor Indicating Code Letters

Code Letter	Kilovolt-Amperes per Horsepower with Locked Rotor
A	0–3.14
B	3.15–3.54
C	3.55–3.99
D	4.0–4.49
E	4.5–4.99
F	5.0–5.59
G	5.6–6.29
H	6.3–7.09
J	7.1–7.99
K	8.0–8.99
L	9.0–9.99
M	10.0–11.19
N	11.2–12.49
P	12.5–13.99
R	14.0–15.99
S	16.0–17.99
T	18.0–19.99
U	20.0–22.39
V	22.4 and up

Figure 5-7. NEC Table 430.7(B) Locked-Rotor Indicating Code Letters is the focus of this section.

NEC **Table 430.7(B)** lists a range of values for locked-rotor currents for motors marked with a particular code letter. For example, a motor marked with code letter F has a locked-rotor kVA per horsepower

Problem 5-43

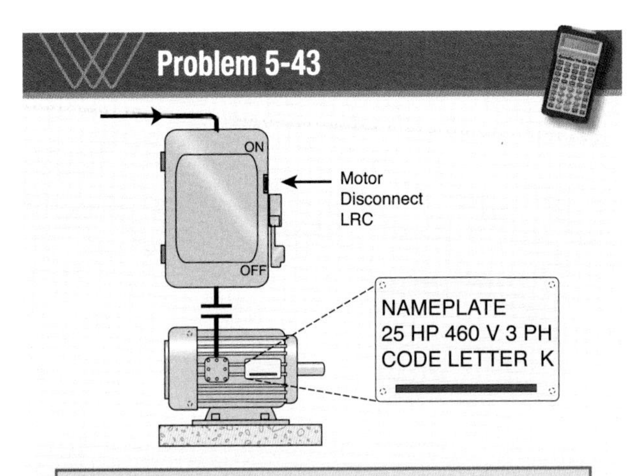

A 25-hp, 460-volt, 3-phase motor with a code letter K is to be installed.

1. What is the lowest LRC for the motor?
2. What is the average LRC for the motor?
3. What is the maximum LRC for the motor?

Solution – Calculation 1

Lowest LRC
Table 430.7(B)
Lowest kVA per hp
Code letter K = 8.0

$$LRC = \frac{(kVA\ per\ hp \times 1{,}000 \times hp)}{(E \times 1.73)}$$

$$= \frac{(8 \times 1{,}000 \times 25)}{(460 \times 1.73)}$$

$$= 251.32 \text{ amps}$$

Answer: 251.32 amperes lowest LRC

Solution – Calculation 2

Average LRC
Table 430.7(B)
Code letter K ranges 8.0 to 8.99
Lowest = 8.0 kVA per hp
Max. = 8.99 kVA per hp

$$Avg. = \frac{(max. + low)}{2}$$

$$= \frac{16.99}{2}$$

$$= 8.495 \text{ kVA per hp}$$

$$LRC = \frac{(kVA\ per\ hp \times 1{,}000 \times hp)}{(E \times 1.73)}$$

$$= \frac{(8.495 \times 1{,}000 \times 25)}{(460 \times 1.73)}$$

$$= 266.87 \text{ amps}$$

Answer: 266.87 amperes average LRC

Solution – Calculation 3

Max. LRC
Table 430.7(B)
Max. kVA per hp
Code letter K = 8.99

$$LRC = \frac{(kVA\ per\ hp \times 1{,}000 \times hp)}{(E \times 1.73)}$$

$$= \frac{(8.99 \times 1{,}000 \times 25)}{(460 \times 1.73)}$$

$$= 282.42 \text{ amps}$$

Answer: 282.42 amperes maximum LRC

ranging from 5.0 to 5.59 kVA per horsepower. This table can be used to calculate three values as illustrated in the following examples:

1. Calculating the lowest value
   Read directly from table lowest value listed.
   Example: Code letter C = 3.55 kVA
2. Calculating the maximum value
   Read directly from table highest value listed.
   Example: Code letter C = 3.99 kVA
3. Calculating the average value
   Lowest value plus highest value divided by two.
   Example: Code letter C

$$\frac{(3.55 + 3.99)}{2} = 3.77 \text{ kVA}$$

5.4.3 Locked-Rotor Current Equations

The following equations are used for calculating the locked-rotor current for motors with a code letter. When working with LRC, do not round off kVA to whole numbers.

Single-Phase LRC

$$LRC = \frac{(\textit{Locked-Rotor kVA per hp}) \times 1{,}000 \text{ hp}}{E}$$

Three-Phase LRC

$$LRC = \frac{(\textit{Locked-Rotor kVA per hp}) \times 1{,}000 \text{ hp}}{E \times 1.732}$$

LRC = locked-rotor current expressed in amperes (*I*)

Locked-rotor kVA per hp = value from Table 430.7(B)

1,000 = value used to convert kVA to volt amperes

hp = the horsepower of the motor under consideration

E = the voltage of the motor circuit under consideration

Note the similarity between the LRC equation above and the following kVA power equation:

Single-Phase

$$I = \frac{kVA \times 1{,}000}{E}$$

Three-Phase

$$I = \frac{kVA \times 1{,}000}{E \times 1.732}$$

5.4.4 Table 430.251(A) and Table 430.251(B), Conversion Tables for Locked-Rotor Current

Note the term "conversion table" in the title of **Table 430.251(A)** and **Table 430.251(B)**. This term indicates the LRC is calculated and the table is used to convert the calculated LRC to the required horsepower-rated motor-circuit switch. **Table 430.251(A)** for single-phase motors and **Table 430.251(B)** for 3-phase motors, as shown in **Figure 5-8** and **Figure 5-9**, are the Locked-Rotor Current (LRC) conversion tables for calculating the required horsepower rated motor-circuit switch when the LRC is not given on the motor nameplate. After the calculations for LRC are made using **Table 430.7(B)**, the calculated LRC is converted to the horsepower rating needed for the motor disconnecting means by the use of **Table 430.251(A)** or **Table 430.251(B)**.

Table 430.251(A) Conversion Table of Single-Phase Locked-Rotor Currents for Selection of Disconnecting Means and Controllers as Determined from Horsepower and Voltage Rating

For use only with 430.110, 440.12, 440.41, and 455.8(C).

	Maximum Locked-Rotor Current in Amperes, Single Phase		
Rated Horsepower	115 Volts	208 Volts	230 Volts
½	58.8	32.5	29.4
¾	82.8	45.8	41.4
1	96	53	48
1½	120	66	60
2	144	80	72
3	204	113	102
5	336	186	168
7½	480	265	240
10	1000	332	300

Figure 5-8. NEC Table 430.251(A).

The starting current or locked-rotor current for a standard motor is generally considered to be six times or 600% of the full-load current. **Table 430.251(A)** for single-phase motors and **Table 430.251(B)** for design letters B, C, and D motors are based upon the full-load current given in **Table 430.248** for single-phase motors and **Table 430.250** for 3-phase motors. The values listed are approximately six times the motor full-load current values given in the full-load current tables. The actual calculations show the

Table 430.251(B) Conversion Table of Polyphase Design B, C, and D Maximum Locked-Rotor Currents for Selection of Disconnecting Means and Controllers as Determined from Horsepower and Voltage Rating and Design Letter
For use only with 430.110, 440.12, 440.41 and 455.8(C).

Rated Horsepower	Maximum Motor Locked-Rotor Current in Amperes, Two- and Three-Phase, Design B, C, and D*					
	115 Volts	200 Volts	208 Volts	230 Volts	460 Volts	575 Volts
	B, C, D	B, C, D	B, C, D	B, C, D	B, C, D	B, C, D
½	40	23	22.1	20	10	8
¾	50	28.8	27.6	25	12.5	10
1	60	34.5	33	30	15	12
1 ½	80	46	44	40	20	16
2	100	57.5	55	50	25	20
3	—	73.6	71	64	32	25.6
5	—	105.8	102	92	46	36.8
7 ½	—	146	140	127	63.5	50.8
10	—	186.3	179	162	81	64.8
15	—	267	257	232	116	93
20	—	334	321	290	145	116
25	—	420	404	365	183	146
30	—	500	481	435	218	174
40	—	667	641	580	290	232
50	—	834	802	725	363	290
60	—	1001	962	870	435	348
75	—	1248	1200	1085	543	434
100	—	1668	1603	1450	725	580
125	—	2087	2007	1815	908	726
150	—	2496	2400	2170	1085	868
200	—	3335	3207	2900	1450	1160
250	—	—	—	—	1825	1460
300	—	—	—	—	2200	1760
350	—	—	—	—	2550	2040
400	—	—	—	—	2900	2320
450	—	—	—	—	3250	2600
500	—	—	—	—	3625	2900

*Design A motors are not limited to a maximum starting current or locked rotor current.

Figure 5-9. NEC Table 430.251(B).

values to be a little less than six times the full-load current. **Problem 5-44** illustrates the relationship between **Table 430.7(B)** and **Table 430.251(B)** and six times the full-load current of **Table 430.250**.

A horsepower-rated disconnect switch indicates the amount of locked rotor current it is capable of interrupting safely.

5.4.5 Calculating Motor-Circuit Switch Horsepower for Motors Marked with Code Letters

5.4.5.1 General Information - As long as there is no code letter, or the code letter is between A and F, the horsepower rating of the disconnect switch is most often listed to, and matches the horsepower rating of the motor. **Problem 5-45** uses a motor with the lowest possible kVA per hp, i.e. code letter A. For this permitted, but certainly uncommon motor, the calculated LRC shows a lower horsepower rated disconnecting means than the horsepower rating which the motor normally requires. However, the *Code* does not specifically permit adjustments as such, except for combination loads in **430.109(C)**.

Problem 5-44

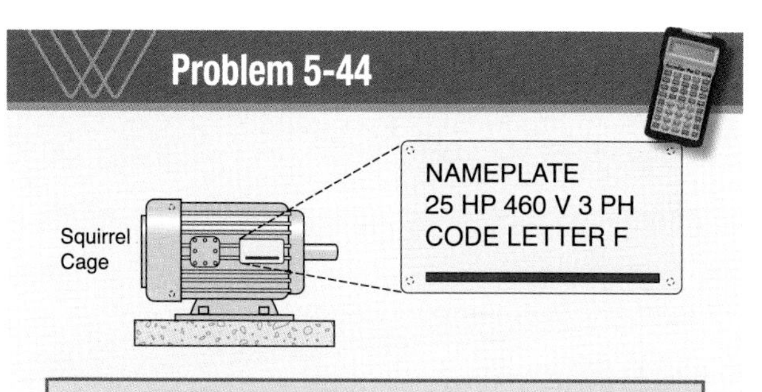

Using a 3-phase, 460-volt, 25-hp, squirrel-cage motor (other than a Design B energy-efficient motor), with a code letter F for the following:

1. What is the LRC rating of the motor using Table 430.251(B), Column 460 volts, B, C, D?
2. What is six times the FLC of Table 430.250?
3. What is the calculated maximum LRC using Table 430.7(B)?

Solution – Calculation 1
Table 430.251(B)
25 hp at 460 volts B, C, D column = 183 amps
Answer: 183 amperes LRC

Solution – Calculation 2
Table 430.250, 25 hp at 460 volts
FLC = 34 amps
= 6 × FLC
= 6 × 34
Answer: 204 amperes LRC

Solution – Calculation 3
Table 430.7(B), Max. kVA per hp
Code letter F = 5.59 kVA per hp

$$\text{LRC} = \frac{(kVA\ per\ hp \times 1{,}000 \times hp)}{(E \times 1.73)}$$

$$= \frac{(5.59 \times 1{,}000 \times 25)}{(460 \times 1.73)}$$

= 175.61 amps

Answer: 175.61 amperes maximum LRC

Comment
Selecting the hp rating of a switch from Table 430.251(B), using the calculated 175.61 amperes, a 25 hp switch is required.

Problem 5-45

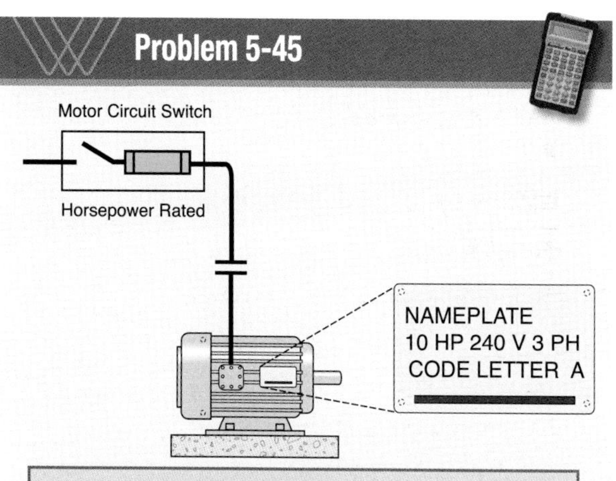

Determine the horsepower rating of a motor-circuit switch used as the disconnecting means for a 10-hp, 240-volt, 3-phase motor, code letter A, using the maximum LRC.

Solution
Table 430.7(B)
Max. kVA per hp
Code letter A = 3.14 kVA

$$\text{LRC} = \frac{(kVA\ per\ hp \times 1{,}000 \times hp)}{(E \times 1.73)}$$

$$= \frac{(3.14 \times 1{,}000 \times 10)}{(240 \times 1.73)}$$

= 75.63 amps
Table 430.251(B)
230 volt column
75.63 amps = 5 hp
Answer: 5 hp motor-circuit switch

Comment
Although this proves that a 5 hp motor circuit switch can safely disconnect this specific 10 hp motor from the circuit, 430.109(A)(1) must be applied.

Problem 5-46

Determine the horsepower rating of a motor-circuit switch used as a motor disconnecting means for a 10-hp, 240-volt, 3-phase motor, code letter F using maximum LRC.

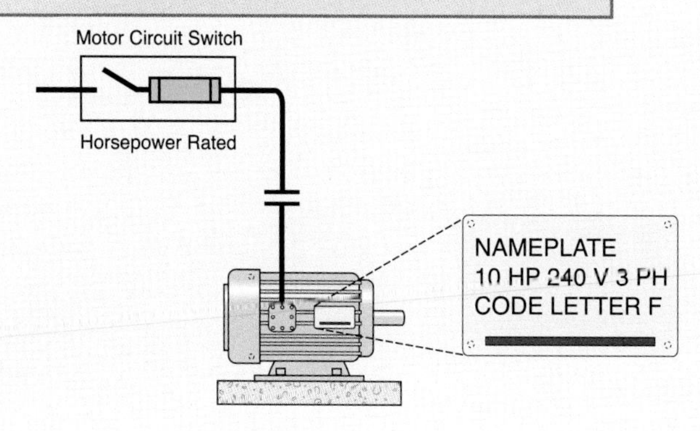

Solution
Table 430.7(B)
Max. kVA per hp
Code letter F = 5.59 kVA

$$\text{LRC} = \frac{(kVA\ per\ hp \times 1{,}000 \times hp)}{(E \times 1.73)}$$

$$= \frac{(5.59 \times 1{,}000 \times 10)}{(240 \times 1.73)}$$

= 134.63 amps
Table 430.251(B)
230 volt column
134.63 amps = 10 hp
Answer: 10 hp motor-circuit switch

Comment
Notice that the calculated LRC requires a motor-circuit switch with the same horsepower rating as the motor when the code letter is F.

Information

As the kVA per hp gets higher, it could result in a disconnecting means rated at a higher horsepower than the horsepower rating of the motor. This can be determined by the calculated locked-rotor current.

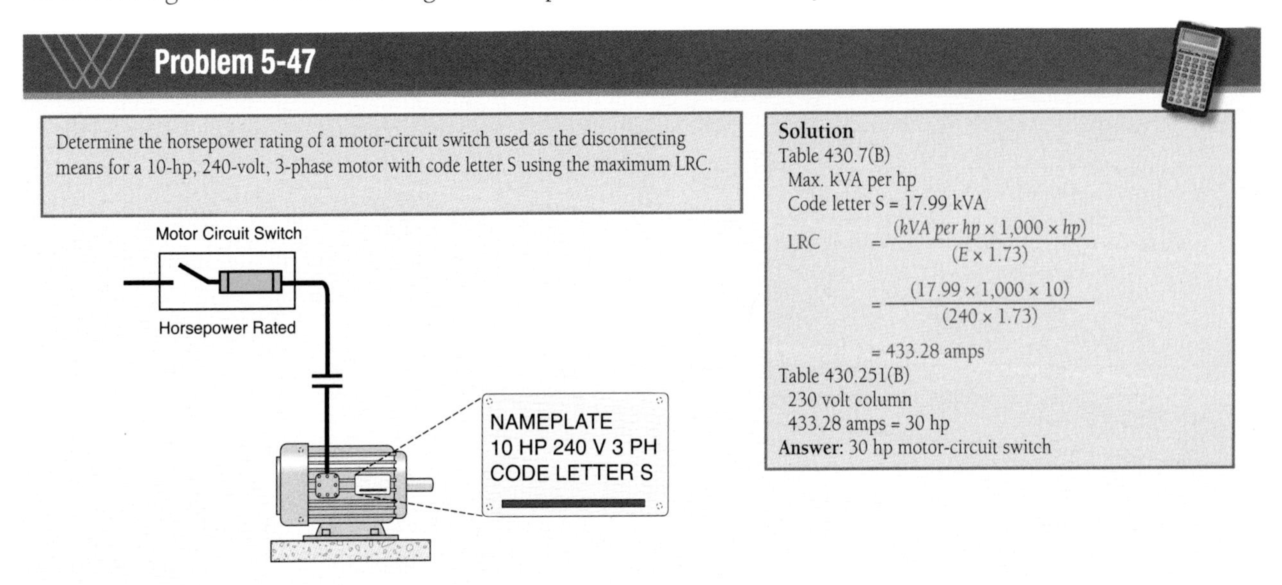

Problem 5-47

Determine the horsepower rating of a motor-circuit switch used as the disconnecting means for a 10-hp, 240-volt, 3-phase motor with code letter S using the maximum LRC.

Solution
Table 430.7(B)
Max. kVA per hp
Code letter S = 17.99 kVA

$$\text{LRC} = \frac{(kVA\ per\ hp \times 1{,}000 \times hp)}{(E \times 1.73)}$$

$$= \frac{(17.99 \times 1{,}000 \times 10)}{(240 \times 1.73)}$$

$$= 433.28 \text{ amps}$$

Table 430.251(B)
230 volt column
433.28 amps = 30 hp
Answer: 30 hp motor-circuit switch

Information

When one motor-circuit switch is used as the disconnecting means for two or more motors, **430.110(C)(1)** requires the horsepower rating of the motor disconnecting means to be equal to the sum of the locked-rotor currents of the motors as listed in **Table 430.251(A)** or **Table 430.251(B)**. **Problem 5-48** demonstrates the calculations necessary to determine the rating of the single disconnecting means from which two motors are supplied. Since both motors are three-phase, **Table 430.251(B)** is used to determine the motor locked-rotor current.

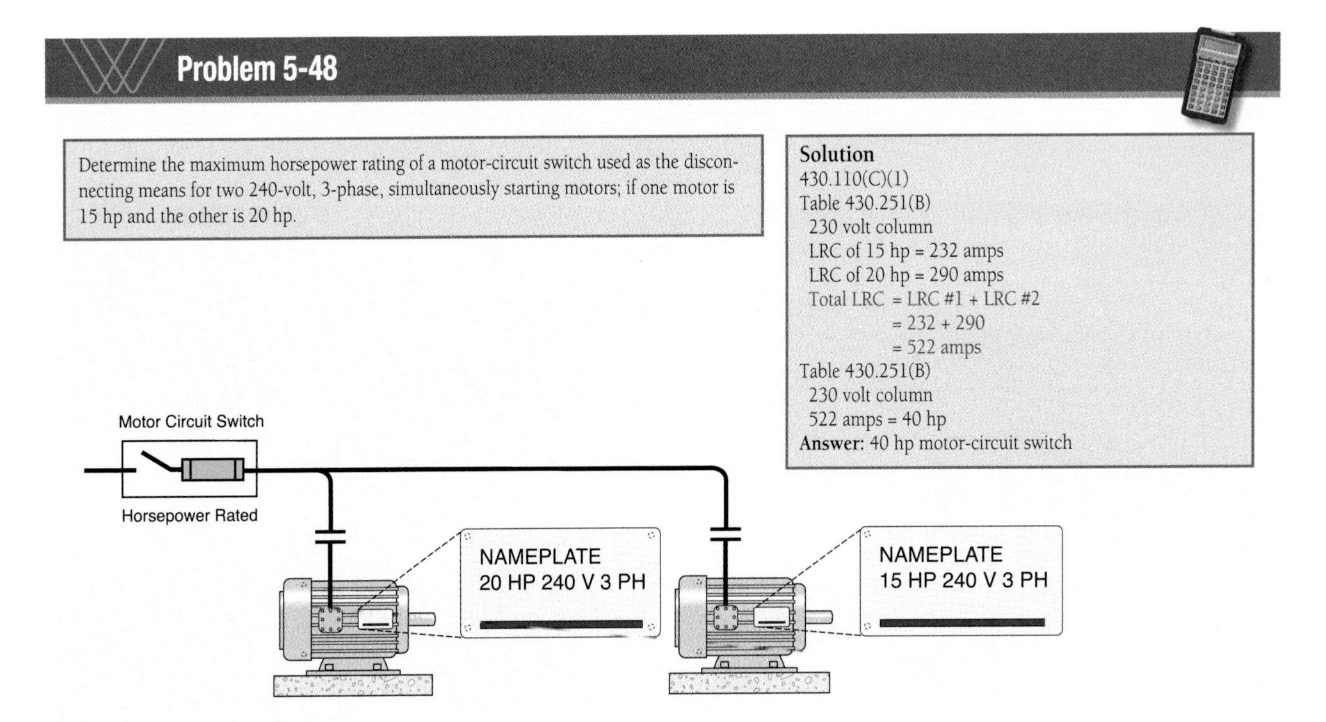

Problem 5-48

Determine the maximum horsepower rating of a motor-circuit switch used as the disconnecting means for two 240-volt, 3-phase, simultaneously starting motors; if one motor is 15 hp and the other is 20 hp.

Solution
430.110(C)(1)
Table 430.251(B)
230 volt column
LRC of 15 hp = 232 amps
LRC of 20 hp = 290 amps
Total LRC = LRC #1 + LRC #2
= 232 + 290
= 522 amps
Table 430.251(B)
230 volt column
522 amps = 40 hp
Answer: 40 hp motor-circuit switch

5.4.5.2 Disconnecting Means for Combination Loads - When combination loads of motors and other than motor loads, such as a motor and a resistance heater load, are controlled by the same disconnecting means, **430.110(C)(1)** and **440.12(C)(1)** require the disconnecting means to be equal to the LRC of the motor plus the full-load current rating of the heating or other loads. **Problem 5-49** demonstrates the calculations necessary to determine the rating of the single disconnecting means from which one motor and one large electric heater is supplied. Since the motor is single-phase, **Table 430.251(A)** is used to determine the motor locked-rotor current and Ohm's Law is used to calculate the full-load current of the heater.

5.4.6 Circuit Breaker as Motor Disconnecting Means for Other Than Design B Energy-Efficient Motors

The circuit breaker and the molded-case switch are not required to be horsepower rated. Reason: locked-rotor current is considered to be six times the full-load current rating of the motor. Circuit breakers and the molded-case switches are listed in the UL Materials Directory as having been tested for six times the device's rating and identified for motor disconnecting means use as required by **Section 430.109**.

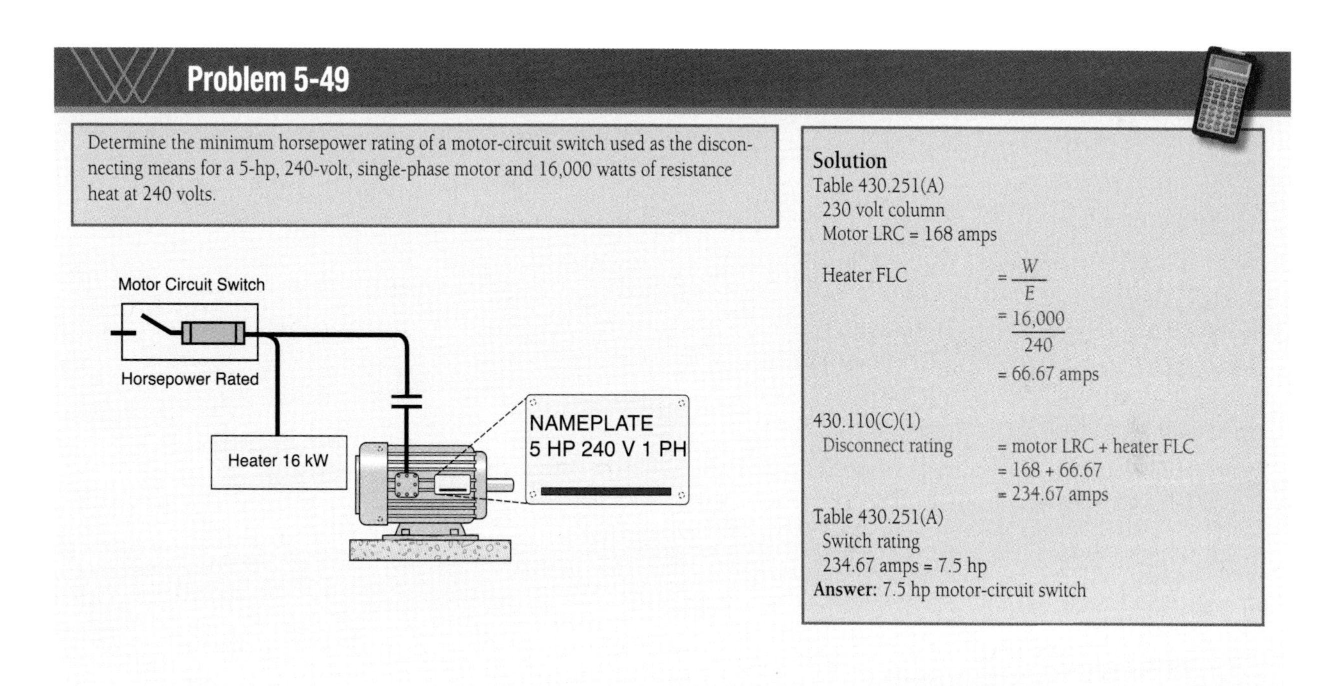

Determine the minimum horsepower rating of a motor-circuit switch used as the disconnecting means for a 5-hp, 240-volt, single-phase motor and 16,000 watts of resistance heat at 240 volts.

Solution

Table 430.251(A)
230 volt column
Motor LRC = 168 amps

Heater FLC $= \frac{W}{E} = \frac{16{,}000}{240} = 66.67$ amps

430.110(C)(1)
Disconnect rating = motor LRC + heater FLC
= 168 + 66.67
= 234.67 amps

Table 430.251(A)
Switch rating
234.67 amps = 7.5 hp

Answer: 7.5 hp motor-circuit switch

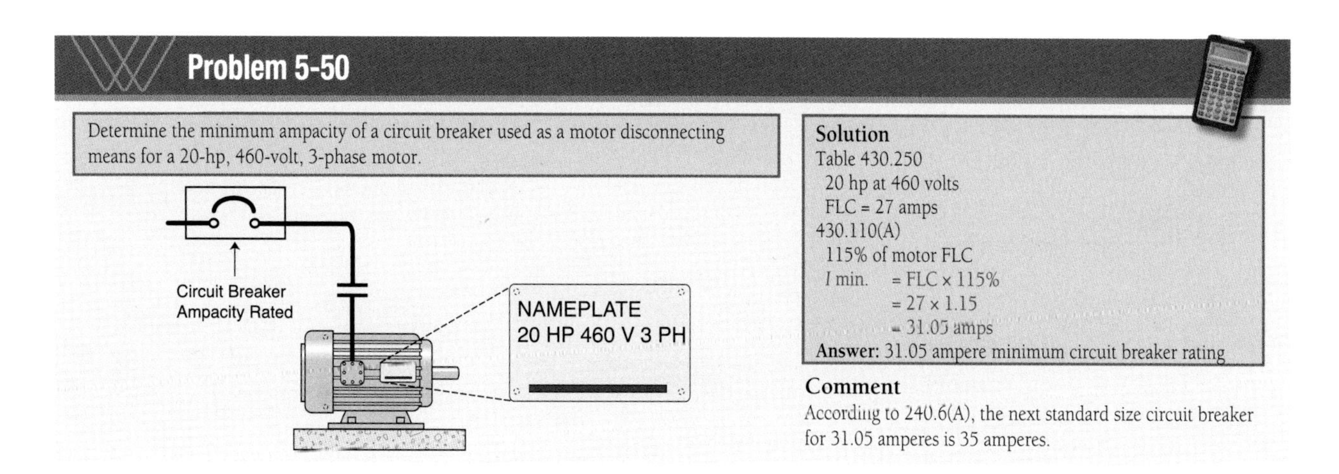

Determine the minimum ampacity of a circuit breaker used as a motor disconnecting means for a 20-hp, 460-volt, 3-phase motor.

Solution

Table 430.250
20 hp at 460 volts
FLC = 27 amps
430.110(A)
115% of motor FLC
I min. = FLC × 115%
= 27 × 1.15
= 31.05 amps

Answer: 31.05 ampere minimum circuit breaker rating

Comment

According to 240.6(A), the next standard size circuit breaker for 31.05 amperes is 35 amperes.

430.110(A) requires the motor disconnecting means to have an ampacity of at least 115% of the motor full-load current and is used for calculating the required ampere rating of the circuit breaker or the molded-case switch. The following is an example of how the horsepower rating is already built into a circuit breaker or a molded-case switch. **See Figure 5-10.**

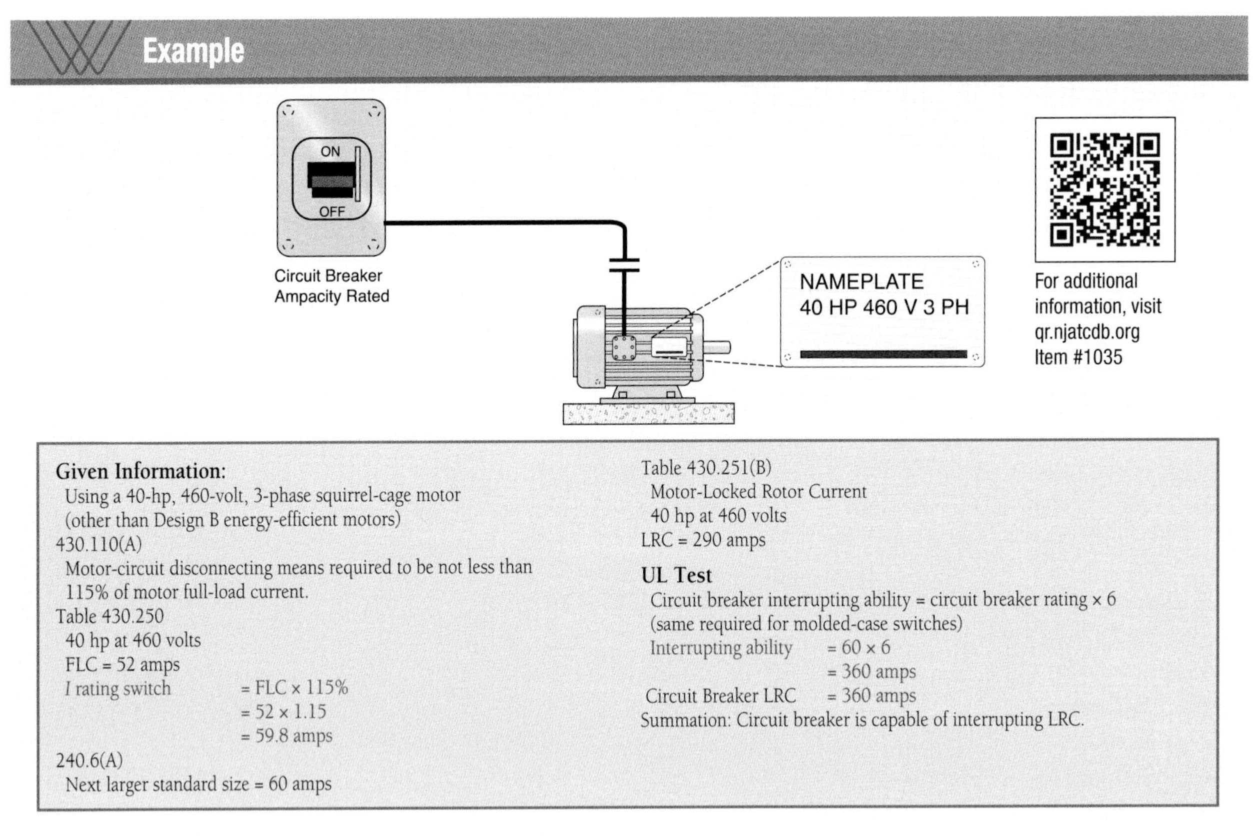

Given Information:
Using a 40-hp, 460-volt, 3-phase squirrel-cage motor (other than Design B energy-efficient motors)
430.110(A)
Motor-circuit disconnecting means required to be not less than 115% of motor full-load current.
Table 430.250
40 hp at 460 volts
FLC = 52 amps
I rating switch = FLC × 115%
= 52 × 1.15
= 59.8 amps
240.6(A)
Next larger standard size = 60 amps

Table 430.251(B)
Motor-Locked Rotor Current
40 hp at 460 volts
LRC = 290 amps

UL Test
Circuit breaker interrupting ability = circuit breaker rating × 6
(same required for molded-case switches)
Interrupting ability = 60 × 6
= 360 amps
Circuit Breaker LRC = 360 amps
Summation: Circuit breaker is capable of interrupting LRC.

Figure 5-10. The focus of this example is the testing and listing requirements of a circuit breaker.

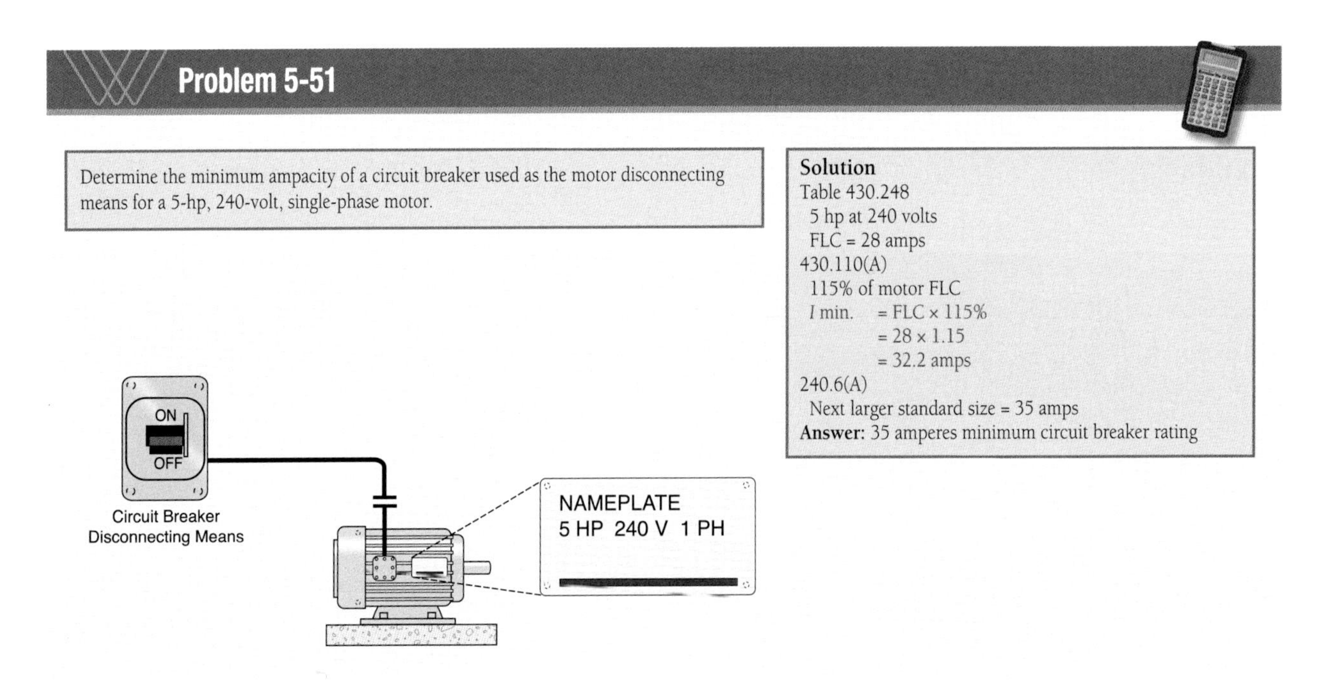

Determine the minimum ampacity of a circuit breaker used as the motor disconnecting means for a 5-hp, 240-volt, single-phase motor.

Solution
Table 430.248
5 hp at 240 volts
FLC = 28 amps
430.110(A)
115% of motor FLC
I min. = FLC × 115%
= 28 × 1.15
= 32.2 amps
240.6(A)
Next larger standard size = 35 amps
Answer: 35 amperes minimum circuit breaker rating

Problem 5-52

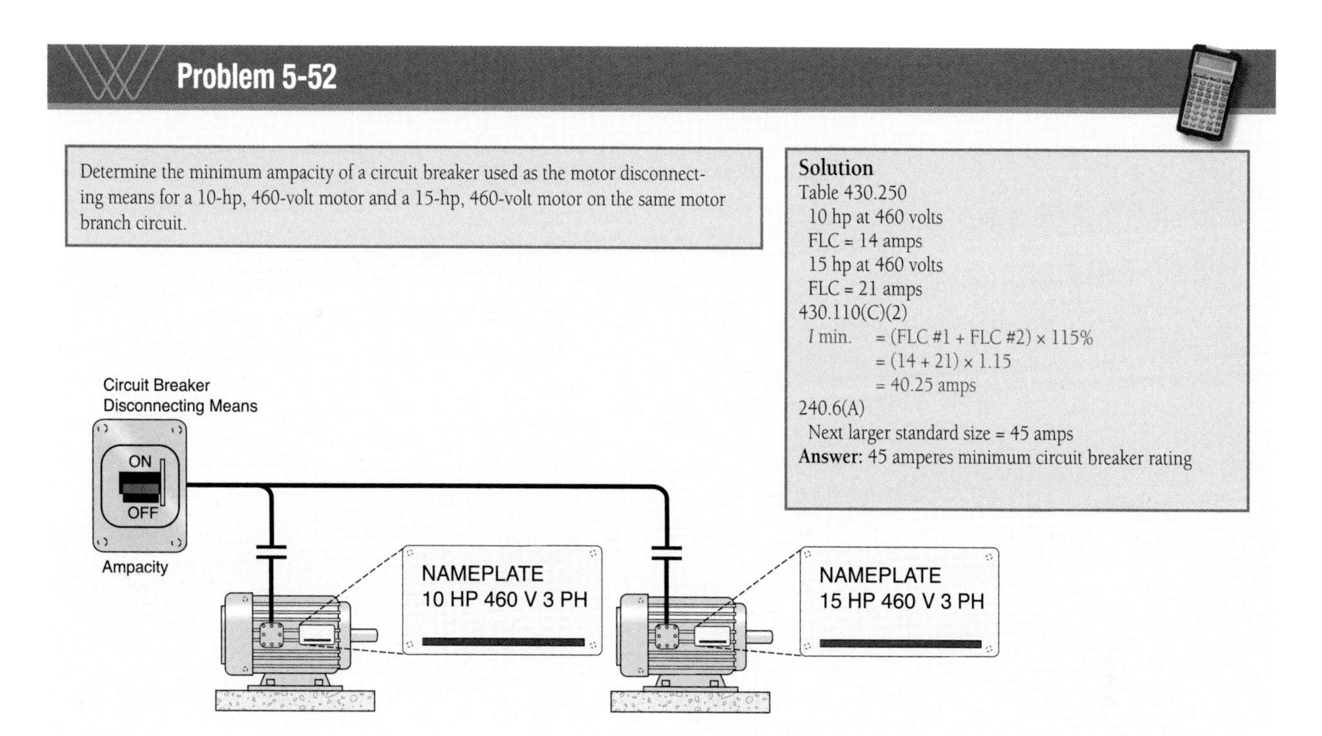

Determine the minimum ampacity of a circuit breaker used as the motor disconnecting means for a 10-hp, 460-volt motor and a 15-hp, 460-volt motor on the same motor branch circuit.

Solution

Table 430.250

10 hp at 460 volts
FLC = 14 amps
15 hp at 460 volts
FLC = 21 amps

430.110(C)(2)

I min. = (FLC #1 + FLC #2) x 115%
= (14 + 21) x 1.15
= 40.25 amps

240.6(A)

Next larger standard size = 45 amps

Answer: 45 amperes minimum circuit breaker rating

Information

When two motors are installed on the same branch circuit, **430.110(C)(1)** and **430.110(C)(2)** require the minimum ampacity of the motor disconnecting means to be 115% of the total of the motor full-load currents.

The full-load motor currents are not determined by using the motor nameplate information. Rather, the full-load current of each motor is determined using **Table 430.248** through **Table 430.250**.

Information

When the same circuit breaker is used as the disconnecting means for a motor and heating load, or other load type, the disconnecting means is required to be 115% of the full-load current of the motor plus the heating or other load.

Problem 5-53

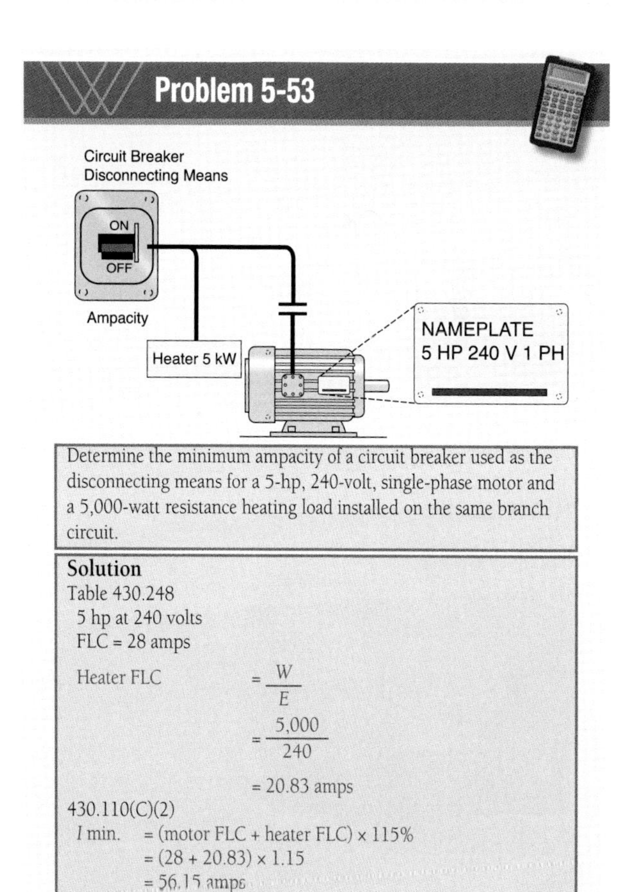

Determine the minimum ampacity of a circuit breaker used as the disconnecting means for a 5-hp, 240-volt, single-phase motor and a 5,000-watt resistance heating load installed on the same branch circuit.

Solution

Table 430.248

5 hp at 240 volts
FLC = 28 amps

Heater FLC $= \frac{W}{E}$

$= \frac{5{,}000}{240}$

= 20.83 amps

430.110(C)(2)

I min. = (motor FLC + heater FLC) x 115%
= (28 + 20.83) x 1.15
= 56.15 amps

240.6(A)

Next larger standard size = 60 amps

Answer: 56.15 amperes minimum circuit breaker rating

5.4.7 Molded-Case Switch for Other Than Design B Energy-Efficient Motors

A molded-case switch looks like a circuit breaker and is similarly rated in amperes, but no overcurrent protection is built into it. It is tested by UL for six times its ampacity rating. Therefore, it is identified to be used as a disconnecting means for motor circuits.

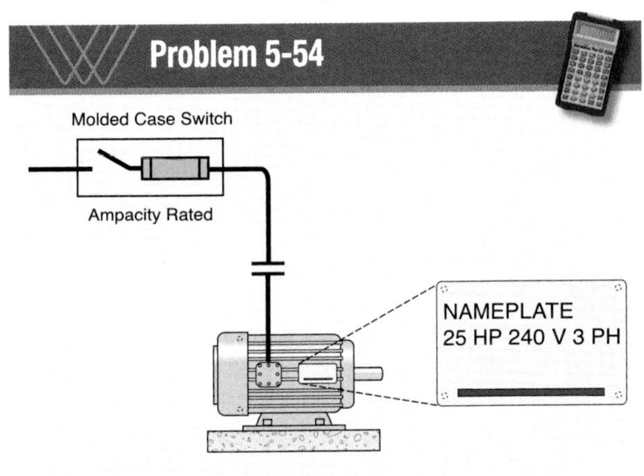

Determine the minimum ampacity rating for a molded-case switch used as a motor disconnecting means for a 25-hp, 240-volt, 3-phase motor.

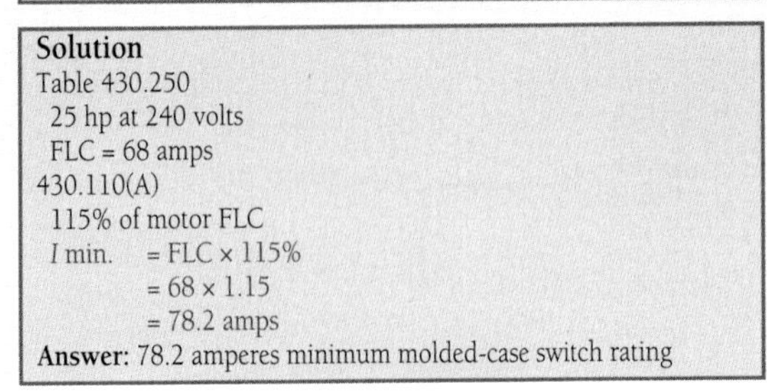

Solution
Table 430.250
25 hp at 240 volts
FLC = 68 amps
430.110(A)
115% of motor FLC
I min. = FLC × 115%
= 68 × 1.15
= 78.2 amps
Answer: 78.2 amperes minimum molded-case switch rating

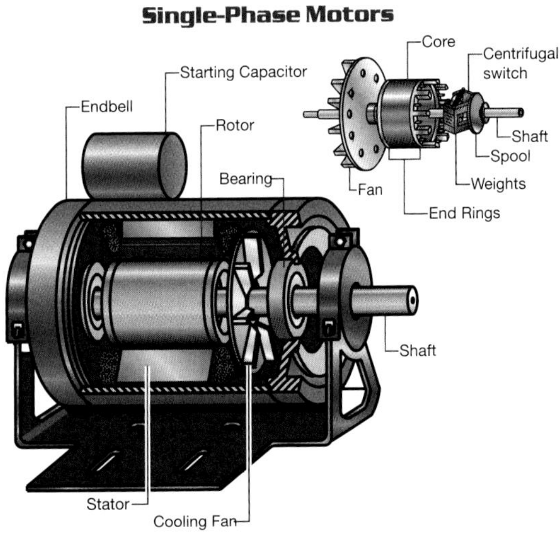

Table 430.248 contains full-load current values for single-phase AC motors running at usual speeds and with normal torque characteristics.

5.4.8 Combination Ampere and Horsepower Rating for Other Than Design B Energy-Efficient Motors

Problem 5-55 is as fundamental as the first problem dealing with sizing the disconnecting means for a motor circuit (for example, **Problem 5-43**). **Section 430.109** points out the different types of motor disconnects. Each disconnecting device is required to be "listed." **Section 430.110** identifies ampere rating calculation methods used to determine the disconnect size. Provided the motor code letter is not an issue, a listed disconnecting means rated and marked for the horsepower and voltage of a motor is all that is necessary for sizing the motor disconnect.

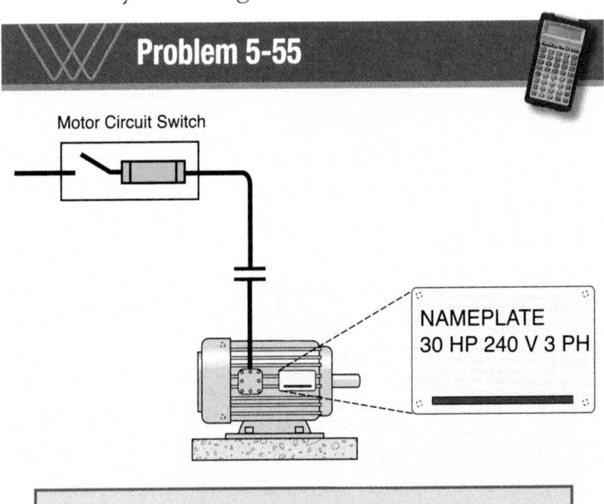

Determine the minimum ampacity and minimum horsepower rating for a motor-circuit switch used for the disconnecting means for a 30-hp, 240-volt, 3-phase motor.

Solution
Table 430.250
30 hp at 240 volts
FLC = 80 amps
430.110(A)
115% of FLC
I min. = FLC × 115%
= 80 × 1.15
= 92 amps
No code letter
Switch hp = Motor hp
Switch hp = 30 hp
Answer: 92 amperes and 30 hp minimum

In summary, a motor disconnecting means has three responsibilities:

1. To safely carry the motor full-load current under normal operation
2. To be capable of being opened under load
3. To be capable of being opened under locked-rotor conditions.

5.5 Air-Conditioning and Refrigerating Equipment Motors

A hermetic refrigerant motor-compressor is a combination of a motor and a compressor enclosed in the same housing, the shaft sealed with seals, and the motor operating in a refrigerant. Therefore, hermetically sealed motors have different operating characteristics. These characteristics vary with the manufacturer. The manufacturer establishes the rated-load current (RLC) and the branch-circuit selection current (BCSC) for their particular motor.

The rated-load current is the current resulting from the operation of the hermetically sealed motor at its rated load, rated voltage, and rated frequency for the load served.

The branch-circuit selection current (BCSC) is always equal to or greater than the rated-load current. Therefore, it is used in calculations for selecting the motor branch-circuit conductors, short-circuit and ground-fault protection, overload protection, controller and disconnecting means size.

Normally, the nameplate of a compressor-type motor has the minimum branch-circuit amperes, maximum branch-circuit fuse rating, circuit breaker rating, voltage rating, number of phases, rated-load current, and locked-rotor current. Therefore, there is seldom use for these calculations in the field.

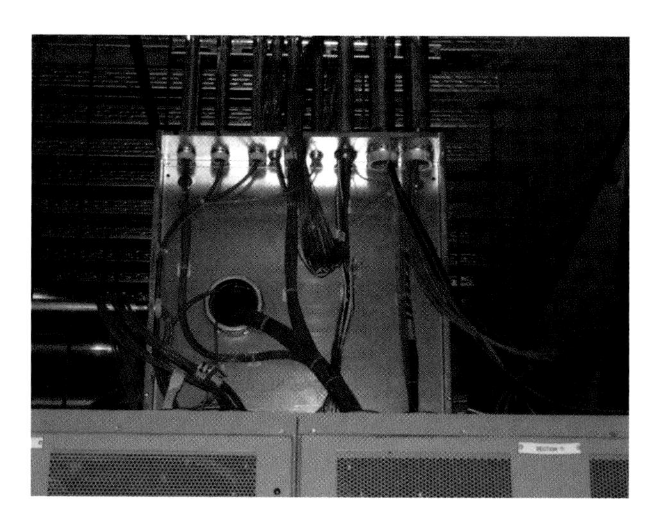

The branch circuit selection current (BCSC) is used to determine the minimum size of the motor branch-circuit conductors.

5.5.1 Branch-Circuit Conductor Sizing

Section 440.32 requires the conductors supplying a motor compressor to be 125% of the rated-load current (RLC) or the branch-circuit selection current (BCSC), whichever is the larger. For **Problem 5-56** and **Problem 5-57**, use **Table 310.15(B)(16)**, 75°C, copper, THWN conductors in accordance with **110.14(C)(1)**.

Problem 5-56

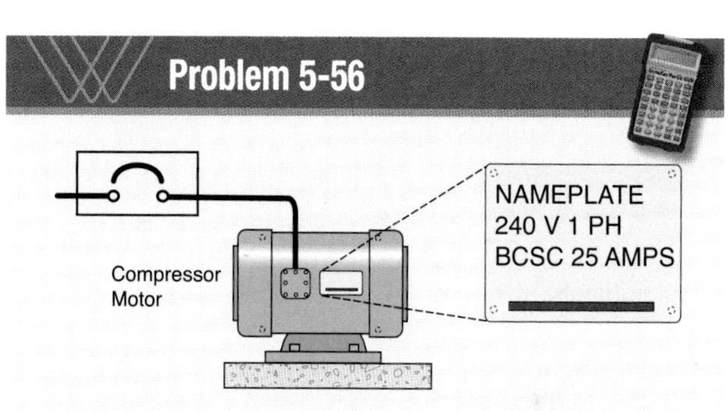

Determine the minimum branch-circuit conductor ampacity and size (THWN copper) for a single-phase, 240-volt compressor motor with a branch-circuit selection current (BCSC) of 25 amperes. Select conductors based upon Table 310.15(B)(16), 75°C, copper, THWN conductors in accordance with 110.14(C)(1)(a)(4).

Solution – Calculation 1
Branch-circuit conductor ampacity
440.32
Ampacity = BCSC × 125%
= 25 × 1.25
= 31.25 amps
Answer: 31.25 amperes

Solution – Calculation 2
Branch-circuit conductor size
Table 310.15(B)(16)
75°C, copper column
31.25 amps = 10 AWG
Answer: 10 AWG THWN copper

Problem 5-57

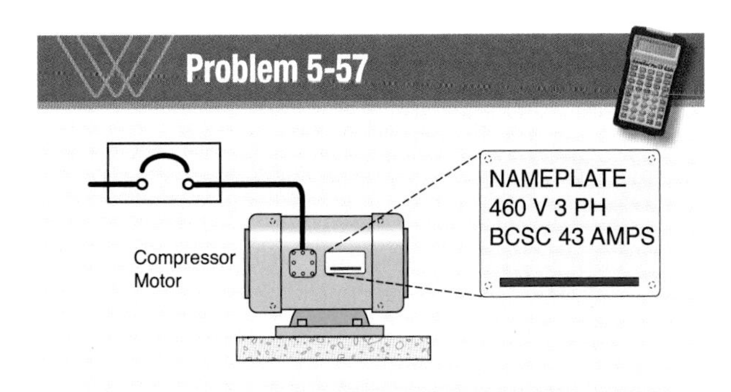

Determine the minimum branch-circuit conductor ampacity and size for a 3-phase, 460-volt compressor motor, with a branch-circuit selection current of 43 amperes. Select conductors based upon Table 310.15(B)(16), 75°C, copper, THWN conductors in accordance with 110.14(C)(1)(a)(4).

Solution – Calculation 1
Branch-circuit conductor ampacity
440.32
Ampacity = BCSC × 125%
= 43 × 1.25
= 53.75 amps
Answer: 53.75 amperes

Solution – Calculation 2
Branch-circuit conductor size
Table 310.15(B)(16)
75°C, copper column
53.75 amps = 6 AWG
Answer: 6 AWG THWN copper

Information

When two or more hermetically sealed motors are on the same branch circuit, the branch-circuit conductors are required to be 125% of the largest motor branch-circuit selection current or the rated-load current, whichever is higher, plus the branch-circuit selection current of the remaining motors.

5.5.2 Overload Calculations

440.52(A) covers the calculations for sizing the motor overload protection. This calculation uses the rated-load current (RLC) of the compressor. Where overload relays are applied, the RLC is multiplied by 140%. Where fuses or inverse time circuit breakers are applied, the RLC is multiplied by 125%. This sizing provides sufficient time delay to permit normal starting of the compressor.

Problem 5-59

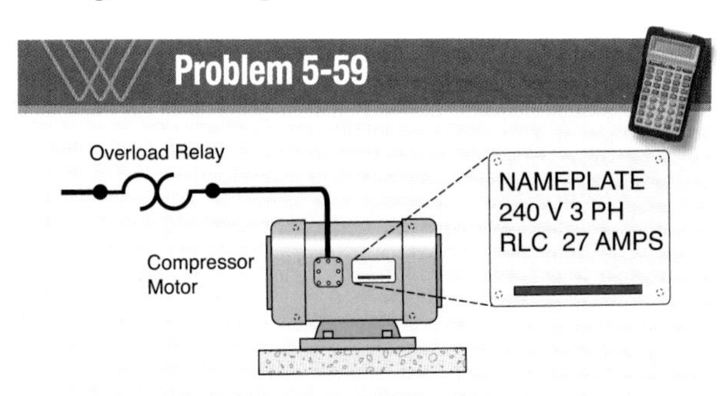

Determine the maximum overload relay protection permitted for a 240-volt, 3-phase hermetically sealed compressor-type motor with a rated-load current (RLC) of 27 amperes.

Solution
440.52(A)(1)
OL relay protection = RLC × 140%
= 27 × 1.40
= 37.8 amps
Answer: 37.8 amperes

Problem 5-58

Determine the motor branch-circuit conductor ampacity and size when two 240-volt, 3-phase, hermetically sealed motors with branch-circuit selection currents of 56 amperes are installed on the same branch circuit. Select conductors based upon Table 310.15(B)(16), 75°C, copper, XHHW-2 conductors in accordance with 110.14(C)(1)(a)(4).

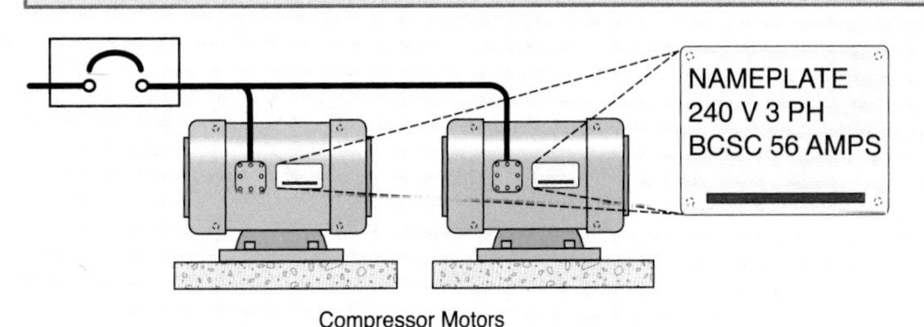

Solution – Calculation 1
Branch circuit conductor ampacity
440.33
Ampacity = (largest BCSC × 125%) + others
= (56 × 1.25) + 56
= 126 amps
Answer: 126 amperes

Solution – Calculation 2
Branch circuit conductor size
Table 310.15(B)(16)
75°C, copper column
126 amps = 1 AWG
Answer: 1 AWG XHHW-2 copper

5.5.3 Motor Branch-Circuit Short-Circuit and Ground-Fault Protection

Basic requirement of **440.22(A)** requires the branch-circuit short-circuit and ground-fault protection not to exceed 175% of the hermetically sealed motor rated-load current or branch-circuit selection current; whichever is the larger.

When starting current is a problem, the branch-circuit short-circuit and ground-fault protection must not exceed 225% of the hermetically sealed motor rated-load current or branch-circuit selection current; whichever is greater.

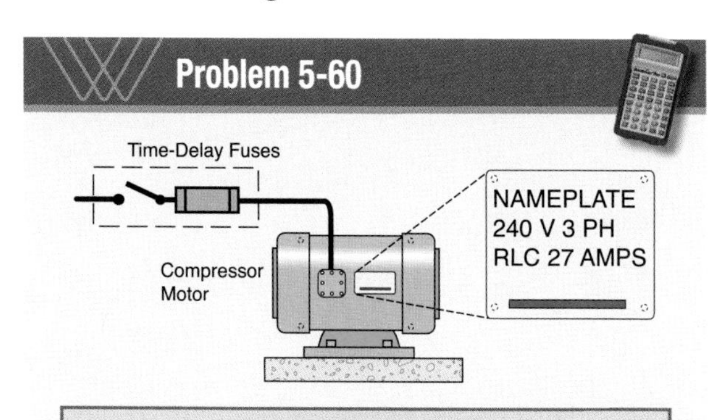

Determine the maximum overload protection for a 240-volt, 3-phase, hermetically sealed compressor-type motor with a rated-load current (RLC) of 27 amperes using time-delay fuses.

Solution
440.52(B)(3)
Not to exceed 125%
OL protection (fuse) = RLC x 125%
= 27 x 1.25
= 33.75 amps
240.6(A)
Next smaller standard size = 30 amps
Answer: 30 amperes

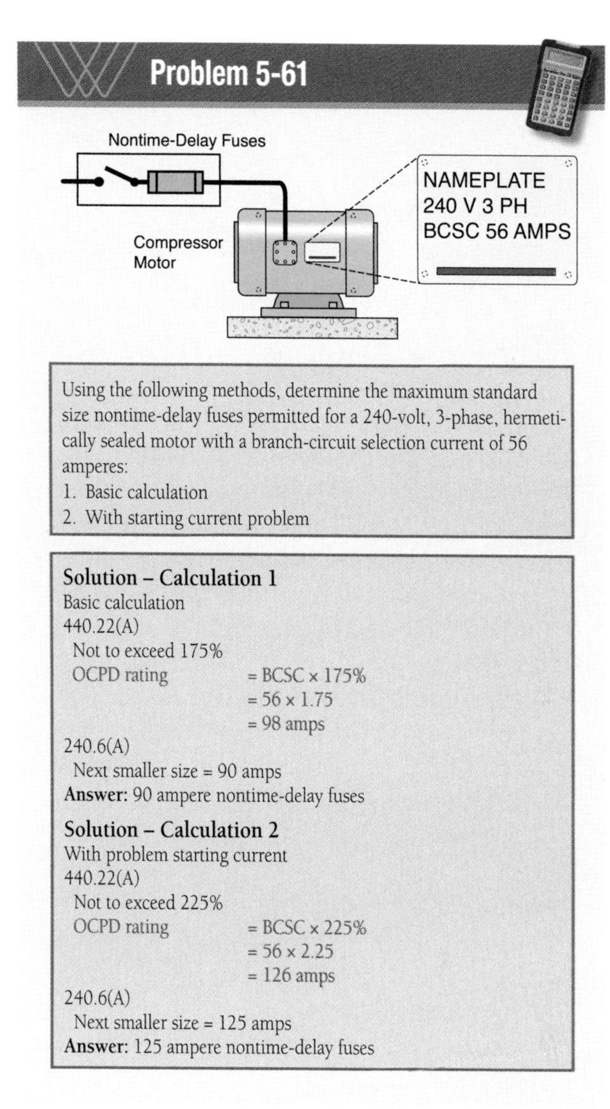

Using the following methods, determine the maximum standard size nontime-delay fuses permitted for a 240-volt, 3-phase, hermetically sealed motor with a branch-circuit selection current of 56 amperes:

1. Basic calculation
2. With starting current problem

Solution – Calculation 1
Basic calculation
440.22(A)
Not to exceed 175%
OCPD rating = BCSC x 175%
= 56 x 1.75
= 98 amps
240.6(A)
Next smaller size = 90 amps
Answer: 90 ampere nontime-delay fuses

Solution – Calculation 2
With problem starting current
440.22(A)
Not to exceed 225%
OCPD rating = BCSC x 225%
= 56 x 2.25
= 126 amps
240.6(A)
Next smaller size = 125 amps
Answer: 125 ampere nontime-delay fuses

Connection to the equipment grounding conductor is required to be accomplished as specified in Part VI of Article 250.

The provisions of Article 440 apply to electric motor-driven air-conditioning and refrigerating equipment and the related branch circuits and controllers.

Definitions and Terms

Continuous Duty - Operation at a substantially constant load for an indefinitely long time.

Continuous Duty (Motor) - A motor which can continue to operate within the temperature limits, after it has reached normal operating temperature .

Control Circuit - The circuit of a control apparatus or system that carries the electrical signals directing the performance of the cotnroller, but does not carry the main power current.

Feeder - All circuit conductors between the service equipment, the source of a separately derived system, or other power supply source and the final branch-circuit overcurrent device.

Full-Load Current (Motor) - The current drawn from the line when the motor is operating at full load torque and full load speed at rated frequency and voltage. The *Code* specifies when the table value is used and when the motor nameplate value is used.

Locked-Rotor Current - Measured current with the rotor locked and with rated voltage and frequency applied to the motor. This is the current seen when starting the motor and load.

Motor Branch Circuit - The circuit conductors between the final overcurrent device protecting the circuit and the motor.

Motor Circuit Switch - A switch rated in horsepower capable of interrupting the maximum operating overload current of a motor of the same horsepower rating as the switch at the rated voltage.

Motor Controller - A controller is any switch or device that is normally used to start and stop a motor by making and breaking the motor circuit current.

Overcurrrent - Any current in excess of the rated current of equipment or the ampacity of a conductor. It may result from overload, short circuit, or ground fault.

Overload - Operation of equipment in excess of normal, full-load rating, or of a conductor in excess of rated ampacity that, when it persists for a sufficient length of time, would cause damage or dangerous overheating. A fault, such as a short circuit or ground fault, is not an overload.

Service Factor (SF) - A measure of the reserve margin built into a motor. Motors rated over 1.0 SF have more than normal margin, and are used where unusual conditions such as occasional high or low voltage, momentary overloads, etc., are likely to occur.

Temperature Rise - The amount by which a motor, operating under rated conditions, is hotter than its surroundings. On most motors, manufacturers have replaced the rise rating on the motor nameplate with a listing of the ambient temperature rating, insulation class, and service factor.

Thermal Protector - An inherent overheating protective device which is responsive to motor temperature and which, when properly applied to a motor, protects the motor against dangerous overheating due to overload or failure to start. This protection is available with either manual reset or automatic reset.

Summary

- Generally, the ampacity of a motor-branch circuit must be sized to at least 125% of the motor full-load current (FLC) as found in the motor tables of Part XIV in **Article 430**.
- This requirement (125% of the full-load current from the *NEC* tables) applies to the continuous-duty motors; which is the most common motor.
- Other less common motors, such as low-speed, multiple-speed, periodic and varying duty, torque motors, and a few others are calculated at 125% of the nameplate values instead of the table values of full-load current.
- Overcurrent protection of motors circuits is divided between two separate and distinct overcurrent protection devices: the branch-circuit short-circuit and ground-fault protective devices and overload protection device(s).
- Branch-circuit short-circuit and ground-fault protective devices are located near the beginning of the circuit and provide circuit and equipment protection from abnormally high values of current.
- Common branch-circuit short-circuit and ground-fault protective devices are sized at 175% for fuses and 250% for circuit breakers.
- Branch-circuit short-circuit and ground-fault protective devices are sized according to **Section 430.52**.
- Operation of this device causes automatic disconnection of the motor from its source of supply.
- Overload protection prevents individual motor overheating, and so the devices used are closely aligned with the individual motor nameplate current being protected.
- Overload protection devices are located near the end of the circuit and provide motor protection against smaller values of current that exceed nominal values of running current.
- Overload protection devices are sized according to **Section 430.32**.
- Operation of the overload device causes automatic disconnection of the motor from its source of supply.
- Motor-circuit switches must be robust enough to safely interrupt the full-load current (FLC) of the motor being disconnected.
- Motor-circuit switches must also be robust enough to safely interrupt the locked-rotor current (LRC) of a stalled or jammed motor.
- Operation of either device must cause safe and predictable automatic disconnection of the motor from its power supply.
- Circuit breakers used to disconnect motors are already rated for 600% full-load current disconnection in accordance with their listing.
- Air-conditioning and refrigerating equipment motors covered in **Article 440** have many of the same requirements found in **Article 430 Motors**.

Voltage Drop

Equations for resistance and voltage drop are based upon values of resistance from **Chapter 9, Table 8** of the *NEC* and equations based upon Ohm's Law. Various theory textbooks cover the subjects of wire resistance and voltage drop using various conductor temperature ratings, but few focus on the same conductor temperature ratings of **Table 310.15(B)(16)** and **Table 310.104(A)** of the *NEC*. The specific conductor temperature focus is 75°C. Other temperatures will be dealt with by using specific voltage drop adjustment factors described within this chapter. Most voltage drop calculations in the trade are made simple by using only DC values. Near the end of this chapter, there are two example problems presented which use AC circuit voltage equations related to **Chapter 9, Table 9** impedance values.

Objectives

- Calculate the resistance of various lengths of wire based upon the values of **Chapter 9, Table 8** of the *Code*.
- Calculate the voltage drop on a branch circuit using the values from **Chapter 9, Table 8** of the *Code*.
- Adjust voltage drop calculations for use with other than 75°C insulated conductors.
- Make calculations based upon the voltage drop equation when the wire size, ampacity, or length of a circuit is unknown.
- Apply various voltage drop equations to both single-phase and 3-phase circuits.
- Calculate the AC voltage drop of branch circuits and motor circuits using the values from **Chapter 9, Table 9** of the *Code*.

Chapter 6

Table of Contents

6.1 Chapter 9, Table 8 for Resistance of Wire

The voltage drop in an electrical circuit depends on the resistance which an electrical conductor offers to the flow of current in the circuit. The resistance of wire will change according to the temperature.

Ampacity **Table 310.15(B)(16)** lists three insulation temperatures: 60°C, 75°C, and 90°C. These are the insulation temperature ratings recognized by the *Code* for conductors rated 0-2000 volts. **Chapter 9, Table 8** lists the DC resistance of wire at 75°C. This would apply only to those conductors listed in the 75°C column of **Table 310.15(B)(16)**. **Chapter 9, Table 8** is the key for *Code* calculations of wire resistance and voltage drop. The resistance values in **Table 8** are given as DC resistance values. For all practical purposes, these same values are used for AC conductors and will give a reasonably accurate calculation of the voltage drop. **Chapter 9, Table 9** gives AC resistance and reactance for AC circuits with a power factor of 85%. The values in this table are based upon the Neher-McGrath conductor resistance calculations. **Table 9** is one source for computer-based electrical calculation programs used by design engineers for commercial and industrial installations where a conservative calculation is desired. However, **Table 8** is generally used for field-based calculations to determine approximate voltage drop in AC circuits.

The heading for the right-hand column of **Table 8** is "Direct-Current Resistance at 75°C (167°F)." The subheadings list copper and aluminum. Copper has two subheadings: uncoated and coated. When electrical conductors are insulated with natural rubber, a reaction occurs between the sulfur in the vulcanized rubber and the copper. Electrical conductors were coated with tin to protect against that reaction. Since there is no reaction between other insulation and copper or aluminum, wire using plastic insulation is not coated. All 600-volt building wire installed today has non-rubber based insulation and is uncoated.

Note the number of decimal places used to indicate the ohms per thousand feet. It indicates that decimals for ohms per 1,000 feet should not be rounded off when calculating voltage drop. Conversely, volts, current, and length round off to the second decimal place and circular mils round off to whole numbers.

6.2 Resistance Textbook Equation: *Code* Book Equation

Theory textbook equation for resistance:

$$R = \frac{k \times L}{A}$$

Or:

$$R = \frac{k \times L}{\text{cmil}}$$

Where:

R = resistance of the wire

k = resistance per mil foot of the material under consideration at a given temperature; usually based on 20°C

L = length of the wire

A = area of the wire expressed in circular mils

cmil = circular mil area of the wire

When the textbook equation is converted to *Code* book values, the following apply:

1. The resistance of the wire will increase as k is increased because the temperature of the conductor under consideration is increased from 20°C to 75°C. The temperature used is different.
2. The resistance of the wire will decrease as the cmil area is increased. The constant k remains the same.
3. The resistance of the wire is increased as the length of the conductor increases. The constant k remains the same.

Industry	Avg. Downtime Cost ($/hr)
Brokerage	$6,450,000
Credit Card	$2,600,000
Pay-Per-View	$150,000
Home Shopping	$113,000
Catalog Sales	$90,000
Airline Reservations	$90,000
Telephone Ticket Sales	$72,000
Cellular Communications	$41,000
Package Shipping	$28,000
ATM Fees	$14,400

Average cost of downtime for different industries

This is why the IBEW and NECA care about power quality.

6.3 Resistance Equation Based upon *Code* Book Values

This equation is used for calculating the resistance of wire using the *Code* resistance values given in **Chapter 9, Table 8**:

$$R = \frac{DC\ resistance \times L}{1{,}000}$$

Where:

R = resistance of the wire

DC = direct current

DC *resistance* = resistance of 1,000 ft of wire (as given in Chapter 9, Table 8)

L = length of the wire in feet

1,000 = converting ohms per 1,000 ft to ohms per foot

Direct-current resistance in the equation contains the *k* and the cmil area expressed in the theory textbook equation and is based upon a temperature of 75°C.

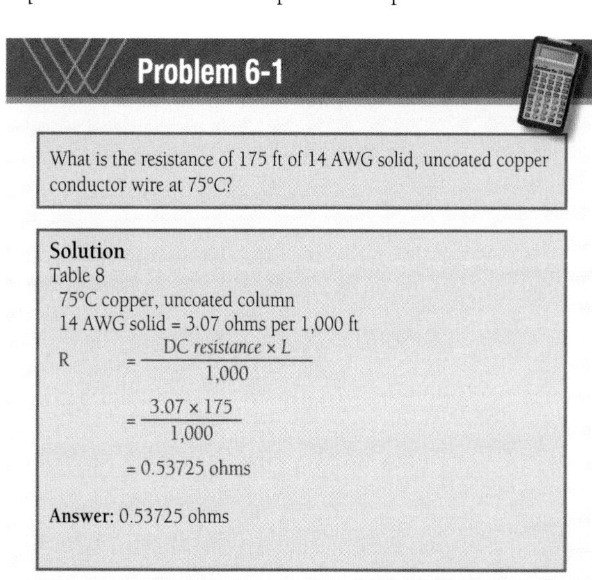

Problem 6-1

What is the resistance of 175 ft of 14 AWG solid, uncoated copper conductor wire at 75°C?

Solution

Table 8

75°C copper, uncoated column

14 AWG solid = 3.07 ohms per 1,000 ft

$$R = \frac{DC\ resistance \times L}{1{,}000}$$

$$= \frac{3.07 \times 175}{1{,}000}$$

$$= 0.53725 \text{ ohms}$$

Answer: 0.53725 ohms

Problem 6-2

What is the resistance of 200 ft of 14 AWG THWN solid copper conductor wire?

Solution

Table 310.104(A) or Table 310.15(B)(16)

THWN = 75°C wire

THWN indicates thermoplastic insulation; therefore, it is uncoated.

Table 8

75°C copper, uncoated column

14 AWG solid = 3.07 ohms per 1,000 ft

$$R = \frac{DC\ resistance \times L}{1{,}000}$$

$$= \frac{3.07 \times 200}{1{,}000}$$

$$= 0.614 \text{ ohms}$$

Answer: 0.614 ohms

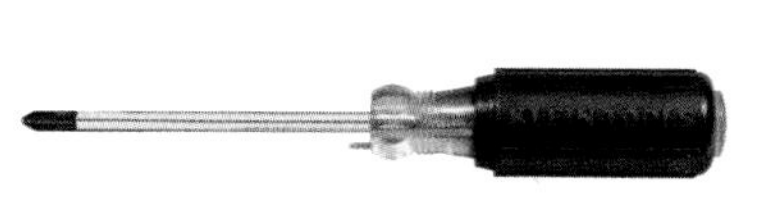

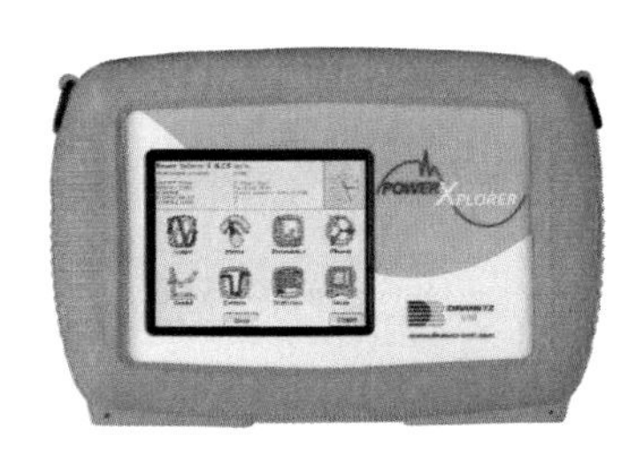

Knowledge of and experience with each tool enables the user to select the correct tool and use the correct procedure for determining a power quality source or event.

Financial loss statistics according to the *NJATC Power Quality Analysis* textbook:

1. $50 billion per year in the USA is lost as a result of power quality breakdown.
2. A manufacturing company lost more than $3 million in one day in Silicon Valley when the "lights went out."
3. "A voltage sag in a paper mill can waste a whole day of production - $250,000 loss."
4. Costs of power outages at Carnegie Mellon University are estimated at $5 million to $15 million annually.

6.4 Resistance Equation Correction Factor

The resistance of wire is directly proportional to the ambient heat. As the maximum operating temperature is decreased, the resistance of a conductor will decrease. The resistance of fully heated 60°C conductors will be less than the resistance of fully heated 75°C conductors.

Table 8 is based upon maximum operating temperature of conductors rated at 75°C and correlates very nicely with the *NEC* 75°C insulated conductors of **Table 310.104(A)**. Note the differences among 60°C, 75°C, and 90°C insulation ratings. There is a 15° change each time. Therefore, there is a correction factor which can be used for the 15° change in temperature. The correction factor is 1.05 for either copper or aluminum.

Problem 6-3

What is the resistance of 1,000 ft of 4 AWG uncoated copper conductor wire with 60°C insulation?

Solution
Table 8
75°C copper, uncoated column
1,000 ft of 4 AWG = 0.308 ohms
60°C is less than 75°C; resistance will decrease, so divide

$$\text{DC resistance at 60°C} = \frac{\text{DC resistance at 75°C}}{\textit{correction factor}} = \frac{0.308}{1.05} = 0.2933 \text{ ohms}$$

Answer: 0.2933 ohms

Problem 6-4

What is the resistance of 1,000 ft of 2 AWG aluminum conductor wire with 90°C insulation?

Solution
Table 8
75°C aluminum column
1,000 ft of 2 AWG = 0.319 ohms
90°C is more than 75°C; resistance will increase, so multiply

$$\text{DC resistance at 90°C} = \text{DC resistance at 75°C} \times 1.05 = 0.319 \times 1.05 = 0.33495 \text{ ohms}$$

Answer: 0.33495 ohms

6.5 Equation for Changing Resistance from 75°C to 60°C Wire

When the temperature drops, the resistance drops. Therefore, the 75°C equation is divided by the correction factor of 1.05.

$$R = \frac{\text{DC } \textit{resistance} \times L}{1{,}000 \times 1.05}$$

When the temperature is increased, the resistance will increase. Therefore, the 75°C equation is multiplied by the correction factor of 1.05.

$$R = \frac{\text{DC } \textit{resistance} \times L \times 1.05}{1{,}000}$$

Problem 6-5

What is the resistance of 400 ft of 90°C insulated, uncoated 6 AWG copper wire?

Solution
Table 8
75°C copper, uncoated column
1,000 ft of 6 AWG = 0.491 ohms

$$R = \frac{\text{DC } \textit{resistance} \times L \times 1.05}{1{,}000} = \frac{0.491 \times 400 \times 1.05}{1{,}000} = 0.20622 \text{ ohms}$$

Answer: 0.20622 ohms

Problem 6-6

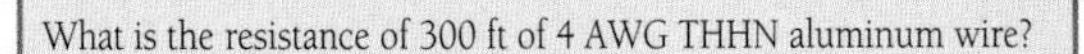

What is the resistance of 300 ft of 4 AWG THHN aluminum wire?

Solution
Table 310.104(A) or Table 310.15(B)(16)
THHN = 90°C wire
Table 8
75°C aluminum column
1,000 ft of 4 AWG = 0.508 ohms

$$R = \frac{\text{DC } \textit{resistance} \times L \times 1.05}{1{,}000} = \frac{0.508 \times 300 \times 1.05}{1{,}000} = 0.16002 \text{ ohms}$$

Answer: 0.16002 ohms

6.6 Developing the Voltage Drop Equation

Voltage is the amount of pressure necessary to force a given amount of current through the circuit conductor's resistance, although no productive work is done. It is measured in volts and identified with the symbol E_d or V_d.

The following is the development of the voltage drop equation using resistance values given in **Chapter 9, Table 8**:

Basic Ohm's Law Equation:

$$E_d = I \times R$$

Resistance Equation:

$$R = \frac{\text{DC } \textit{resistance} \times L}{1{,}000}$$

R in both equations stands for the same thing, resistance. Therefore, the R equation can be substituted for the R in the basic Ohm's Law equation (substituting equal for equal).

Substitute the resistance equation for the R in the basic Ohm's Law equation to solve for voltage drop.

$$E_d = \frac{I \times \text{DC } \textit{resistance} \times L}{1{,}000}$$

Because two conductors are needed to power an electrical load, the distance to the load is doubled. Hence, the length L is multiplied by 2 in the equation for single-phase circuits, and 1.73 is used for 3-phase circuits.

Appropriate lockout or tagout devices must be applied to equipment in order to properly isolate all energy sources.

6.6.1 Voltage Drop Equation

Single-Phase

$$V_d = \frac{\text{DC } \textit{resistance} \times I \times 2L}{1{,}000}$$

Three-Phase

$$V_d = \frac{\text{DC } \textit{resistance} \times I \times 1.73L}{1{,}000}$$

Where:

V_d = voltage drop expressed in volts

I = current in the conductor

DC *resistance* = the resistance value per 1,000 ft as listed in Table 8

$2L$ = twice the length of the circuit from the supply point to the single-phase load

$1.73L$ = length of the circuit from the supply point to the 3-phase load

1,000 = dividing by 1,000 converts the ohms per 1,000 ft to ohms per foot

The difference between these two formulas is that single-phase uses $2L$ where as 3-phase uses $1.73L$.

Problem 6-7

Calculate the voltage drop on a 60 ft, 120-volt, single-phase branch circuit of 14 AWG THWN uncoated solid copper conductor wire carrying 12 amperes.

Solution

Table 310.104(A) or Table 310.15(B)(16)

THWN = 75°C

Table 8

75°C copper, uncoated column

14 AWG = 3.07 ohms per 1,000 ft

$$V_d = \frac{\text{DC } \textit{resistance} \times I \times 2L}{1{,}000}$$

$$= \frac{3.07 \times 12 \times 2 \times 60}{1{,}000}$$

$$= 4.42 \text{ volts}$$

Answer: 4.42 volts

6.6.2 Voltage Drop and Line Loss

Voltage drop results in a loss of power in the line called "line loss" because it does no useful work. It is expressed in watts (W).

Single-Phase

$$W = V_d \times I$$

Three-Phase

$$W = V_d \times I \times 1.73$$

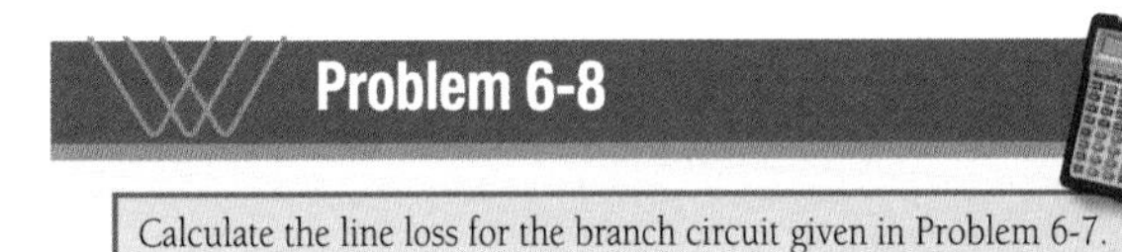

Problem 6-8

Calculate the line loss for the branch circuit given in Problem 6-7.

Solution

$W = V_d \times I$
$= 4.42 \times 12$
$= 53.04$ watts

Answer: 53.04 watts

Problem 6-9

Calculate the voltage drop on a 100 ft, 480-volt, 3-phase branch circuit carrying 60 amperes and using 6 AWG THWN copper conductors.

Solution

Table 310.104(A) or Table 310.15(B)(16)
THWN = 75°C uncoated
Table 8
75°C copper uncoated
6 AWG = 0.491 ohms per 1,000 ft

$$V_d = \frac{\text{DC resistance} \times I \times 1.73L}{1{,}000}$$

$$= \frac{0.491 \times 60 \times 1.73 \times 100}{1{,}000}$$

$= 5.097$ volts

Answer: 5.097 volts

Problem 6-10

Calculate the line loss for the branch circuit given in Problem 6-9.

Solution

$W = V_d \times I \times 1.73$
$= 5.097 \times 60 \times 1.73$
$= 529.07$ watts

Answer: 529.07 watts

6.6.3 Voltage Drop in Aluminum Conductors

The basic voltage drop equation applies to copper or aluminum. The DC resistance per 1,000 ft changes, but the equation remains the same.

Problem 6-11

Calculate the voltage drop on a 3-phase, 240-volt branch circuit of 1 AWG THWN aluminum carrying 90 amperes a distance of 150 ft.

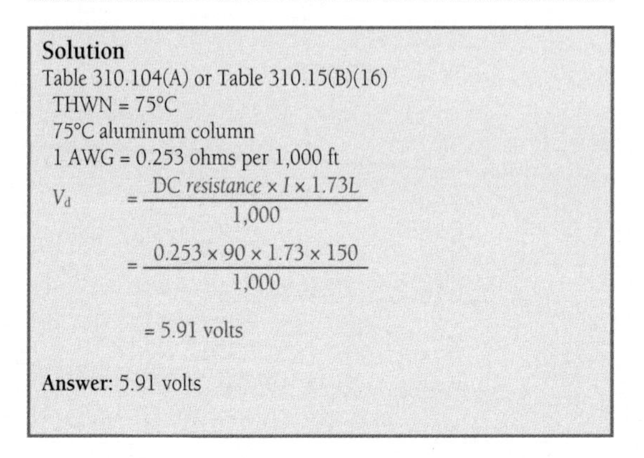

Solution

Table 310.104(A) or Table 310.15(B)(16)
THWN = 75°C
75°C aluminum column
1 AWG = 0.253 ohms per 1,000 ft

$$V_d = \frac{\text{DC resistance} \times I \times 1.73L}{1{,}000}$$

$$= \frac{0.253 \times 90 \times 1.73 \times 150}{1{,}000}$$

$= 5.91$ volts

Answer: 5.91 volts

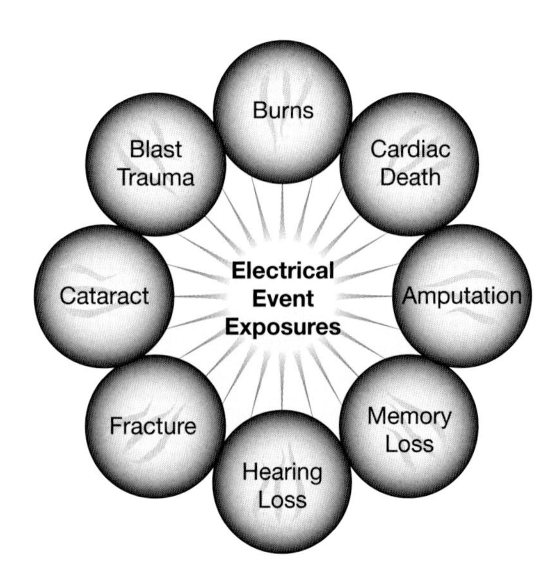

The PPE requirements of NFPA 70E do not address protection from physical trauma other than exposure to the thermal effects of arc flash. These circles identify the different types of trauma that result from injuries related to arc flash and arc blast.

6.6.4 Voltage Drop and Temperature Correction Factor

Table 310.15(B)(16) has three temperature ratings and **Table 8** has one 75°C rating. Therefore, the temperature correction factor also is used in conjunction with the voltage drop equation.

For 60°C wire

$$V_d = \frac{\text{DC } \textit{resistance} \times I \times 2L}{1{,}000 \times 1.05}$$

For 90°C wire:

$$V_d = \frac{\text{DC } \textit{resistance} \times I \times 2L \times 1.05}{1{,}000}$$

For 3-phase, $2L$ is changed to $1.73L$.

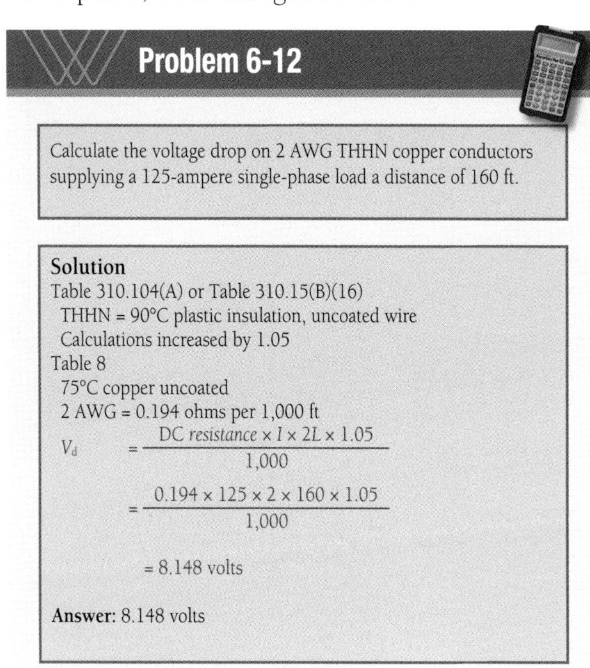

Problem 6-12

Calculate the voltage drop on 2 AWG THHN copper conductors supplying a 125-ampere single-phase load a distance of 160 ft.

Solution
Table 310.104(A) or Table 310.15(B)(16)
THHN = 90°C plastic insulation, uncoated wire
Calculations increased by 1.05
Table 8
75°C copper uncoated
2 AWG = 0.194 ohms per 1,000 ft

$$V_d = \frac{\text{DC } \textit{resistance} \times I \times 2L \times 1.05}{1{,}000}$$

$$= \frac{0.194 \times 125 \times 2 \times 160 \times 1.05}{1{,}000}$$

$$= 8.148 \text{ volts}$$

Answer: 8.148 volts

6.7 Using the Voltage Drop Equation

210.19(A), Informational Note No. 4 states that voltage drop for branch circuits is recommended not to exceed 3% where reasonable efficiency is expected. This requires calculating the wire size in circular mil area in order to limit the voltage drop.

6.7.1 Equations for Selecting Wire Size

When the ampacity, length, and desired voltage drop are known, the equation is transposed to solve for the minimum circular mil area of the wire.

Single-Phase

$$\text{cmil} = \frac{k \times I \times 2L}{V_d}$$

Three-phase

$$\text{cmil} = \frac{k \times I \times 1.73L}{V_d}$$

Where:

cmil = circular mil area of 75°C wire as listed in Table 8

k = approximate resistance value per mil foot

I = line current

$2L$ = twice the length of a single-phase circuit

$1.73L$ = length of a 3-phase circuit

V_d = desired voltage drop

If equipment instructions include a minimum voltage, voltage drop calculations should always be done to ensure compliance with **110.3(B)**.

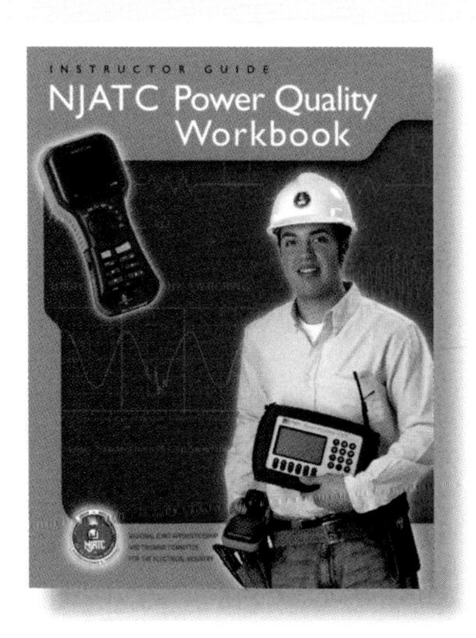

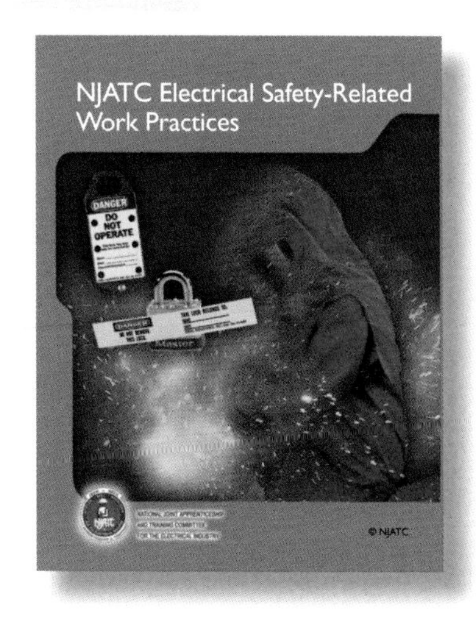

6.7.2 Developing the Values of *k* for Use with Table 8

The approximate value of *k* is used when calculating wire size, length of conductor, or maximum amperes when trying to remain within the *Code* recommended voltage drop percentage.

The approximate value of *k* used in the theory textbook equation is 10.4 for copper and 17 for aluminum at 20°C. **Table 8** is set up for 75°C; therefore, *k* must be a different value. The following *k* values are applicable to the **Table 8** based upon 75°C.

For copper:

k for 75°C = 12.9

For aluminum:

k for 75°C = 21.2

The following two examples illustrate how these two values are calculated:

For additional information, visit qr.njatcdb.org Item #1036

Examples

Determining *k* for Copper

Using this circuit, first solve for voltage drop
300 ft, 6 AWG copper uncoated, 60 amps, single-phase

Table 8

DC resistance, 6 AWG uncoated copper = 0.491 ohms per 1,000 ft

$$V_d = \frac{\text{DC resistance} \times I \times 2L}{1{,}000}$$

$$= \frac{0.491 \times 60 \times 2 \times 300}{1{,}000}$$

$$= 17.676 \text{ volts}$$

Textbook equation:

$$V_d = \frac{k \times I \times 2L}{cmil}$$

Solving for *k*:

$$k = \frac{V_d \times cmil}{I \times 2L}$$

Using values from this circuit, next solve for *k* value for copper

Table 8

6 AWG copper = 26,240 cmil

$$k = \frac{V_d \times cmil}{I \times 2L}$$

$$= \frac{17.676 \times 26{,}240}{60 \times 2 \times 300}$$

$$= 12.883 \text{ or } 12.9 \text{ ohms}$$

Answer: *k* = 12.9 for copper

Determining *k* for Aluminum

Using this circuit, first solve for voltage drop
300 ft, 6 AWG aluminum, 60 amps, single-phase

Table 8

DC resistance, 6 AWG aluminum = 0.808 ohms per 1,000 ft

$$V_d = \frac{\text{DC resistance} \times I \times 2L}{1{,}000}$$

$$= \frac{0.808 \times 60 \times 2 \times 300}{1{,}000}$$

$$= 29.088 \text{ volts}$$

Using values from this circuit, next solve for *k* value for aluminum

Table 8

6 AWG aluminum = 26,240 cmil

$$k = \frac{V_d \times cmil}{I \times 2L}$$

$$= \frac{29.088 \times 26{,}240}{60 \times 2 \times 300}$$

$$= 21.201 \text{ or } 21.2 \text{ ohms}$$

Answer: *k* = 21.2 for aluminum

The Success in the Workplace series explores critical conduct issues through interactive video scenarios. Each program in the series highlights the attitudes that lead to a successful career in the electrical industry.

Problem 6-13

Determine the size of THWN copper conductors needed for a branch circuit supplying a 200-ampere, single-phase, 240-volt load a distance of 150 ft, with the voltage drop held within the recommended *Code* limits where reasonable efficiency is expected. Does this wire size comply with Table 310.15(B)(16)?

Solution

210.19(A)(1) Informational Note No. 4

3% V_d recommended

$V_d = V \text{ supply} \times 3\%$

$= 240 \times 0.03$

$= 7.2 \text{ volts}$

$$\text{cmil} = \frac{k \times I \times 2L}{V_d}$$

$$= \frac{12.9 \times 200 \times 2 \times 150}{7.2}$$

$= 107{,}500$ cmil minimum

Table 8

Select next larger wire size

133,100 cmils = 2/0 AWG copper

Table 310.15(B)(16) at 200 amps requires 3/0 AWG THWN copper

Answer: 3/0 AWG THWN, copper

Problem 6-14

Determine the THWN aluminum wire size needed for branch conductors supplying a 115-ampere, 208-volt, 3-phase load a distance of 130 ft with the voltage drop held within the *Code* recommendations where reasonable efficiency is expected.

Solution

210.19(A)(1) Informational Note No. 4

3% V_d recommended

$V_d = V \text{ supply} \times 3\%$

$= 208 \times 0.03$

$= 6.24 \text{ volts}$

$$\text{cmil} = \frac{k \times I \times 1.73L}{V_d}$$

$$= \frac{21.2 \times 115 \times 1.73 \times 130}{6.24}$$

$= 87{,}870$ cmil

Table 8

87,870 cmil = 1/0 AWG aluminum

Answer: 1/0 AWG aluminum

What single electrical activity was identified as resulting in 24% of all electrical related injuries?

Troubleshooting

6.7.3 Solving for Length (*L*)

When the wire size, voltage drop, and ampacity are known, the equation can be transposed to solve for length (L).

Single-Phase

$$L = \frac{\text{cmil} \times V_d}{2\,k \times I}$$

Three-phase

$$L = \frac{\text{cmil} \times V_d}{1.73\,k \times I}$$

Problem 6-15

A 120 volt branch circuit is to be installed with two 6 AWG THWN aluminum conductors serving a load with 50 amperes of resistance. Determine the maximum length of the conductors when the voltage drop is held within a 3% limit.

Solution

210.19(A)(1) Informational Note No. 4

3% V_d recommended

$V_d = V \text{ supply} \times 3\%$

$= 120 \times 0.03$

$= 3.6 \text{ volts}$

Table 8

cmil column

6 AWG aluminum = 26,240 cmil

$$L = \frac{\text{cmil} \times V_d}{2k \times I}$$

$$= \frac{26{,}240 \times 3.6}{2 \times 21.2 \times 50}$$

$= 44.56$ ft

Answer: 44.56 ft

This is one example of a catastrophic arc flash.

Problem 6-16

A 480-volt branch circuit is to be installed with three 1/0 AWG THWN copper conductors serving a 140-ampere, 3-phase load. Determine the maximum length of the conductors, with the voltage drop not exceeding 3%.

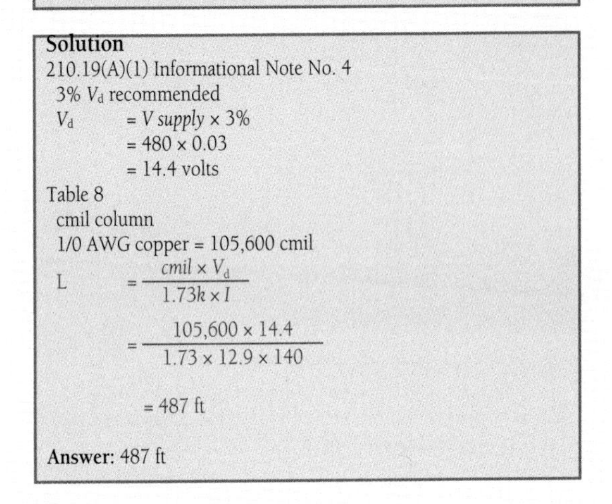

Solution

210.19(A)(1) Informational Note No. 4

3% V_d recommended

V_d = $V\ supply \times 3\%$

= 480×0.03

= 14.4 volts

Table 8

cmil column

1/0 AWG copper = 105,600 cmil

$$L = \frac{cmil \times V_d}{1.73k \times I}$$

$$= \frac{105{,}600 \times 14.4}{1.73 \times 12.9 \times 140}$$

= 487 ft

Answer: 487 ft

6.7.4 Solving for Current (*I*)

The same equation can be transposed to solve for the maximum current when the wire size, length, and desired voltage drop are known.

Single-Phase

$$I = \frac{\text{cmil} \times V_d}{k \times 2L}$$

Three-phase

$$I = \frac{\text{cmil} \times V_d}{k \times 1.73L}$$

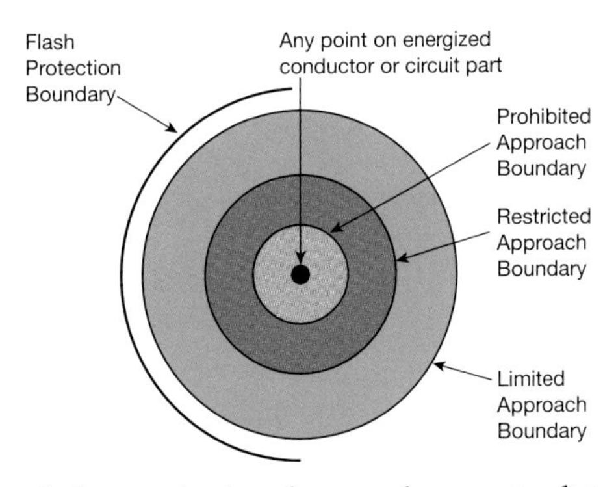

An arc flash protection boundary may be a greater distance from the exposed energized electrical conductors or circuit parts than the limited approach boundary.

Problem 6-17

Determine the maximum load for a 220-volt, 3-phase circuit with 10 AWG THWN stranded copper conductors installed a distance of 160 ft with the voltage drop held within *Code* recommendations.

Solution

210.19(A)(1) Informational Note No. 4

3% V_d recommended

V_d = $V\ supply \times 3\%$

= 220×0.03

= 6.6 volts

Table 8

cmil column

10 AWG copper stranded = 10,380 cmil

$$I = \frac{cmil \times V_d}{k \times 1.73L}$$

$$= \frac{10{,}380 \times 6.6}{12.9 \times 1.73 \times 160}$$

= 19.19 amps

Answer: 19.19 amperes

Problem 6-18

Determine the maximum load for a 115-volt, single-phase circuit with 8 AWG THWN solid aluminum conductors installed a distance of 120 ft, with the voltage drop not exceeding 3%.

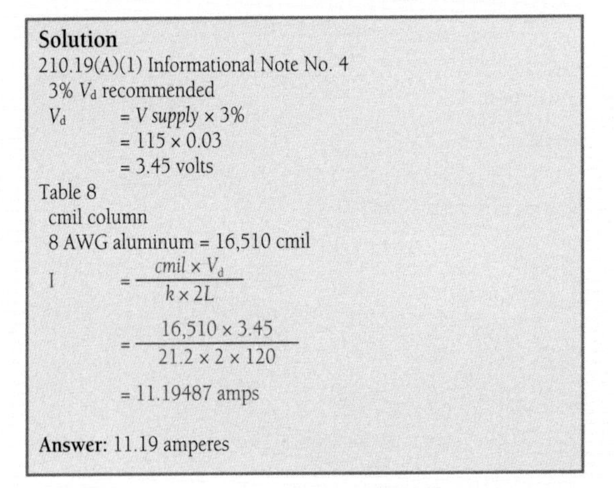

Solution

210.19(A)(1) Informational Note No. 4

3% V_d recommended

V_d = $V\ supply \times 3\%$

= 115×0.03

= 3.45 volts

Table 8

cmil column

8 AWG aluminum = 16,510 cmil

$$I = \frac{cmil \times V_d}{k \times 2L}$$

$$= \frac{16{,}510 \times 3.45}{21.2 \times 2 \times 120}$$

= 11.19487 amps

Answer: 11.19 amperes

In summary, the voltage drop equation can be used to select the wire size, determine k and using Table 8, calculate the circuit length, and find the circuit amperes.

6.8 Voltage Drop for Feeders and Branch Circuits

According to **210.19(A) Informational Note No. 4**, the total voltage drop on a branch circuit and feeder must not exceed 5% where reasonable efficiency is expected.

Problem 6-19

An installation on a 480-volt 3-phase system consists a 4/0 THWN copper feeder supplying 225 amperes to a control center located 300 ft from the service. A 3-phase branch circuit of 2 AWG THWN copper carries 105 amperes to a 3-phase load located 125 ft from the control center.

1. What is the voltage drop on the feeder?
2. What is the voltage drop on the branch circuit (BC)?
3. What is the total voltage drop to the load?
4. What is the voltage drop percentage for feeder and branch circuit?

Solution – Calculation 1
Feeder voltage drop
Table 8
cmil column
4/0 AWG = 211,600 cmil

$$V_d = \frac{k \times I \times 1.73L}{cmil}$$

$$= \frac{12.9 \times 225 \times 1.73 \times 300}{211{,}600}$$

= 7.12 volts

Answer : 7.12 volts feeder voltage drop

Solution – Calculation 2
Branch circuit voltage
Table 8
cmil column
2 AWG = 66,360

$$V_d = \frac{k \times I \times 2L}{cmil}$$

$$= \frac{12.9 \times 105 \times 1.73 \times 125}{66{,}360}$$

= 4.41 volts

Answer: 4.41 volts

Solution – Calculation 3
Total circuit voltage

V_d = V_d (feeder) + V_d (BC)
= 7.12 + 4.41

Answer = 11.53 volts

Solution – Calculation 4
Percent for total circuit

$$V_d = \frac{V_d}{E\ (supply)}$$

$$= \frac{11.53}{480}$$

= 0.024 or 2.4%

Answer: 2.4% for the total circuit

6.9 Using Alternate Methods to Solve for Voltage Drop

An alternate method can also be used for calculating voltage drop.

6.9.1 Varying the Value of *k*

Instead of the correction factor of 1.05 being used, the value of *k* is given. The given value will take into consideration the material of the conductor and the temperature.

For example: *k* for copper could possibly be any number between 12.0 and 14.0; *k* for aluminum could possibly be between 20.0 and 23.0.

Problem 6-20

What is the voltage drop on a single-phase, 240-volt circuit carrying 80 amperes on a 4 AWG copper conductor a distance of 200 ft? (k = 13.0)

Solution
Table 8
cmil column
4 AWG = 41,740 cmil

$$V_d = \frac{k \times I \times 2L}{cmil}$$

$$= \frac{13.0 \times 80 \times 2 \times 200}{41{,}740}$$

= 9.96 volts

Answer: 9.96 volts

Problem 6-21

What is the voltage drop on a 3-phase, 240-volt circuit carrying 45 amperes on an 8 AWG aluminum conductor a distance of 175 ft? (k = 21.2)

Solution
Table 8
cmil column
8 AWG = 16,510 cmil

$$V_d = \frac{k \times I \times 1.73L}{cmil}$$

$$= \frac{21.2 \times 45 \times 1.73 \times 175}{16{,}510}$$

= 17.49 volts

Answer: 17.49 volts

Information

When k has a given value, the equation can be transposed to solve for conductor size, maximum current, or length for a given or recommended voltage drop.

Single-Phase

$$\text{cmil} = \frac{k \times I \times 2L}{V_d}$$

Three-phase

$$\text{cmil} = \frac{k \times I \times 1.73L}{V_d}$$

Problem 6-22

Determine the correct wire size for a THWN copper conductor needed to carry a single-phase, 60- ampere load a distance of 200 ft and not exceed a 3% drop for a 240 volt circuit. (k = 12.8)

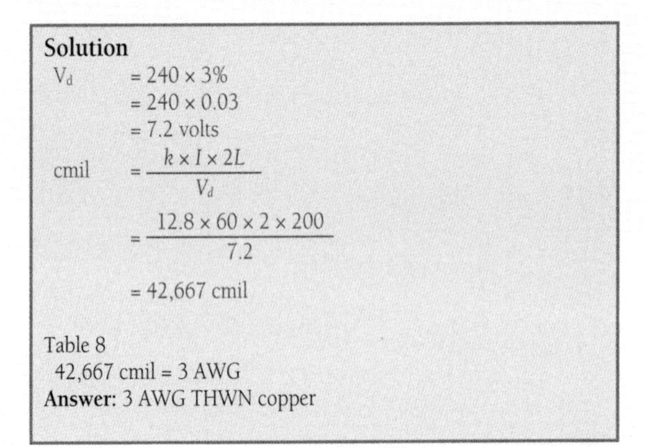

Solution

$V_d = 240 \times 3\%$
$= 240 \times 0.03$
$= 7.2$ volts

$\text{cmil} = \frac{k \times I \times 2L}{V_d}$
$= \frac{12.8 \times 60 \times 2 \times 200}{7.2}$
$= 42{,}667$ cmil

Table 8
42,667 cmil = 3 AWG
Answer: 3 AWG THWN copper

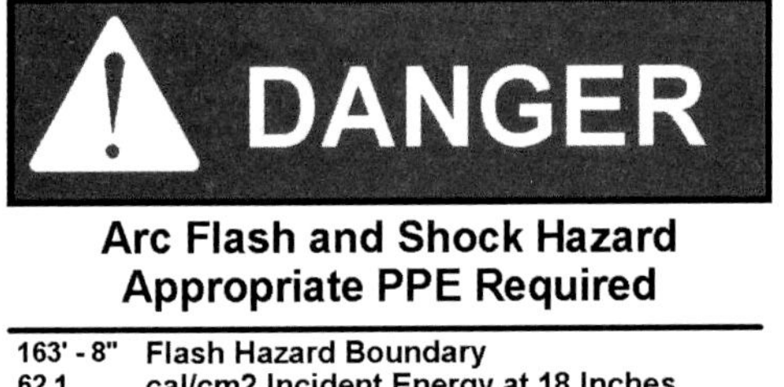

An example of an arc flash and shock warning label

6.9.2 Using Chapter 9, Table 9

Chapter 9, Table 9 may also be used to perform AC voltage drop calculations. This table provides a comprehensive list of impedance (Z) values for AC circuits at 85% power factor using 75°C insulated copper or aluminum conductors installed in PVC, steel, or aluminum conduits. According to the table footnote, adjustments may be made for different power factors as well. The final voltage drop calculated values results are approximate, as are most previous voltage drop calculations.

Note that the header of **Table 9** indicates that the values given are either ohms to neutral per kilometer or ohms to neutral per 1,000 feet. Care must be exercised to use consistent units of measure. English units (feet or ft) are the choice for *Code* calculations.

To find the impedance of a circuit whose length is known in feet, use:

$$Z = \frac{\text{length (ft)} \times \text{table value}}{1{,}000 \text{ (ft)}}$$

Voltage drop is calculated with Ohm's Law:

$$V_d = I \times Z$$

Most applications will require a calculation of voltage drop using phase-to-phase voltage. To do so, multiply the phase-to-neutral voltage drop by the square root of three (rounded to 1.732 in **Problem 6-23** and **Problem 6-24**).

Personal Protective Equipment (PPE) is required when working on energized circuits. The level of PPE is determined by the possible exposed hazard.

Problem 6-23

A feeder has a 320-ampere continuous load. The system source is 480 volts, 3-phase at the supply 400-ampere circuit breaker. The feeder is a 3 in. steel conduit with three 500 kcmil THWN copper conductors operating at their rating of 75°C. The circuit length is 300 ft, and the power factor is 85%. Using Table 9, determine the approximate voltage drop of this circuit.

Solution – Calculation 1
Line-to-neutral circuit impedance (*Z*)
Table 9
Effective *Z* at 0.85 PF for uncoated copper wires
Steel conduit column
500 kcmil copper wire = 0.050 ohms per 1,000 ft

$$Z = \frac{\textit{circuit length} \text{ (ft)} \times \textit{table value} \text{ (ohms)}}{1{,}000 \text{ ft}}$$

$$= \frac{300 \times 0.050}{1{,}000}$$

= 0.015 ohms (line-to-neutral)

Answer: 0.015 ohms line-to-neutral circuit impedance (*Z*)

Solution – Calculation 2
Line-to-neutral voltage drop (V_d)
$V_d = I \times Z$
= 320 × 0.015
= 4.8 volts

Answer: 4.8 volts line-to-neutral voltage drop

Solution – Calculation 3
Line-to-line voltage drop (V_d)
V_d (line-to-line) = Vd (line-to-neutral) × 1.732
= 4.80 × 1.732
= 8.31 volt drop (line-to-line)

Answer: 8.31 volts line-to-line voltage drop

Solution – Calculation 4
Voltage drop as a percentage of the circuit voltage

$$Vd\% = \frac{V_d}{\textit{circuit voltage}} \times 100$$

$$= \frac{8.31}{480} \times 100$$

= 1.73%

Answer: 1.73% voltage drop

Solution – Calculation 5
Actual voltage present at the load
Voltage at load = *source voltage – voltage drop*
= 480 – 8.31
= 471.69 volts

Answer: 471.69 volts

Problem 6-24

A motor circuit supplies a 50-hp, 208-volt, 3-phase motor with an 85% power factor and draws a 143-ampere load. The system source is 208 volts, 3-phase at the motor control center and the motor circuit is protected by a 300-ampere circuit breaker. The feeder is a 2 in. steel conduit with three 3/0 AWG THWN copper conductors operating at their rating of 75°C. The circuit length is 150 ft. Using Table 9, determine the approximate voltage drop of this circuit.

Solution – Calculation 1
Line-to-neutral circuit impedance (*Z*)
Table 9
Effective *Z* at 0.85 PF for uncoated copper wires
Steel conduit column
3/0 AWG copper wire = 0.094 ohms per 1,000 ft

$$Z = \frac{\textit{circuit length} \text{ (ft)} \times \textit{table value} \text{ (ohms)}}{1{,}000 \text{ ft}}$$

$$= \frac{150 \times 0.094}{1{,}000}$$

= 0.0141 ohms (line-to-neutral)

Answer: 0.0141 ohms line-to-neutral circuit impedance (*Z*)

Solution – Calculation 2
Line-to-neutral voltage drop (V_d)
$V_d = I \times Z$
= 143 amps × 0.0141 ohms
= 2.016 volts

Answer: 2.016 volts line-to-neutral voltage drop

Solution – Calculation 3
Line-to-line voltage drop (V_d)
V_d (line-to-line) = V_d (line to neutral) × 1.732
= 2.016 volts × 1.732
= 3.49 volts

Answer: 3.49 volts line-to-line voltage drop

Solution – Calculation 4
Voltage drop as a percentage of the circuit voltage

$$V_d\% = \frac{V_d}{\textit{circuit voltage}} \times 100$$

$$= \frac{3.49}{208} \times 100$$

= 1.68%

Answer: 1.68% voltage drop

Solution – Calculation 5
Actual voltage present at the motor
Voltage at motor = *source voltage – voltage drop*
= 208 – 3.49
= 204.51 volts

Answer: 204.51 volts

Definitions and Terms

Conductor, Aluminum - According to **310.106(B)**, both solid and stranded aluminum conductors in sizes 12 AWG through 1,000 kcmil are required to be composed of AA-8000 series electrical grade aluminum alloy.

Conductor, Copper - Annealed copper is the material generally used as a conductor in insulated wire and cable. Only soft drawn copper is used for general inside wiring.

Conductor, Copper Coated - When annealed copper is directly covered by rubber insulation, the copper is coated and covered usually with tin or a lead alloy. This is done for the mutual protection of the copper and the rubber. The manufacture of rubber insulated conductors is generally considered as obsolete today. Rubber insulated (with coated) conductors are limited to existing installations only.

Voltage Drop - The difference between the value of source voltage and load voltage.

Summary

- Most calculations are based upon a basic DC circuit model using Ohm's Law, thereby keeping the math straight forward and relatively easy to learn.
- All circular conductors have a resistance, the values of which are clearly presented in **Chapter 9, Table 8**.
- As the temperature rating of a specific circuit conductor increases, its resistance also increases.
- There is a small difference in resistance between solid and stranded wires. But, there is a larger difference in resistance between copper and aluminum wires.
- The actual resistance of solid and stranded, copper and aluminum conductors used in building wire circuits are calculated using the basic Ohm's Law equation and the general values of resistance from **Chapter 9, Table 8**.
- Correction factors are developed and used to permit the resistance calculation of other *NEC* insulated conductors with differing temperature ratings such as 60°C and 90°C.
- After appropriate formulas are developed and common *NEC* parameters of acceptable voltage drop are explained, satisfactory minimum size circuit conductors are selected.
- Voltage drop solutions using the DC model are applied to a variety of 2-wire, 3–wire and 3-phase branch circuits, and feeder circuits.
- AC voltage drop solutions are demonstrated and solved using **Chapter 9, Table 9**.
- Both AC and DC circuits can achieve a reasonable efficiency by keeping voltage drop low.

Appliances

The fundamental calculation methods of **Article 220** are used in determining branch-circuit, feeder, and service loads as they relate to appliances. Also, specific calculation requirements of **Article 422** will be used as necessary. To a large extent, most of the calculations deal with household type appliances, but some calculations may be applied to commercial appliances.

Specific appliance calculations include electric ranges, electric ovens, counter-mounted cook units, household electric clothes dryers, storage-type electric water heaters, and a few other basic appliances. **Section 220.3** and **Table 220.3** of the *Code* provides a more detailed list of specialized applications that are in addition to, or modifications of, those within.

Branch circuits for household ranges must comply with **210.19(A)(3)**. Load calculations for household ranges are permitted according to **Table 220.55** and the accompanying **Note 4**. Load calculations for household dryers must comply with **Section 220.54**, but are permitted to use the demand factors of **Table 220.54**. For other than household kitchen equipment, the requirements of **Section 220.56** must be applied to commercial kitchen equipment and the demand factors of **Table 220.56** are permitted to be used for these calculations. Feeder and service neutral loads related to appliances are permitted to be reduced in accordance with **Section 220.61**. Both single-phase and 3-phase feeder demand calculations are demonstrated. Most feeder calculations are 120/240 volt single-phase, 3-wire, unless they are specifically noted to be 3-phase.

Objectives

- Calculate branch-circuit loads for household appliances such as ranges, wall-mounted ovens, counter-mounted cooking units, clothes dryers, storage-type water heaters, and kitchen waste disposers.
- Determine the proper size branch-circuit conductor for an individual appliance branch circuit.
- Determine the proper size overcurrent protective device for an individual appliance branch circuit.
- Calculate feeder demand loads and ampacities for appliance loads.
- Calculate neutral loads and conductor sizes for appliance loads.
- Calculate appliance service and feeder demand loads for one-family and multifamily dwelling units.
- Calculate appliance loads, branch-circuit conductor sizes, and overcurrent protective device sizes for commercial appliances.

Chapter 7

Table of Contents

7.1 Range Loads–Feeder Demands–Dwelling Units with 120/240 Volt, Single-Phase Service

Section 220.55 and **Table 220.55** are used for calculating loads for household electric ranges and other cooking appliances installed in dwelling units. This section and table cover the installation of any of the following household cooking appliances having a rating of over $1\frac{3}{4}$ kW:

1. Household electric ranges
2. Wall-mounted electric ovens
3. Counter-mounted cooking units

A rating of $1\frac{3}{4}$ or less would be considered an electrical appliance and not subject to this section or **Table 220.55**. Other household cooking appliances rated $1\frac{3}{4}$ kW or over include a warming oven, a broiler, a microwave oven, or similar cooking appliances.

The calculations derived from the use of **Table 220.55** are generally reserved for feeder loads. However, according to **Table 220.55 Note 4**, it is permissible to calculate the branch-circuit load of one range according to **Table 220.55**. Additional permissions are contained in **Note 4** as well.

220.61(B)(1) permits the feeder demand on the neutral to be 70% of the feeder demand on the ungrounded conductors (line or phase conductors).

For range load calculations, kVA and kW are considered to be the same. All feeder demands will be calculated for line and neutral conductors and will be given in VA.

The title of **Table 220.55** contains one instruction regarding the use of the table: Column C is to be used in all cases except as otherwise permitted in **Note 3**.

For the following illustrated problems, the service is 120/240 volts, 3-wire, single-phase, unless otherwise noted in the illustration. The use of single-phase ranges on a 3-phase system will be explained later in this chapter.

7.1.1 Ranges Not over 12 kW and of Unequal Rating

When the ranges are of unequal rating and no range is rated over 12 kW, the number of ranges can be counted

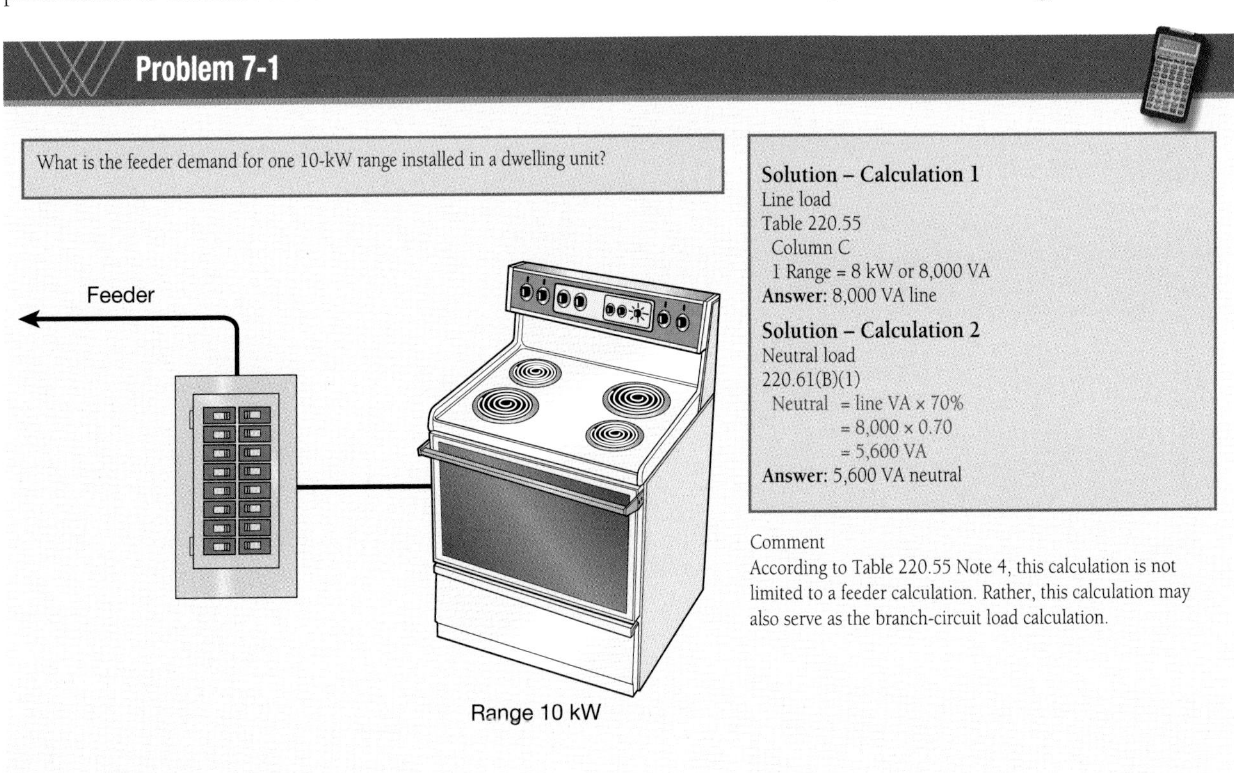

Problem 7-1

What is the feeder demand for one 10-kW range installed in a dwelling unit?

Solution – Calculation 1
Line load
Table 220.55
Column C
1 Range = 8 kW or 8,000 VA
Answer: 8,000 VA line

Solution – Calculation 2
Neutral load
220.61(B)(1)
Neutral = line VA × 70%
= 8,000 × 0.70
= 5,600 VA
Answer: 5,600 VA neutral

Comment
According to Table 220.55 Note 4, this calculation is not limited to a feeder calculation. Rather, this calculation may also serve as the branch-circuit load calculation.

and read from Column C. This is permitted by **Table 220.55 Note 2**. **Note 2** requires that when certain ranges of unequal ratings are installed, all ranges less than 12 kW are required to be counted as 12 kW ranges.

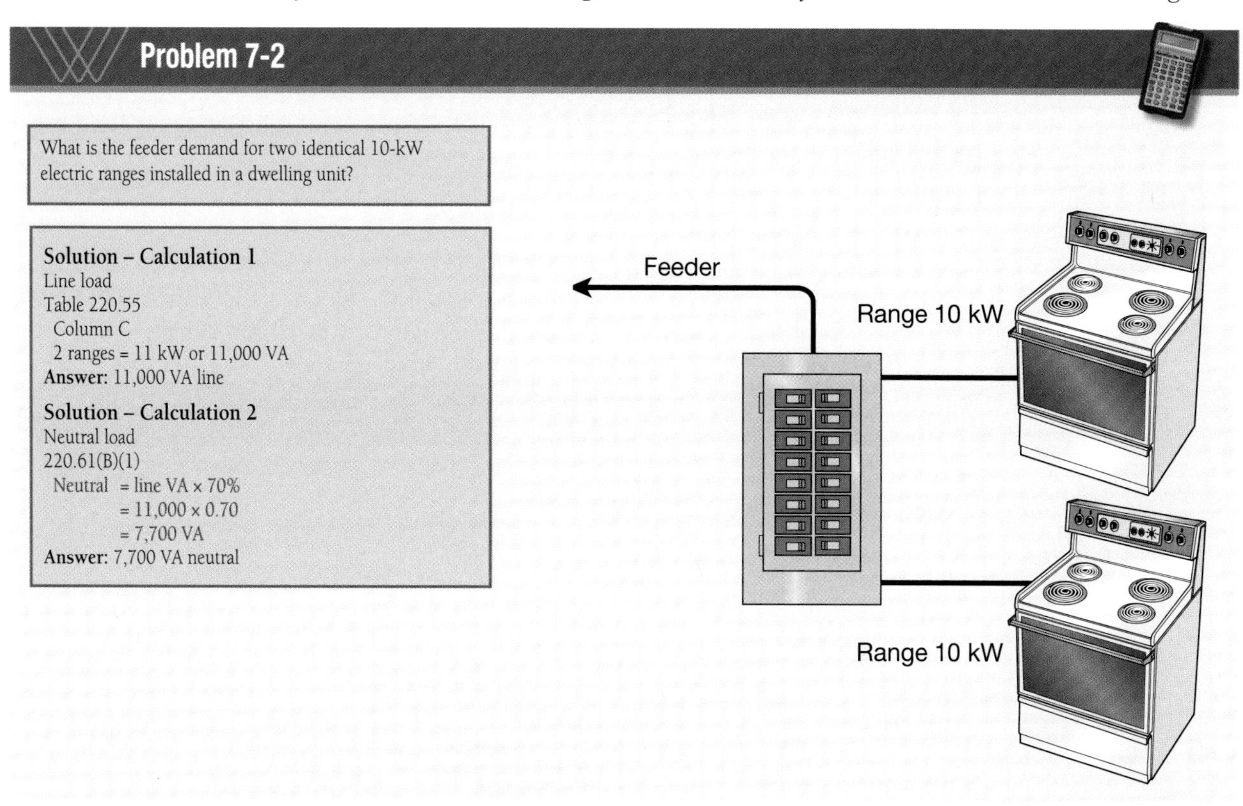

Problem 7-2

What is the feeder demand for two identical 10-kW electric ranges installed in a dwelling unit?

Solution – Calculation 1
Line load
Table 220.55
Column C
2 ranges = 11 kW or 11,000 VA
Answer: 11,000 VA line

Solution – Calculation 2
Neutral load
220.61(B)(1)
Neutral = line VA × 70%
= 11,000 × 0.70
= 7,700 VA
Answer: 7,700 VA neutral

Problem 7-3

What is the feeder demand for one 10-kW range and one 12-kW range installed in a dwelling unit?

Solution – Calculation 1
Line load
Table 220.55 Note 2
Two ranges unequal
Both between 8¾ kW and 12 kW
Treat as equal ranges
Table 220.55
Column C
2 ranges = 11 kW, or 11,000 VA
Answer: 11,000 VA line

Solution – Calculation 2
Neutral load
220.61(B)(1)
Neutral = line VA × 70%
= 11,000 × 0.70
= 7,700 VA
Answer: 7,700 VA neutral

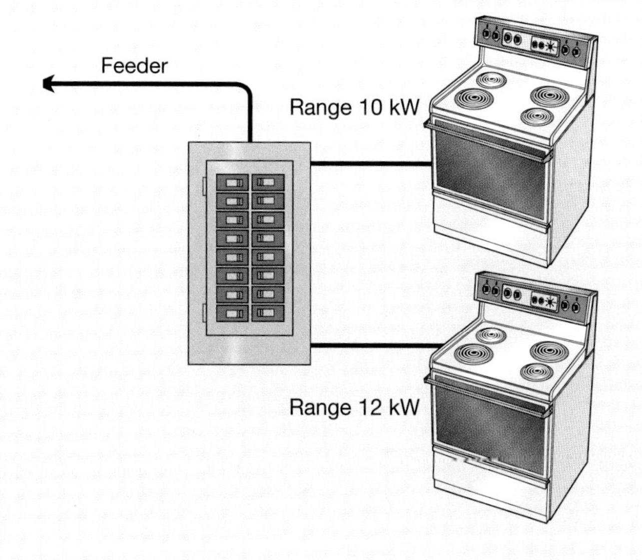

7.1.2 All Household Cooking Appliances over 1¾ kW through 8¾ kW

The following problem is based upon **Table 220.55 Note 3**. **Note 3** contains four instructions:

1. **Note 3** may be used in lieu of Column C. The word *may* means that Column C does not have to be used (both ways must be checked when trying to establish the smallest demand).
2. Add nameplate ratings of all cooking appliances rated between 1¾ kW and less than 3½ kW and multiply by percentage in Column A.
3. Add nameplate ratings of all cooking appliances rated from 3½ kW through 8¾ kW and multiply by percentage in Column B.
4. When the ranges are of unequal rating, and no range is rated over 12 kW, the number of ranges can just be counted and read from Column C.

Reason: **Note 2** requires that when ranges of unequal rating are installed, all ranges less than 12 kW are required to be counted as 12 kW ranges.

Problem 7-4

What is the feeder demand for two electric wall-mounted ovens installed in a dwelling unit, one at 2½ kW and the other at 3 kW?

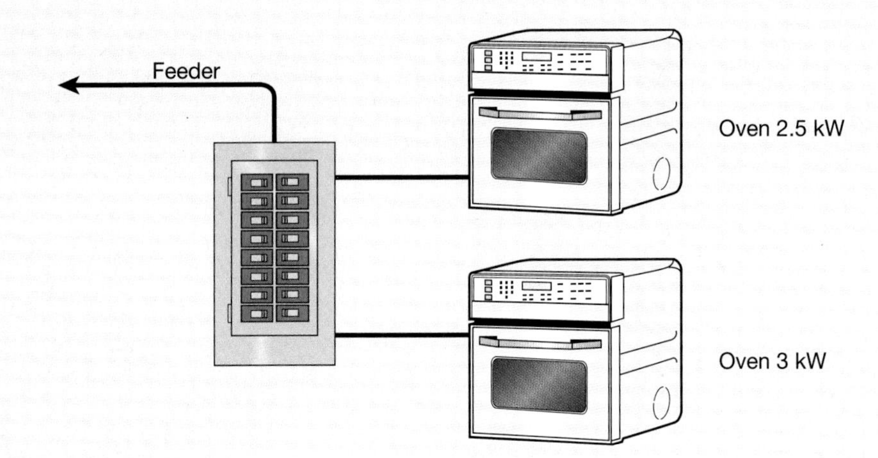

Solution – Calculation 1
Line load
Table 220.55 Note 3
Establish smallest demand
Column A calculation
Add nameplate ratings
2.5 + 3 = 5.5 kW
Table 220.55
Column A
2 units = 75%
5.5 kW x 0.75 = 4.125 kW
Column C calculation

Table 220.55
Column C
2 units = 11 kW
4.125 kW is smaller than 11 kW; use smallest demand
Line = 4.125 kW or 4,125 VA
Answer: 4,125 VA line

Solution – Calculation 2
Neutral load
220.61(B)(1)
Neutral = line VA x 70%
= 4,125 x 0.70
= 2,888 VA
Answer: 2,888 VA neutral

Information

An option given by **Note 3** permits the use of either Column B or Column C. When the minimum answer is requested, both options must be calculated and the smaller used.

Problem 7-5

What is the feeder demand for three 8-kW electric ranges installed in a dwelling unit?

Feeder

Range 8 kW

Range 8 kW

Range 8 kW

Solution – Calculation 1
Line load
Table 220.55 Note 3
Establish smallest demand
Column B calculation
Add nameplate ratings
8 + 8 + 8 = 24 kW
Table 220.55
Column B
3 ranges = 55%
24 kW × 0.55 = 13.2 kW or 13,200 VA
Column C calculation

Table 220.55
Column C
3 ranges under 12 kW = 14 kW
13.2 kW is smaller than 14 kW; use smallest demand
Answer: 13,200 VA line

Solution – Calculation 2
Neutral load
220.61(B)(1)
Neutral = line VA × 70%
= 13,200 × 0.70
= 9,240 VA
Answer: 9,240 VA neutral

Comment
Note 3 is permitted in lieu of Column C. As a comparison, if the three ranges had been read from Column C, the feeder demand would have been 14 kW. A lower demand load would be calculated by the use of Columns A or B when Note 3 is applied.

Information

Where the rating of the cooking appliances falls under both Columns A and B for a given number of appliances, calculate the demand factors for each column, then and add them together.

Problem 7-6

What are the line and neutral feeder demands for three 3-kW ovens and three 8-kW ranges on the same feeder?

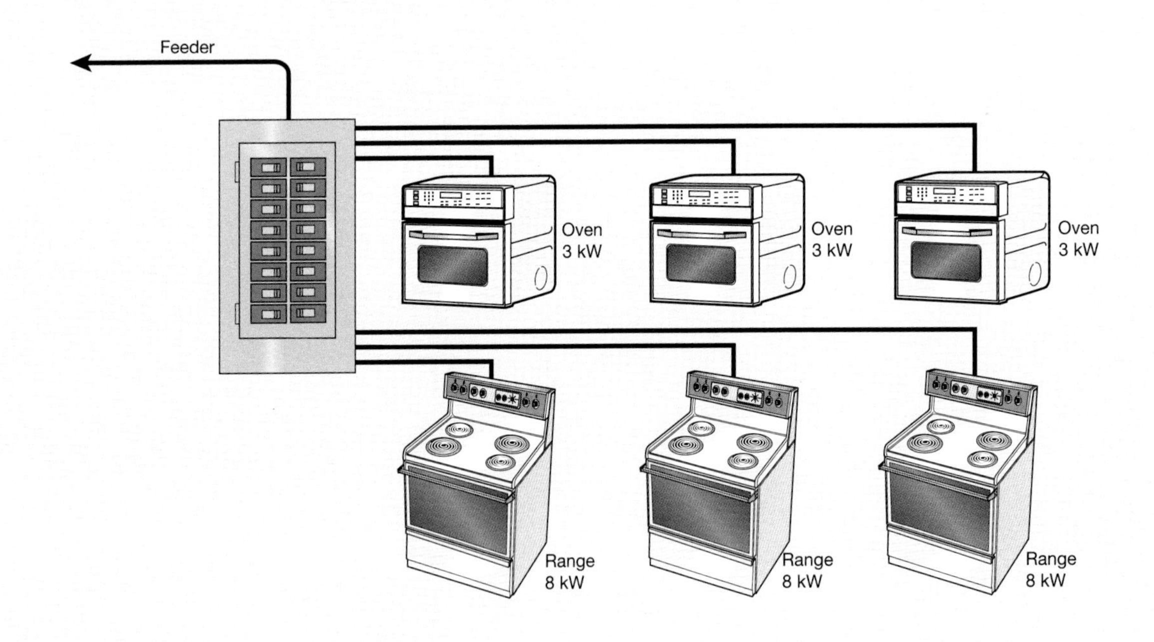

Solution – Calculation 1
3 kW oven line load
Table 220.55
Column A
3 units = 70%
3 × 3 kW × 0.70 = 6.3 kW
8 kW range line load
Table 220.55
Column B
3 units = 55%
3 × 8 kW × 0.55 = 13.2 kW
Line = 6.3 + 13.2
= 19.5 kW or 19,500 VA
Answer: 19,500 VA line

Solution – Calculation 2
Neutral load
220.61(B)(1)
Neutral = line VA × 70%
= 19,500 × 0.70
= 13,650 VA
Answer: 13,650 VA neutral

Comment
The use of Note 3 and Columns A and B is used for a multifamily dwelling where a large number of electrical cooking appliances rated 1¾ kW though 8¾ kW are to be installed.

7.1.3 Ranges in Multifamily Dwellings

Table 220.55 is applicable to range loads installed in multifamily dwellings.

Problem 7-7

A 28-unit multifamily dwelling has one 10-kW range installed in each unit. What is the feeder demand for the complex?

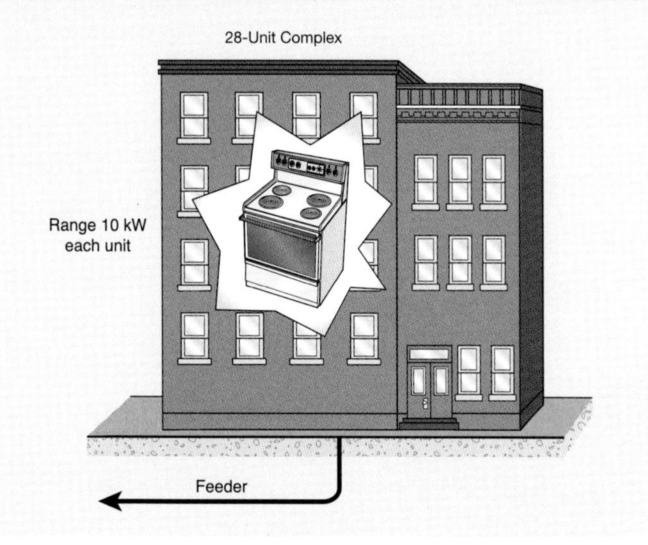

Solution – Calculation 1
Line load
Table 220.55
Column C
26–30 ranges = 15 kW + 1 kW for each range
Total kW = 15 kW + (28 × 1)
= 43 kW or 43,000 VA
Answer: 43,000 VA line

Solution – Calculation 2
Neutral load
220.61(B)(1)
Neutral = line VA × 70%
= 43,000 × 0.70
= 30,100 VA
Answer: 30,100 VA neutral

Problem 7-8

A 48-unit multifamily dwelling has a 12-kW range installed in each apartment. What is the feeder demand for the complex?

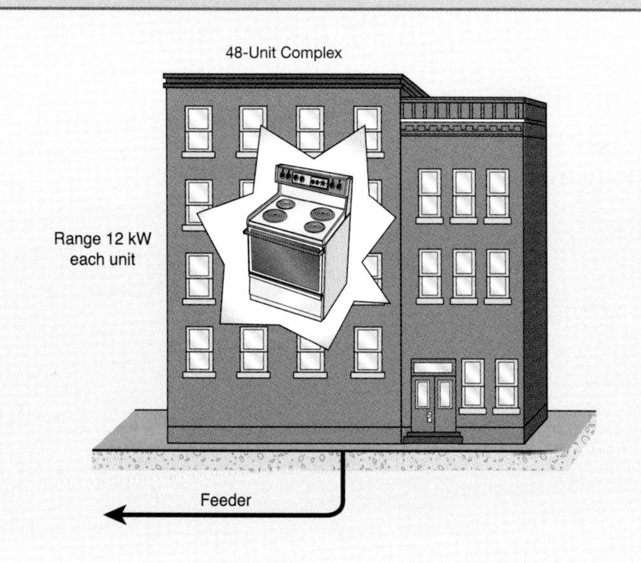

Solution – Calculation 1
Line load
Table 220.55
Column C
41–50 ranges = 25 kW + $\frac{3}{4}$ kW for each range
Total kW = 25 kW + (48 × 0.75)
= 25 + 36
= 61 kW or 61,000 VA
Answer: 61,000 VA line

Solution – Calculation 2
Neutral load
220.61(B)(1)
Neutral = line VA × 70%
= 61,000 × 0.70
= 42,700 VA
Answer: 42,700 VA neutral

7.1.4 Ranges Rated over 12 kW and Less Than 27 kW

Table 220.55 Note 1 covers individual ranges and ranges over 12 kW and less than 27 kW. The demand in Column C for the number of ranges is increased by 5% for each kW over 12 kW. According to **220.5(B)**, calculations are permitted to be rounded to the nearest whole ampere, with the decimal fraction smaller than 0.5 dropped. Generally, these solutions do not round or drop fractions less than 0.5 until the end of the solution or for the answer.

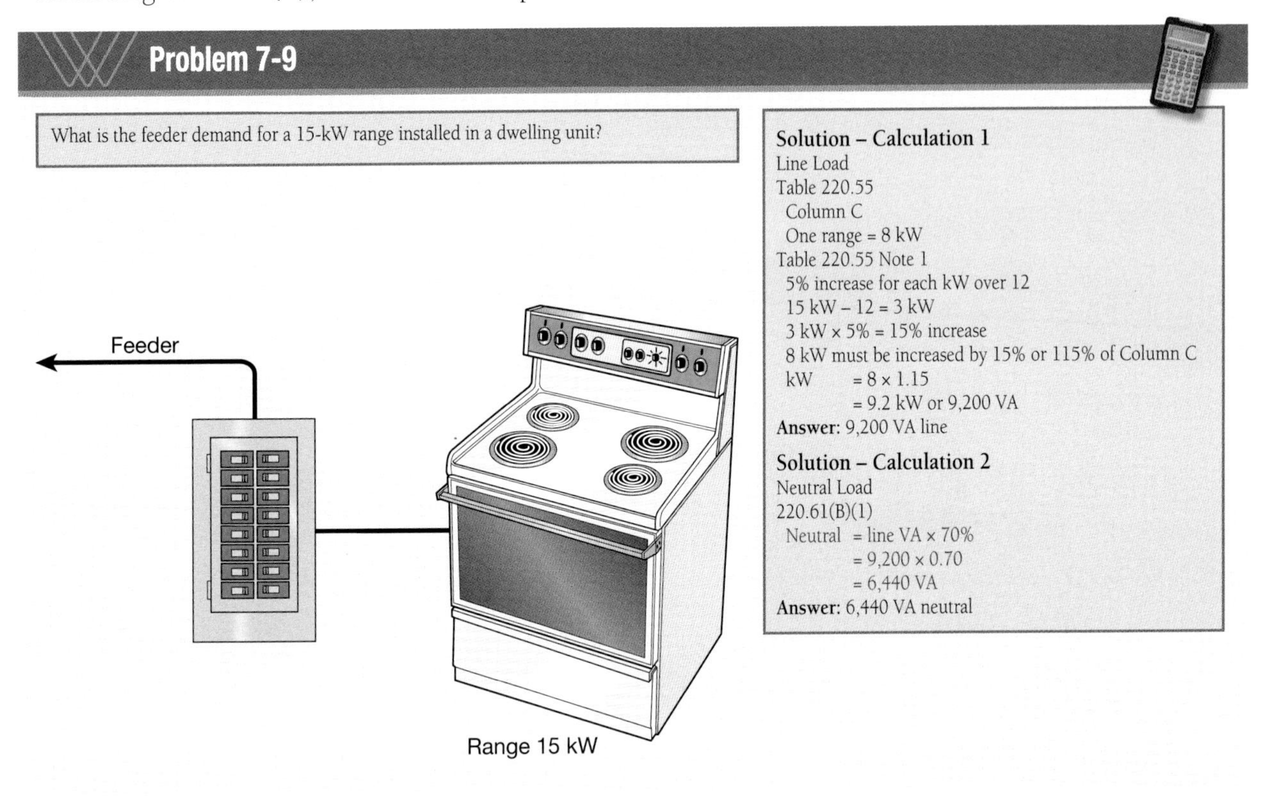

Problem 7-9

What is the feeder demand for a 15-kW range installed in a dwelling unit?

Solution – Calculation 1
Line Load
Table 220.55
Column C
One range = 8 kW
Table 220.55 Note 1
5% increase for each kW over 12
15 kW – 12 = 3 kW
3 kW x 5% = 15% increase
8 kW must be increased by 15% or 115% of Column C
kW = 8 x 1.15
= 9.2 kW or 9,200 VA
Answer: 9,200 VA line

Solution – Calculation 2
Neutral Load
220.61(B)(1)
Neutral = line VA x 70%
= 9,200 x 0.70
= 6,440 VA
Answer: 6,440 VA neutral

Problem 7-10

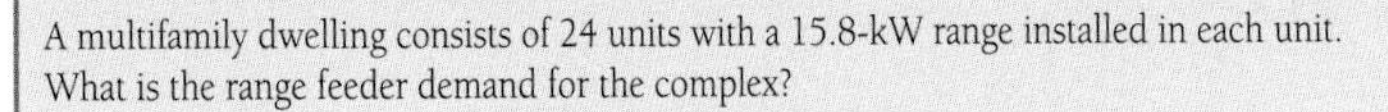
A multifamily dwelling consists of 24 units with a 15.8-kW range installed in each unit. What is the range feeder demand for the complex?

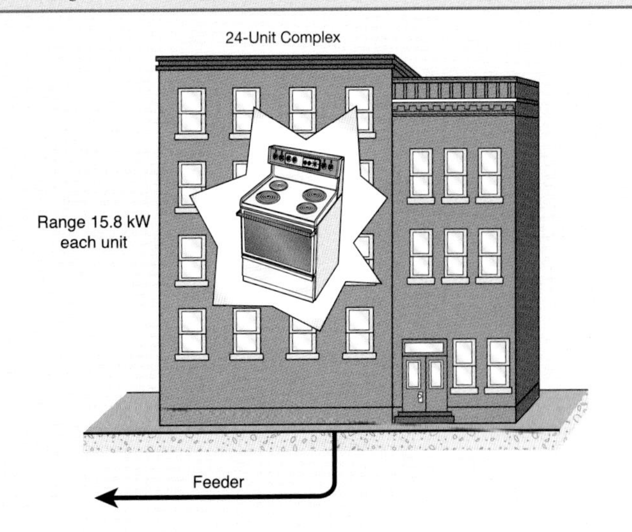

Solution – Calculation 1
Line Load
Table 220.55
Column C
24 ranges = 39 kW
Table 220.55 Note 1
5% increase for each kW (or major fraction) over 12
15.8 kW – 12 = 3.8 kW
3.8 = 4.0
4.0 x 5% = 20% increase
39 kW must be increased by 20%, or 120% of Column C
kW = 39 x 1.20
= 46.8 kW or 46,800 VA
Answer: 46,800 VA line

Solution – Calculation 2
Neutral Load
220.61(B)(1)
Neutral = line VA x 70%
= 46,800 x 0.70
= 32,760 VA
Answer: 32,760 VA neutral

Information

Table 220.55 Note 2 covers the installation of a number of ranges with unequal ratings of 8¾ kW to 27 kW. An average value of kW ratings is calculated. Use 12 kW for any range less than 12 kW. Read the demand for the total number of ranges in Column C. When the average value exceeds 12 kW, increase the Column C demand by 5% for each kW or major fraction thereof.

Problem 7-11

A 12-unit multifamily installation consists of four 10-kW ranges, four 14-kW ranges, and four 16-kW ranges. What is the feeder demand for the complex?

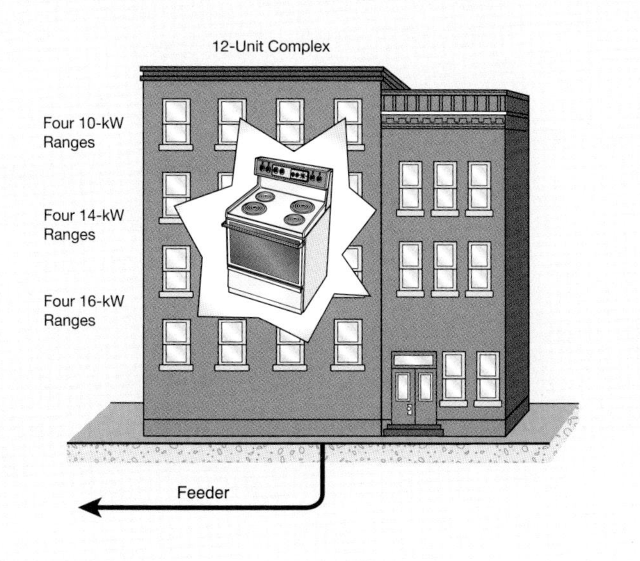

Solution – Calculation 1
Line Load
Table 220.55
Column C
12 ranges = 27 kW
Table 220.55, Note 2
Use 12 kW for the 10 kW ranges
4 x 12 kW = 48
4 x 14 kW = 56
4 x 16 kW = 64
12 ranges = 168 kW

$$\text{Average} = \frac{168 \text{ kW}}{12 \text{ ranges}}$$
= 14 kW

Table 220.55 Note 1
5% increase for each kW over 12
14 kW – 12 = 2 kW
2 kW x 5% = 10% increase
27 kW must be increased by 10%, or 110% of Column C
Total kW = Column C x 110%
= 27 x 1.10
= 29.7 kW or 29,700 VA
Answer: 29,700 VA line

Solution – Calculation 2
Neutral Load
220.61(B)(1)
Neutral = line VA x 70%
= 29,700 x 0.70
= 20,790 VA
Answer: 20,790 VA neutral

7.2 Developing an Equation for Single-Phase Ranges on a 3-Phase System

It is not unusual for single-phase ranges to be installed on 3-phase systems, especially in multi-family dwellings where the service is 120/208 volts, 3-phase, 4-wire and a single-phase, 120/208 volt, 3-wire feeder is installed to each unit. An equation can be developed for calculating the 3-phase load for single-phase ranges installed on a 3-phase system.

Q. *When a 3-phase, 4-wire system has 5 kW of lighting installed between phase A and the neutral, and 5 kW of lighting installed between phase B and the neutral, how much lighting needs to be installed between phase C and the neutral to keep the system balanced?*

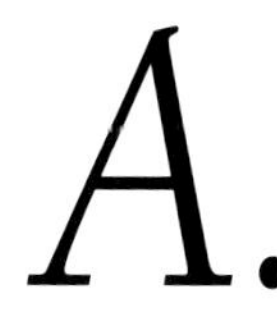

It is not difficult to calculate the answer of 5 kW.

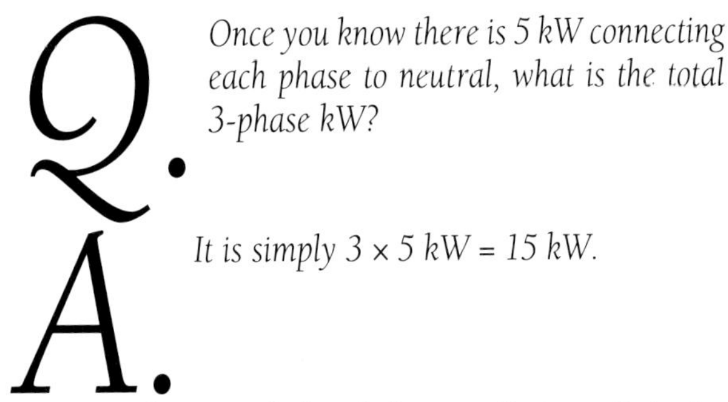

Q. *Once you know there is 5 kW connecting each phase to neutral, what is the total 3-phase kW?*

A. *It is simply 3 × 5 kW = 15 kW.*

This same theory is behind the installation of single-phase ranges on a 3-phase system. On a 10-kW range, each phase is (for the most part) a balanced load. However, **Table 220.55** has a built-in derating factor and it takes a few more steps to take advantage of **Table 220.55**. The range loads will balance out at less than 10 kW per phase.

The following sequence uses three single-phase 10-kW ranges installed on a 120/208 volt, 3-phase, 4-wire system. Evenly divided, there will be one range installed on each phase to neutral.

Step 1: **Placing single-phase ranges on a single-phase, 120/240 volt, 3-wire system.** First, look at how a single-phase installation of a 120/240 volt, single-phase, 3-wire balanced system would look in the graphic shown in **Figure 7-1**. This is the system that **Table 220.55** is built upon.

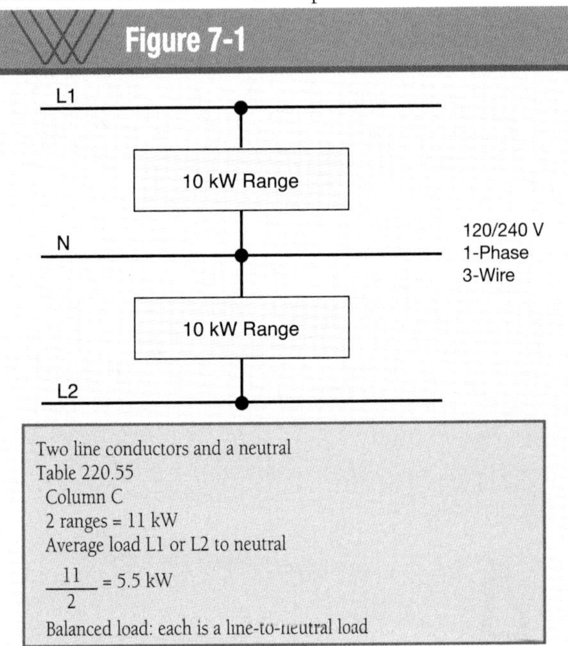

Figure 7-1. Two 10-kW ranges connected to a balanced 120/240 single-phase, 3-wire system.

When three 10-kW, single-phase ranges are installed on a 3-phase system, there will be the same balanced load as illustrated with the lighting load. Just as with the lighting load, 3 × 5 kW lighting equals a 15-kW 3-phase load, or when 3 × 5.5 kW equals 16.5 kW is the 3-phase load for three 10-kW, single-phase ranges as is shown later.

Step 2: **Putting Two Single-Phase Ranges on Two Phases of a 3-Phase System.** Even though the voltage might change, there is no change in the kW load for 3-phase, range-load calculations.
Calculations begin with two ranges and two phases as shown in the 120/208 volt, single-phase, 3-wire in **Figure 7-2**.

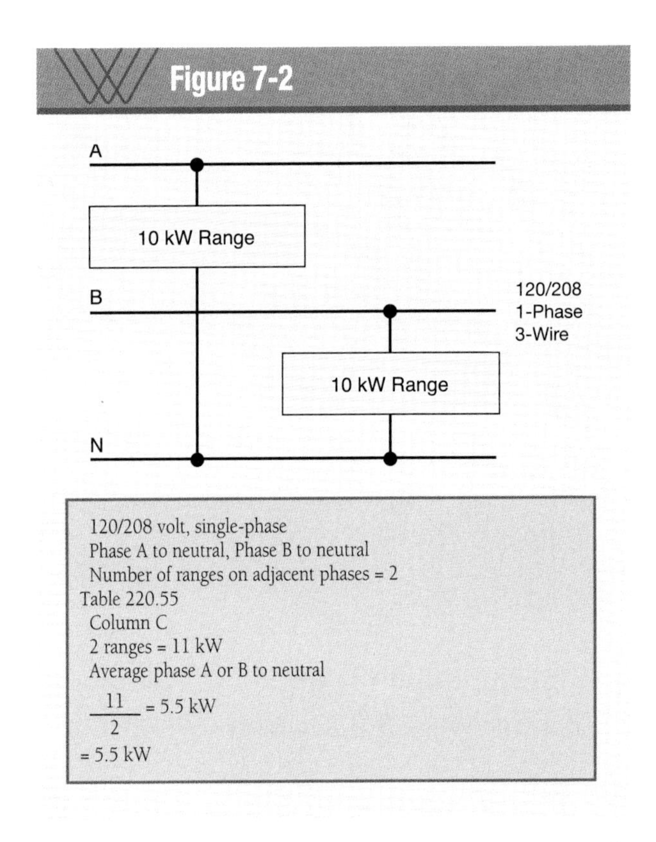

Figure 7-2. Two 10-kW ranges connected to a 120/208 V single-phase, 3-wire system.

Step 3: **Adding a Third Range and a Third Phase.** Calculations continue with three ranges and three phases as shown in the 120/208 volt, 3-phase, 4-wire system in **Figure 7-3**.

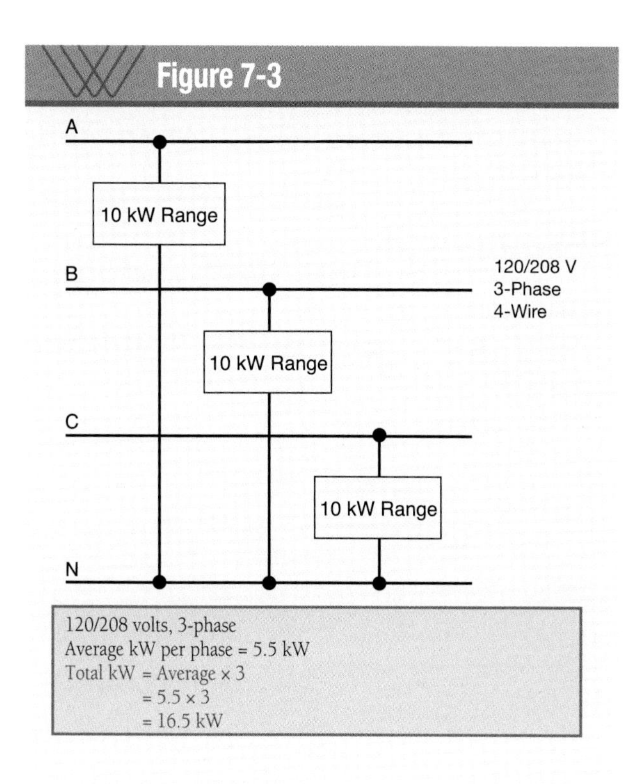

Figure 7-3. Three 10-kW ranges connected to a 120/208 V 3-phase, 4-wire system.

Summary: When single-phase ranges are connected on 3-phases:

Step 1: Evenly divide the number of ranges on the three phases.
Step 2: Calculate the number of ranges on adjacent phases.
Step 3: Read the kW from **Table 220.55 Column C** for the number of ranges between adjacent phases.
Step 4: Calculate the average kW for each phase to neutral.
Step 5: Multiply the average by 3 for the total 3-phase kW load.

Simplified equation:

$$\text{kW (3-phase)} = \frac{\text{kW (Column C)} \times 3}{2}$$

Where:

kW for 3-phase = total 3-phase load

kW (Column C) = kW for ranges on adjacent phases according to **Table 220.55 Column C**

3 = adjustment for 3-phase

2 = adjustment for averaging **Table 220.55 Column C** kW for 2-phase

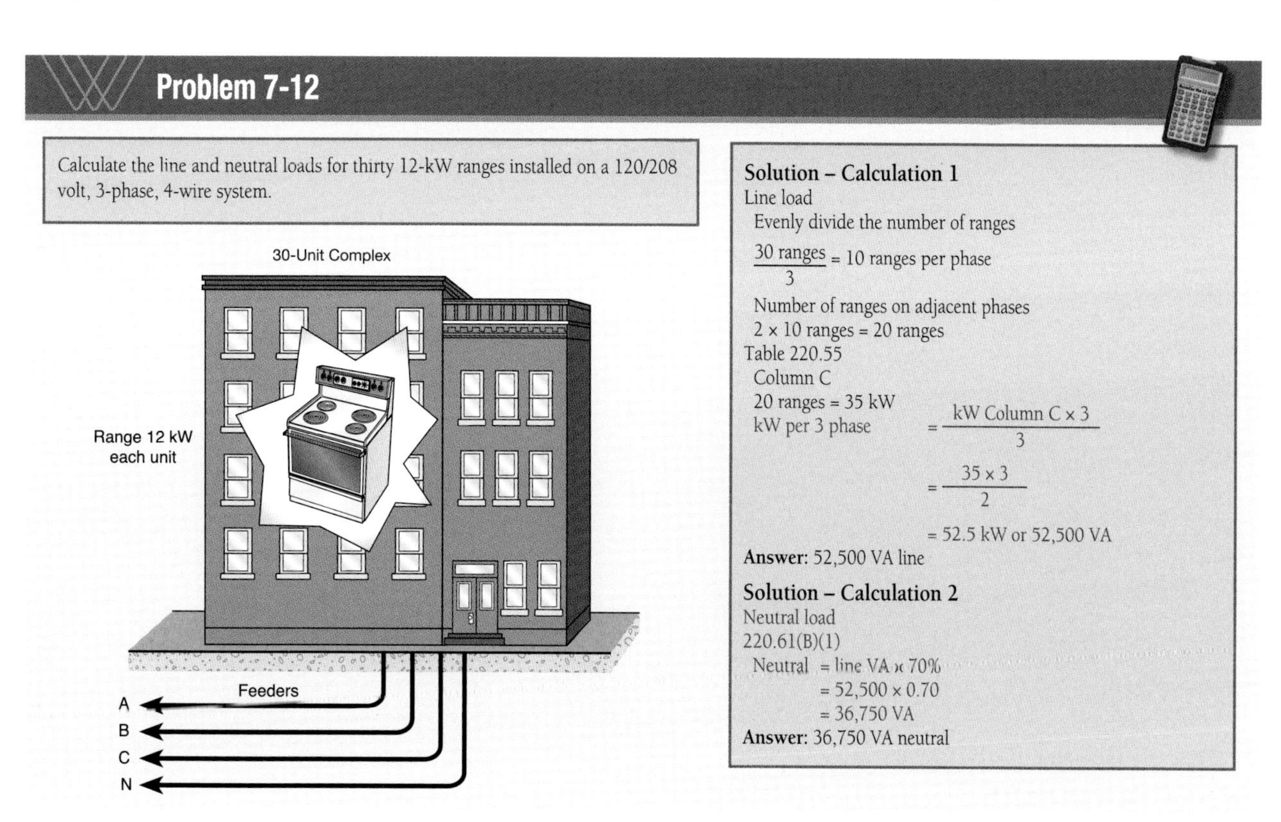

Calculate the line and neutral loads for thirty 12-kW ranges installed on a 120/208 volt, 3-phase, 4-wire system.

Solution – Calculation 1
Line load
Evenly divide the number of ranges
$\frac{30 \text{ ranges}}{3}$ = 10 ranges per phase
Number of ranges on adjacent phases
2 × 10 ranges = 20 ranges
Table 220.55
Column C
20 ranges = 35 kW
kW per 3 phase $= \frac{\text{kW Column C} \times 3}{3}$
$= \frac{35 \times 3}{2}$
= 52.5 kW or 52,500 VA
Answer: 52,500 VA line

Solution – Calculation 2
Neutral load
220.61(B)(1)
Neutral = line VA × 70%
= 52,500 × 0.70
= 36,750 VA
Answer: 36,750 VA neutral

7.3 Branch-Circuits for Range Loads

Up to this point, all calculations have been directed at feeders. **Table 220.55 Note 4** permits the use of **Table 220.55** for calculating branch-circuit loads for household electrical cooking appliances. **Note 4** covers several installations:

1. For the branch-circuit load for one range, use **Table 220.55**.
2. The branch-circuit load for one wall-mounted oven is the nameplate rating of the oven.
3. The branch-circuit load for one counter-mounted cooking unit is the nameplate rating of the cooking unit.
4. For the branch-circuit load for a counter-mounted cooking unit and not more than two wall-mounted ovens, all on the same branch circuit and located in the same room, add nameplate ratings and calculate as one range, using **Table 220.55**.

Problem 7-13

What is the branch-circuit load for one 16-kW range installed in a one-family dwelling unit?

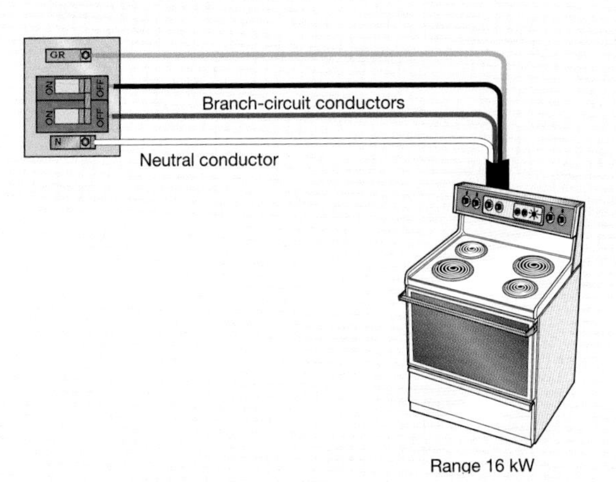

Solution – Calculation 1
Branch-circuit conductor load
Table 220.55 Note 4
  Use Table 220.55 for branch-circuit loads
Table 220.55
  Column C
  1 range = 8 kW
Table 220.55 Note 1
  5% increase for each kW over 12
  16 kW – 12 = 4 kW
  4 kW × 5% = 20% increase
  4 kW must be increased by 20%, or 120% of Column C
  Total kW = Column C × 120%
  = 8 × 1.20
  = 9.6 kW or 9,600 VA
Answer: 9,600 VA line

Solution – Calculation 2
Neutral conductor load
210.19(A)(3) Exception No. 2
  Range size exceeds 8,750 VA
  Neutral = line VA × 70%
  = 9,600 × 0.70
  = 6,720 VA
Answer: 6,720 VA neutral

Problem 7-14

What is the branch-circuit load for one 5-kW counter-mounted cooking unit?

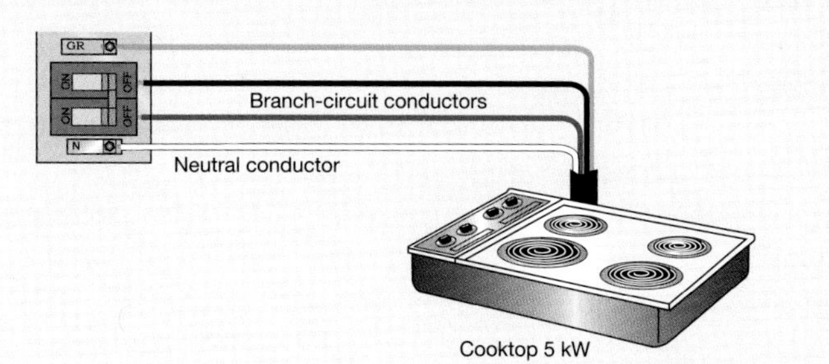

Solution – Calculation 1
Branch-circuit conductor load
Table 220.55 Note 4
Use nameplate rating
Line = 5 kW or 5,000 VA
Answer: 5,000 VA line

Solution – Calculation 2
Neutral conductor load
210.19(A)(3) Exception No. 2
Does not apply since 5,000 VA is less than 8,750 VA
Neutral conductor load = 5,000 VA
Answer: 5,000 VA neutral

Problem 7-15

What is the branch-circuit load for two 4-kW wall-mounted ovens and a 5-kW counter-mounted cooking unit installed on the same branch circuit in the same room of a one-family dwelling unit?

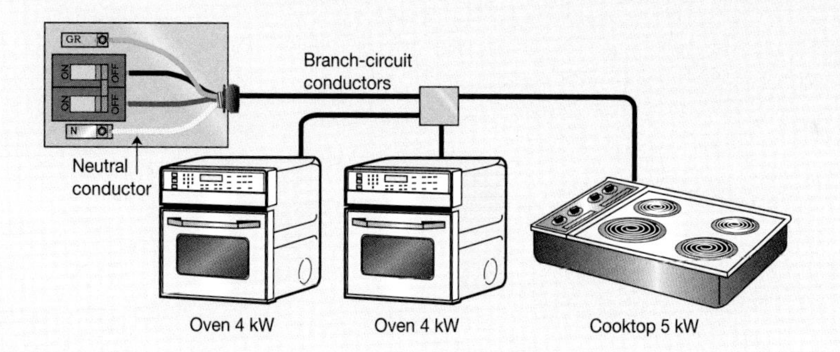

Solution – Calculation 1
Branch-circuit conductor load
Table 220.55 Note 4
Same circuit, same room, not over 2 ovens
Add and calculate as one range
4 kW + 4 kW + 5 kW = 13 kW
Table 220.55
Column C
One range = 8 kW
Table 220.55 Note 1
5% increase for each kW over 12
13 kW – 12 = 1 kW
1 kW x 5% = 5% increase
1 kW must be increased by 5%, or 105% of Column C
Total kW = Column C x 105%
= 8 x 105%
= 8.4 kW or 8,400 VA
Answer: 8,400 VA line

Solution – Calculation 2
Neutral conductor load
210.19(A)(3) Exception No. 2
Does not apply since 8,400 VA is less than 8,750 VA
Neutral conductor load = 8,400 VA
Answer: 8,400 VA neutral

7.4 Range Loads–Branch-Circuit Conductors

Section 210.19 sets the minimum ampacity and size of a branch circuit. **210.19(A)(3)**, in turn, establishes the minimum conductor size permitted for branch-circuit conductors supplying household ranges and cooking appliances.

210.19(A)(3) sets the minimum branch-circuit rating for an 8¾ kW or larger range at not less than 40 amperes. This means the overcurrent protective device is a minimum of 40 amperes. Therefore, the branch-circuit conductors must have an ampacity of at least 40 amperes.

210.19(A)(3) Exception No. 2 permits the neutral of a 3-wire branch circuit for 8¾ kW or larger household cooking appliances to be 70% of the branch-circuit rating. However, it also limits the neutral to not be smaller than a 10 AWG wire.

Problem 7-16

What is the minimum size of THWN copper branch-circuit conductors and neutral conductor permitted for one 120/240 volt, 10-kW electric range installed in a one-family dwelling unit? What is the matching size circuit breaker for the conductors selected? What is the minimum size copper equipment grounding conductor?

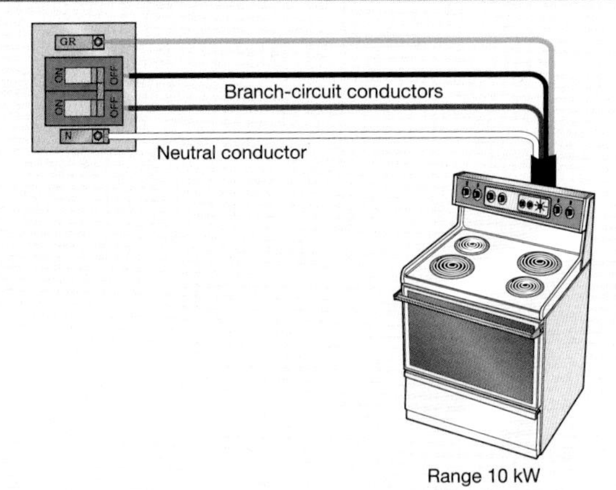

Solution – Calculation 1
Branch-circuit conductors
Table 220.55 Note 4
Use Table 220.55 for branch-circuit loads
Table 220.55
One Range = 8 kW
Table 310.15(B)(16)
75°C copper THWN column
33.33 amps = 10 AWG THWN
240.4(D)(7)
Maximum overcurrent for 10 AWG = 30 amps
210.19(A)(3)
Does not meet minimum of 40 amps
Next larger standard size conductor = 8 AWG THWN
Table 310.15(B)(16)
8 AWG THWN = 50 amps
Answer: 8 AWG THWN minimum branch-circuit conductors

Solution – Calculation 2
Neutral conductor
220.61(B)(1)
Neutral = line VA × 70%
= 33.33 × 0.70
= 23.33 amps
Table 310.15(B)(16)
75°C copper THWN column
23.33 amps = 12 AWG THWN
210.19(A)(3) Exception No. 2
Does not meet minimum of 10 AWG for neutral
Next larger standard size conductor = 10 AWG neutral
Answer: 10 AWG THWN minimum neutral conductor

Solution – Calculation 3
Circuit breaker
210.19(A)(3)
Minimum rating of 40 amps
Answer: 40 ampere 2-pole circuit breaker

Solution – Calculation 4
Equipment Grounding Conductor
250.122 and Table 250.122
40 amp CB = 10 AWG copper
Answer: 10 AWG copper

Article 422 Appliances.

The *NEC* definition of a vending machine in **Section 422.2** provides the *Code* user with the necessary information to determine whether an appliance is indeed a vending machine. For example, a coin-operated washing machine, while it is an appliance, is not a vending machine because it does not dispense products or merchandise. Typical vending machines include those that dispense soda, candy, food, cigarettes, lottery tickets, coffee, and the like.

Problem 7-17

What is the minimum ampacity and size of THHN copper branch-circuit and neutral conductors permitted for one 120/240 volt, 17.6-kW range installed in a one-family dwelling unit?

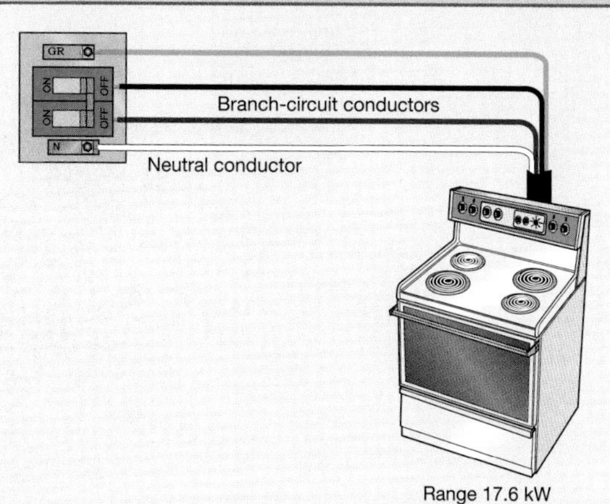

Solution – Calculation 1

Branch-circuit conductors

Table 220.55 Note 4

Use Table 220.55 for branch-circuit loads

Table 220.55

One range = 8 kW

Table 220.55 Note 1

5% increase for each kW over 12

17.6 kW – 12 = 5.6 kW

220.5(B)

Round to the nearest whole number

5.6 = 6 kW

6 kW x 5% = 30% increase

8 kW must be increased by 30% or 130% of Column C

kW = 8 x 1.30

= 10.4 kW or 10,400 VA

$$I = \frac{\text{kW} \times 1{,}000}{E}$$

$$= \frac{10.4 \times 1{,}000}{240}$$

= 43.33 amps

110.14(C)(1)

Not to exceed Table 310.15(B)(16)

Table 310.15(B)(16)

75°C copper THHN column

43.33 amps = 8 AWG THHN

210.19(A)(3)

Fulfills minimum requirement of 40 amps

Answer: 8 AWG THHN minimum branch-circuit conductors

Solution – Calculation 2

Neutral conductor

220.61(B)(1)

Neutral = line VA x 70%

= 43.33 x 0.70

= 30.33 amps

Table 310.15(B)(16)

30.33 amps = 10 AWG THHN

210.19(A)(3) Exception No. 2

Fulfills minimum requirement of 10 AWG for neutral

Answer: 10 AWG THHN minimum neutral conductor

7.5 Household Cooking Equipment in Schools

Table 220.55 Note 5 permits the use of **Table 220.55** and its notes to be used for calculating feeder loads for household cooking appliances installed and used for instructional programs.

Problem 7-18

Calculate the feeder demand for six 16-kW household electric ranges to be installed in a high school for instructional programs.

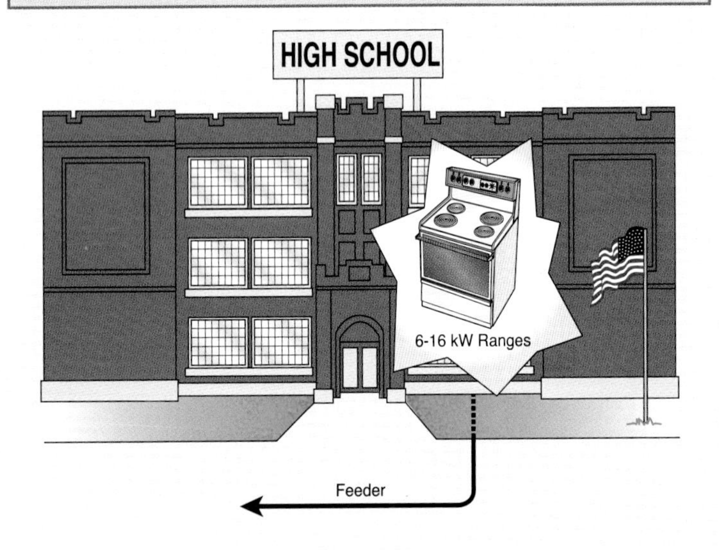

Solution – Calculation 1
Line load
Table 220.55 Note 1
5% increase for each kW over 12
16 kW – 12 = 4 kW
4 kW x 5% = 20% increase
4 kW must be increased by 20% or 120% of Column C
Table 220.55
Column C
6 ranges = 21 kW
Total kW = 21 x 1.20
= 25.2 kW or 25,200 VA
Answer: 25,200 VA line

Solution – Calculation 2
Neutral load
The ranges are household electric ranges, so 220.61(B)(1) applies
220.61(B)(1)
Neutral = line VA x 70%
= 25,200 x 0.70
= 17,640 VA
Answer: 17,640 VA neutral

Problem 7-19

Calculate the lower feeder demand for four 12-kW ranges and four 4-kW wall-mounted ovens to be installed in a high school cooking lab for instructional programs.

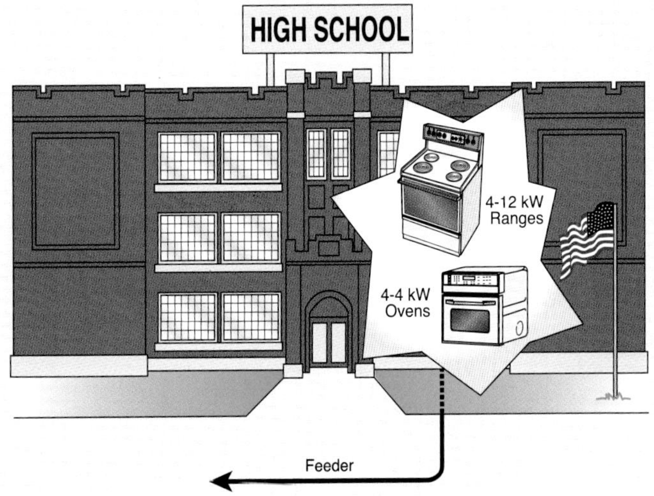

Comment
Note 3 is permitted to be used in lieu of Column C. Therefore, to calculate the lowest feeder demand, calculate both with and without Column C and use the lowest demand. In lieu of indicates a choice.

Solution – Calculation 1
Lowest feeder demand
Using Note 3
Table 220.55 Note 3
4 ovens x 4 kW = 16 kW
Table 220.55
Column B calculation
4 units = 50% demand
16 kW x 0.50 = 8 kW
Table 220.55
Column C
4 ranges = 17 kW
8 kW oven load + 17 kW range = 25 kW
Column C calculation
4 ovens + 4 ranges = 8 units under 12 kW
Table 220.55
Column C
8 units = 23 kW
23 kW is less than 25 kW; use lower demand
Answer: 23,000 VA line

Solution – Calculation 2
Neutral load
220.61(B)(1)
Neutral = line VA x 70%
= 23,000 x 0.70
= 16,100 VA
Answer: 16,100 VA neutral

7.6 Commercial Cooking Appliances

Commercial cooking appliances are covered by **Section 220.56** and **Table 220.56**.

Included in commercial cooking appliances are the following:

- Electrical cooking equipment
- Dishwashers
- Booster heaters
- Water heaters
- Other kitchen equipment, including warming ovens, mixing machines, food processors, etc.

For additional information, visit qr.njatcdb.org Item #1037

Section 220.56 applies to all the preceding kitchen equipment, provided it is controlled by a thermostat or is used in an intermittent duty fashion.

Section 220.56 does not apply to space heating, ventilating, and air-conditioning equipment.

The feeder demand is not permitted to be less than the sum of two largest kitchen equipment loads.

220.61(B)(1) permits the 70% derating factor for the neutral and applies to household electric ranges only. It does not apply to commercial cooking equipment.

Problem 7-20

Calculate the feeder load for three 16-kW ranges, one 7.5-kW water heater, and one 4-kW dishwasher installed in a restaurant kitchen.

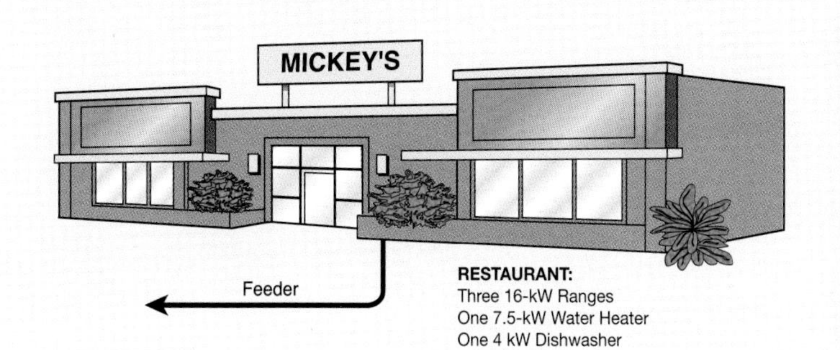

Solution
Section 220.56 and Table 220.56
3 × 16 kW range = 48.0
1 × 7.5 kW range = 7.5
1 × 4 kW dishwasher = 4.0
Total load of 5 units = 59.5 kW
Table 220.56
5 units = 70% demand
59.5 × 0.70 = 41.65 kW
Section 220.56
Demand not to be less than the sum of the two largest loads
Two largest = 16 + 16
= 32 kW
41.65 kW is larger than 32 kW; use largest load
Answer: 41.6 kW or 41,650 VA for line and neutral feeder demand

Problem 7-21

Calculate the feeder load for two 24-kW ranges, one 10-kW water heater, one 3-kW dishwasher with a 2-kW booster heater, and two 2-kW warming ovens to be installed in a restaurant kitchen.

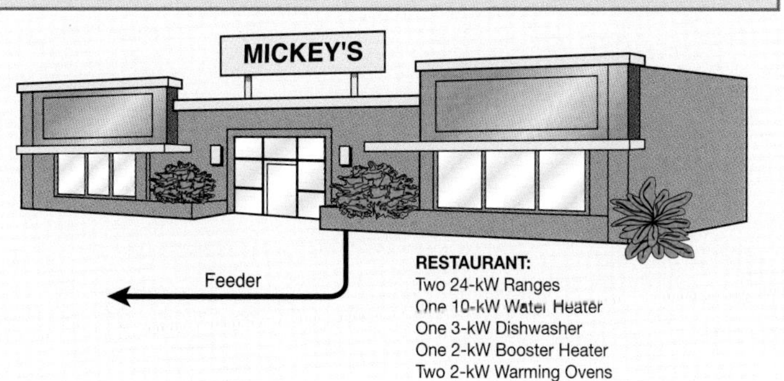

Solution
Section 220.56 and Table 220.56
2 × 24 kW ranges = 48
1 × 10 kW water heater = 10
1 × 3 kW dishwasher = 3
1 × 2 kW booster heater = 2
2 × 2 kW warming oven = 4
Total load of 7 units = 67 kW
Table 220.56
6 units and more = 65% demand
67 kW × 0.65 = 43.55 kW
Section 220.56
Demand not to be less than the sum of the two largest loads
Two largest loads = 48 kW
Calculated 43.55 kW is less than 48 kW; use largest load
Answer: 48 kW or 48,000 VA for line and neutral feeder demands

7.7 Branch Circuits Serving an Electric Clothes Dryer

The branch circuit loads for an electric clothes dryer is covered by **Part II** of **Article 220**. **220.14(B)** permits a dryer load to be calculated according to **Section 220.54** (located in **II. Feeders**). The branch-circuit load shall be either 5,000 VA or the nameplate rating; whichever is larger. This is the basic method used for the following calculation.

Electric clothes dryers are not considered continuous loads.

Problem 7-22

What is the minimum size of Type MC cable with 90°C copper conductors permitted to supply a 4.5-kW, 120/240 volt, single-phase, 3-wire electric dryer installed in a one-family dwelling unit? What size overcurrent protective device using 75°C wire terminations will protect this circuit?

Solution – Calculation 1
Circuit conductors
220.14(B)
Load to be calculated using Section 220.54
Section 220.54
4.5 kW dryer must be calculated at 5 kW

$$I = \frac{kW \times 1{,}000}{E}$$

$$= \frac{5 \times 1{,}000}{240}$$

$$= 20.83 \text{ amps}$$

220.5(B)
Round to the nearest whole
20.83 amps = 21 amps
110.14(C)(1)(a)(3)
Use higher 75°C temperature conductor to match circuit breaker temperature rating of 75°C
Table 310.15(B)(16)
75°C copper column
Double asterick note: use 240.4(D)(7)
240.4(D)(7)
21 amps = 10 AWG copper
Answer: 10-3 Type MC Cable (copper) with ground

Solution – Calculation 2
Overcurrent protective device
240.4(D)(7)
10 AWG = 30 amps
Answer: 30 ampere 2-pole overcurrent protective device

7.8 Other Appliances

The fundamental requirement for appliance branch circuits is that they are required to carry the appliance current without overheating under their conditions of use. And for individual appliances, the branch circuit rating cannot be less than the rating marked on the appliance. **Article 422** covers appliances used in any occupancy. For motor operated appliances, **Article 430 Motors** is to be used to determine the branch circuit rating unless there are specific requirements for motor operated appliances within **Article 422**.

7.8.1 Branch Circuits Serving an Electric Water Heater

Storage-type water heaters are covered by **Section 422.13**. This section requires that these water heaters be considered a continuous load for sizing the branch circuit conductors and overcurrent protection. **210.19(A)** for conductor sizing and **210.20(A)** for overcurrent protection are used for sizing these branch circuits.

Problem 7-23

What is the minimum size of Type NM, nonmetallic sheathed cable using copper branch-circuit conductors to supply a 50 gallon storage-type water heater? This water heater is furnished with standard 240 volt AC, single-phase non-simultaneous wiring, with separate 4,500 watt upper and lower heating elements. What size circuit breaker with 60/75°C wire terminations will protect this circuit?

Solution – Calculation 1
Type NM cable size

$$I = \frac{W}{E}$$

$$= \frac{4{,}500}{240}$$

$$= 18.75 \text{ amps}$$

Section 422.13
Considered a continuous load
210.19(A)(1)

$$I = \textit{continuous load} \times 125\%$$
$$= 18.75 \times 1.25$$
$$= 23.44 \text{ amps}$$

Section 334.80
NM cable not to exceed ampacity of 60°C rated conductor
Table 310.15(B)(16)
60°C copper column
Double asterisk note: use 240.4(D)(7)
240.4(D)(7)
23.44 amps = 10 AWG
Answer: 10-2 Type NM cable (copper) with ground

Solution – Calculation 2
Circuit breaker
240.4(D)(7)
10 AWG = 30 amps
Answer: 30 ampere 2-pole CB

Comment
Non-simultaneous means these elements are NOT operating at the same time, and therefore only one 4,500 watt element is the actual load at any given time.

7.8.2 Branch Circuits Serving a Kitchen In-Sink Waste Disposer

According to **422.60**, the voltage and current ratings are marked on the in-sink waste disposer. These ratings are permitted to be used to select the conductor size and the overcurrent protection. Also, **422.11(G)** requires a in-sink waste disposal motor to be equipped with an overload protective device, usually of the integral type. See **430.32(A)(2)**.

Since the overcurrent protective device size for the motor of the waste disposer is sometimes not marked on the product nameplate, **422.11(A)** is used for this calculation. This section requires the use of **Section 240.4**. Therefore, **240.4(D) Small Conductors** is used to "match" the ampacity of the conductor size to the selected overcurrent protective device.

Problem 7-24

A household kitchen waste disposer is supplied by an individual branch circuit. The waste disposer nameplate indicates it is a single-phase, ¾ hp, 120 volt, 8.1 amperes, intermittent-time rated motor and equipped with manual reset overload protection. It does not have a recommended wire size or overcurrent device rating on the nameplate or in the product specification sheet. What is the minimum size of nonmetallic sheathed cable using copper branch-circuit conductors to supply this kitchen waste disposer? What size circuit breaker with 60/75°C wire terminations will protect this circuit? What is the minimum size AC general-use snap switch permitted to be used to disconnect the waste disposer?

Solution – Calculation 1
Type NM cable size
422.11(A)
Use appliance marked rating
Nameplate = 8.1 amps
334.80
NM cable not to exceed ampacity of 60°C rated conductor
210.3
Minimum branch-circuit rating
Minimum conductor size = 15 amps
Table 310.15(B)(16)
60°C copper column

Solution – Calculation 1 (continued)
Double asterisk note: use 240.4(D)(3)
240.4(D)(3)
15 amps = 14 AWG copper
Answer: 14-2 Type NM cable (copper) with ground

Solution – Calculation 2
Circuit breaker
240.4(D)(3)
14 AWG copper = 15 amps CB
Answer: 15 amp 1-pole CB

Solution – Calculation 3
Disposer disconnect switch using an AC general-use snap switch
404.14(A)(3)
Motor load not to exceed 80% rating of switch
Verify 15 amp switch rating is sufficient
Motor Load = *switch rating* × 80%
= 15 × 0.8
= 12 amps
12 amps = max. motor load for 15 amp switch
8.1 amps is less than 12 amps; use 15 amp switch
Answer: 15 ampere, 120 volt general-use snap switch

7.8.3 Branch Circuits Serving a Unit Electric Heater

Article 424 covers fixed electric space-heating equipment, such as heating cable, unit heaters, boilers, central systems and the like. Fixed electric space heating equipment must be considered a continuous load for determining the branch-circuit conductors and overcurrent protection.

Problem 7-25

What is the minimum size of THWN copper branch-circuit conductors permitted for one 3-kW, 120-volt, ceiling-mounted unit electric heater installed in a one-family attached garage with a nameplate full load current of 26.5 amperes? What size overcurrent protective device with 60/75°C wire terminations will protect this circuit? What is the minimum size copper equipment grounding conductor?

Solution – Calculation 1
Circuit conductors
424.3(B)
Considered a continuous load
210.19(A)(1)
I = *continuous load* × 125%
= 26.5 × 1.25
= 34.125 amps
Table 310.15(B)(16),
75 °C THWN copper column
35 amps = 10 AWG
240.4(D)(7)

Solution – Calculation 1 (continued)
Prohibits a 10 AWG THWN;
34.125 amps load requires an 8 AWG THWN
Answer: 8 AWG THWN circuit conductors

Solution – Calculation 2
Overcurrent protective device (OCPD)
210.20(A)
Table 310.15(B)(16)
75°C copper column
8 AWG THWN permits a 50 amp OCPD
Answer: 50 ampere fuse or circuit breaker

Solution – Calculation 3
Equipment grounding conductor (EGC)
Table 250.122
50 amp OCPD requires a 10 AWG EGC
Answer: 10 AWG equipment grounding conductor

Definitions and Terms

Appliance - Utilization equipment, generally other than industrial, that is normally built in standardized sizes or types and is installed or connected as a unit to perform one or more functions such as clothes washing, air conditioning, food mixing, deep frying, and so forth.

Branch Circuit - The circuit conductors between the final overcurrent device protecting the circuit and the outlet(s).

Branch Circuit, Appliance - A branch circuit that supplies energy to one or more outlets to which appliances are to be connected and that has no permanently connected luminaires that are not a part of an appliance.

Continuous Load - A load where the maximum current is expected to continue for 3 hours or more.

Cooking Unit, Counter Mounted - A cooking appliance designed for mounting in or on a counter and consisting of one or more heating elements, internal wiring, and built-in or mountable controls.

Demand Factor - The ratio of the maximum demand of a system, or part of a system, to the total connected load of a system or the part of the system under consideration.

Dwelling Unit - A single unit, providing complete and independent living facilities for one or more persons, including permanent provisions for living, sleeping, cooking, and sanitation.

Feeder - All circuit conductors between the service equipment, the source of a separately derived system, or other power supply source and the final branch-circuit overcurrent device.

Vending Machine - Any self-service device that dispenses products or merchandise without the necessity of replenishing the device between each vending operation and is designed to require insertion of coin, paper currency, token card, key, or receipt of payment by other means.

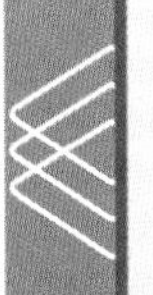

A Variety of 2014 NEC Changes for **Article 422 Appliances.**

New **422.5** requires the device providing GFCI protection within **Article 422** to be readily accessible.

Also, new **422.23** requires tire inflation machines and automotive vacuum machines provided for public use to be protected by a ground-fault circuit-interrupter. And new **422.51(B)** requires vending machines NOT utilizing a cord and plug connection to be connected to a ground-fault circuit-interrupter protected circuit. Revised **422.49** requires cord and plug connected high-pressure spray washers to incorporate factory-installed ground-fault circuit-interrupter protection for personnel at or near the beginning of the supply cord for all 250 volt or less single-phase units and all three-phase equipment rated 208Y/120 volts and 60 amperes or less.

Summary

- Generally, branch-circuit conductors must have an ampacity not less than the maximum load to be served according to **210.19(A)(1)**.
- According to **422.10(A)**, the rating of an individual branch circuit cannot be less than the marked rating of the appliance.
- The general rules for branch-circuit ratings, found in **210.19(A)**, require that where a branch circuit supplies continuous loads (or any combination of continuous and noncontinuous loads) the minimum branch-circuit conductor size must have an allowable ampacity not less than 100% of the noncontinuous load plus 125% of the continuous load.
- **210.19(A)(3)** requires that branch-circuit conductors supplying household ranges, wall-mounted ovens, counter-mounted cooking units, and other household cooking appliances have an ampacity not less than the rating of the branch circuit and not less than the maximum load to be served.
- For ranges of 8¾ kW or more rating, the minimum branch-circuit rating must be at least 40 amperes.
- According to **Table 220.55, Note 4**, the first sentence permits a single appliance load of 8 kW to be used for any household range(s) and cooking appliance rated not over 12 kW.
- The demand factors and loads of **Table 220.55** apply to services, feeders, and branch circuits for household electric ranges, wall-mounted ovens, counter-mounted cooking units, and other household cooking appliances over 1¾ kW but not over 27 kW.
- The demand factors and loads of **Table 220.55** apply to multiple appliances of equal and unequal ratings, single- and 3-phase installations, as well as household cooking appliances used for instructional purposes.
- In many cases, the neutral conductor of a single-phase, three-wire circuit may be reduced to 70% of the phase conductor.
- The minimum load to be used for calculations of electric clothes dryer circuits is the larger of either 5,000 VA or the nameplate rating of each dryer served.
- **Table 220.54** demand factors and loads apply to services, feeders, and branch circuits for household electric clothes dryers.
- Load calculations for household dryers must comply with **220.54**, but are permitted to use the demand factors of **Table 220.54**.
- Where a branch circuit supplies continuous loads (or any combination of continuous and noncontinuous loads) according to **210.20(A)**, the rating of the overcurrent device must not be less than 100% of the noncontinuous load plus 125% of the continuous load.
- Commercial electric cooking equipment, dishwasher booster heaters, water heaters, and other kitchen equipment loads are permitted to be calculated in accordance with the demand factors of **Table 220.56**.
- These demand factors can be applied to all equipment which has either thermostatic control or intermittent use as kitchen equipment.
- A fixed storage-type water heater which has a capacity of 450 L (120 gal) or less must be considered a continuous load for the purposes of sizing branch circuits.

Load Calculations

Generally speaking, one-family dwelling unit load calculations are the simplest of the calculation methods presented in this lesson. These loads are simply converted to volt-amperes (VA), with some loads permitted to be reduced by applicable demand factors. There are two different methods of performing these one-family dwelling unit calculations: the standard method or the optional method. The optional method is reserved for service sizes greater than 100 amperes.

Multifamily dwelling unit calculations in some ways are similar to one-family calculations. In addition to a calculation of each dwelling unit within the structure, there are feeders and their associated demand factors that need to be handled. In addition, the services to multifamily dwelling units are generally larger than one-family dwelling structures. Continuous-duty loads do not generally apply to one-family and multifamily dwellings, unless fixed electric space-heating equipment is installed within the unit(s) or common area.

Objectives

- Calculate one-family and multifamily dwelling unit lighting and appliance loads including the number of branch circuits needed for lighting loads.
- Prepare the heating and air-conditioning loads and multifamily dwelling unit feeder demands for one-family and multifamily dwelling units.
- Determine the feeder demand for service-entrance conductors total for one-family and multifamily dwelling units.
- Select the size of one-family and multifamily dwelling unit services based upon calculations.
- Apply various multifamily dwelling feeder demands including the neutral feeder and service conductor size for one-family and multifamily dwellings.
- Compute branch-circuit general lighting loads and the minimum number of branch circuits required for commercial occupancies.
- Classify branch-circuit and feeder loads as continuous and noncontinuous.
- Solve for the adequate size feeder and service overcurrent protection and conductor sizes for commercial occupancies.
- Calculate the neutral load and select the proper conductor size for commercial occupancies.

Chapter 8

Table of Contents

8.1 General Requirements for Residential Loads

Load calculations can be done in simple or complex form. The complex calculations convert all loads into amperes and allot specific loads to the neutral conductor. The simple method of calculations converts all loads to volt-amperes (VA) and considers all neutral conductor loads to be evenly divided.

Chapter 8 of *Code Calculations* uses the simplified volt-ampere method of calculation and rounds off all loads to a whole number.

For certain equipment, the *Code* permits the neutral conductor to be smaller than the line conductors and this is reflected back into the feeder demand. To avoid overlooking the neutral conductor, neutral calculations will be made following the feeder demand calculations for the ungrounded conductors as each particular item is calculated.

The *Code* recognizes two methods of calculation. One is the standard method, usually called the *long method*, and the other is the optional method; usually referred to as the *short method*.

Although most of these calculations are directed at feeder demand loads, there are opportune times when certain branch-circuit calculations can be used. Special note will be made when this is done.

8.2 Dwelling Unit—Standard Calculation Method

Dwelling unit loads are more commonly called *residential loads*. They include one-family dwellings, two-family dwellings, townhouses, individual apartments, and multifamily apartment complexes. Dwelling unit load can be applied for an entire building or an area of a building. Chapter 8 is directed mainly at "feeder" loads, which, for the dwelling unit, are most often the service-entrance conductors. Some calculations apply to branch-circuit loads.

The *Code* lists requirements for specific loads which might be installed in a dwelling unit. The following outline of these loads can be used as a checklist for calculating residential loads:

- General lighting and general-use receptacle load
- Small appliance load
- Laundry load
- Other fastened-in-place appliance loads
- Cooking appliance load
- Clothes dryer load
- Heating load
- Air-conditioning load
- 25% of largest motor installed
- Other loads

8.2.1 Dwelling Unit—General Lighting and General-Use Receptacle Load

The minimum lighting load is based upon the square footage (ft^2) area of a building using the outside measurements of the building. For dwelling units, the calculated floor area does not include open porches, garages, and unused or unfinished spaces not adaptable for future use. The assigned minimum lighting load is volt-amperes per ft^2 according to **Table 220.12**. A footnote to **Table 220.12** refers to **220.14(J)**, which indicates that the general-purpose receptacles in a dwelling unit are included in the volt-amperes per square foot (VA per ft^2) calculation. Energy code compliance is also permitted in accordance with **220.12, Exception.**

The problems for the dwelling unit under consideration are based upon a home which the Smiths are planning to build. In the planning stages, the process is called *calculating loads*. Initially, all we know is that the house is to be single story, 55 feet wide by 65 feet long excluding open porches and garage areas, and the service will be 120/240 volt, 3-wire. Other things will be added as the planning progresses.

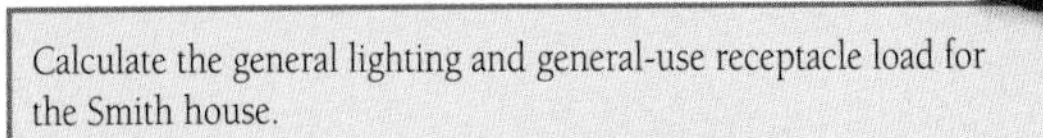

Calculate the general lighting and general-use receptacle load for the Smith house.

Solution

Area = length × width
= 55 × 65
= 3,575 ft^2

Table 220.12

Dwelling unit = 3 VA per ft^2

Load = area × VA per ft^2
= 3,575 × 3
= 10,725

Answer: 10,725 VA

8.2.2 Dwelling Unit—Number of Branch Circuits

A branch-circuit calculation at this point seems appropriate since **210.11(A)** requires that the total calculated lighting load (prior to the application of demand factors) be expressed in volt-amperes and used to calculate the minimum number of lighting branch-circuits needed.

This lighting circuit will cover the general lighting and general-purpose receptacles. The two or more required small-appliance branch circuits and the required laundry circuit(s) are in addition to the lighting branch circuits.

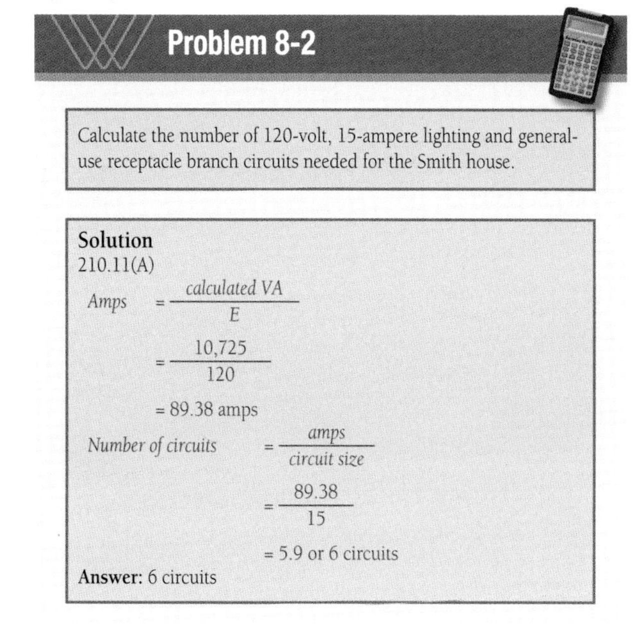

Problem 8-2

Calculate the number of 120-volt, 15-ampere lighting and general-use receptacle branch circuits needed for the Smith house.

Solution

210.11(A)

$$Amps = \frac{calculated\ VA}{E}$$

$$= \frac{10{,}725}{120}$$

$$= 89.38 \text{ amps}$$

$$Number\ of\ circuits = \frac{amps}{circuit\ size}$$

$$= \frac{89.38}{15}$$

$$= 5.9 \text{ or } 6 \text{ circuits}$$

Answer: 6 circuits

8.2.3 Dwelling Unit—Required Branch Circuits

When calculating the number of branch circuits, the calculated load is used before any demand factors are applied. When the number of calculated branch circuits is a fraction of a circuit, the next higher whole number must be used. Example: A calculated quantity of 4.1 actually requires 5 circuits. Had 20-ampere circuits been called for, the calculated ampere load would have been divided by 20.

The Smith House Required number of branch circuits		
210.11(A) General Lighting and Receptacle	15-amp branch circuits	6
210.11(C)(1) Small-Appliance	20-amp branch circuits	2
210.11(C)(2) Laundry	20-amp branch circuit	1
210.11(C)(3) Bathroom	20-amp branch circuit	1
Total branch circuits		10

Small Appliance and Laundry Circuits

The required minimum of two small-appliance circuits for the Smith house will be installed and calculated at 1,500 VA each, according to **220.52(A)**. The required laundry circuit will be installed and also calculated at 1,500 VA according to **220.52(B)**.

All the lighting will not be utilized at the same time; therefore, the *Code* permits the calculated lighting load to be reduced by a percentage, called a demand factor, which is given in **Table 220.42**. The lighting load demand factors for a dwelling unit are as follows:

First 3,000 VA at 100%

3,001 to 120,000 VA at 35%

Over 120,000 VA at 25%

220.52(A) and **220.52(B)** permit the required small appliance and laundry loads to be included with the lighting and general-use receptacle load before the lighting load demand factors are applied. The permitted combined lighting and receptacle loads may seem strange at first, but it serves to greatly simplify dwelling unit calculations, while providing safe and ample feeders and services.

A solid math foundation will allow the student to build an understanding and knowledge to the competency level necessary to be a confident Journeyman Electrical Worker.

Problem 8-3

Calculate the lighting feeder demand for the Smith house.

Solution

Calculated lighting load = 10,725 volt-amperes	
Lighting load total	10,725 VA
Small appliances 2 circuits at 1,500	3,000 VA
Laundry 1 circuit at 1,500	1,500 VA
Total	15,225 VA
Table 220.42	
First 3,000 VA at 100%	3,000 VA
3,001 to 120,000 VA at 35%	
15,225 VA – 3,000 VA = 12,225	
= 12,225 x 0.35	
= 4,279 VA	4,279 VA
Lighting Feeder Demand	7,279 VA

Answer: 7,279 VA ungrounded conductor load
7,279 VA neutral conductor load

Feeder Demand: Clothes Dryer

Section 220.54 requires the minimum amount of feeder demand for a clothes dryer to be not less than 5,000 volt-amperes; or if the nameplate rating is larger than 5,000 volt-amperes, the nameplate rating is to be used. This means that a 4-kW clothes dryer is calculated at 5,000 VA. A 6-kW clothes dryer is calculated at 6,000 VA.

220.61(B)(1) permits the neutral conductor load to be 70% of the calculated load on the ungrounded conductors for electric ranges, wall-mounted ovens, counter-mounted cooking units, and electric dryers.

Problem 8-4

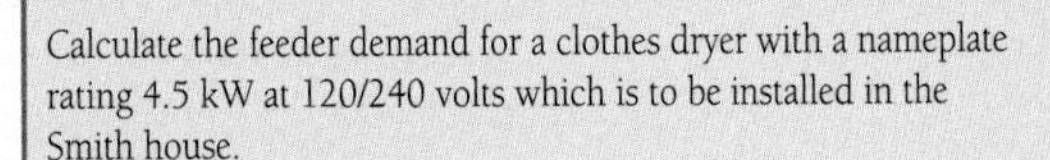

Calculate the feeder demand for a clothes dryer with a nameplate rating 4.5 kW at 120/240 volts which is to be installed in the Smith house.

Solution – Calculation 1
Ungrounded conductors
Section 220.54
Minimum for 1 dryer = 5,000 VA
Answer: 5,000 VA ungrounded conductor load

Solution – Calculation 2
Neutral conductor
220.61(B)(1)
VA *demand*= VA x 70%
= 5,000 x 0.70
= 3,500 VA
Answer: 3,500 VA neutral conductor load

Feeder Demand: Appliance Loads

Many appliances installed in a dwelling unit are fastened in place. These fastened-in-place appliances are calculated at their nameplate rating. However, when four or more of these appliances are installed in the same dwelling unit, a 75% demand factor is permitted by **Section 220.53**.

Electric ranges, clothes dryers, space heating equipment, and air-conditioning equipment are appliances, but are not counted as one of the four appliances; nor is the 75% demand factor applied to these loads. Each of these loads will be calculated individually.

The following list of fastened-in-place appliances count toward the total of four and are then subject to the 75% demand factor.

- Water heater
- Dishwasher
- Disposer
- Trash compactor
- Ventilation hood (fan/light)
- Sump pump

Problem 8-5

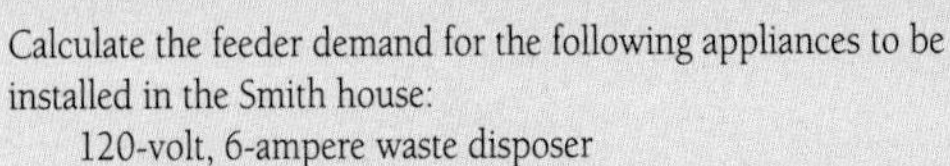

Calculate the feeder demand for the following appliances to be installed in the Smith house:
120-volt, 6-ampere waste disposer
1.5-kW dishwasher at 120 volts
5-kW, 120-volt water heater
120-volt, 4.4-ampere ventilation hood (fan/light)
960-watt, 120-volt trash compactor

Solution

Convert all loads to VA	
VA = *E* x *I* or VA = kW x 1,000	
Disposer	
VA = E x I	
= 120 x 6	
= 720 VA	720 VA
Dishwasher 1.5 kW x 1,000	1,500 VA
Water heater 5 kW x 1,000	5,000 VA
Ventilation hood	
VA = E x I	
= 120 x 4.4	
= 528 VA	528 VA
Compactor 960 Watts	960 VA
Total appliance demand	8,708 VA
Section 220.53	
Demand = VA x 75%	
= 8,708 x 0.75	
= 6,531 VA	

Answer: 6,531 VA ungrounded conductor load
6,531 VA neutral conductor load

Feeder Demand: Range Load

Calculations for range load and other cooking appliances were covered in Chapter 7 of *Code Calculations*. The neutral conductor for range loads is permitted to be reduced by 70%, according to **220.61(B)(1)**.

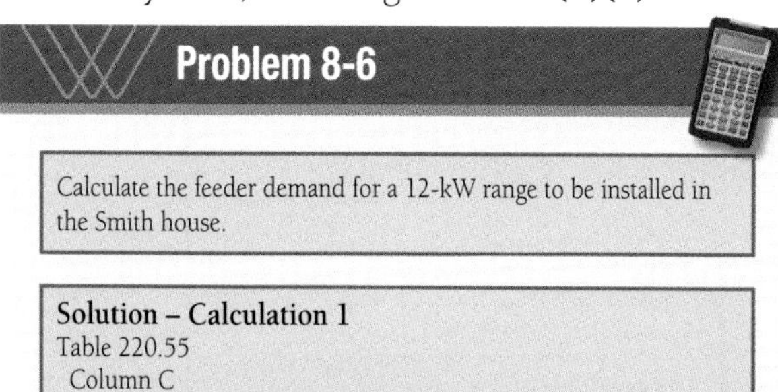

Problem 8-6

Calculate the feeder demand for a 12-kW range to be installed in the Smith house.

Solution – Calculation 1
Table 220.55
Column C
One range 12 kW or less = 8,000 VA
Answer: 8,000 VA ungrounded conductor load

Solution – Calculation 2
220.61(B)(1)
Neutral VA = VA × 70%
= 8,000 VA × 0.70
= 5,600 VA
Answer: 5,600 VA neutral grounded conductor load

Feeder Demand: Heating and Air Conditioning

Where fixed electrical space-heating equipment is installed in a dwelling unit, the loads are calculated at 100%. The same holds true for air-conditioning loads, which are also calculated at 100%. Where desired, **Section 220.60** permits only the largest load of the two noncoincidental loads to be used for calculating feeder demand. Noncoincident loads are loads which are unlikely to be used at the same time.

Courtesy of Raychem Quicknet, Tyco Thermal Controls

The installation of fixed electric space heating cables in a dwelling unit.

Problem 8-7

The Smith house will have four 500-watt strip heaters and two 750-watt bathroom heaters. Each heater is separately controlled and operated at 240 volts. In addition, there is a 7½-hp, 240-volt air-conditioning unit. Calculate the heating and air-conditioning feeder demand and determine the noncoincident load.

Solution – Calculation 1
Heating load
4 strip heaters × 500 watts
= 4 × 500
= 2,000 VA — 2,000 VA
2 heaters × 750 watts
= 2 × 750
= 1,500 VA — 1,500 VA
Total heating load — 3,500 VA
Answer: 3,500 VA heating load

Solution – Calculation 2
Air-conditioning load
Table 430.248
7½ hp at 240 volts FLC = 40 amps
VA = E × I
= 240 × 40
= 9,600 VA
Answer: 9,600 VA air-conditioning load

Solution – Calculation 3
Noncoincident load
Section 220.60
Use larger load
9,600 VA is larger than 3,500 VA
Use air-conditioning load
Answer: 9,600 VA ungrounded conductor load
No neutral grounded conductor load

Feeder Demand: Largest Motor

Section 220.50, by referencing sections in **Article 430** and specifically **Section 430.24**, requires that 125% of the largest motor full-load current (FLC) to be part of the feeder load calculation. Since the motor loads have all been accounted for at 100% load in Calculation 3 of **Problem 8-7**, this calculation must not include 125%, but rather only the remaining 25%. This next calculation is used to determine that 25%. It is used in the summary calculation as well.

Problem 8-8

What is 25% of the largest motor full-load current (FLC) planned for the Smith house?

Solution
The largest motor load is the air-conditioning motor
Section 430.24
VA = Largest load × 25%
= 9,600 × 0.25
= 2,400 VA
Answer: 2,400 VA ungrounded conductor load

8.2.4 Summary Dwelling Unit—Smith House, Standard Calculation Method

The Smith House Summary of Installations	
Area	65 feet by 55 feet
Service	120/240 volts, single-phase, 3-wire
Clothes dryer	4.5 kW
Disposer	120 volts, 6 amperes
Dishwasher	1.5 kW
Water heater	5 kW
Ventilation hood (fan/light)	120 volts, 4.4 amperes
Trash compactor	960 watts
Range	10 kW
Strip heaters	4-500 watts, 240 volts each
Heaters	2-750 watts, 240 volts each
Air-conditioning motor	7½ hp, 240 volts
25% of largest motor installed	

Summary: Standard Calculation Method for One Dwelling Unit

According to **Article 220, Part III, Section 220.40**, the calculated load of a feeder or service cannot be less than the sum of the branch-circuit loads determined by Part II, after any applicable demand factors allowed by Part III have been applied.

This summary step is arranged to provide two solutions using the standard calculation method: the total feeder demand in VA for ungrounded or line conductors and a separate total feeder demand in VA for the grounded conductor. Additional summary calculation examples are available for review using **Informative Annex D** of the *NEC*.

A one-family dwelling with an attached and a detached garage.

Problem 8-9

Summarize the total feeder demand for the Smith house using the standard calculations.

Solution

Area = *length* × *width*
= 65 × 55
= 3,575 ft²
VA per ft² dwelling = 3 VA per ft²

Feeder Demand			**VA Ungrounded Conductors**	**VA Neutral Conductor**
Lighting				
Calculated lighting load = *area* × VA per ft²				
= 3,575 × 3				
= 10,725 VA		10,725 VA		
Small appliances 2 circuits at 1,500		3,000 VA		
Laundry 1 circuit at 1,500		1,500 VA		
Calculated total		15,225 VA		
First 3,000 at 100%	3,000 VA			
3,001 to 120,000 at 35%				
15,225 – 3,000 = 12,225				
= 12,225 × 0.35				
= 4,279	4,279 VA			
Lighting demand	7,279 VA		7,279	7,279
Clothes dryer				
Minimum for 1 dryer = 5,000 VA			5,000	0
Neutral = 5,000 × 70%				
=5,000 × 0.70				
= 3,500			0	3,500
Appliances				
Disposer 120 × 6		720 VA		
Dishwasher 1.5 kW × 1,000		1,500 VA		
Water heater 5 kW × 1,000		5,000 VA		
Ventilation hood 120 × 4.4		528 VA		
Compactor 960 watts		960 VA		
Total appliance demand		8,708 VA		
4 or more = 75%				
= 8,708 × 0.75				
= 6,531 VA			6,531	6,531
Range				
One range 12 kW or less = 8,000 VA			8,000	0
Neutral = VA × 70%				
= 8,000 × 0.70				
= 5,600 VA			0	5,600
Heating				
4 strip heaters × 500 watts				
= 2,000 VA	2,000 VA			
2 heaters × 750 watts				
= 1,500 VA	1,500 VA			
Total heating load	3,500 VA		0	0
Air conditioning				
7½ hp at 240 volts FLC = 40 amps				
VA = *E* × *I*				
= 240 × 40				
= 9,600 VA			9,600	0
Air-conditioning load is larger than heat load				
Air-conditioning motor is largest motor				
= 9,600 VA × 25%				
= 9,600 × 0.25			2,400	0
Total Demand			38,810	22,910

Answer: 38,810 VA ungrounded conductor load
22,910 VA neutral grounded conductor load

Service-Entrance Conductors: Ampacity

Now that everything is calculated in VA, the ampacity of the line and neutral conductors for the service-entrance conductors can be calculated for the Smith House using Ohm's Law and the wire size selected. **220.5(A)** specifies that the voltage to be used in these calculation is the nominal voltage of the system, which is 240 volts in this example.

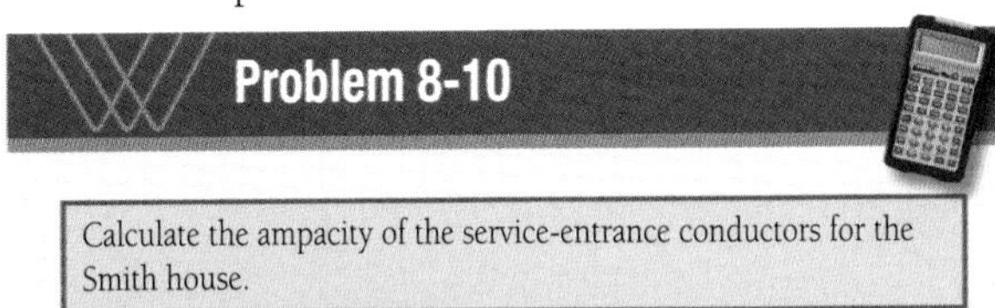

Problem 8-10

Calculate the ampacity of the service-entrance conductors for the Smith house.

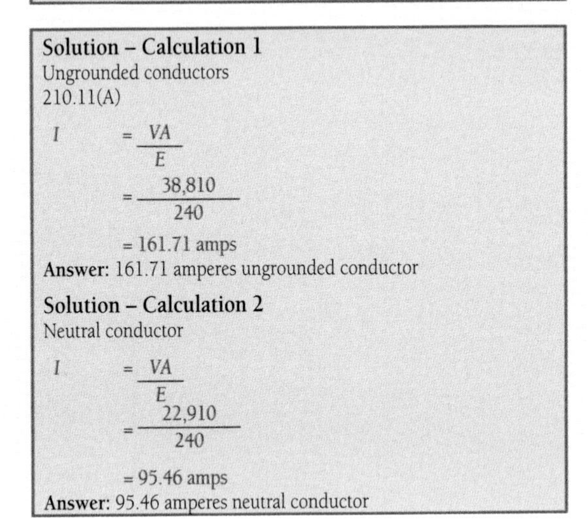

Solution – Calculation 1

Ungrounded conductors

210.11(A)

$$I = \frac{VA}{E}$$

$$= \frac{38{,}810}{240}$$

$$= 161.71 \text{ amps}$$

Answer: 161.71 amperes ungrounded conductor

Solution – Calculation 2

Neutral conductor

$$I = \frac{VA}{E}$$

$$= \frac{22{,}910}{240}$$

$$= 95.46 \text{ amps}$$

Answer: 95.46 amperes neutral conductor

Service-Entrance Conductors: Wire Size

The next step is to select the size of the dwelling unit service-entrance conductors. **Section 310.15(B)(7)** now permits service conductors that supply the entire load 120/1240-volt single-phase dwelling services to have an ampacity not less than 83 percent of the service rating. Service ratings are the same as standard size overcurrent devices from 100 through 400 amperes as found in **240.6(A)**. For the 2014 *NEC*, Annex D provides a new Example D7 to assist user with direct application of this percentage.

The 120/240 volt, 3-wire service fits **310.15(B)(7)**, provided the service is not over 400 amperes. Generally, not many one-family dwelling services are over 400 amperes.

1. Unless the calculated amperage matches a service size, select the next larger standard size service rating from the calculated feeder load amperes.
2. The service (or feeder) conductor supplying the entire load associated with the dwelling is permitted to have an ampacity not less than 83 percent of the standard ampere rating between 100 amp and 400 amperes as found in standard ampere rating of **240.6(A)**.
3. For the neutral conductor, use the amperes found using Ohm's Law and **Table 310.15(B)(16)** for conductor size.

Problem 8-11

What size THWN copper conductors are needed for the 120/240 volt, 3-wire service to the Smith house?

Solution – Calculation 1

Determine ungrounded service conductor size, THWN, copper

Ungrounded conductor load = 161.71 amps

Determine Service Raring: use next larger standard size, 240.6(A)

Next larger standard size Service Rating = 175 amps

Service conductor ampacity = (Service rating) × (310.15(B)(7) factor

Service conductor ampacity = (175 amps) × (0.83)

Service conductor ampacity = 145.25 aamps

Using Table 310.15(B)(16), for THWN, copper

145.25 amp load requires a service conductor size = 1/0 AWG

Answer: 1/0 THWN CU ungrounded conductor

Solution – Calculation 2

Determine neutral service conductor size, THWN, copper

Ungrounded conductor load = 95.46 amps

Using Table 310.15(B)(16), for THWN, copper

95.46 amps requires a 3 AWG

Answer: 3 THWN CU neutral conductor

Comment

The 175-ampere service could use a 200-ampere switch with 175-ampere fuses or a 175-ampere circuit breaker. The standard sizes for fuses and circuit breakers are listed in 240.6(A).

Grounding Electrode Conductor: Wire Size

The size of the grounding electrode conductor is determined by its relationship to the size of the ungrounded conductor used for the service. **Table 250.66** is used to size the grounding electrode conductor.

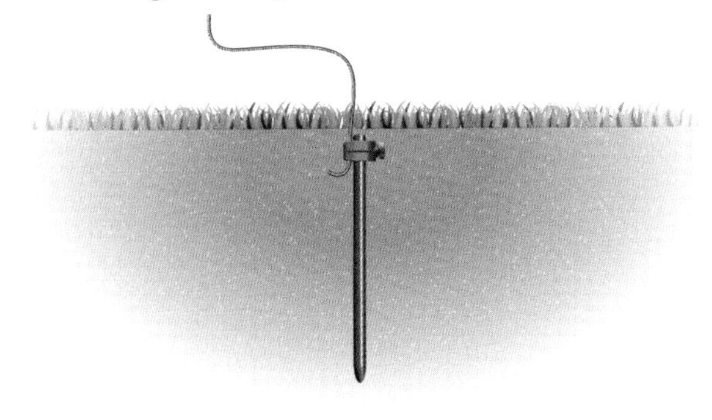

Grounded electrode conductor

Problem 8-12

What minimum size is needed for a copper grounding electrode conductor to a metal underground water pipe grounding electrode for the Smith house?

Solution
Largest ungrounded service-entrance conductor = 1/0 AWG
Table 250.66
1/0 AWG copper = 6 AWG copper
Answer: 6 AWG

Comment
For most residential services, the neutral conductor is not permitted to be smaller than the grounding electrode conductor found in Table 250.66. The minimum size grounded conductor for services is described in detail in 250.24(C)(1).

8.2.5 Dwelling Unit—Smith House, Optional Calculation Method

The optional feeder and service load calculations are found in **Article 220, Part IV**. This section only applies to a dwelling unit having a total connected load with an ampacity of 100 amperes or greater.

Summary: Optional Method for One Dwelling Unit
Using **Section 220.82**, the optional method of calculating feeder demand for a dwelling unit can be summed up in three statements:

1. Convert all loads to VA
2. Calculate the first 10,000 VA at 100% and the balance at 40%.
3. Select the correct method of calculating the heating and air-conditioning feeder demand from **220.82(C)**.

Problem 8-13

Calculate the feeder demand for the Smith house using the optional method.

Solution

Area = length × width	
= 65 × 55	
= 3,575 ft²	
VA per ft² dwelling = 3 VA per ft²	
Lighting	
Calculated lighting load = area × VA per ft²	
= 3,575 × 3	
= 10,725 VA	10,725 VA
Small appliances 2 circuits at 1,500	3,000 VA
Laundry 1 circuit at 1,500	1,500 VA
Appliances	
Clothes dryer 4.5 kW × 1,000	4,500 VA
Disposer 120 × 6	720 VA
Dishwasher 1.5 kW × 1,000	1,500 VA
Water heater 5 kW × 1,000	5,000 VA
Ventilation hood 120 × 4.4	528 VA
Compactor 960 watts	960 VA
Household cooking	
Range 12 kW (at nameplate)	12,000 VA
Total of nameplate ratings	40,433 VA

First 10,000 at 100%		10,000 VA
Balance at 40%		
40,433 – 10,000 = 30,433 VA		
30,433 VA × 0.40		
=12,173 VA		12,173 VA
Heating		
220.82(C)(5)		
40% demand factor		
4 strip heaters × 500 watts		
= 2,000 VA	2,000 VA	
2 heaters × 750 watts		
= 1,500 VA	1,500 VA	
Subtotal	3,500 VA	
6 units at 40%		
3,500 VA × 0.40 = 1,400 VA		
Heating load not used		
Air conditioning		
220.82(C)(1)		
7½ hp at 240 volts FLC = 40 amps		
VA = E × I		
= 240 × 40		
= 9,600 VA		
Air-conditioning load is the largest load		9,600 VA
Total feeder demand optional method		31,773 VA
Answer: 31,773 VA total feeder demand		

Using the Optional Method for Service-Entrance Conductor Ampacity, Line Size, Neutral Size, and Grounding Electrode Conductor Size
As long as a service fulfils the requirements of **310.15(B)(7)**, a factor of 0.83 is permitted to be used to reduce the size of the ungrounded service conductor for the Optional Method for calculating line size service-entrance conductor ampacity. This 2014 *NEC* change effects both Solution - Calculation 2 and 3 in Problem 8-11 without changing the answer(s) (from previous editions of this problem.) The *NEC* provides an example of the change within Annex D, Example D7.

When the service does not fall within the parameters of **310.15(B)(7)**, the line conductors must be sized directly from **Table 310.15(B)(16)** without further reduction. However, **220.30(A)** still permits the neutral conductor to be calculated according to **220.61(B)**, which is the same method by which the neutral was calculated using the previous standard calculation method.

8.3 Multifamily Dwellings

Multifamily dwelling are permitted to be calculated according to the standard method of Part III or the optional method of Part IV within **Article 220**. The standard methods will be used in **8.3.1 Introduction to Multifamily Dwellings** and in **8.3.2 Homestead Apartments**. Later in 8.3.3, the optional calculations according to Part IV of **Article 220** will be used.

Problem 8-14

Calculate the following using the optional calculation method:
1. Ampacity of service-entrance conductors
2. Minimum size service
3. Minimum size THWN copper service-entrance conductors
4. Minimum size neutral for service
5. Minimum size copper grounding electrode conductor using a metal water pipe grounding electrode

Solution – Calculation 1
Ampacity of service-entrance conductors

$$I = \frac{VA}{E}$$

$$= \frac{31{,}773}{240}$$

= 132 amps

Answer: 132 amperes

Solution – Calculation 2
Determine minimum service size
Min. calculated service size = 132 amps
240.6(A), next larger standard size OCPD
Minimum service size = 150 amps

Solution – Calculation 3
Determine minimum conductor size
Section 310.15(B)(7)(1)
150 amps × 0.83 = 124.5 amps
Table 310.15(B)(16) = 1 AWG

Answer: 1 AWG THWN CU

Solution – Calculation 4
Neutral for service
220.61(A)
Maximum unbalance

Lighting load		
$3{,}575 \text{ ft}^2 \times 3 \text{ VA per ft}^2$	10,725 VA	
Small appliances 2 circuits at 1,500	3,000 VA	
Laundry 1 circuit at 1,500	1,500 VA	
Subtotal	15,225 VA	
3,000 VA at 100%		3,000 VA
Balance at 35%		
15,225 VA – 3,000 VA = 12,225		
12,225 × 0.35		
= 4,279 VA		4,279 VA
Appliances		
Household range at 12 kW (8 kW at 70%)		5,600 VA
Clothes dryer at 4.5 kW (5 kVA at 70%)		3,500 VA
Disposer 120 volts × 6		720 VA
Dishwasher 1.5 kW × 1,000		1,500 VA
Ventilation hood 120 volts × 4.4		528 VA
Compactor 960 watts		960 VA
Total		20,087 VA

$$I = \frac{VA}{E}$$

$$= \frac{20{,}087}{240}$$

= 83.7 amps

Table 310.15(B)(16)
83.7 amps = 4 AWG THWN copper

Answer: 4 AWG THWN neutral

Solution – Calculation 5
Grounding electrode conductor
Table 250.66
1 AWG copper = 6 AWG copper

Answer: 6 AWG copper

8.3.1 Introduction to Multifamily Dwellings

The same standard calculation method used to calculate a one-family dwelling is used to calculate the feeder to each individual apartment of a multifamily dwelling. **310.15(B)(7)** is applicable to a 120/240 volt, 3-wire feeder installed to each apartment, when the service for the complex is 120/240 volts, 3-wire, rated 400 amperes or less. However, when the apartment complex is supplied with 120/208 volts, a 3-phase, 4-wire feeder and a 120/208 volt, single-phase, 3-wire feeder are installed to each apartment, **310.15(B)(7)** is not applicable.

The sequence of calculations for a multifamily dwelling follows the calculation for a one-family dwelling very closely. However, a few additional demand factors apply to a multifamily dwelling. The following outline of these loads can be used as a checklist for calculating specific loads:

- Lighting load
- Small appliance loads
- Laundry load
- Other fastened-in-place appliance loads
- Cooking appliance load
- Clothes dryer load
- Heating load
- Air-conditioning load
- Other loads common to premises
- 25% of largest motor installed

The following sequence of problems pertains to the planning of the Homestead Apartments, a multifamily dwelling. The basic plan calls for eighteen 1,400 ft^2 units, with a 120/240 volt, single-phase, 3-wire service. The lighting is incandescent, and there are no plans for general laundry facilities on the premises.

Lighting Load

The Homestead Apartments consists of 18 separate apartments. The apartment lighting load is based on the area of each apartment which is 1,400 ft^2. As a reminder, **220.12(J)** indicates the general-use receptacles in one-family, two-family and multifamily dwellings are included in the general lighting load calculation. In addition to lighting within each apartment, there is also outdoor security lighting connected to the house load and not associated with any apartment load. The security lighting load is a separate calculation which will appear later in this section.

Problem 8-15

Calculate the incandescent lighting load for the Homestead Apartments.

Solution

Area = ft^2 of each unit × number of units
= 1,400 × 18
= 25,200 ft^2

Table 220.12

Dwelling unit = 3 VA per ft^2

Lighting load = area × VA per ft^2
= 25,200 × 3
= 75,600 VA

Answer: 75,600 VA calculated lighting load

Feeder Demand: Lighting, Small Appliance, and Laundry Circuit

Two small appliance branch circuits are required for each unit. There are no general laundry facilities on the premises. Therefore, the laundry branch-circuit load must be calculated for each unit according to **210.11(C)(2)**. Both of these loads are permitted to be added to the calculated lighting load for the complex, and the demand factors of **Table 220.42** are then applied.

Problem 8-16

Calculate the feeder demand for the lighting, small appliance, and laundry loads for the Homestead Apartments.

Solution

Calculated lighting load	75,600 VA
Small appliances load	
= 2 circuits × 18 units × 1,500 VA	
= 2 × 18 × 1,500	
= 54,000 VA	54,000 VA
Laundry circuit load	
= 1 circuit × 18 units × 1,500	
= 1 × 18 × 1,500	
= 27,000 VA	27,000 VA
Subtotal	156,600 VA
Table 220.42	
Apply demand factors	
First 3,000 VA at 100%	3,000 VA
3,001 to 120,000 VA at 35%	
120,000 − 3,001 = 117,000	
117,000 × 0.35	
= 40,950 VA	40,950 VA
Remaining at 25%	
156,600 − 3,000 − 117,000 = 36,600	
36,600 × 0.25	
= 9,150 VA	9,150 VA
Feeder demand for lighting	53,100 VA

Answer: 53,100 VA ungrounded and neutral conductor load

According to *LED Lighting Explained* published by Philips Color Kinetics, LED light sources deliver high-quality white, colored, or color-changing light while consuming far less energy than conventional sources.
www.ledlightingexplained.com

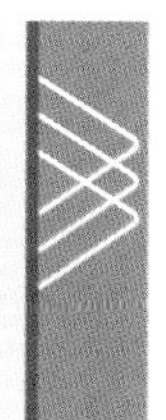

Feeder Demand: Appliance Loads

The fastened-in-place appliances installed in each apartment are additive for the multifamily complex and easily meet the four required appliances for the 75% reduction of **Section 220.53**.

Problem 8-17

Each unit of the Homestead Apartments will have one 2-kW, 120-volt dishwasher, one 3-kW, 120-volt water heater, and a 6.2-ampere, 120-volt waste disposer. Calculate the fastened-in-place appliance load for the Homestead Apartments.

Solution

VA = kW × 1,000

Dishwasher

2 kW × 1,000 = 2,000 VA

2,000 VA × 18 units = 36,000 VA — 36,000 VA

Water heater

3 kW × 1,000 = 3,000 VA

3,000 VA × 18 units = 54,000 VA — 54,000 VA

Disposer

Amps × volts = VA

6.2 × 120 = 744 VA

744 VA × 18 units = 13,392 VA — 13,392 VA

Total — 103,392 VA

Section 220.53

Over 4 appliances = 75% reduction

103,392 × 0.75 = 77,544 VA

Answer: 77,544 VA ungrounded and neutral conductor load

Feeder Demand: Dryer Loads

When a number of household dryers are installed, **Table 220.54** permits a demand factor according to the number of household dryers installed.

One example of a multifamily complex where a number of appliances and dryers are often used.

Problem 8-18

Plans are to install one 5-kW, 120/240 volt dryer in each unit of the Homestead Apartments. Calculate the feeder demand for the dryer load.

Solution – Calculation 1

Line load

VA = Dryer kW × 1,000

= 5 × 1,000

= 5,000 VA

Total number of dryers = 18

Table 220.54

12-23 dryers = 47% – 1% for each dryer exceeding 11

18 – 11 = 7

Demand factor = 47% – 7%

= 40%

Load = *number of dryers* × VA × 40%

= 18 × 5,000 × 0.40

= 36,000 VA

Answer: 36,000 VA ungrounded conductor load

Solution – Calculation 2

Neutral conductor load

220.61(B)

Neutral VA = VA × 70%

= 36,000 × 0.70

= 25,200 VA

Answer: 25,200 VA neutral conductor load

Feeder Demand: Range Loads

Calculations for range load and other cooking appliances were covered in Chapter 7 of *Code Calculations*. The neutral conductor for range loads is permitted to be reduced by 70%, according to **220.61(B)(1)**.

Problem 8-19

Plans are to install one 9-kW, 120/240 volt household electric range in each unit of the Homestead Apartments. Calculate the feeder demand for the ranges.

Solution – Calculation 1

Ungrounded conductor load

Table 220.55

Column C

18 ranges = 33 kW or 33,000 VA

Answer: 33,000 VA ungrounded conductor load

Solution – Calculation 2

Neutral conductor load

220.61(B)

Neutral VA = VA × 70%

= 33,000 × 0.70

= 23,100 VA

Answer: 23,100 VA neutral conductor load

Feeder Demand: Heating and Air-Conditioning Loads

Where air-conditioning loads are installed in a dwelling unit, the loads are calculated at 100%.

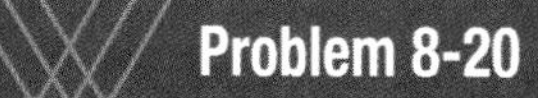

Problem 8-20

Plans call for one 3-hp, 240-volt, single-phase air conditioner in each unit of the Homestead Apartments. Calculate the feeder demand for the air-conditioning load.

Solution
Table 430.248
3 hp = 17 amps
VA = E x I x *number of units*
= 240 x 17 x 18
= 73,440 VA
Answer: 73,440 VA

Feeder Demand: 25% of Largest Motor

The calculation of the largest motor at 25% is also required for a multifamily dwelling. In this case, it is one of the eighteen 3-hp air conditioner motors.

Problem 8-21

What is 25% of the largest motor?

Solution
Largest motor
Air conditioner motor
240 volts and 17 amps
VA = E x I
= 240 x 17
= 4,080 VA
VA *demand*= 4,080 x 0.25
= 1,020 VA
Answer: 1,020 VA ungrounded conductor load

Feeder Demand: Other Loads

Other loads that might be added are calculated at nameplate rating.

Courtesy of NECA

Other loads may include a self-contained hot tub.

Problem 8-22

Plans are to have six 500-watt, 120-volt incandescent outdoor luminaires installed on the premises of the Homestead Apartments for security lighting. Calculate the feeder demand for the security lighting.

Solution
Load = 6 x 500 watts
= 3,000 VA
Answer: 3,000 VA ungrounded and neutral conductor load

8.3.2 Homestead Apartments – Using the Standard Method

The following is a summary of the Homestead Apartments:

The Homestead Apartments	
FOR THE COMPLEX:	
Security lighting	Six 500-watt outdoor luminaires
Service to complex	120/240 volts, single-phase, 3-wire
Size	18 units, each unit = 1,400 ft^2
WITHIN EACH UNIT:	
Dishwasher	2 kW, 120 volts
Water heater	3 kW, 120 volts
Disposer:	6.2 amps, 120 volts
Dryer:	5 kW, 120/240 volts
Range	9 kW, 120/240 volts
Air conditioner	3 hp, 240 volts

Feeder Demand: Summary Using the Standard Calculation Method

According to **Article 220, Part II, Section 220.40**, the calculated load of a feeder or service cannot be less than the sum of the branch-circuit loads determined by Part II, after any applicable demand factors allowed by Part III have been applied.

This summary step is arranged to provide two solutions using the standard calculation method: the total feeder demand in VA for ungrounded or line conductors and a separate total feeder demand in VA for the grounded conductor. Additional summary calculations are available for review using **Informative Annex D** of the 2011 *NEC*.

Problem 8-23

Calculate the feeder load for the Homestead Apartments using the standard method.

Solution

Area = ft² × *number of units*
= 1,400 × 18
= 25,200 ft²

Lighting load = *area* × VA per ft²
= 25,200 × 3
= 75,600 VA

Lighting Demand	**VA Ungrounded Conductors**		**VA Neutral Conductor**
Section 220.42			
Apply demand factor			
Calculated lighting load	75,600 VA		
2 small appliance branch circuits			
= 2 circuits × 18 units × 1,500			
= 54,000 VA	54,000 VA		
1 laundry branch circuit			
= 1 circuit × 18 units × 1,500			
= 27,000 VA	27,000 VA		
Subtotal	156,600 VA		
First 3,000 at 100%	3,000 VA		
Next 117,000 at 35%			
= 117,000 x 0.35			
= 40,950 VA	40,950 VA		
Remainder of 36,600 at 25%			
= 36,600 x 0.25			
= 9,150 VA	9,150 VA		
Lighting feeder demand	53,100 VA	53,100	53,100
Appliance loads			
Dishwasher 2 kW × 18 36,000 VA			
Water heater 3 kW × 18	54,000 VA		
Disposer 6.2 × 120 × 18	13,392 VA		
Subtotal	103,392 VA		
Over 4 appliances at 75%			
= 103,392 × 0.75			
= 77,544 VA		77,544	77,544
Dryer Feeder Demand			
Table 220.54			
18 dryers = demand factor 40%			
= 18 × 5 kW × 0.40			
= 36,000 VA		36,000	0
Neutral = 36,000 × 70%			
= 36,000 x 0.70			
= 25,200 VA		0	25,200
Household Cooking Feeder Demand			
Table 220.55			
18 ranges = 33 kW × 1,000		33,000	0
Neutral = 33,000 × 70%			
33,000 × 0.70			
= 23,100 VA		0	23,100
Air-Conditioning Demand			
3 hp, 240 volt FLC = 17 amps			
VA = 240 × 17			
= 4,080 VA			
4,080 VA × 18 motors		73,440	0
25% of largest motor			
4,080 VA × 0.25			
= 1,020 VA		1,020	0
Other Loads			
Security lighting 6 × 500 W		3,000	3,000
Total Demand		277,104	181,944

Answer: 277,104 VA ungrounded conductor load
181,944 VA neutral grounded conductor load

Information

Once the total feeder demands are calculated, several other items can be calculated. The following problems take a look as some of these items. They include the ungrounded and grounded conductor ampacity, the corresponding conductor sizes, and the conduit size(s).

Service-Entrance Conductors: Ampacity

The ampacity calculation is performed by dividing the VA by the voltage. However, **220.61(B)(2)** permits additional derating factors for the neutral conductor when:

1. The calculated neutral conductor current is over 200 amperes, and
2. The neutral conductor load is other than nonlinear loads such as electric discharge lighting.

Since discharge lighting is not present, further derating according to **220.61(B)(2)** is used.

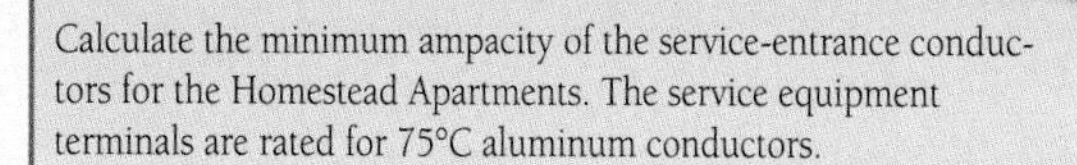

Problem 8-24

Calculate the minimum ampacity of the service-entrance conductors for the Homestead Apartments. The service equipment terminals are rated for 75°C aluminum conductors.

Solution – Calculation 1
Ungrounded service-entrance conductors
210.11(A)

$$I = \frac{VA}{E}$$

$$= \frac{277{,}104}{240}$$

$$= 1{,}154.6 \text{ amps}$$

Answer: 1,154.6 amperes ungrounded service-entrance conductors

Solution – Calculation 2
Neutral conductor

$$I = \frac{VA}{E}$$

$$= \frac{181{,}944}{240}$$

$$= 758.1 \text{ amps}$$

220.61(B)(2)
Derating of neutral

Total neutral calculated amps	758
First 200 amps at 100%	200
Remaining 558 amps at 70%	
558 × 0.70 = 391	391
Total neutral ampacity	591 amps

Answer: 591 amperes neutral service-entrance conductors

Service-Entrance Conductors: Wire Size

The next step is to select the proper conductor size using parallel aluminum service-entrance conductors. The calculated load is divided by the number of conductors (per phase) connected in parallel. **310.10(H)** restricts the minimum size parallel conductor permitted to 1/0 AWG (with few exceptions).

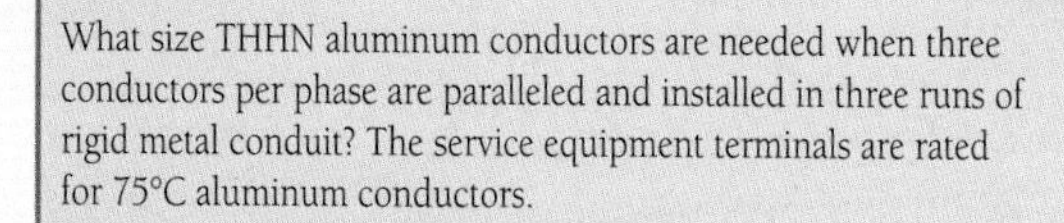

Problem 8-25

What size THHN aluminum conductors are needed when three conductors per phase are paralleled and installed in three runs of rigid metal conduit? The service equipment terminals are rated for 75°C aluminum conductors.

Solution – Calculation 1
Ungrounded service-entrance conductors
210.11(A)

$$\textit{Single conductor amps} = \frac{\textit{total amps}}{3}$$

$$= \frac{1{,}155}{3}$$

$$= 385 \text{ amps}$$

Table 310.15(B)(16)
75°C THHN aluminum column
385 amps = 750 kcmil
Answer: Three 750 kcmil THHN aluminum conductors per phase

Solution – Calculation 2
Service-entrance neutral conductor

$$\textit{Single conductor amps} = \frac{\textit{total amps}}{3}$$

$$= \frac{591}{3}$$

$$= 197 \text{ amps}$$

Table 310.15(B)(16)
75°C THHN aluminum column
197 amps = 250 kcmil
Answer: 250 kcmil THHN aluminum

Comment
The line and neutral conductors are all over 1/0 AWG and are permitted to be paralleled by 310.10(H)(1). When the neutral calculations result in a conductor smaller than 1/0 AWG, the minimum 1/0 AWG size takes precedence.

Electrical continuity of at service equipment must be ensured. This is one prescribed method of bonding three service conduits with parallel conductors.

Service-Entrance Conductors: Conduit Size

Chapter 9, Tables 1, 4, and **5A** are used to determine conduit size if a raceway contains mixed size aluminum conductors. **Table 5A** is used to calculate the total area of wire present in the conduit. Then, using the rigid metal conduit (RMC) portion of **Table 4**, the total calculated square inches of wire fill is applied to the 40% column such that the total fill does not exceed the total area of the selected conduit.

Problem 8-26

What is the minimum size rigid metal conduit (RMC) needed where compact aluminum conductors are installed in three parallel conduit sets?

Solution

Chapter 9 Table 5A

750 kcmil THHN aluminum = 0.9076 $in.^2$

0.9076 $in.^2 \times 2$ = 1.8152 $in.^2$ — 1.8152 $in.^2$

250 kcmil THHN aluminum = 0.3525 $in.^2$

0.3525 $in.^2 \times 1$ = 0.3525 $in.^2$ — <u>0.3525 $in.^2$</u>

Total area of 3 conductors — 2.1677 $in.^2$

Chapter 9 Table 4 Article 344 (RMC)

Over two wires, 40% fill column

2.1677 $in.^2$ = 3 in. RMC

Answer: 3 in. RMC

8.3.3 Homestead Apartments – Optional Calculations

Optional calculations for multifamily dwellings are permitted to be used according to **Article 220, Part IV, Section 220.84**, instead of Part III. This section and associated table basically permits the use of additional demand factors based upon the number of dwelling units per service or feeder. Before using this option, verify that all of the qualifications are available.

Summary: Optional Calculation Method for Multifamily Dwellings

The following specific qualifications must be met before a multifamily dwelling unit can qualify for the optional calculation:

1. Each dwelling unit is supplied with one feeder
2. Each dwelling unit is equipped with electric cooking
3. Each unit has air conditioning, electric heat, or both

For additional information, visit qr.njatcdb.org Item #1038

The optional method of calculating the feeder demand for a multifamily dwelling unit is based upon **Section 220.84** and **Table 220.84**. All loads common to each unit are calculated at nameplate rating. The demand factor of **Table 220.84** is applied using the following steps:

1. Calculate the general lighting load and general-use receptacles at 3 VA per ft^2.
2. Calculate the small appliance load at 1,500 VA per circuit.
3. Calculate the laundry circuit at 1,500 VA.
4. The nameplate rating of the following:
   a. All appliances fastened in place, permanently connected, or located to be on a specific circuit
   b. Ranges, wall-mounted ovens, counter-mounted cooking units
   c. Clothes dryers not connected to the laundry circuit in item 3.
5. Calculate the nameplate amperes or VA rating of permanently-connected motors not included in item 4.
6. The larger of heating or air-conditioning loads or fixed electric heating load
7. Total the loads and apply the demand factor for the number of units as given in **Table 220.84**
8. Total the house loads according to **220.84(B)**.

The term *house load* refers to all public electrical loads within a multifamily which are not directly associated with a particular dwelling unit within the property.

From the quantity of SER cables in view, this is most likely a corridor ceiling just outside the main electrical room of a large multifamily dwelling.

Problem 8-27

Calculate the feeder demand for the Homestead Apartments using the optional method.

Solution

One Dwelling Unit

Lighting 1,400 ft² × 3	4,200 VA
Small appliances 2 circuits at 1,500	3,000 VA
Laundry 1 circuit at 1,500	1,500 VA
Dishwasher 2 kW × 1,000	2,000 VA
Water heater 3 kW× 1,000	3,000 VA
Disposer 6.2 amps × 120	744 VA
Dryer 5 kW× 1,000	5,000 VA
Range 9 kW× 1,000	9,000 VA

Heating and air conditioning
3 hp, 240 volts FLC = 71 amps
VA = E × I × number of units
= 240 × 17 × 1
= 4,080 VA — 4,080 VA

Subtotal — 32,524 VA

Total= 18 units × 32,524 VA
= 585,432 VA

Table 220.84
18 units at 38%
= 585,432 × 0.38
= 222,464 VA — 222,464 VA

Security lighting (house load) — 3,000 VA

Total — 225,464 VA

Answer: 225,464 VA demand using the optional method

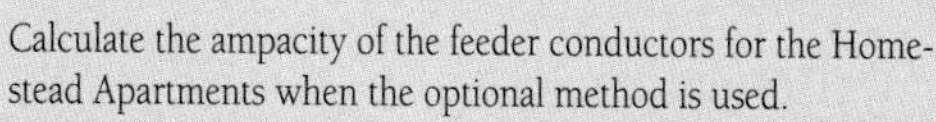

Problem 8-28

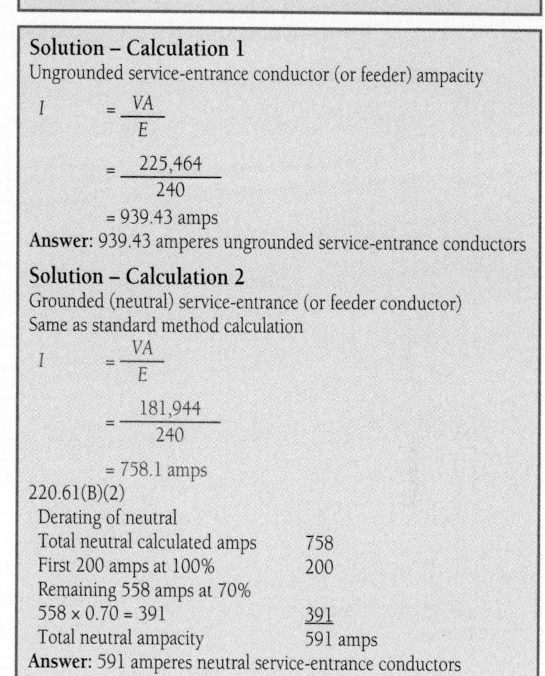

Calculate the ampacity of the feeder conductors for the Homestead Apartments when the optional method is used.

Solution – Calculation 1

Ungrounded service-entrance conductor (or feeder) ampacity

$$I = \frac{VA}{E}$$

$$= \frac{225,464}{240}$$

$$= 939.43 \text{ amps}$$

Answer: 939.43 amperes ungrounded service-entrance conductors

Solution – Calculation 2

Grounded (neutral) service-entrance (or feeder conductor)
Same as standard method calculation

$$I = \frac{VA}{E}$$

$$= \frac{181,944}{240}$$

$$= 758.1 \text{ amps}$$

220.61(B)(2)
Derating of neutral

Total neutral calculated amps	758
First 200 amps at 100%	200
Remaining 558 amps at 70%	
558 × 0.70 = 391	391
Total neutral ampacity	591 amps

Answer: 591 amperes neutral service-entrance conductors

Service-Entrance Conductors: Size and Ampacity
220.84(A)(3) indicates the use of **220.61(B)** for calculating neutral conductors. This permits the application of a 70% demand factor on the neutral load. It also indicates that the optional method is used for ungrounded feeder and service-entrance conductors, while the standard method is used for neutral feeder and service-entrance conductors.

An example of stackable apartment meter sockets and feeder disconnect circuit breakers, all supplied from a service disconnect.

8.4 Commercial Buildings

The calculations for commercial buildings are somewhat more restrictive since there are fewer demand factors to apply and does not have an optional method of calculating the feeder load. However, the basic system is very much like the residential in that everything is converted to volt-amperes and all loads are considered to be balanced.

It is important to understand that commercial buildings often have continuous loads. Continuous loads are loads which continue for three hours or more. Continuous loads must be accounted for during load calculations specifically when overcurrent devices and conductor sizes are calculated and determined.

8.4.1 Variety Store with Warehouse

The first commercial building to plan is a variety store. The store area has a 60 foot frontage and is 150 feet deep, with an additional 5,000 ft² of storage warehouse area in the basement. The service will be 120/208 volts, 3-phase, 4-wire. Plans call for the general lighting to be 120-volt fluorescent throughout.

Feeder Demand: General Lighting

Table 220.12 lists the lighting VA per ft² for different types of locations by occupancy. More than one of these locations can be located in the same building. When they are, each area is calculated separately with the VA per ft² listed in **Table 220.12**.

Problem 8-29

Calculate the lighting load for the variety store. The general store area fluorescent lighting is considered a continuous load since it will be on for three hours or more. The storage area lighting is considered a noncontinuous lighting load.

Solution – Calculation 1
General store area (continuous load)
Area = *length* × *width*
= 150 × 60
= 9,000 ft²
Table 220.12
Store lighting = 3 VA per ft²
Lighting load = *area* × VA
= 9,000 × 3
= 27,000 VA
Answer: 27,000 VA continuous load

Solution – Calculation 2
Storage area (noncontinuous)
Table 220.12
Storage lighting = 1/4 VA per ft²
VA = *area* × VA per ft²
= 5,000 × 0.25
=1,250 VA
Answer: 1,250 VA noncontinuous load

Solution – Calculation 3
Neutral conductor load

Continuous load	27,000 VA
Noncontinuous load	1,250 VA
Total	28,250 VA

Answer: 28,250 VA neutral conductor load

Feeder Demand: Receptacle Loads

The number of general-purpose duplex receptacles in a store is not regulated. The receptacles are not factored in with the lighting load as they are for a dwelling unit. Each general-purpose duplex receptacle is required to be calculated at 180 volt-amperes, according to **220.14(I)**. When receptacles are installed in nondwelling locations and calculated at 180 VA per receptacle, the demand factor of **Table 220.44** is used. The demand factor is applied as follows:

1. The first 10,000 VA are calculated at 100%
2. The balance is calculated at 50%

Problem 8-30

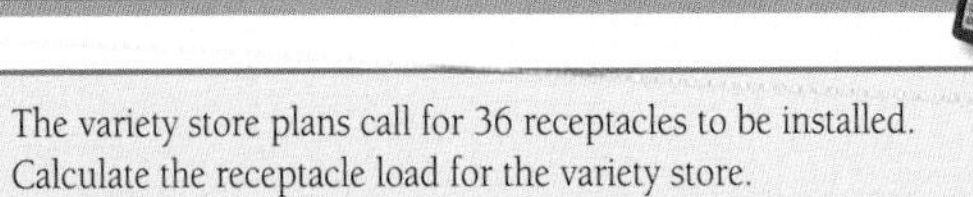

The variety store plans call for 36 receptacles to be installed. Calculate the receptacle load for the variety store.

Solution – Calculation 1
220.14(I)
VA load = *number of receptacles* × 180 VA
= 36 × 180
= 6,480 VA
Answer: 6,480 VA noncontinuous load

Solution – Calculation 2
Table 220.44
Receptacle load is less than 10,000 VA = 100% load
Answer: 6,480 VA neutral conductor load

Feeder Demand: Sign Circuit

Almost every store needs an electric sign. The *Code* anticipates this need and in **600.5(A)** requires a 20-ampere branch circuit to be installed inside or outside at the front of the store for other than neon signs. **220.14(F)** requires the circuit to be calculated at 1,200 volt-amperes. This is in addition to other receptacle loads.

Problem 8-31

Calculate the feeder demand for the required 120-volt, 20-ampere sign circuit. The sign circuit is considered to be a continuous load according to 600.5(B).

Solution
600.5(A)
One (20-ampere) receptacle at 1,200 VA
Answer: 1,200 VA continuous and neutral conductor load

An example of a neon tubing type sign.

Feeder Demand: Show Window Lighting

Stores like to display their merchandise in well-lighted show windows. **220.43(A)** requires a special calculation of 200 VA per linear foot of show window.

Problem 8-32

Plans call for 120 volt lighting in two 20-ft show windows for the variety store. Calculate the load for the show window lighting. The show window lighting is considered to be a continuous load.

Solution

VA load = *number of show windows* × *length* × 200 VA
= 2 × 20 × 200
= 8,000 VA

Answer: 8,000 VA continuous and neutral conductor load

Feeder Demand: Multioutlet Assembly

Multioutlet assemblies are very handy for use with 120 volt electrical appliances, lamps, etc., especially in retail stores. The *Code* allots a special load to multioutlet assemblies according to the likelihood that the load will be used simultaneously, according to **220.14(H)**:

1. Light use: 180 VA per five feet of multioutlet assembly
2. Heavy use: 180 VA per foot of multioutlet assembly

Problem 8-33

Plans call for 30 ft of multioutlet assembly to be installed in the variety store for light use. Calculate the feeder demand for the multioutlet assembly. This load is considered to be non-continuous.

Solution

Ungrounded service-entrance conductor (or feeder) ampacity

$$Units = \frac{length}{5\text{ ft}} = \frac{30}{5} = 6 \text{ units}$$

VA load = units × 180 VA
= 6 × 180
= 1,080 VA

Answer: 1,080 VA noncontinuous and neutral conductor load

Feeder Demand: Other Loads—Commercial Store

There are numerous other loads which could be used in a store. Other loads are calculated at 100% of the nameplate rating.

Problem 8-34

A 10-kW, 208-volt, 3-phase, 80-gallon storage-type water heater is to be installed in the variety store. Calculate the water heater load.

Solution

One water heater at nameplate rating
10 kW = 10,000 VA

Answer: 10,000 VA noncontinuous load
Neutral conductor load = 0

Comment

The line and neutral conductors are all over 1/0 AWG and are permitted to be paralleled by 310.10(H)(1). When the neutral calculations result in a conductor smaller than 1/0 AWG, the minimum 1/0 AWG size takes precedence.

Feeder Demand: Small Motor Loads

Single-phase motor loads are calculated using the full-load current tables given in **Article 430** when the horsepower is given. If the horsepower is not given, but the full-load current is given, the full-load current is used. For continuous and noncontinuous loads, motor calculation should follow the structure as well as the requirements of **Section 430.24**. **Problem 8-35** follows **430.24(2)** since none of these motor loads will be considered the largest motor load.

Problem 8-35

Three 1/3-hp, 120-volt, single-phase motors are to be installed for ventilation purposes in the variety store. Calculate the full-load current (FLC) of the feeder demand for the single-phase motors.

Solution

Calculate at FLC rating
Table 430.248
1/3 hp at 120 volts FLC = 7.2 amps
VA = E × I × 3
= 120 × 7.2 × 3
= 2,592 VA

Answer: 2,592 VA noncontinuous and neutral conductor load

Feeder Demand: Air Conditioning

When there is both electric heat and air conditioning, the noncoincidental load regulation is applied. When no electric heat is indicated, air conditioning is calculated at 100%.

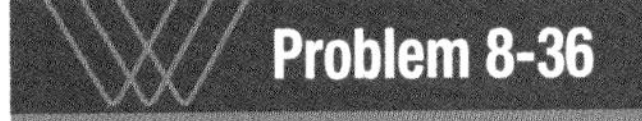

A 3-phase, 20-hp, 208-volt motor is used for air conditioning of the variety store. Calculate the feeder demand for the air conditioning.

Solution
Table 430.250
20 hp at 208 volts FLC = 59.4 amps
VA = E x I x 1.73
= 208 x 59.4 x 1.73
= 21,374 VA
Answer: 21,374 VA noncontinuous load
Neutral conductor load = 0

Comment
This calculation follows 430.24(2) so that it can be used with continuous and noncontinuous load calculations.

Feeder Demand: Largest Motor

Clarified for the 2011 *Code*, **Section 430.24** requires feeders which supply several motors or a motor and other loads have an ampacity not less than the sum of the following:

1. 125% of the FLC rating of the highest rated motor
2. Sum of the FLC ratings of all other motors in the group
3. 100% of the noncontinuous non-motor load
4. 125% of the continuous non-motor load

Since this change was a clarification as far as motors are concerned, the largest motor continues to multiplied by a factor of 25% and added as a separate line item. A sample calculation performed in this manner can be found in **Annex D, Example D3(A)**.

Courtesy of Baldor Electric Company

An example of a 20-hp 3-phase motor

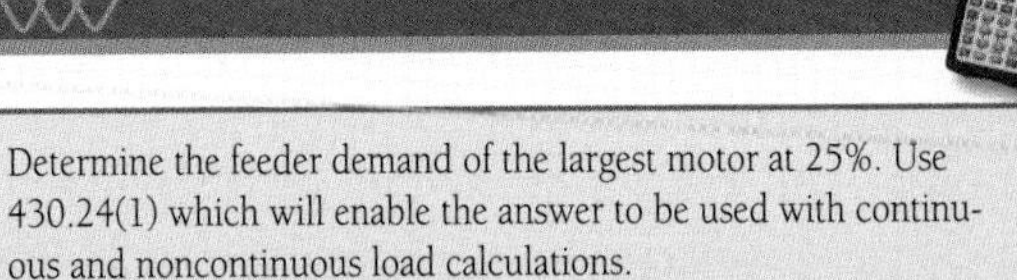

Determine the feeder demand of the largest motor at 25%. Use 430.24(1) which will enable the answer to be used with continuous and noncontinuous load calculations.

Solution
Air-conditioning motor is largest
Demand = VA x 25%
= 21,374 x 0.25
= 5,344 VA
Answer: Noncontinuous load = 5,344 VA
Neutral conductor load = 0

8.4.2 Variety Store Calculation Summary

Calculations for commercial buildings are viewed as more restrictive due to fewer applicable demand factors. Basic commercial calculations are very straightforward.

Since most commercial buildings have continuous loads, these loads must be accounted for during load calculations specifically when overcurrent devices and conductor sizes are calculated and determined. This summary consists of three individual calculations: total noncontinuous, continuous, and neutral loads.

However, the basic summary is similar to a residential summary, all loads are converted to volt-amperes and are considered balanced loads.

Variety Store Summary Feeder Demand	
Service	120/208 volts, 3-phase, 4-wire
Store area	60 feet x 150 feet
Basement storage area	5,000 ft²
Discharge lighting will be used	
General-purpose duplex receptacles	Thirty-six 120 volt
Sign circuit	One 20 amp, 120 volt, required
Show windows	Two 20 feet
Multioutlet assembly, light use	30 feet 120 volt
Water heater	One 10 kW at 208 volts, 3-phase
Single-phase motors	Three 1/3 hp, 120 volt
Motor	One 20 hp, 208 volt, 3-phase
25% of largest motor	

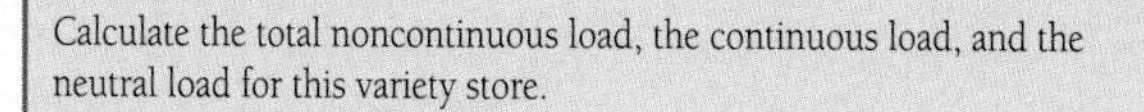

Problem 8-38

Calculate the total noncontinuous load, the continuous load, and the neutral load for this variety store.

Solution – Calculation 1

Noncontinuous loads		**Noncontinuous load VA**
Lighting storage area		
Load	= *area* × VA per ft²	
	= 5,000 × 1/4	
	= 5,000 × 0.25	
	= 1,250 VA	1,250 VA
General-purpose receptacles		
Load	= *number* × 180 VA	
	= 36 × 180	
	= 6,480 VA	6,480 VA
Multioutlet assembly		
180 VA per 5 ft light use		
Units	= $\frac{length}{5\text{ ft}}$	
	= $\frac{30}{5}$	
	= 6 units of 5 ft	
Load	= *number* × 180 VA	
	= 6 × 180 VA	
	= 1,080 VA	1,080 VA
Water heater		
One at 10 kW, 208 volts, 3-phase		
VA	= kW × 1,000	
	= 10 x 1,000	
	= 10,000 VA	10,000 VA
Single-phase motors		
Table 430.248		
1/3 hp at 120 volts		
FLC = 7.2		
Load	= *E* × *I* × 3	
	= 120 × 7.2 × 3	
	= 2,592 VA	2,592 VA
Air conditioning		
Table 430.250		
20 hp, 208 volt, 3-phase		
FLC = 59.4 amps		
VA	= *E* × *I* × 1.73	
	= 208 × 59.4 × 1.73	
	= 21,374 VA	21,374 VA
Largest motor x 25%		
	= 21,374 × 0.25	
	= 5,344 VA	5,344 VA
Noncontinuous load subtotal		48,120 VA

Answer: 48,120 VA Noncontinuous load

Solution - Calculation 2

Continuous loads		**Continuous load VA**
Store area lighting		
Area	= *length* x *width*	
	= 150 × 60	
	= 9,000 ft²	
Load	= *area* × VA per ft²	
	= 9,000 × 3	
	= 27,000 VA	27,000 VA
Required sign receptacle		
1 circuit at 1,200 VA		1,200 VA
Show window lighting		
Load	= *length* × 200 VA per ft	
	= 2 × 20 × 200	
	= 8,000 VA	8,000 VA
Continuous load subtotal		36,200 VA

Answer: 36,200 VA Continuous load

Solution - Calculation 3

Neutral loads		**Neutral load VA**
Store area lighting		
Area	= *length* × *width*	
	= 150 × 60	
	= 9,000 ft²	
Load	= *area* × VA per ft²	
	= 9,000 × 3	
	= 27,000 VA	27,000 VA
Note: Store lighting is fluorescent lighting		
Lighting storage area		
Load	= *area* × VA per ft²	
	= 5,000 × 1/4	
	= 5,000 × 0.25	
	= 1,250 VA	1,250 VA
General-purpose receptacles		
Load	= *number* × 180 VA	
	= 36 × 180	
	= 6,480 VA	6,480 VA
Required sign receptacle		
1 circuit at 1,200 VA		1,200 VA
Show window lighting		
Load	= *length* × 200 VA per ft	
	= 2 × 20 × 200	
	= 8,000 VA	8,000 VA
Multioutlet assembly		
180 VA per 5 ft light use		
Units	= $\frac{length}{5\text{ ft}}$	
	= $\frac{30}{5}$	
	= 6 units of 5 ft	
Load	= *number* × 180 VA	
	= 6 × 180	
	= 1,080 VA	1,080 VA
Single-phase motors		
Table 430.248		
1/3 hp at 120 volt		
FLC = 7.2		
Load	= *E* × *I* × 3	
	= 120 × 7.2 × 3	
	= 2,592 VA	2,592 VA
Neutral loads subtotal		47,602 VA

Answer: 47,602 VA Neutral loads

Service-Entrance Conductors: Ampacity and Overcurrent Protection

Once the total volt-ampere demand is calculated, the next step is to calculate the ampacity and size of the service-entrance conductors in accordance with **230.42(A)**. Consider this installation to be 3-phase, 4-wire with all four conductors in one conduit. Determining the overcurrent protection will follow the same basic method except without considering **Table 310.15(B)(3)(a)**.

Note that all the lighting in the variety store is fluorescent, which is discharge lighting, which can result in nonlinear load and cause harmonic currents in the neutral conductor. Therefore, **310.15(B)(4)(c)** requires the neutral to be counted as a conductor when the neutral load is the majority of discharge lighting. There are four conductors in the conduit. Therefore, the adjustment factors of **310.15(B)(2)(a)** will apply. No temperature is indicated, so it is assumed to be within the range of ordinary ambient, or 86°F.

Problem 8-39

Calculate the ampacity of the service conductors for the variety store and determine the minimum size of the overcurrent protection for this service.

Solution - Calculation 1

Service overcurrent protection

230.42(A)

Noncontinuous loads	48,120 VA
Continuous loads = 36,200 VA	
125% of the continuous loads	
36,200 × 1.25 = 45,250 VA	45,250 VA
Total load	93,370 VA

Ungrounded conductors

$$I = \frac{VA}{E \times 1.73}$$

$$= \frac{93{,}370}{208 \times 1.73}$$

$$= 259 \text{ amps}$$

240.4(B)

Next higher rating

240.6(A)

300 amp overcurrent protective device

Answer: 300 ampere overcurrent protective device

Solution - Calculation 2

Service conductor ampacity

230.42(A)

Noncontinuous loads	48,120 VA
Continuous loads = 36,200 VA	
125% of the continuous loads	
36,200 VA × 1.25 = 45,250 VA	45,250 VA
Total load	93,370 VA

Ungrounded conductors

$$I = \frac{VA}{E \times 1.73}$$

$$= \frac{93{,}370}{208 \times 1.73}$$

$$= 259 \text{ amps}$$

310.15(B)(5)(c)

Neutral load is majority of discharge lighting

Neutral counts as a current-carrying conductor for conduit fill Table 310.15(B)(3)(a)

Ampacity correction for fill 4 conductors = 80%

$$Ampacity = \frac{I}{80\%}$$

$$= \frac{259}{0.80}$$

$$= 324 \text{ amps}$$

Answer: 324 amperes ungrounded conductors

Solution - Calculation 3

Neutral conductor load

$$I = \frac{VA}{E \times 1.73}$$

$$= \frac{47{,}602}{208 \times 1.73}$$

$$= 132 \text{ amps}$$

310.15(B)(5)(c)

Neutral load is majority of discharge lighting

Neutral counts as a current-carrying conductor for conduit fill, use Table 310.15(B)(3)(a)

Ampacity correction for fill 4 conductors = 80%

$$Ampacity = \frac{I}{80\%}$$

$$= \frac{132}{0.80}$$

$$= 165 \text{ amps}$$

Answer: 165 amperes neutral conductor

Comment

The conductors are required to be a larger size than the overcurrent device because there are four current-carrying conductors in the conduit.

Service-Entrance Conductors and a Bare Neutral: Size

According to **Section 230.41**, service-entrance conductors are required to be insulated conductors. However, grounded service-entrance conductors are permitted to be bare according to the exception following this section.

Problem 8-40

The variety store service is to consist of single THWN copper conductors for the line conductors and a bare neutral conductor installed in rigid metal conduit. 230.41 Exception permits a bare neutral conductor. What is the minimum size service-entrance conductors required?

Solution – Calculation 1
Ungrounded conductors
Table 310.15(B)(16)
324 amps THWN copper = 400 kcmil
Answer: 400 kcmil ungrounded conductors

Solution – Calculation 2
Neutral conductor
Neutral is bare, but its ampacity is the same as line conductor insulation.
Table 310.15(B)(16)
165 amps THWN copper = 2/0 AWG
Answer: 2/0 AWG bare neutral conductor

Service-Entrance Conductors: Conduit Size

Determining the minimum conduit size using bare conductors is different from previous conduit size calculations because bare conductor dimensions are less than insulated conductors. Bare conductor dimensions are found in **Chapter 9, Table 8**, whereas **Chapter 9, Table 5** is used for insulated conductor dimensions.

As Smart Grid technology grows, the common electric meters will see major changes. 2011 NEC changes to Section 230.82 will help pave the way for the these future advancements in power management via utility customer electric meters.

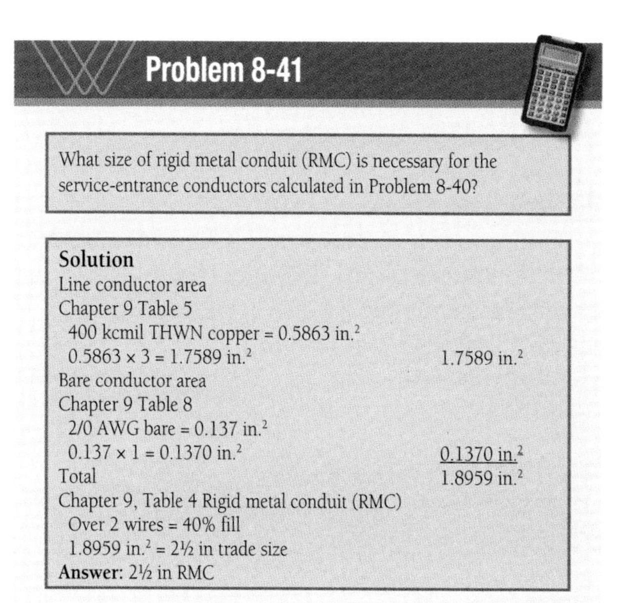

Problem 8-41

What size of rigid metal conduit (RMC) is necessary for the service-entrance conductors calculated in Problem 8-40?

Solution
Line conductor area
Chapter 9 Table 5
400 kcmil THWN copper = 0.5863 in.^{2}
0.5863 × 3 = 1.7589 in.^{2} — 1.7589 in.^{2}
Bare conductor area
Chapter 9 Table 8
2/0 AWG bare = 0.137 in.^{2}
0.137 × 1 = 0.1370 in.^{2} — 0.1370 in.^{2}
Total — 1.8959 in.^{2}
Chapter 9, Table 4 Rigid metal conduit (RMC)
Over 2 wires = 40% fill
1.8959 in.^{2} = 2½ in trade size
Answer: 2½ in RMC

8.5 Office Buildings

Generally speaking, the general term *office building* is broad classification of commercial property usually designated as a business occupancy and designed with spaces to be used for offices.

8.5.1 Method of Feeder Demand Calculation

There are multiple methods of calculating the lighting and receptacle loads for office buildings. Often, an office building is built before there are tenants. It is simply a shell of a building, often referred to as a speculative or "spec" type building. Without any tenants, no receptacles have been planned. Other office buildings are built with tenants in mind.

Within **Table 220.12 General Lighting Loads by Occupancy**, in the row for Office Buildings, Footnote[b] points to **220.14(K)**. This section accommodates either method of providing for receptacles. In practice, this section requires that the larger of either an actual count of receptacles or an allowance of an additional 1 VA per ft^2 be used.

For the Tower Office Building, a value of 1 VA per ft^2 for receptacle loads will be used in addition to the **Table 220.12** value of 3½ VA per ft^2 for lighting loads.

Multiple calculations can be made for the Tower Office Building. This building will consist of five floors, with an outside dimension of 125 feet by 88 feet. The service is 277/480 volts, 3-phase, 4-wire, and the lighting is fluorescent throughout.

Feeder Demand: Lighting without Including Receptacles

Determining the general lighting load in VA for an office building without considering receptacle load is done by applying the VA per ft^2 of **Table 200.12** to the total building area.

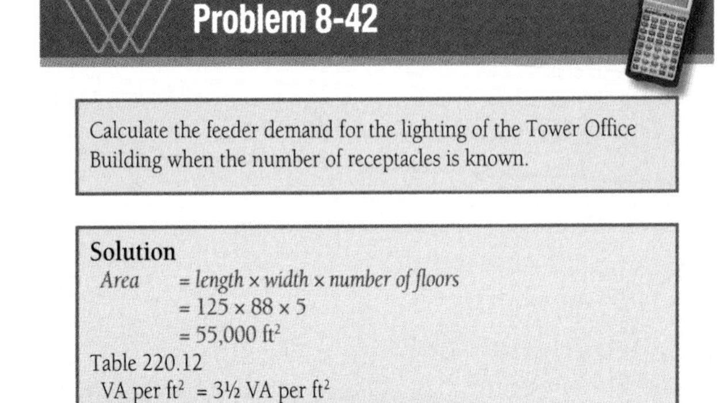

Problem 8-42

Calculate the feeder demand for the lighting of the Tower Office Building when the number of receptacles is known.

Solution

Area = *length × width × number of floors*
= 125 × 88 × 5
= 55,000 ft^2

Table 220.12

VA per ft^2 = 3½ VA per ft^2

Lighting load = *area* × VA per ft^2
= 55,000 × 3.5
= 192,500 VA

Answer: 192,500 VA

Calculating the Required Number of Lighting Circuits

Office building lighting is considered a continuous load. For this reason, **210.20(A)** requires the branch-circuit overcurrent protective device, which is also the branch-circuit rating, to be 125% of the connected continuous load. **Section 215.2** requires the same for feeders.

When calculating the number of branch circuits needed, 125% for continuous loads is used before the number of circuits is calculated. The 15- or 20-ampere branch circuits used are equivalent to the ratings for the overcurrent protective devices selected.

Typical office building lighting.

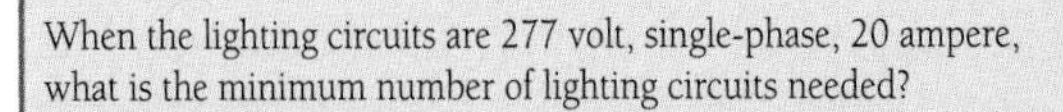

Problem 8-43

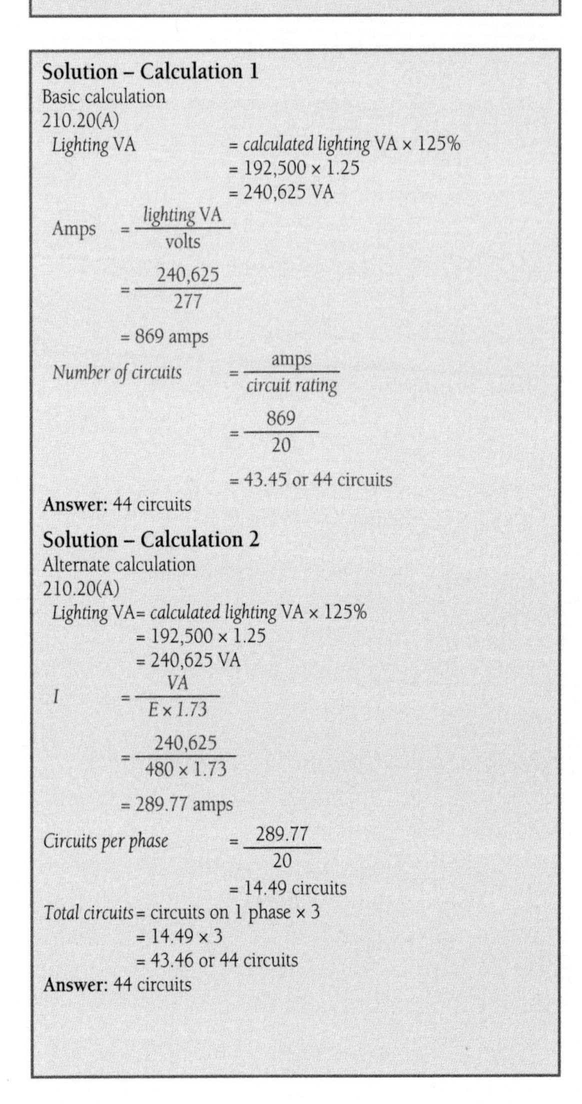

When the lighting circuits are 277 volt, single-phase, 20 ampere, what is the minimum number of lighting circuits needed?

Solution – Calculation 1

Basic calculation

210.20(A)

Lighting VA = *calculated lighting* VA × 125%
= 192,500 × 1.25
= 240,625 VA

$$\text{Amps} = \frac{\textit{lighting VA}}{\text{volts}} = \frac{240{,}625}{277} = 869 \text{ amps}$$

$$\textit{Number of circuits} = \frac{\text{amps}}{\textit{circuit rating}} = \frac{869}{20} = 43.45 \text{ or } 44 \text{ circuits}$$

Answer: 44 circuits

Solution – Calculation 2

Alternate calculation

210.20(A)

Lighting VA= *calculated lighting* VA × 125%
= 192,500 × 1.25
= 240,625 VA

$$I = \frac{VA}{E \times 1.73} = \frac{240{,}625}{480 \times 1.73} = 289.77 \text{ amps}$$

$$\textit{Circuits per phase} = \frac{289.77}{20} = 14.49 \text{ circuits}$$

Total circuits = circuits on 1 phase × 3
= 14.49 × 3
= 43.46 or 44 circuits

Answer: 44 circuits

Feeder Demand: Including Receptacles with Lighting

Determining the general lighting load and the receptacle load in VA for an office building and is done by following the requirements of both **Table 220.12** and **Section 220.14**. The total VA per ft^2 used in the calculation is the sum of the value of **Table 220.12** plus the value or **220.14(K)**. This calculation results in adding 1 VA per ft^2 for receptacles to 3½ VA of **Table 220** with a total value of 4½ VA per ft^2.

Problem 8-44

Calculate the lighting and receptacle load for the Tower Office Building when the number of receptacles is not known.

Solution
Area = 55,000 ft^2
Table 220.12
VA per ft^2 = 3½ VA per ft^2
220.14(K):
Increase VA by 1 VA = 4½ VA per ft^2
Lighting load = area × VA per ft^2
= 55,000 × 4.5
= 247,500 VA
Answer: 247,500 VA

Feeder Demand: Number of Receptacles Known

According to **220.14(K)**, receptacles for banks and office building receptacle loads must be calculated according to the larger of either the 180 VA method used in **220.14(I)** or the 1 VA per ft^2 method used in **220.14(K)**.

Problem 8-45

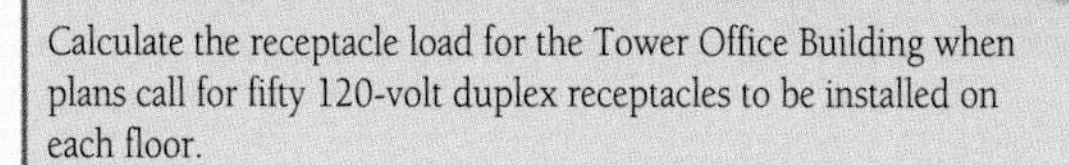

Calculate the receptacle load for the Tower Office Building when plans call for fifty 120-volt duplex receptacles to be installed on each floor.

Solution – Calculation 1
180 VA method
Number of receptacles = 50 per floor × 5 floors
= 250
220.14(I)
Receptacles calculated at 180 VA each
Load = number of receptacles × 180 VA
= 250 × 180
= 45,000 VA
Table 220.44
Apply demand factors
Total calculated load = 45,000 VA

First 10,000 at 100%	10,000 VA
Remaining 35,000 VA at 50%	
= 35,000 × 0.50	
=17,500 VA	17,500 VA
Total	27,500 VA

Answer: 27,500 VA

Solution – Calculation 2
One VA per ft^2 method
220.14(K)
One VA per ft^2 = 55,000 VA
Table 220.44 does not apply
Answer: 55,000 VA

Solution – Calculation 3
220.14(K)
Use larger load
55,000 VA is larger than 27,500 VA
Use One VA per ft^2 method
Answer: 55,000 VA

Sizing a Transformer for 120 Volt Receptacles

When the general distribution system is 480/277 volts, a transformer is needed to furnish the 120 volts for the general-purpose receptacle load. When the receptacle load is calculated in volt-amperes and transformers are rated in kVA, the calculated receptacle volt-ampere load can be divided by 1,000 to determine the minimum kVA transformer rating needed.

Problem 8-46

The receptacle load will be supplied by a transformer with a 3-phase, 480-volt primary and a 120/208 volt, 3-phase, 4-wire secondary. Calculate the minimum kVA rating of the transformer.

Solution
Receptacle load = 55,000 VA

$$kVA = \frac{VA}{1,000}$$

$$= \frac{55,000}{1,000}$$

= 55 kVA
Answer: 55 kVA

Comment
The answer of a 55 kVA transformer is not a standard size 3-phase transformer. Therefore, any size larger would be permitted.

Feeder Demand: Motor Loads

Three-phase motor loads are calculated using the full-load current **Table 430.250**. Motor calculations follow the requirements of **Section 430.24** as well as the *NEC* **Informative Annex D, Example D3(a)**.

Three large motors used for air-handling purposes in a commercial building under construction.

Problem 8-47

The following motors are to be installed in the Tower Office Building:
Two 5 hp, 460 volt, 3-phase
Two 10 hp, 460 volt, 3-phase
Two 50 hp, 460 volt, 3-phase.
Calculate the VA feeder demand for the motor loads.

Solution
Table 430.250
5 hp, 460 volt FLC = 7.6 amps
10 hp, 460 volt FLC= 14 amps
50 hp, 460 volt FLC = 65 amps

VA	= E × I × 1.73 × *number of motors*	
VA (5 hp)	= 460 × 7.6 × 1.73 × 2	
	= 12,096	12,096
VA (10 hp)	= 460 × 14 × 1.73 × 2	
	= 22,282	22,282
VA (50 hp)	= 460 × 65 × 1.73 × 2	
	= 103,454	103,454
25% of largest		
VA hp	= 460 × 65 × 1.73 × 0.25	
	= 12,932	12,932
Total motor VA		150,764

Answer: 150,764 VA

8.5.2 Office Building Calculation Summary

This summary of the office building calculation follows the same procedures as previously outlined.

Total Feeder Demand for Office Building, Number of Receptacles Unknown

The general lighting load for the office building is calculated by using **Table 220.12**. The receptacles are calculated using **220.14(K)(2)**. The solution for **Problem 8-48** could combine the office lighting load of 3½ VA per ft² with the office receptacle's load of 1 VA per ft², but, because continuous and noncontinuous loads are being calculated, the calculations should remain separate. The motor load calculations follow **430.24(1)** and **430.24(2)**.

Problem 8-48

Calculate the total feeder demand based upon continuous and noncontinuous loads for the Tower Office Building when the number of receptacles is not known.
Summary of Tower Office Building:
5 floors, 125 ft by 88 ft
Service: 480/277 volts, 3-phase, 4-wire
Discharge lighting used throughout
Motor loads: two 5 hp; two 10 hp, and two 50 hp; all of which are 460 volts, 3-phase

Solution

		Noncontinuous Load	Continuous Load
Building total area	= *length* × *width* × floors		
	= 125 × 88 × 5		
	= 55,000 ft²		
Lighting VA	= *total area* × 3½ VA per ft²		
	= 55,000 × 3.5		
	= 192,500 VA	0	192,500 VA
Receptacles VA	= *total area* × 1 VA per ft²		
	= 55,000 × 1		
	= 55,000 VA	55,000 VA	0
Motor loads calculated in Problem 8-47			
Two 5 hp, 460 volts	12,096 VA		
Two 10 hp, 460 volts	22,282 VA		
Two 50 hp, 460 volts	103,454 VA		
Total motor load	137,832 VA	137,832 VA	0
25% of largest (50 hp)			
VA hp	= *largest motor load* × 25%		
	= 51,727 × 0.25		
	= 12,932 VA	12,932 VA	0
Totals		205,764 VA	192,500 VA

Answer: 205,764 VA noncontinuous load
192,500 VA continuous load

Ampacity of Neutral Conductor

220.61(B)(2) permits the size of the neutral conductor to be decreased to 70% after the first 200 amperes is calculated at 100%. However, the ampacity of the neutral is not permitted to be reduced for the portion of the load that supplies electric discharge lighting or other harmonic loads.

Problem 8-49

Calculate the ampacity of the neutral conductor for the following calculated office building loads when the service is 120/208 volts, 3-phase, 4-wire:

Fluorescent lighting load	204,060 VA
Receptacle load (after demand factor applied)	65,000 VA
Incandescent lighting load	150,000 VA
120-volt motor load	30,000 VA

Solution

Fluorescent lighting load

$$I = \frac{VA}{E \times 1.73}$$

$$= \frac{204{,}060}{208 \times 1.73}$$

= 567 amps

Other loads = receptacle + incandescent + motors

VA = 65,000 + 150,000 + 30,000

= 245,000 VA

$$I = \frac{VA}{E \times 1.73}$$

$$= \frac{245{,}000}{208 \times 1.73}$$

= 681 amps

200 amperes at 100%	200 amps
681 amps – 200 amps = 481 amps	
481 amperes at 70%	
= 481 × 0.70	
= 337 amps	337 amps
100% fluorescent (nonlinear loads)	567 amps
Total amps neutral conductor	1,104 amps

Answer: 1,104 amperes neutral conductor load

Definitions and Terms

Appliance - Utilization equipment, generally other than industrial, that is normally built in standardized sizes or types and is installed or connected as a unit to perform one or more functions such as clothes washing, air conditioning, food mixing, deep frying, and so forth.

Building - A structure that stands alone or that is cut off from adjoining structures by fire walls with all openings therein protected by approved fire doors.

Demand Factor - The ratio of the maximum demand of a system, or part of a system, to the total connected load of a system or the part of the system under consideration.

Dwelling, One-family - A building that consists solely of one dwelling unit.

Dwelling, Multifamily - A building that contains three or more dwelling units.

Dwelling Unit - A single unit, providing complete and independent living facilities for one or more persons, including permanent provisions for living, sleeping, cooking, and sanitation.

Feeder - All circuit conductors between the service equipment, the source of a separately derived system, or other power supply source and the final branch-circuit overcurrent device.

Grounding Electrode Conductor - A conductor used to connect the system grounded conductor or the equipment to a grounding electrode or to a point on the grounding electrode system.

Neutral Conductor - The conductor connected to the neutral point of a system that is intended to carry current under normal conditions.

Neutral Point - The common point on a wye-connection in a polyphase system or midpoint on a single-phase, 3-wire system, or midpoint of a single-phase portion of a 3-phase delta system, or a midpoint of a 3-wire, direct-current system.

Noncoincident Loads - Two electrical loads are considered to be noncoincident loads where the two electrical loads are unlikely to operate at the same time (220.60).

Service - The conductors and equipment for delivering electric energy from the serving utility to the wiring system of the premises served.

Service Conductors - The conductors from the service point to the service disconnecting means.

Summary

- **Article 210, Part III** contains a description and quantity of required outlets.
- **Article 220** contains most of the load calculation requirements for branch circuits in **Part II**, and the feeder, and service load calculation requirements in **Part III**.
- **Annex D** also provides a variety of acceptable step-by-step load calculations. Even the simplest of the load calculations for a one-family dwelling often exceeds twenty four separate steps.
- For accurate calculations and the elimination of errors, load calculations should always be done following the step-by-step methods as they are presented in this chapter.
- Many different demand factors for dwelling unit calculations are permitted to be used.
- Typical residential items and references for the standard method for branch-circuit, feeder, and service load calculation include the following steps:

  » Lighting—**Table 220.12**
  » Small appliance—**220.52(A)**
  » Clothes dryer—**Section 220.54**
  » Appliances—**Section 220.53**
  » Electric range—**Section 220.55**
  » Heating—**Section 220.51** and **Section 220.60**
  » Air conditioning—**Section 220.60**
  » 25% of largest motor—**Section 220.50** and **Section 430.24**
  » Application for various demand factor—**Article 220, Parts II, III,** and **IV**.

- For *Code*-based commercial building load calculations, it is always best to set up three separate tables of loads in the following order:

  » Noncontinuous loads
  » Continuous loads
  » Neutral loads

- The first two tables (noncontinuous and continuous loads) are then combined using 125% of the continuous and 100% of the noncontinuous loads.
- The sum of these two loads can then be used to calculate and determine the overcurrent protective device and the conductor size before adjustment or correction factors are applied.
- Setting up the noncontinuous table of loads should proceed according to the "*Code* required" order: noncontinuous loads, motors loads (including 25% of the largest), and loads carrying full current for less than three hours.
- Next, set up the continuous table of loads which should proceed according to the "*Code* required" order: continuous loads and loads carrying full current for three hours of more. Many specific appliances and dedicated office machines fall into this category.
- Next, set up a neutral load table, determining and listing all of the phase-to-neutral loads. The many references to sections within each Chapter 8 problem can be used as a load calculation guide. In addition, use **Annex D** of the *NEC* as another source of methods to perform load calculations.

Transformer Overcurrent Protection

Protection techniques of power transformers, their supply conductors, and their load conductors consist of basic overcurrent and overload protection. Calculations to determine and properly size this protection are broken down into four basic general categories.

The first category is general overcurrent protection afforded to the windings of a transformer. Calculations based upon **Table 450.3(B)** are used to demonstrate how to properly apply and select the maximum size overcurrent protection as 'primary only' and 'primary plus secondary' protection for a given transformer. Each of the three individual primary categories and each secondary category will be explored and calculated, as permitted.

The second category is the maximum overcurrent protective device along with the correct conductor size. Calculations and selection are based upon the load(s) served by the secondary and the primary circuit of a transformer with the use of **Table 450.3(B),** along with more complex sections of overcurrent protection within **240.21(B) Feeder Taps** and **240.21(C) Transformer Secondary Conductors**. The seven specific tap rules are studied in depth.

The third category is the overcurrent protective device and the proper size conductor for transformers equipped with coordinated thermal overload protection by the manufacturer.

The fourth category is dedicated transformers used in fire pump circuits. Special calculations based upon **Article 695** are used to correctly size a transformer and apply short-circuit protection to a fire pump circuit.

This study of transformers is limited to 1000 volts, nominal, or less.

Objectives

- Calculate overcurrent protection for primary and secondary windings of a transformer.
- Select standard size overcurrent protective devices.
- Calculate and select primary and secondary transformer conductor sizes.
- Calculate and select the primary overcurrent devices for single-phase transformers with a two-wire secondary.
- Apply the 10 foot and 25 foot tap rule to transformer secondary conductors.
- Apply the feeder tap rule which supplies a transformer (primary plus secondary not over 25 feet long).

Chapter 9

Table of Contents

9.1 Protecting Transformer Windings (1000 Volts, Nominal, or Less)

9.1.1 General

The study of overcurrent protection of transformers begins with a look at the transformers used for power and lighting with an operating voltage of 1000 volts, nominal, or less. Such transformers are used as premises distribution transformers, often in conjunction with lighting distribution panels.

The following primary sections of the *Code* will be used:

Table 450.3(B) - This table is directed at the overcurrent protection of transformer windings for transformers operating at 1000 volts, nominal, or less.
Table 450.3(B), Note 1 - This note permits overcurrent protection, where limited to 125% of the transformer full-load current, to be increased to a higher rating that does not exceed the next higher standard size overcurrent device.
240.6(A) - This section includes standard ampere ratings of fuses and circuit breakers used for selecting the proper final transformer size and conductor overcurrent protective device.
Table 450.3(B), Note 2 - This note permits overcurrent protection on the secondary of a transformer to be not more than six circuit breakers or sets of fuses grouped at one location.

Primary overcurrent protection can be responsive to secondary short circuits, but is not always responsive to secondary overcurrent caused by an unbalanced condition of a single-phase, 3-wire system or a 3-phase, 4-wire system. The *Code* calls attention to this in **240.21(C)(1)**, in which it states in part:

> **(1) Protection by Primary Overcurrent Device.** Conductors supplied by the secondary side of a single-phase transformer having a 2-wire (single voltage) secondary, or a three-phase, delta-delta connected transformer having a 3-wire (single voltage) secondary, shall be permitted to be protected by overcurrent protection provided on the primary (supply) side of the transformer provided...

The *Code* does not give a specific method for sizing the primary or secondary conductors of transformers 1000 volts, nominal, or less, other than as required by the following general requirements for electrical conductors:

Section 215.2 - requires feeder conductors to have sufficient ampacity to supply the load served.
Section 240.4 - requires conductors to be protected against overcurrent according to their ampacities.

Table 450.3(B) requires transformer primary and secondary windings to be protected against overcurrents according to the ampacity rating of the transformer windings. The ampacity rating of transformer windings is based upon the current rating of the transformer.

The sizing of primary or secondary conductors and the primary or secondary overcurrent protection need to be considered together. The reason is that the overcurrent device can be used to protect the transformer windings and the primary and secondary circuit conductors. This will be illustrated in several of the problems to follow.

240.6(A) is divided into the following three separate parts:

1. Standard ampere ratings for fuses and inverse-time circuit breakers
2. Additional standard ratings for fuses of only 1, 3, 6, 10, and 601 amperes
3. The use of nonstandard fixed ampere rated fuses and circuit breakers which could have any ampacity rating

The type of overcurrent protective device required for transformers is not specified. The overcurrent protective device can be time-delay or nontime-delay fuses, or adjustable or nonadjustable circuit breakers. Thermal overload protective devices are also permitted for use as transformer winding overcurrent protection.

When a transformer is energized, there is an inrush of current. The inrush of current is similar to the starting current of a motor. It will last only until a counter-EMF is built up in the primary of the transformer, which takes only a short period of time. This initial current surge is called inrush current, charging current, or excitation current. The inrush current can be as much as eight to ten times the rated primary current. A time-delay type overcurrent protective device can be used to compensate for the inrush current.

Symbol Identification

Where transformer overcurrent protection examples are presented in this text, the following standard references and symbols will be used:

Symbol Identification	
OCPD	Overcurrent protective device
$OCPD_{pri}$	Overcurrent protective device, primary side of transformer
$OCPD_{sec}$	Overcurrent protective device, secondary side of transformer
I_{pri}	Primary current
E_{pri}	Primary voltage
I_{sec}	Secondary current
E_{sec}	Secondary voltage
kVA	Kilovolt amperes
LRC	Locked-rotor current
CB	Circuit breaker

All conductors used in the illustrations within this chapter will be THWN copper, unless otherwise noted.

Table 310.15(B)(16) is used for selecting the ampacity of conductors.

Equations

Transformer primary and secondary current ratings can be taken from tables or can be calculated using transformer power equations:

Single-phase

$$I = \frac{kVA \times 1{,}000}{E}$$

Three-phase

$$I = \frac{kVA \times 1{,}000}{E \times 1.73}$$

9.1.2 Primary Overcurrent Protection Only

Table 450.3(B) includes multiple transformer overcurrent protection schemes based upon the percentage of a transformer's rated current. Each part of this table and its accompanying notes will be looked at individually.

Notice that the table focuses only on the protection of the transformer windings and excludes the primary or secondary feeder conductor protection. Consider the overcurrent protection to be individually located at the transformer.

9.1.2.1 Primary Protection, 9 Amperes or More - **Table 450.3(B)** requires each 1000 volt, nominal, or less transformer primary with a primary current of 9 amperes or more to be protected by an individual overcurrent device on the primary side of the transformer. Where the protection method is primary only protection, the overcurrent protective device must be rated or set at not more than 125% of the rate primary current (I_{pri}). The equation is stated as follows:

$$OCPD_{pri} = I_{pri} \times 125\%$$

Table 450.3(B) Note 1 permits a higher rating which does not exceed the next higher standard size overcurrent device to be used for the primary only protection method.

Although **Note 1** is similar to the permission given in **240.4(B)**, it is not appropriate to use the permission of **240.4(B)** because the *Code* clearly states in two separate places that **240.4(B)** should not be used: **240.21(B) Tap Conductors** and **240.21(C) Transformer Secondary Conductors**. The discipline to use **Note 1** and not to use **240.4(B)** will provide more accurate calculations while improving *Code* compliance.

The second protection method is primary and secondary protection. For this method, the overcurrent protective device is allowed to be rated or set at not more than 250% of the rated primary current (I_{pri}).

$$\text{Maximum } OCPD_{pri} = I_{pri} \times 250\%$$

Note 1 of the table is not available for this protection method. Where **Note 1** is not specifically referenced for the particular calculation, going to the next higher standard size overcurrent protective device is not an option and the selection of the next lower standard size overcurrent protective device will be necessary.

Problem 9-1

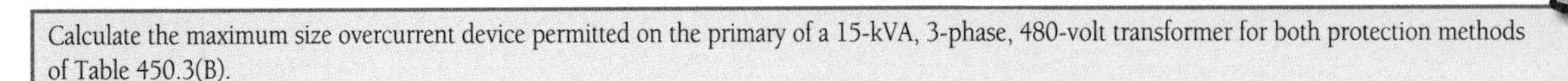
Calculate the maximum size overcurrent device permitted on the primary of a 15-kVA, 3-phase, 480-volt transformer for both protection methods of Table 450.3(B).

Solution – Calculation 1
Primary only protection
3-phase

$$I = \frac{kVA \times 1{,}000}{E_{pri} \times 1.73}$$

$$= \frac{15 \times 1{,}000}{480 \times 1.73}$$

$= 18.06$ amps

Table 450.3(B)
Currents of 9 amperes or more column
Primary only protection = 125%
Max. $OCPD_{pri}$ $= I_{pri} \times 125\%$
$= 18.06 \times 1.25$
$= 22.58$ amps

Table 450.3(B) Note 1
Next larger std. size permitted
240.6(A)
Next larger = 25 amps
Answer: 25 ampere Max. $OCPD_{pri}$

Solution – Calculation 2
Primary and secondary protection
$I_{pri} = 18.06$ amps
Table 450.3(B)
Currents of 9 amperes or more column
Primary and secondary protection = 250%
Max. $OCPD_{pri}$ $= I_{pri} \times 250\%$
$= 18.06 \times 2.50$
$= 45.15$ amps

Table 450.3(B) Note 1
Does not apply, use next smaller
240.6(A)
Next smaller = 45 amps
Answer: 45 ampere max. $OCPD_{pri}$

Comment
The dual solution calculations from Problem 9-1 demonstrate a stark difference in the maximum overcurrent protection permitted for a 15-kVA transformer. Using Solution - Calculation 1, the transformer primary OCPD max is limited to 25 amperes whereas using Solution - Calculation 2, the transformer primary OCPD max is limited to 45 amperes. For a 150-kVA transformer, the differences could be as large as 250 amps versus 500 amps. The primary only protection demonstrated by Solution - Calculation 1 is often recommended for transformers with stable loads and other loads without high inrush currents. The primary protection demonstrated by Solution - Calculation 2 is often recommended for transformers with varying loads or loads with high inrush currents.

For additional information, visit qr.njatcdb.org Item #1039

The front cover of the transformer is removed to expose the wiring terminal for view and connection. Notice that the low voltage secondary wiring uses parallel conductors.

9.1.2.2 Primary Protection, Less Than 9 Amperes - Table 450.3(B) requires each 1000 volt, nominal, or less transformer primary with a primary current of less than 9 amperes to be protected by an individual overcurrent device on the primary side of the transformer rated or set at not more than 167% of the rate primary current (I_{pri}). **Note 1** is not applicable.

The reason for the increase in the permitted sizing of the protective device is that the transformer "charging current" for the smaller transformers is proportionally larger for the ampere rating of the transformer.

Information

Some transformers are designed for specific applications, have single winding primaries or single winding secondaries. Most transformers, however, have dual windings on both the primary and the secondary.

Problem 9-2

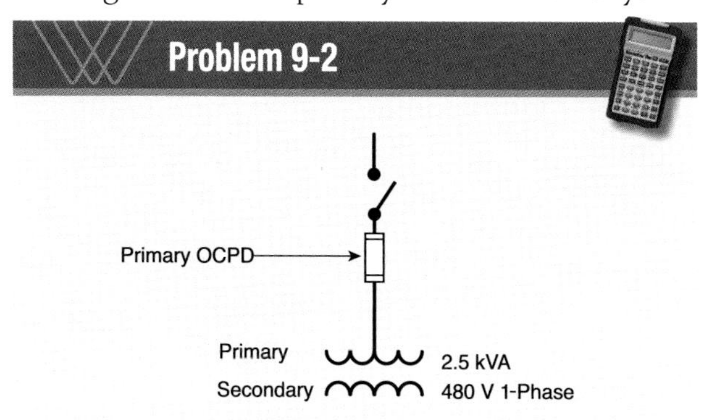

Calculate the maximum primary overcurrent protection permitted for the primary of a 2.5-kVA single-phase, 480-volt transformer where fuses are used and secondary overcurrent protection for the transformer is not desired.

Solution

Primary overcurrent protection

$I_{pri} = \frac{kVA \times 1{,}000}{E_{pri}}$

$= \frac{2.5 \times 1{,}000}{480}$

= 5.2 amps

Table 450.3(B)

Currents less than 9 amperes column

Primary only protection = 167%

Max. $OCPD_{pri}$ = I_{pri} × 1.67

= 5.2 × 1.67

= 8.68 amps

Table 450.3(B) Note 1

Does not apply, any less permitted

240.6(A)

Next smaller = 6 amps

Answer: Fuse size of 8.68 amperes or less

9.1.2.3 Primary Protection, Less Than 2 Amperes - Table 450.3(B) requires each 1000 volt, nominal, or less transformer primary with a primary current of less than 2 amperes to be protected by an individual overcurrent device on the primary side of the transformer rated or set at not more than 300% of the rated primary current (I_{pri}). **Note 1** is not applicable.

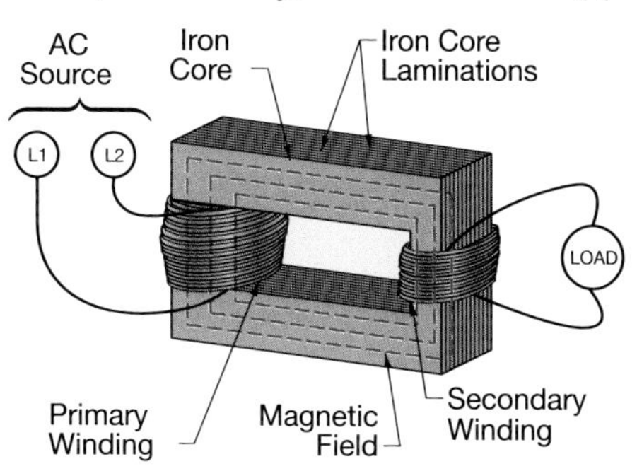

The primary winding is connected to the AC input and the secondary is connected to the load.

Problem 9-3

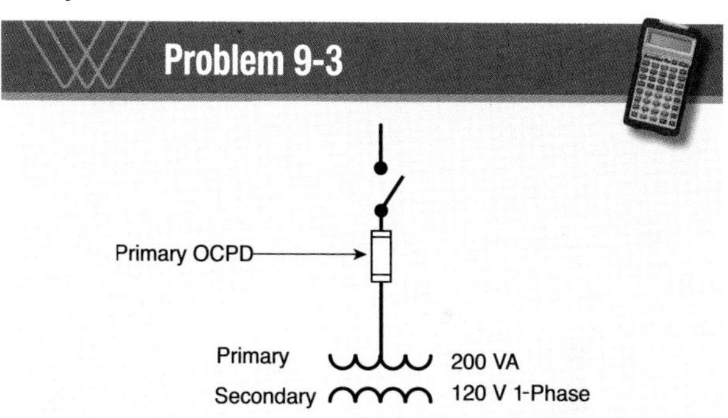

Calculate the maximum overcurrent protection using the primary only method of protection for a 200-VA, single-phase, 120-volt primary transformer when fuses are used.

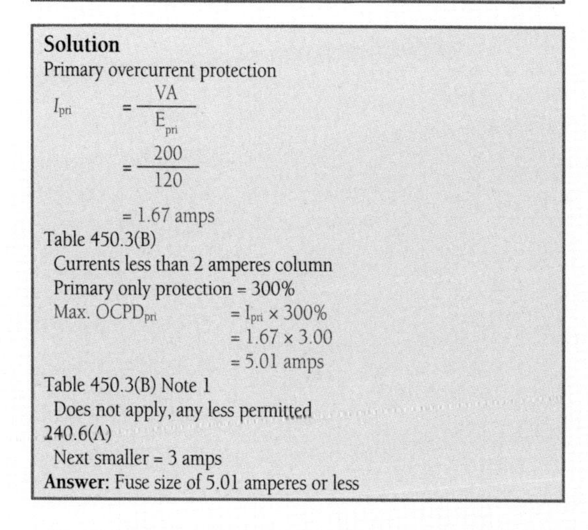

Solution

Primary overcurrent protection

$I_{pri} = \frac{VA}{E_{pri}}$

$= \frac{200}{120}$

= 1.67 amps

Table 450.3(B)

Currents less than 2 amperes column

Primary only protection = 300%

Max. $OCPD_{pri}$ = I_{pri} × 300%

= 1.67 × 3.00

= 5.01 amps

Table 450.3(B) Note 1

Does not apply, any less permitted

240.6(A)

Next smaller = 3 amps

Answer: Fuse size of 5.01 amperes or less

9.1.2.4 Overcurrent Protection Primary Windings and Primary Feeder Conductors - Provided the overcurrent device protects the primary transformer windings according to **Table 450.3(B)**, the overcurrent device protecting the primary of a transformer may be located other than at the transformer.

This means that the overcurrent device protecting the circuit conductors supplying the transformer can also be used to protect the primary windings of the transformer. This is a case of the overcurrent device doing double duty.

1. The overcurrent protective device must protect the primary of the transformer according to **Table 450.3(B)**.
2. The overcurrent device must protect the conductors according to **Section 240.4**.

Problem 9-4 is an excellent example of how the overcurrent protective device is responsible for the protection of the transformer windings and the circuit conductors, and both must be considered. And to the contrary, there is more risk of nuisance tripping where a transformer is fully loaded since the OCPD is less than 125% of the primary full-load current.

Problem 9-4

The primary of a 45-kVA, 440-volt, 3-phase transformer is supplied with THWN copper conductors. Calculate the ampacity of the overcurrent protective device for the primary conductors and the primary of the transformer. The method of protection is selected to be primary only.

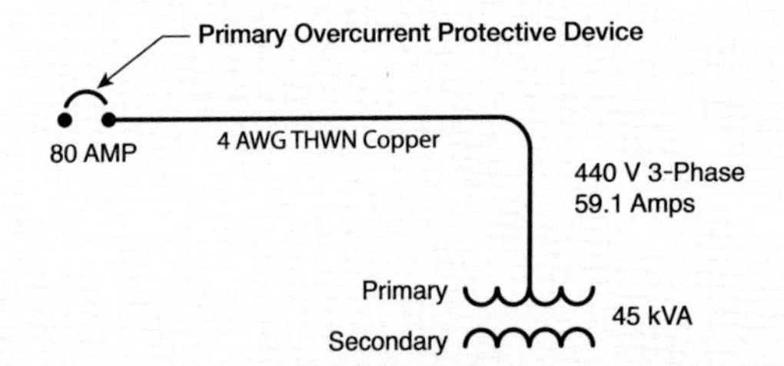

Solution – Calculation 1
Primary only protection
3-phase

$$I = \frac{\text{kVA} \times 1{,}000}{E_{pri} \times 1.73}$$

$$= \frac{45 \times 1{,}000}{440 \times 1.73}$$

$= 59.1$ amps

Table 450.3(B)
Currents of 9 amperes or more column
Primary only protection = 125%
Max. $OCPD_{pri}$ $= I_{pri} \times 125\%$
$= 59.1 \times 1.25$
$= 73.87$ amps
Table 450.3(B) Note 1
Next larger std. size permitted
240.6(A)
Next larger = 80 amps
Max. $OCPD_{pri}$ = 80 amps OCPD
Table 310.15(B)(16)
THWN copper column
80 amps = 4 AWG THWN
4 AWG THWN = 85 amps
Answer: 80 ampere $OCPD_{pri}$ with 4 AWG THWN copper

Solution – Calculation 2
Primary overcurrent protection
$I_{pri} = 59.1$ amps
Table 310.15(B)(16)
THWN copper column
59.1 amps = 6 AWG THWN
6 AWG THWN copper = 65 amps
240.4(A)
Next larger std. size permitted
240.6(A)
Next larger = 70 amps
Table 450.3(B)
Currents of 9 amperes or more column
Primary only protection = 125%
Max. $OCPD_{pri}$ $= I_{pri} \times 125\%$
$= 59.1 \times 1.25$
$= 73.87$ amps
Table 450.3(B) Note 1
Next larger std. size permitted
240.6(A)
Next larger = 80 amps
Select 6 AWG THWN Copper with 70 amp OCPD
Reason: Neither method exceeds 125% of I_{pri}
Answer: 70 ampere $OCPD_{pri}$ with 6 AWG THWN copper
Selection of Answers:
4 AWG THWN copper with 80 ampere $OCPD_{pri}$
6 AWG THWN copper with 70 ampere $OCPD_{pri}$

Comment
Notice that when a 6 AWG conductor is used with a 70-ampere overcurrent device, both the conductor and the transformer are protected. No transformer capacity is being lost by using a conductor one size smaller and keeping the conductor and the primary windings protected. When the smaller overcurrent protection is used, less transformer overload current will be permitted before the primary overcurrent protective device operates.

9.1.3 Secondary Overcurrent Protection

9.1.3.1 Secondary Protection, 9 Amperes or More - Table 450.3(B) requires each 1000 volt, nominal, or less transformer secondary with a secondary current of 9 amperes or more to be protected by an individual overcurrent device on the secondary side of the transformer rated or set at not more than 125% of the rated secondary current (I_{sec}). **Note 1** of the table permits a higher rating which does not exceed the next higher standard size overcurrent device to be used for the transformer overcurrent protection. According to **Note 1**, using the next higher standard size OCPD is permitted.

9.1.3.2 Secondary Protection, Less Than 9 Amperes - Table 450.3(B) requires each 1000 volt or less transformer with a secondary current of less than 9 amperes to be protected by an individual overcurrent device on the secondary side of the transformer rated or set at not more than 167% of the rated secondary current (I_{sec}). **Note 1** is not applicable.

Information

Small transformers are often more vital to the successful operation of a building than realized. Small transformers are responsible for building automation systems, mass notification and fire alarm systems, elevator control, and a myriad of other very important electrical systems.

Problem 9-5

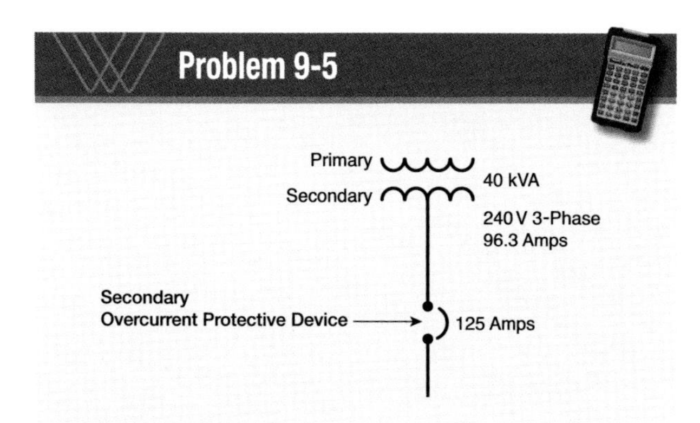

Calculate the maximum secondary overcurrent protection for a 40-kVA, 3-phase transformer with a secondary voltage of 240 volts.

Solution

Transformer secondary current

3-phase

$$I_{sec} = \frac{\text{kVA} \times 1{,}000}{E_{sec} \times 1.73}$$

$$= \frac{40 \times 1{,}000}{240 \times 1.73}$$

$$= 96.3 \text{ amps}$$

Secondary overcurrent protection

Table 450.3(B)

Currents of 9 amperes or more column

Primary and secondary protection = 125%

Max. OCPDsec = Isec × 125%

= 96.3 × 1.25

= 120.37 amps

Table 450.3(B) Note 1

Next larger std. size permitted

240.6(A)

Next larger = 125 amps

Answer: 125 amperes max. $OCPD_{sec}$

Problem 9-6

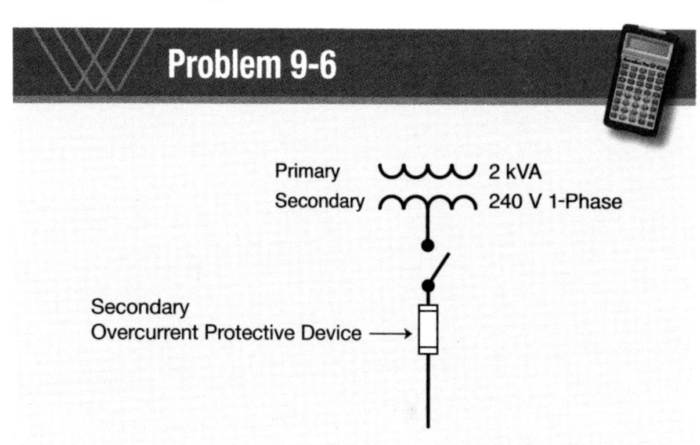

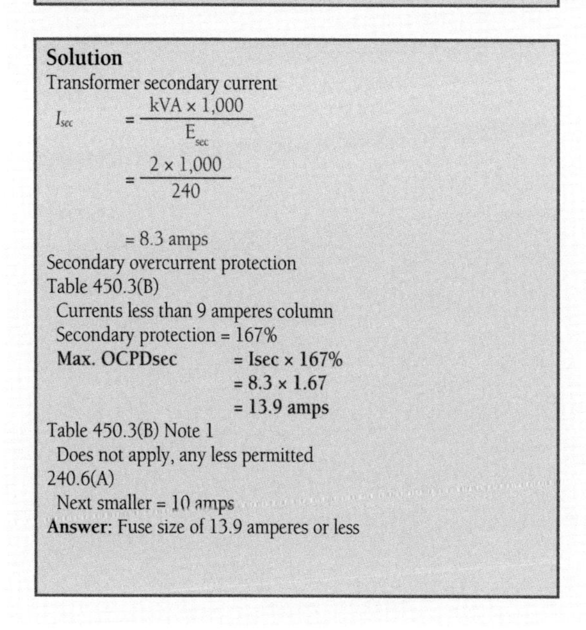

Calculate the maximum secondary overcurrent fuse protection for a single-phase, 2-kVA transformer with a secondary voltage of 240 volts.

Solution

Transformer secondary current

$$I_{sec} = \frac{\text{kVA} \times 1{,}000}{E_{sec}}$$

$$= \frac{2 \times 1{,}000}{240}$$

$$= 8.3 \text{ amps}$$

Secondary overcurrent protection

Table 450.3(B)

Currents less than 9 amperes column

Secondary protection = 167%

Max. OCPDsec = Isec × 167%

= 8.3 × 1.67

= 13.9 amps

Table 450.3(B) Note 1

Does not apply, any less permitted

240.6(A)

Next smaller = 10 amps

Answer: Fuse size of 13.9 amperes or less

9.1.4 Primary Protection at 250% and Secondary Protection at 125%

Two parts of **Table 450.3(B)** are considered in this situation. One part applies to the overcurrent protection on the secondary side of the transformer. The second part applies to the overcurrent protection on the primary side of the transformer.

1. The secondary overcurrent current protection is sized according to the 125% rule.
2. Then the overcurrent device on the primary side may be set at not over 250% of the primary current.

This indicates that the overcurrent protective device used for circuit conductor overcurrent protection is also permitted to be used for transformer primary winding overcurrent protection, provided it does not exceed 250% of the transformer primary current. When the overcurrent device on the secondary opens or trips, only the transformer primary winding charging current will flow. Therefore, the secondary overcurrent device limits the current on the primary windings.

Problem 9-7

Calculate the maximum primary and secondary overcurrent protection and size of the primary THWN copper conductors.

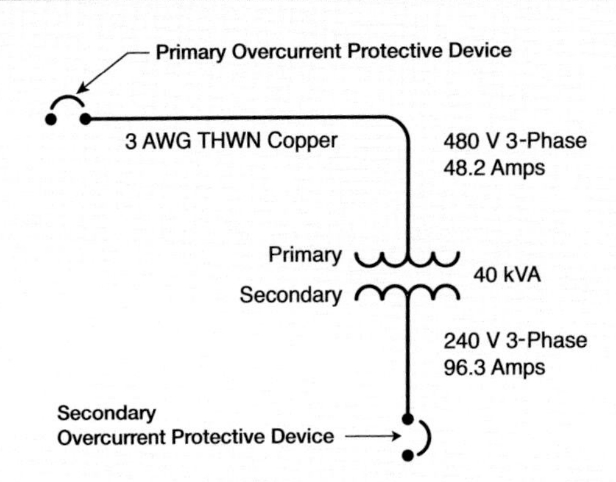

Solution – Calculation 1
Secondary overcurrent protection
I_{sec} = 96.3 amps
Table 450.3(B)
Currents of 9 amperes or more column
Secondary protection = 125%
Max. $OCPD_{sec}$ $= I_{sec} \times 125\%$
$= 96.3 \times 1.25$
$= 120.37$ amps
Table 450.3(B) Note 1
Next larger std. size permitted
240.6(A)
Next larger = 125 amps
Answer: 125 amperes Max. $OCPD_{sec}$

Solution – Calculation 2
Primary overcurrent protection
I_{pri} = 48.2 amps
Table 450.3(B)
Currents of 9 amperes or more column
Primary and secondary protection = 250%

$OCPD_{pri}$ $= I_{pri} \times 250\%$
$= 48.2 \times 2.50$
$= 120.5$ amps
Table 450.3(B) Note 1
Does not apply, any less permitted
240.6(A)
Next smaller = 110 amps
Answer: 110 amperes max. $OCPD_{pri}$

Solution – Calculation 3
Primary conductor size
$OCPD_{pri}$ = 110 amps
Match conductor to OCPD
Table 310.15(B)(16)
THWN copper column
110 amps = 2 AWG THWN
2 AWG THWN = 115 amps
Therefore, it is protected at 110 amps
Answer: 2 AWG THWN copper

9.1.5 Secondary Protection Using Multiple Overcurrent Devices

As pointed out in **Note 2** of **Table 450.3(B)**, multiple overcurrent protective devices including fuses and circuit breakers are allowed to be substituted for an individual or single overcurrent protective device where the individual device is required for secondary protection according to **450.3(B)**.

Table 450.3(B) Note 2 permits overcurrent protection for the secondary of a transformer to consist of the following:

- Up to six overcurrent devices
- Grouped in one location
- Total device ratings not exceeding the allowed rating of a single overcurrent device

For example, if the calculated size of a circuit breaker or fuse used for the transformer secondary overcurrent protection is 200 amperes, it would also be permissible to use two 100-ampere or four 50-ampere overcurrent protective devices.

Problem 9-8

What is the maximum number of 50-ampere circuit breakers permitted to be used as overcurrent protection on the secondary of a 480/208 volt, 3-phase, 75-kVA transformer? The transformer primary is protected by an overcurrent device set at not more than 250% of the rated transformer current.

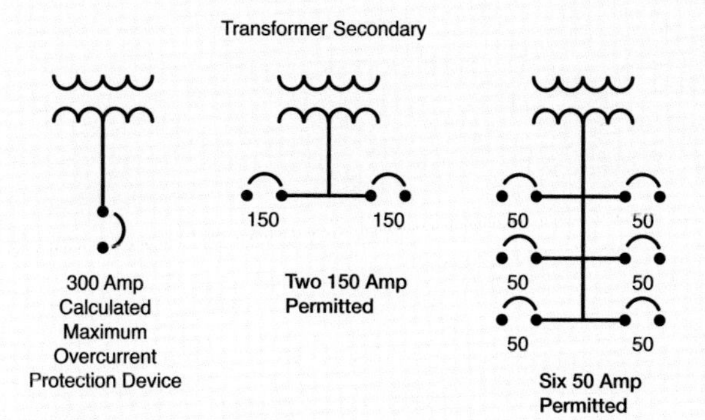

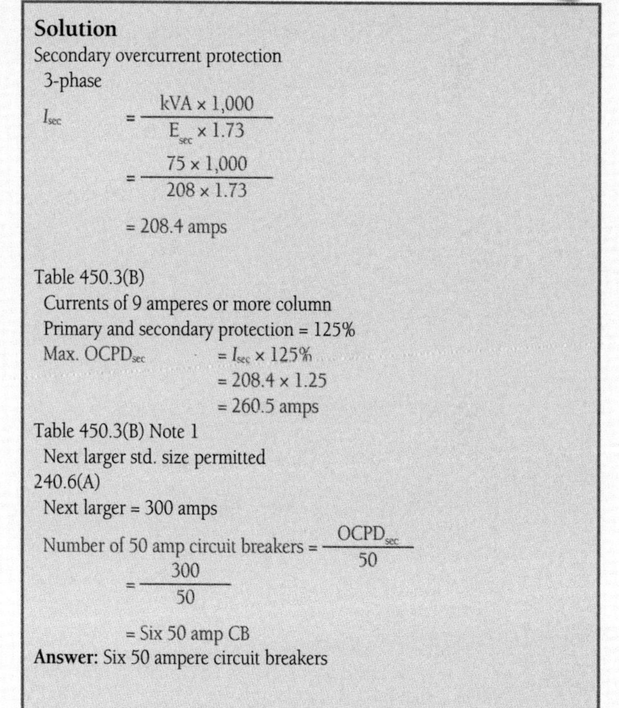

Solution

Secondary overcurrent protection

3-phase

$$I_{sec} = \frac{kVA \times 1{,}000}{E_{sec} \times 1.73}$$

$$= \frac{75 \times 1{,}000}{208 \times 1.73}$$

= 208.4 amps

Table 450.3(B)

Currents of 9 amperes or more column

Primary and secondary protection = 125%

Max. $OCPD_{sec}$ = I_{sec} × 125%

= 208.4 × 1.25

= 260.5 amps

Table 450.3(B) Note 1

Next larger std. size permitted

240.6(A)

Next larger = 300 amps

$$\text{Number of 50 amp circuit breakers} = \frac{OCPD_{sec}}{50}$$

$$= \frac{300}{50}$$

= Six 50 amp CB

Answer: Six 50 ampere circuit breakers

The required disconnect for this transformer is not located within sight of the transformer location. Where a transformer disconnect is remotely located and not within sight, the transformer must be field marked with the location of the disconnecting means. In addition, this remote disconnecting means must be lockable in accordance with 110.25.

9.1.6 Transformer Thermal Overload Protection

Thermal overload protection is sometimes built into the transformer by the manufacturer. It is also called self-contained protection (SCP). The thermal overload unit is sensitive to the heat generated within the transformer which could damage the windings of the transformer. In case of dangerous overheating within the transformer, the thermal overload operates very much like the overload heaters in a magnetic switch and opens the control circuit to the overcurrent device protecting the circuit to the primary windings of the transformer. The primary feeder overcurrent protection is sized according to the impedance (Z) of the transformer. The impedance of a transformer is marked on the nameplate of the transformer. The following values are listed in **Note 3** of **Table 450.3(B)**:

Where Z is 6% or less

$$\text{Max. OCPD}_{pri} = I_{pri} \times 6$$

Where Z is over 6%, but less than 10%

$$\text{Max. OCPD}_{pri} = I_{pri} \times 4$$

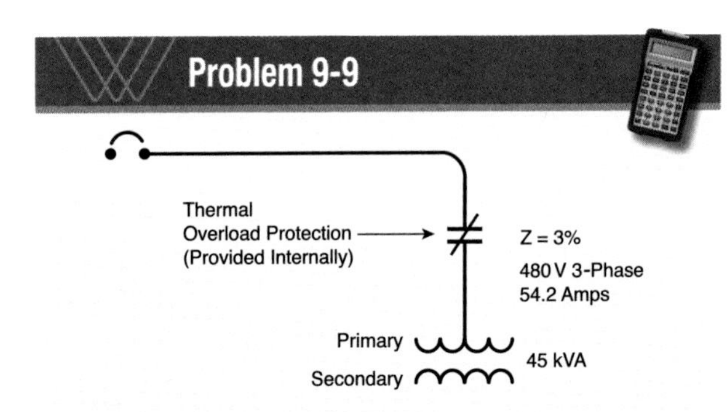

Calculate the maximum overcurrent protection permitted for the transformer feeder with built-in thermal overload protection in the primary of the transformer. The transformer impedance is 3%.

Solution
Primary overcurrent protection
I_{pri} = 54.2 amps
Table 450.3(B) Note 3
Transformer Z is 6% or less
Max. OCPD_{pri} = $I_{pri} \times 6$
= 54.2 × 6
= 325.2 amps
Table 450.3(B) Note 1
Does not apply, any less permitted
240.6(A)
Next smaller = 300 amps
Answer: 300 ampere max. OCPD_{pri} or less

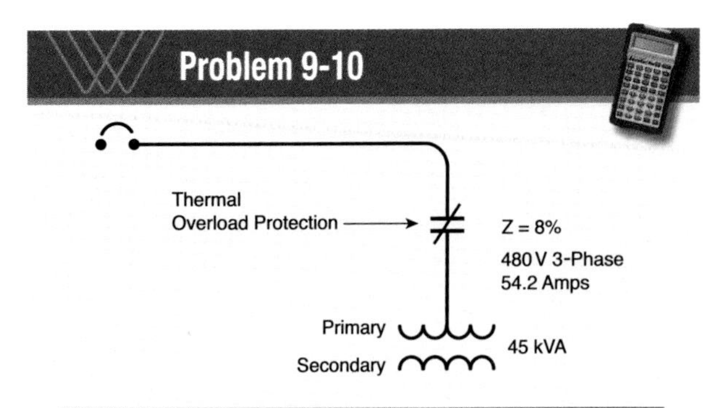

Calculate the feeder overcurrent protection permitted for the transformer with built-in thermal overload protection for the primary with an 8% impedance rating.

Solution
Primary overcurrent protection
I_{pri} = 54 amps
Table 450.3(B) Note 3
Transformer Z is over 6%, but less than 10%
Max. OCPD_{pri} = $I_{pri} \times 4$
= 54.2 × 4
= 216.8 amps
Table 450.3(B) Note 1
Does not apply, any less permitted
240.6(A)
Next smaller = 200 amps
Answer: 200 ampere max. OCPD_{pri} or less

9.1.7 Summary

The first consideration is for overcurrent protection of the transformer primary and secondary windings with a voltage rating of 1000 volts, nominal, or less. This can be thought of as being located at the transformer.

The following is a summary of **Table 450.3(B)**:

1. When the primary current rating is 9 amperes or more, overcurrent protection is permitted to be 125% of the rated primary current.
   Overcurrent device selection - Note 1, use the next higher standard size.
2. When the primary current rating is less than 9 amperes, overcurrent protection is permitted to be 167% of the primary current.
   Overcurrent device selection - Use the 167% value or any lower size.
3. When the primary current rating is less than 2 amperes, overcurrent protection is permitted to be 300% of the primary current.
   Overcurrent device selection - Use the 300% value or any lower size.
4. When the overcurrent protection for the primary windings is located away from the transformer,

the overcurrent device must protect both the transformer primary winding and the conductors supplying that transformer primary.

5. When the secondary current rating is 9 amperes or more, overcurrent protection is permitted to be 125% of the rated secondary current.
   Overcurrent device selection - Note 1, use the next higher standard size.
6. When the secondary current is less than 9 amperes, overcurrent protection is permitted to be up to 167% of the rated secondary current.
   Overcurrent device selection - Use the 167% value or any lower size.
7. When the secondary windings are protected by the 125% rule and the overcurrent protection on the primary protects the conductors and the primary winding, overcurrent protection is permitted to be 250% of the rated primary current.
   Overcurrent device selection, secondary - Note 1, use the next higher standard size.
   Overcurrent device selection, primary - Use the 250% value or any lower size.
8. The maximum of six overcurrent devices are permitted for protection of the secondary, provided the total ampacity of the six overcurrent devices does not exceed the use of a single device.
9. Feeder protection with thermal overload and 6% or less impedance requires not more than six times primary current.
   Overcurrent device selection, primary feeder - Use the 600% value or any lower size.
10. Feeder protection with thermal overload and more than 6%, but less than 10% requires not more than four times primary current.
    Overcurrent device selection, primary feeder - Use the 400% value or any lower size.

9.2 240.21(B) Feeder Taps and 240.21(C) Transformer Secondary Conductors

The location of the overcurrent protection for conductors, as required by **Section 240.21**, may be applied in conjunction with transformer overcurrent protection.

The basic rule of **Section 240.21** is that an overcurrent protective device is required to be connected at the point where a conductor receives its supply.

The primary of a transformer receives its supply from branch-circuit conductors, or it may be tapped from a feeder. Basically, the overcurrent protection for the conductors, tapped from the feeder, supplying a transformer must be located at the point the feeder is tapped. Some permissive rules in this section permit otherwise.

The secondary of a transformer is the source of supply for the secondary conductors. Basically, the overcurrent protection for the secondary conductors would be located at the transformer. The permissive rules in this section permit otherwise.

There are, however, several subsections to the basic rule, several of which can be directly applied to overcurrent protection of transformers. But, a number of specific requirements must be met before the conductors can be installed without overcurrent protection at the point at which the conductor receives its supply. Each of the permissive rules will be looked at individually as they apply to the primary or secondary of the transformer.

9.2.1 240.21(C)(1) Primary Protection, Including Secondary Protection

240.21(C)(1) applies to the secondary conductors of a transformer. The first statement in the regulation stipulates that the transformer secondary is not permitted to be protected by the primary overcurrent device unless it is one of the following transformer connections:

- Single-phase, 2-wire, single-voltage secondary
- Delta-delta, 3-phase, 3-wire, single-voltage secondary

Secondary conductors, as such, are not limited in length. However, when the primary overcurrent device is used to also protect the secondary, the following are required:

1. The primary overcurrent must be in accordance with **Section 450.3.**
2. The primary overcurrent protection must not exceed the value determined by multiplying the ampacity of the secondary conductors by the secondary-to-primary voltage ratio.

Examples of application:

1. Single-phase, 2-wire, 480-volt primary with a single-phase, 2-wire, 120-volt ungrounded secondary
2. Single-phase, 2-wire, 480-volt primary with a single-phase, 2-wire, 120-volt secondary with one conductor grounded
3. Single-phase, 2-wire, 120-volt primary with a single-phase, 2-wire, 120-volt ungrounded secondary used as an isolation transformer
4. Delta, 3-phase, 480-volt primary with a 240-volt, 3-phase, 3-wire delta secondary

Problem 9-11

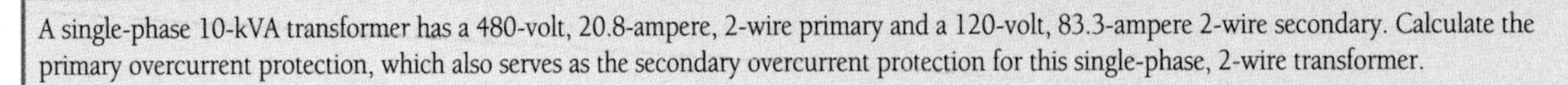

A single-phase 10-kVA transformer has a 480-volt, 20.8-ampere, 2-wire primary and a 120-volt, 83.3-ampere 2-wire secondary. Calculate the primary overcurrent protection, which also serves as the secondary overcurrent protection for this single-phase, 2-wire transformer.

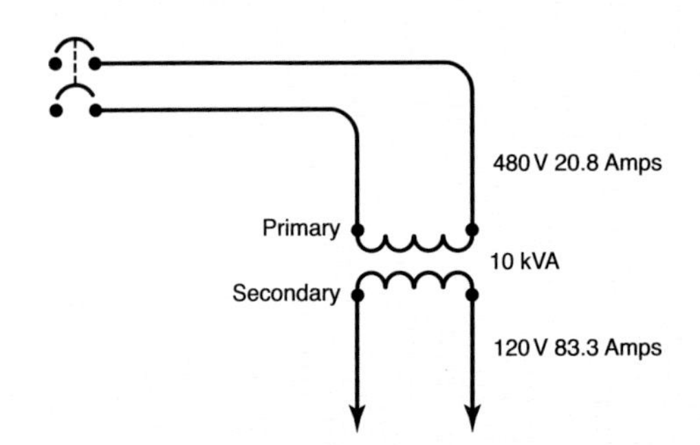

Solution - Calculation 1
Primary overcurrent protection
Table 450.3(B)
Currents of 9 amperes or more column
Primary protection only = 125%
Max. $OCPD_{pri}$ = $I_{pri} \times 125\%$
= 20.8 x 1.25
= 26 amps
Table 450.3(B) Note 1
Next larger std. size permitted
240.6(A)
Next larger = 30 amps
Answer: 30 ampere $OCPD_{pri}$
Note: The use of a 25 ampere $OCPD_{pri}$ is not prohibited as an alternate (less than maximum) solution.

Solution - Calculation 2
Secondary conductor ampacity
I_{sec} = 83.3 amps
Table 310.15(B)(16)
THWN copper column
83.3 amps = 4 AWG THWN
4 AWG THWN = 85 amps ampacity
Answer: Ampacity of 85 amperes

Solution - Calculation 3
Secondary-to-primary voltage ratio

$$V_{ratio} = \frac{E_{sec}}{E_{pri}}$$

$$= \frac{120}{480}$$

$$= 0.25$$

Answer: 0.25 voltage ratio

Solution - Calculation 4
Verification No. 1
Confirm acceptable secondary conductor size
4 AWG THWN = 85 amps
$OCPD_{pri} = I_{sec} \times V_{ratio}$
= 85 x 0.25
= 21.25 amps
$OCPD_{pri}$ (30 amps or 25 amps) rating exceeds 21.25 amps
Verification No. 1 fails
Verification No. 2
Increase size secondary conductors
Table 310.15(B)(16)
THWN copper column
3 AWG THWN = 100 amps
$OCPD_{pri} = I_{sec} \times V_{ratio}$
= 100 x 0.25
= 25 amps
$OCPD_{pri}$ (25 amps) rating does not exceed 25 amps
Verification No. 2 is acceptable installation
$OCPD_{pri}$ 25 amps, 10 AWG on primary and 3 AWG on secondary
Verification No. 3
Increase secondary conductors to 1 AWG
Table 310.15(B)(16)
THWN copper column
1 AWG THWN = 130 amps
$OCPD_{pri} = I_{sec} \times V_{ratio}$
= 130 x 0.25
= 33 amps
Use 30 amp $OCPD_{pri}$
30 amps $OCPD_{pri}$, 10 AWG THWN copper primary conductors, and 1 AWG THWN copper secondary conductors
Verification No. 3 is an acceptable installation
Answer: $OCPD_{pri}$ 25 amps, 10 AWG on primary and 3 AWG THWN copper on secondary
Answer: $OCPD_{pri}$ 30 amperes, 10 AWG on primary and 1 AWG THWN copper on secondary

240.21(C)(1) cannot be applied to a transformer with a single-phase, 2-wire, 480-volt or 277-volt primary and a 120/240 volt, 3-wire secondary; as commonly used for 120-volt receptacles with a 277/480 volt system. Nor can the rule be applied to a 480-volt delta primary and a 120/208 volt, 4-wire wye secondary. Summarizing the calculation requirements of **240.21(C)(1)**, the following two rules apply:

1. Rating of the primary OCPD cannot exceed requirements of **Table 450.3(B)**.
2. Rating of the primary OCPD cannot exceed the value of the secondary conductor ampacity multiplied by the voltage ratio of the secondary voltage divided by the primary voltage.

9.2.2 Ten Foot Feeder Tap Rule—Transformer Primary Conductors 240.21(B)(1)

The basic rule for overcurrent protection of conductors requires the overcurrent protective device to be located at the point the conductor receives its supply. **Section 240.21** lists several permissive rules for the installation of tap conductors without overcurrent protection. **240.21(B)(1) Feeder Taps** permits a 10 foot tap from a feeder without overcurrent protection on the primary side of the transformer, provided all of the following conditions are met:

1. The tap conductors are not over 10 feet in length
2. The ampacity of the tap conductors is not less than:
   a. The combined calculated loads on the circuit supplied by the tap, and
   b. The current rating of the equipment containing an overcurrent device supplied
3. The tap conductors do not extend beyond the transformer, switchgear, panelboard or disconnecting means they supply
4. Except at the point of connection, the conductors are protected and enclosed in a raceway
5. For field installations, if the tap conductors leave the enclosure where the tap is made, the ampacity of the tap conductors cannot be less than one-tenth of the rating of the overcurrent device protecting the feeder conductors

According to **240.21(B)**, the provisions of **240.4(B)** do not apply to feeder tap conductors. Also, the small conductor requirements of **240.4(D)** do not apply since there are specifically mentioned in **240.4(E)(3)**.

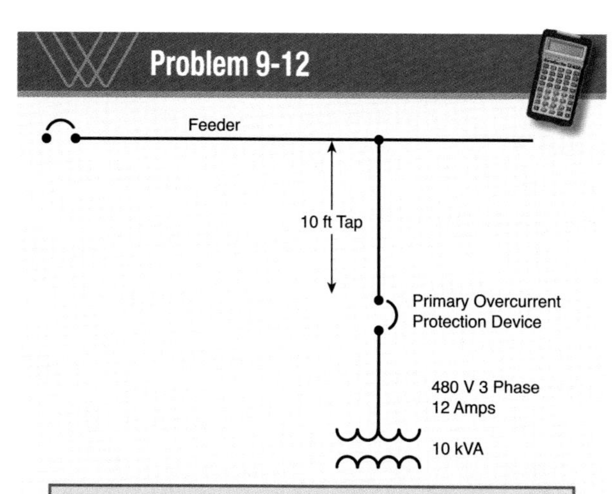

Calculate the minimum size tap conductor using THWN copper, and the maximum setting of the overcurrent protective device on the primary feeder for a 10-kVA, 480-volt, 3-phase transformer with a full-load primary current of 12 amperes.

Solution – Calculation 1
Primary overcurrent protection
Table 450.3(B)
Currents of 9 amperes or more column
Primary protection = 125%
Max. $OCPD_{pri}$ $= I_{pri} \times 125\%$
$= 12 \times 1.25$
$= 15$ amps
15 amps is a standard size
Answer: 15 amp Max. $OCPD_{pri}$

Solution – Calculation 2
Primary conductor size
Tap conductor (TC) amps $= I_{pri}$ amps
$= 12$ amps
Table 310.15(B)(16)
THWN copper column
12 amps = 14 AWG THWN copper
240.21(B) indicates that 240.4(D) does not apply
240.4(E) applies
14 AWG THWN = 20 amps
Answer: 14 AWG THWN copper

Solution – Calculation 3
Feeder CB ratio compliance
240.21(B)(1)(4)
Feeder CB cannot exceed ten times the tap conductor ampacity
Tap conductor ampacity
14 AWG THWN = 20 amps
Feeder OCPD max $= 20 \text{ amps} \times 10$
$= 200$
Feeder CB cannot exceed 200 amps
Answer: Feeder OCPD cannot exceed 200 amperes

Comment
Without the limitation of the "one-tenth" rule, in 240.21(B)(1)(4), a small conductor could be severely damaged by high let-through current of a larger overcurrent protective device. For example, if an insulated 14 AWG conductor were on the load side of a 400-ampere circuit breaker, the insulated 14 AWG conductor could "see" a maximum of 400 amperes for a long period of time without ever tripping. Four hundred amperes on an insulated 14 AWG conductor will far exceed the damage curve of that conductor and may cause a fire. So, for the case of a 400-ampere feeder overcurrent device, instead of an insulated 14 AWG conductor, at least an insulated 8 AWG (40 amp) conductor would be necessary.

9.2.3 Ten Foot Transformer Tap Rule—Transformer Secondary Conductors

240.21(C)(2) covers transformer secondary conductors not over 10 feet long and permits them to be without overcurrent protection on the secondary side of the transformer provided all of the following conditions are met:

1. The length of secondary conductors is limited to 10 feet
2. The ampacity of the secondary conductors is not less than:
   a. The combined calculated load served
   b. The rating of the equipment containing an overcurrent device or the overcurrent protective device at its termination
3. The tap conductors do not extend beyond the panel or equipment supplied
4. The conductors are protected and enclosed in a raceway
5. For field installations, if the secondary conductors leave the enclosure where the connections are made, the rating of the overcurrent device protecting the primary of the transformer, multiplied by the "primary to secondary transformer voltage ratio," cannot exceed 10 times the ampacity of the secondary conductor

Note that the secondary conductors of a transformer are not considered tap conductors. The secondary conductors should be considered a feeder originating at the secondary of the transformer.

The 10 foot secondary conductor rule requires secondary conductors to terminate on a set of fuses or a circuit breaker. The 10 foot tap rule no longer allows conductors to land directly onto main lug only–type switchboards or panelboards. Now, all other secondary tap rules require conductors to terminate on a single set of fuses or a single circuit breaker.

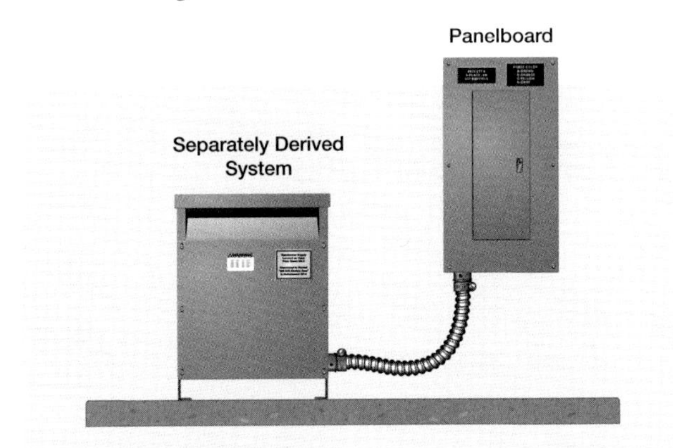

A separately derived system supplying a main circuit breaker panelboard using the ten-foot transformer tap rule and complying with 408.36

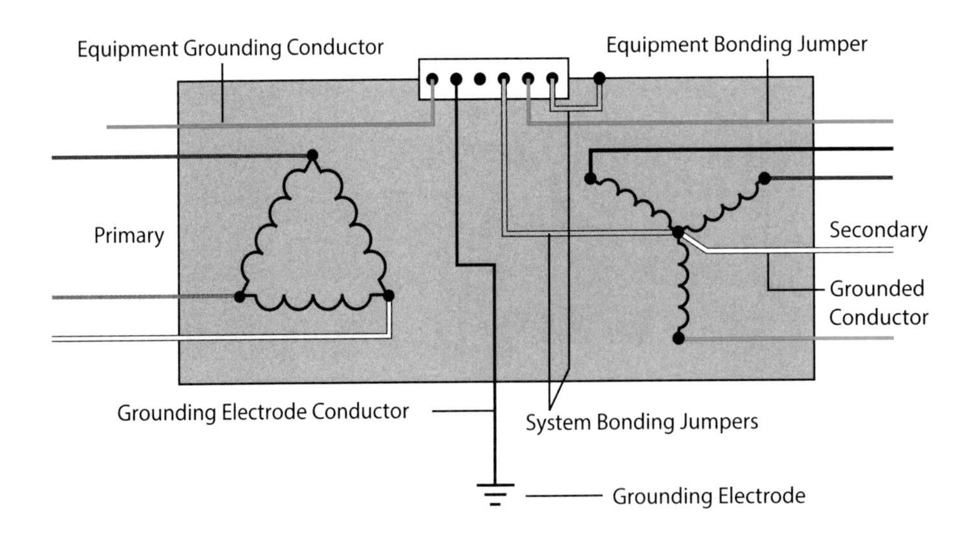

An example of wiring diagram for a three-phase, delta-wye transformer used as separately derived system.

Problem 9-13

The following calculation illustrates an alternate or different use of the 10 foot secondary conductor rule of 240.21(C)(2). The drawing shows two separate sets of 10 foot secondary conductors used on the secondary side of a 120/208 volt, 3-phase, 4-wire, 112.5-kVA transformer. One set of secondary conductors supplies a panelboard protected by a 300-ampere main circuit breaker. Another set of secondary conductors supplies a 60-ampere fusible disconnect switch. Both transformer primary and secondary overcurrent protection is to be used according to Table 450.3(B).

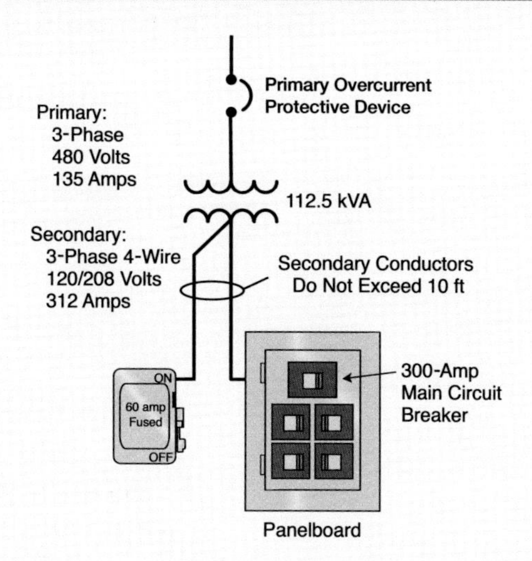

Solution – Calculation 1
Primary overcurrent protection
I_{pri} = 135 amps
Table 450.3(B)
Currents of 9 amperes or more column
Primary and secondary protection = 250%
Max. $OCPD_{pri}$ = I_{pri} × 250%
= 135 × 2.5
= 337.5 amps
Table 450.3(B) Note 1
Does not apply, use next smaller
240.6(A)
Next smaller = 300 amp fuse or circuit breaker
Answer: 300 ampere max. $OCPD_{pri}$

Solution – Calculation 2
Primary conductor size
Match conductors to 300 amp $OCPD_{pri}$
Table 310.15(B)(16)
THWN copper column
350 kcmil THWN copper = 310 amps
Answer: 350 kcmil THWN copper per phase
Answer: 400 amperes

Solution – Calculation 3
Secondary conductors (300 amp circuit)
240.21(C)(2)
300 amp $OCPD_{sec}$
240.4(B) not permitted
300 amp minimum ampacity
Table 310.15(B)(16)
THWN copper column
350 kcmil THWN copper = 310 amps
Answer: 350 kcmil THWN copper per phase and neutral

Solution – Calculation 4
Secondary overcurrent protection
60 amp fused circuit
215.2(A)(1)
Maximum calculated load is 60 amps
Maximum OCPD is 60 amps
Answer: 60 ampere max. $OCPD_{sec}$

Solution – Calculation 5
Secondary conductors
60 amp circuit
240.21(C)(2)(4)
Max. ratio for field installation
Max. ratio = 1/10 × $OCPD_{pri}$
= 0.10 × 300
= 30 amps
Calculate primary-to-secondary voltage ratio

$$V_{ratio} = \frac{E_{pri}}{E_{sec}} = \frac{480}{208} = 2.31$$

$OCPD_{pri} \times V_{ratio}$
30 amps × 2.31 = 69 amps
Table 310.15(B)(16)
4 AWG THWN copper rated 85 amps
Answer: 4 AWG THWN copper (60 ampere circuit)

Solution – Calculation 6
Table 450.3(B) Note 2
Verify maximum secondary overcurrent protection is not exceeded
Step 1
Calculate max. value of a single $OCPD_{sec}$
I_{sec} = 312 amps
Table 450.3(B)
Currents of 9 amperes or more column
Secondary protection = 125%
$OCPD_{sec}$ = I_{sec} × 125%
= 312 × 1.25
= 390 amps
Table 450.3(B) Note 1
Next larger std. size permitted
240.6(A)
Next larger = 400 amps
Answer: 400 amperes
Step 2
Verify total max. $OCPD_{sec}$ does not exceed 400 amps by adding both $OCPD_{sec}$
300 amp $OCPD_{sec}$ + 60 amp $OCPD_{sec}$ = 360 amps
Answer: Total max. $OCPD_{sec}$ not exceeded

Comment
For a configuration such as this example, a load calculation for these two combined secondary circuits must be done according to Article 220. In addition, all loads must be calculated at 100% of the noncontinuous load and 125% of the continuous load according to 215.2(A)(1), and, finally, the total calculated load cannot exceed the maximum permitted load on the selected transformer.

9.2.4 Ten Foot Transformer Tap Rule—Supplying a Panelboard

When the 10 ft tap rule is used to supply a panelboard, **Section 408.36** requires an overcurrent device to protect the panelboard. According to this section, the overcurrent protective device has to be located on any point in the feeder circuit or within the panelboard itself.

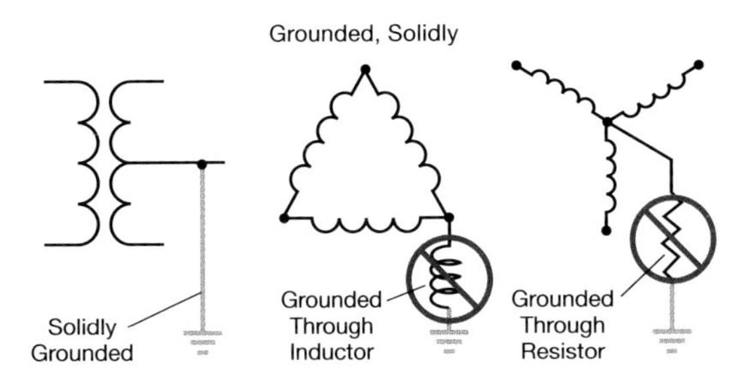

This is a graphic representation of the definition of the NEC term "solidly grounded" as used in transformer installations.

Problem 9-14

Calculate the overcurrent protection and conductor sizes permitted for the 75-kVA transformer using the 10 foot secondary conductor rule to supply a panelboard.

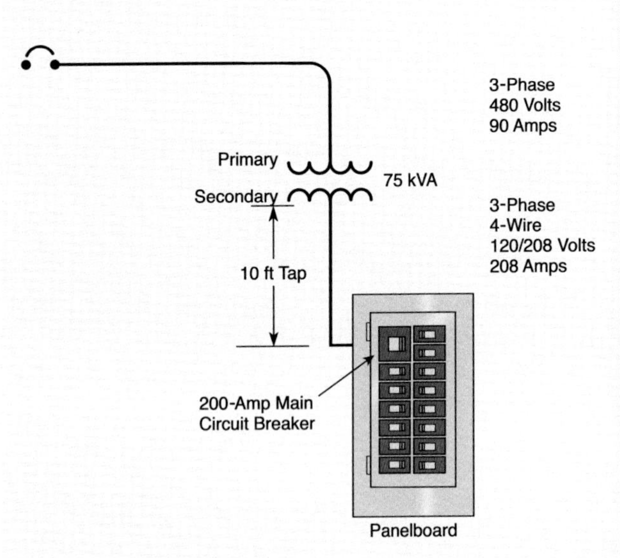

Solution – Calculation 1

Primary overcurrent protection
I_{pri} = 90 amps
Table 450.3(B)
Currents of 9 amperes or more column
Primary protection = 125%
Max. $OCPD_{pri}$ $= I_{pri} \times 125\%$
$= 90 \times 1.25$
$= 113$ amps
Table 450.3(B) Note 1
Next larger std. size permitted
240.6(A)
Next larger = 125 amps
Answer: 125 amperes max. $OCPD_{pri}$

Solution – Calculation 2

Primary conductors
Table 310.15(B)(16)
THWN copper column
113 amps = 2 AWG THWN copper
2 AWG THWN = 115 amps
240.4(B)
Next larger higher rating permitted
OCPD for conductor = 125 amps
Table 310.15(B)(16)
THWN copper column
125 amps = 2 AWG THWN copper
Answer: 2 AWG THWN copper
The 125 ampere overcurrent protection will protect the conductors supplying the transformer and the primary windings of the transformer.

Solution – Calculation 3

Secondary overcurrent protection
I_{sec} = 208 amps
Table 450.3(B) and 408.36
Currents of 9 amperes or more column
Secondary protection = 125%
$OCPD_{sec} = I_{sec} \times 125\%$
$= 208 \times 1.25$
$= 260$ amps
Table 450.3(B) Note 1
Next larger std. size permitted
240.6(A)
Next larger = 300 amps
Max. $OCPD_{sec}$ = 300 amps
But 200 amp panel is less than secondary FLC
200 amp $OCPD_{sec}$ is the max. permitted
Answer: 200 ampere max. $OCPD_{sec}$

Solution – Calculation 4

Secondary conductors
$OCPD_{sec}$ = 200 amps
Table 310.15(B)(16)
THWN copper column
200 amps = 3/0 AWG THWN copper
Answer: 3/0 AWG THWN copper

Comment
In this particular case, the 200-ampere main overcurrent protection in the panelboard has three responsibilities:
1. Protect the transformer secondary windings
2. Limit the current on the 3/0 AWG 10 foot secondary conductors to 200 amperes
3. Protect the 200-ampere panelboard

9.2.5 Twenty-Five Foot Feeder Tap Rule—Transformer Primary 240.21(B)(2)

240.21(B)(2) **Feeder Taps** is a permissive rule permitting a maximum conductor tap length of 25 feet without overcurrent protection at the point where the conductor receives its supply. This permissive rule is applicable to the primary of a transformer when all of the following conditions are met:

1. The tap is not over 25 feet in length
2. The ampacity of the tap conductors is at least one-third the rating of the overcurrent device protecting the conductors
3. The tap conductors terminate in a single set of fuses or a circuit breaker, which will limit the current on the tap conductors
4. The tap conductors are protected from physical damage

There is more than one solution for these calculations. The final solution should be always be verified for compliance with all of the *Code* text.

Problem 9-15

A 300-ampere feeder is tapped to supply a 75-kVA transformer. The total length of the tap leading to the primary circuit breaker does not exceed 25 feet. Calculate the minimum THWN copper conductor size of the tap conductor leading to the 75-kVA transformer. Also, calculate the size of the primary circuit breaker protecting the transformer using the primary only protection rule of Table 450.3(B).

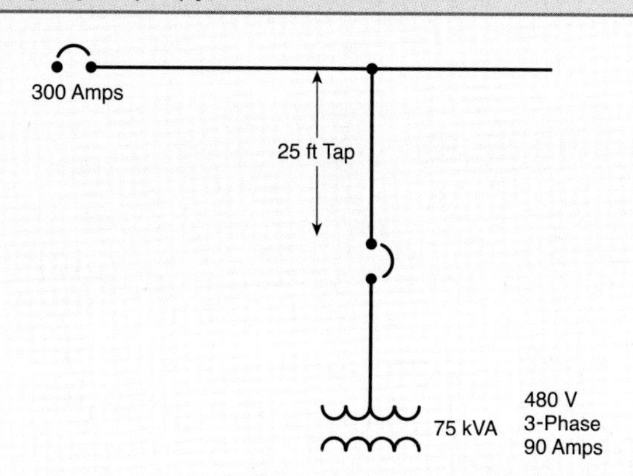

Solution – Calculation 1
Primary overcurrent protection at transformer
Table 450.3(B)
Currents of 9 amperes or more column
Primary protection = 125%
$OCPD_{pri} = I_{pri} \times 125\%$
$= 90 \times 1.25$
$= 113$ amps
Table 450.3(B) Note 1
Next larger std. size permitted
240.6(A)
Next larger = 125 amps
$OCPD_{pri}$ = 125 amp CB
Answer: 125 ampere CB

Solution – Calculation 2
Primary tap conductor size
125 amp CB
Table 310.15(B)(16)
2 AWG THWN copper or 1 AWG THW copper
Answer: 2 AWG THWN copper wire rated 115 amps

Solution – Calculation 3
Verify primary tap conductor size
240.21(B)(2)(1)
Not less than 1/3 rating of feeder CB
$\frac{300}{3} = 100$ amps
Primary tap conductor must be 100 amps or greater
Primary tap conductor = 2 AWG THWN copper
Answer: 2 AWG THWN copper is sufficient

9.2.6 Twenty-Five Foot Transformer Tap Rule—Transformer Secondary Conductors 240.21(C)(6)

240.21(C)(6) is what is commonly referred to as the 25 foot rule for secondary conductors of a transformer. Like the others, this rule is comprised of several parts:

1. Secondary conductors are not over 25 feet long
2. The secondary conductor ampacity is not less than primary-to-secondary voltage ratio multiplied by 1⁄3 the rating of the primary OCPD
3. Secondary conductors terminate in a single circuit breaker or set of fuses which limits the current to the ampacity of the secondary conductor
4. Secondary conductors are protected from physical damage, such as being enclosed in an approved raceway

For additional information, visit qr.njatcdb.org Item #1040

A power supply to a sign panelboard via a transformer using 240.21(C)(6).

Problem 9-16

A 3-phase 80-kVA transformer with a full-load secondary current of 193 amperes supplies a 240-volt, 3-phase, 3-wire panelboard equipped with a 250-ampere main circuit breaker. The transformer secondary conductors supplying the panelboard do not exceed 25 feet in length. Determine if the 250-ampere circuit breaker complies with Table 450.3(B) secondary protection and calculate the size of the copper THWN secondary conductors supplying the transformer.

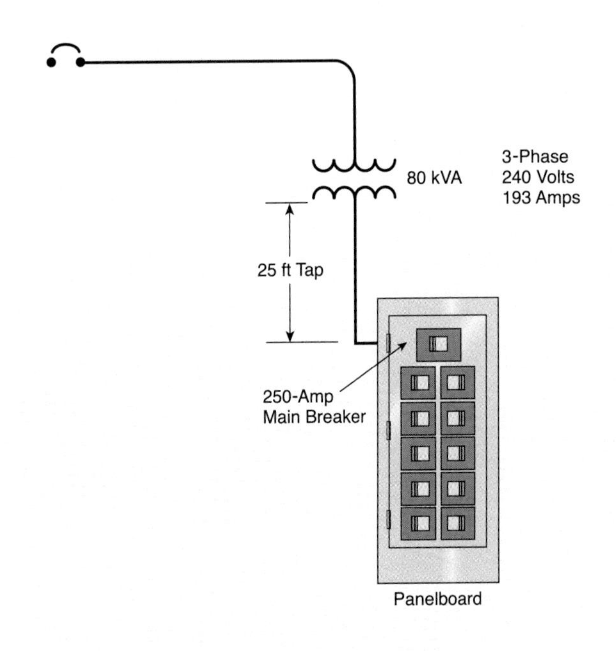

Solution – Calculation 1

Secondary overcurrent protection

$$I_{sec} = \frac{kVA \times 1,000}{E_{sec} \times 1.73}$$

$$= \frac{80 \times 1,000}{240 \times 1.73}$$

= 193 amps

Table 450.3(B)

Currents of 9 amperes or more column

Secondary protection = 125%

Max. $OCPD_{sec}$ $= I_{sec} \times 125\%$

$= 193 \times 1.25$

$= 241.25$ amps

Table 450.3(B) Note 1

Next larger std. size permitted

240.6(A)

Next larger = 250 amps

Answer: 250 amperes max. $OCPD_{sec}$

Solution – Calculation 2

Secondary conductors

Table 310.15(B)(16)

THWN copper column

250 amps = 250 kcmil THWN copper

250 kcmil THWN = 255 amps

250 kcmil will carry 193 amp secondary amps

Answer: 250 kcmil THWN copper

9.2.7 Outside—Secondary Conductors 240.21(C)(4)

240.21(C)(4) is a permissive rule for outside secondary conductors and is applicable to the secondary conductors of a transformer installed with the following limitations:

1. Conductors are installed outdoors except at the point of load termination
2. Conductors must be protected from physical damage
3. Conductors are required to be terminated on a single overcurrent device which will limit the load to the ampacity of the conductors
4. The overcurrent device is an integral part of a disconnecting means, or it is adjacent to the disconnecting means
5. The disconnecting means is readily accessible outside of the building, or inside nearest the point of entrance of the conductors, or is installed in accordance with **Section 230.6** at nearest the point of entrance of the conductors

An example of use would be a 480-volt feeder to a second building on the premises where a transformer is set outside the second building. The transformer secondary conductors terminate in a circuit breaker either outside the building or inside the building. Note: **240.21(C)(4)** does not limit the length of outside secondary conductors.

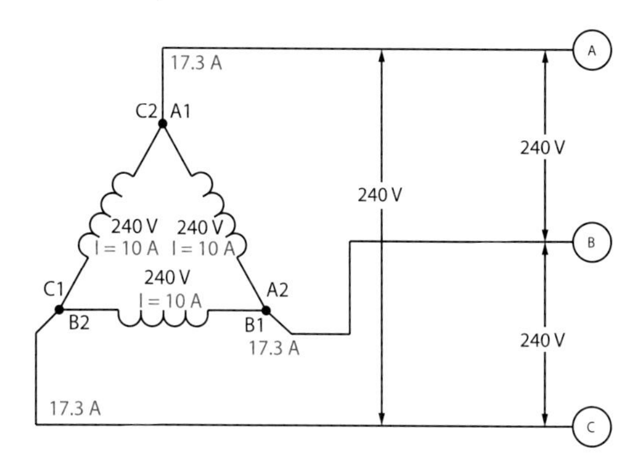

A delta connection has the wires from the ends of each coil connected end-to-end to form a closed loop.

Problem 9-17

What size THWN copper secondary conductors are needed for a 45-kVA, 3-phase, 480-volt primary and a 208/120 volt secondary pad-mounted transformer set outside of a second building? The secondary feeder conductors terminate immediately inside the building, in a single circuit breaker, for secondary overcurrent protection.

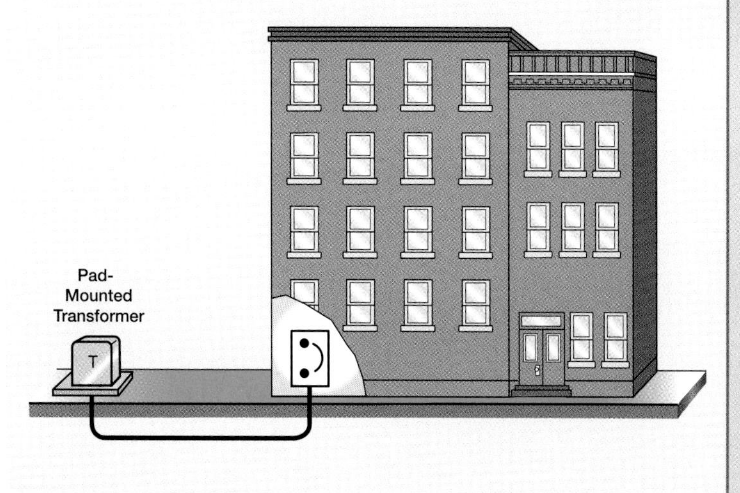

Solution

Secondary conductor size

$$I_{sec} = \frac{\text{kVA} \times 1{,}000}{E_{sec} \times 1.73}$$

$$= \frac{45 \times 1{,}000}{208 \times 1.73}$$

$$= 125 \text{ amps}$$

Table 450.3(B)(2)
Currents of 9 amperes or more column
Secondary protection = 125%
$OCPD_{sec}$ = Isec × 125%
= 125 × 1.25
= 156.25 amps
Table 450.3(B) Note 1
Next larger std. size permitted
240.6(A)
Next larger = 175 amps
Table 310.15(B)(16)
THWN copper column
175 amps = 2/0 AWG THWN

Answer: 2/0 AWG THWN copper

9.3 Dedicated Transformers Used in Fire Pump Circuits

695.5(A) and **695.5(B)** give special attention to fire pump installations with a transformer dedicated to supplying only the power to a fire pump and its related equipment. No overcurrent protection is permitted on the secondary side of the transformer. The primary overcurrent protection rating or setting must be large enough to carry all of the following loads:

1. The locked-rotor current (LRC) of the fire pump motor(s)
2. The locked-rotor current of the pressure maintenance or jockey pump motor(s) when connected to this power supply
3. And 100% of the current(s) of any associated fire pump accessory equipment when connected to this power supply

Problem 9-18

Calculate the minimum primary overcurrent protection for a 480/240 volt dedicated fire pump transformer where it supplies a 240-volt, 3-phase, 25-hp fire pump; a 240-volt, 3-phase, 5-hp pressure maintenance pump; and 3-phase accessory equipment with a full-load rating of 60 amperes.

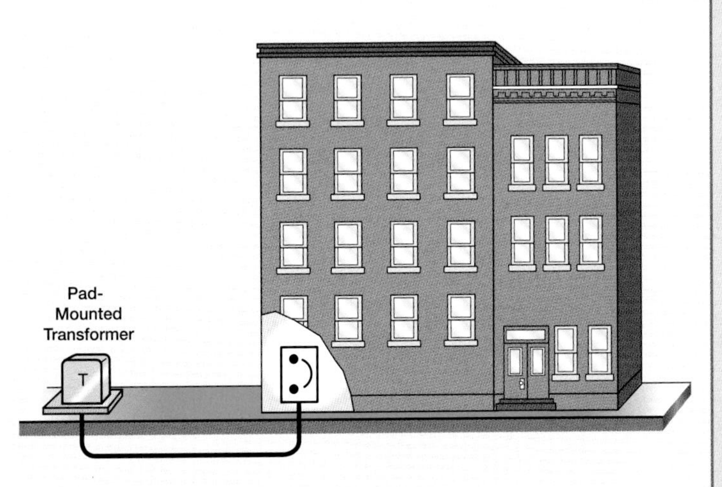

Solution

Primary overcurrent protection

Using locked-rotor current (LRC)

LRC for 3-phase motors

Table 430.251(B)

230 V column

LRC of a 25 hp = 365 amps

LRC of a 5 hp = 92 amps

Min. I_{sec} = LRC motor 1 + LRC motor 2 + I (accessories)

= 365 + 92 + 60

= 517 amps

Transfer I_{sec} (secondary current) to I_{pri} (primary current)

By using secondary-to-primary voltage ratio

$$V_{ratio} = \frac{E_{sec}}{E_{pri}} = \frac{240}{480} = 0.5$$

Min. I_{pri} = Min. $I_{sec} \times V_{ratio}$

= 517 × 0.5

= 258.5 amps

Min. $OCPD_{pri}$ = I_{pri}

= 258.5 amps

240.4(B)

Next larger higher rating permitted

Min. $OCPD_{pri}$ = 300 amps

Answer: 300 ampere min. $OCPD_{pri}$

This page intentionally left blank.

Definitions and Terms

Overcurrent - Any current in excess of the rated current of equipment or the ampacity of a conductor. It may result from overload, short circuit, or ground fault.

Overload - Operation of equipment in excess of normal, full-load rating, or of a conductor in excess of rated ampacity that, when it persists for a sufficient length of time, would cause damage or dangerous overheating. A fault, such as a short circuit or ground fault, is not an overload.

Tap Conductor - A conductor other than a service conductor, that has overcurrent protection ahead of its point of supply that exceeds the value permitted for similar conductors that are protected as described in Section 240.4. The application of tap conductor is limited to Article 240, and does not apply to other areas of the *Code*.

Transformer - An individual transformer, single-phase or polyphase, identified by a single nameplate, unless otherwise indicated in Article 450.

Summary

- Transformers are usually sized based upon three mandatory pieces of information: primary source voltage, secondary load voltage, and kVA of the load served by the transformer.
- Next, the standard size transformer which complements the calculated load can be determined.
- After these items are known, generally the next step is to look up or calculate the full-load currents of the primary and secondary windings for the selected transformer. These serve as the limits for a given size transformer.
- The five fundamental items are: standard kVA size transformer, single- or three-phase, primary voltage and current, secondary voltage and current, as well as the actual load to be served.
- Next, the overcurrent protection for the transformer primary and secondary windings (rated at 600 volts, nominal, or less) can be determined.
  - » These overcurrent devices should first be thought of as being located at the transformer.
  - » The overcurrent devices may also be placed ahead of the transformer feeder.
- **Table 450.3(B)** is used to determine the primary overcurrent protection as follows:
  - » When the primary current rating is 9 amperes or more, overcurrent protection is permitted to be a maximum of 125% of the primary current and using **Note 1**, permission is given to select the next higher standard size device.
  - » When the primary current rating is less than 9 amperes, overcurrent protection is permitted to be a maximum of 167% of the rated primary current; without exception.
  - » When the primary current rating is less than 2 amperes, overcurrent protection is permitted to be 300% of the primary current.
  - » These three sizing requirements are all maximums, but using smaller sizes within reason (such as charging current) is always permitted.

Summary

- » When the overcurrent protection for the primary windings is located away from the transformer, the overcurrent device must protect both the transformer primary winding and the conductors supplying that transformer primary.
- When calculating secondary overcurrent protection, use one of the following methods:
  - » When the secondary current rating is 9 amperes or more, overcurrent protection is permitted to be 125% of the rated secondary current and permission to select the next higher standard size is given using **Note 1**.
  - » When the secondary current is less than 9 amperes, overcurrent protection is permitted to be up to 167% of the rated secondary current. Use the 167% value or any lower size.
  - » Provided the secondary windings are protected by the 125% rule, and provided the overcurrent protection on the primary protects the conductors and the primary winding, then, the primary overcurrent protection is permitted to be up to 250% of the rated primary current.
- The maximum of up to six overcurrent devices are permitted as protection of a transformer secondary, provided the total ampacity of the six overcurrent devices does not exceed that of a single OCPD.
- The value of the primary overcurrent device for a transformer equipped with interior thermal overload protection is calculated according to either of the following:
  - » Where a transformer impedance is 6% or less, the OCPD on the primary must not be greater than six times the primary current. Use the 600% value or any lower size.
  - » Where a transformer impedance is over 6%, but less than 10%, the OCPD on the primary must not be greater than four times the primary current. Use the 400% value or any lower size.
- Since fire pumps are considered sacrificial equipment when called upon to work during a fire, most fire pump circuit overcurrent protective devices are exceedingly larger than normally required on other "nonlife-safety" circuits.
- Where the service or system voltage is different from the utilization voltage of the fire pump motor, transformer(s) protected by disconnecting means and overcurrent protective devices are be permitted to be installed between the system supply and the fire pump controller as follows:
  - » Where a transformer supplies an electric motor driven fire pump, it must be rated at a minimum of 125 percent of the sum of the fire pump motor(s) and pressure maintenance pump(s) motor loads, and 100 percent of the associated fire pump accessory equipment supplied by the transformer.
  - » The primary overcurrent protective device(s) must be selected or set to carry indefinitely the sum of the locked-rotor current of the fire pump motor(s) and the pressure maintenance pump motor(s) and the full-load current of the associated fire pump accessory equipment when connected to this power supply.
  - » The requirement to carry the locked-rotor currents indefinitely shall not apply to conductors or devices other than overcurrent devices in the fire pump motor circuit(s).
  - » Secondary overcurrent protection shall not be permitted. See 695.5(A) and (B).
- Similar requirements exist for a feeder source supplying a fire pump where additional non-fire pump related loads are also served by the transformer. See 695.5(C)

Cable Tray

Cable tray calculations are used to determine the proper cable tray fill using single or multiconductor cables in different types of cable trays for conductors rated 2,000 volts or less. The standard cable tray sizes (based upon inside width) are listed in the first column of **Table 392.22(A)**. According to this table, cable tray sizes are related to both the "... sum of diameters of cables placed within the tray" as well as "...the cross-sectional area of fill in the cable tray." To determine single-conductor cable diameters and cross-sectional areas, use **Table 5** of **Chapter 9** for concentric conductors and **Table 5A** for compact conductors.

The types of cable tray include ladder, ventilated and solid bottom trough, ventilated and solid bottom channel, as well as wire mesh tray.

Fill calculations are somewhat similar to conduit and wireway fill calculations, in that tray fill calculations follow the tables found in **Section 392.22**. When calculating the maximum conductor fill for a circular raceway, the most commonly used column for those calculations is the 40% fill column of **Chapter 9, Table 4**. So, Column 1 of **Table 392.22(A)** can be compared to the 40% column of **Chapter 9, Table 4**.

Ampacity calculations of single and multiconductor cables are generally more liberal than calculations for wiring methods of *NEC* Chapter 3. The ampacity of the conductors installed in cable trays is based upon the installation of the conductors being made according to **392.80(A)(1)** for multiconductor cables and **392.80(B)(1)** for single-conductor cable with few variations. Final ampacity calculations also include **110.14(C)(1)** regarding terminal temperature limitations. The final ampacities for cables in cable trays may require stepping down to the ampacities of **Table 310.15(B)(16)** and the use of the 75°C column to coordinate with temperature limitations of the overcurrent protective devices, switchgear, switchboards, panelboards, and other user equipment.

Objectives

- Calculate cable tray fills for single-conductor and multiconductor cables in various types of cable trays.
- Determine the ampacity of single-conductor and multiconductor cables for given cables, conductors, and cable tray installation arrangements.
- Demonstrate how to calculate cable tray fills for multiconductor control, signal cables, and any mixture of cables.
- For an existing cable tray, determine the presence of additional space within the cable tray and show the necessary steps to place additional cables into the cable tray.

Chapter 10

Table of Contents

10.1 Cable Tray Fill Calculations for Multiconductor Installations

Cable trays are not considered a wiring method. They are a support system for wiring methods. Type TC cable, or tray cable, is just one of the recognized wiring methods which may be supported by cable trays. Many other wiring methods are permitted to be installed in cable trays.

Types of cable tray installations include the following:

1. Ladder
2. Ventilated trough
3. Solid bottom
4. Cable tray with a solid fixed barrier
5. Any of the above with a cover
6. Ventilated and solid channel
7. Wire mesh

For additional information, visit qr.njatcdb.org Item #1041

Each of these trays provides strength for conductor support, physical protection from damage, and ventilation functions to a greater or lesser degree. All ventilated types of cable tray are designed to allow significant airflow around cables to provide dissipation of heat from conductors.

The following is a tabular comparison of the three types of cable trays and their ability to dissipate heat generated within the conductors.

Type of Tray	Ventilation	Heat Dissipation
Ladder	Yes	Yes
Wire Mesh	Yes	Yes
Trough	Yes	Less than Ladder type
Solid Bottom	No	Less than Trough type
Covered (all types)	Limited	Reduced

Ladder, ventilated trough, and wire mesh cable trays allow better circulation of air for better heat dissipation. Therefore, it is understandable that the cable tray fill for a solid bottom tray will be less than that for a ventilated tray.

Whether the cable tray is covered or not covered is not taken into consideration when deciding cable tray fill. It is factored in when calculating the ampacity of the conductors in cable trays.

For all practical purposes, when a solid fixed barrier is installed in a cable tray, it can be considered as two cable trays. Ventilated and solid channel trays are used for cable dropouts, and have a smaller width measurement.

According to **392.10 Uses Permitted**, cable tray installations may be used as a support system for:

1. Service conductors
2. Feeders
3. Branch circuits
4. Communication cables and raceways
5. Control circuits
6. Signaling circuits

A cable tray can be dedicated to any one type of circuit or any combination within the range of 2,000 volts or less. Some combinations will require cable tray barriers. Cable tray installations are not restricted to industrial installations. Rather, cable trays are used in a wide variety of installations in many commercial establishments.

The following is a partial list of the many wiring methods found in **Table 392.10(A)** which are permitted to be supported by cable trays:

1. Armored cable (AC)
2. Communication cables and raceways
3. Class 2 and Class 3 cables
4. Electrical metallic tubing (EMT)
5. Fire alarm cables
6. Flexible metal conduit (FMC)
7. Instrument tray cable (ITC)
8. Intermediate metal conduit (IMC)
9. Liquidtight flexible metal conduit (LFMC)
10. Liquidtight flexible nonmetallic conduit (LFNC)
11. Metal-clad cable (MC)
12. Mineral-insulated, metal-sheathed cable (MI)
13. Network-powered broadband communications cables
14. Optical fiber cables and optical fiber raceways
15. Power and control tray cable (TC)
16. Power-limited tray cable (PLTC)
17. Other factory-assembled multiconductor control, signal, or power cables that are specifically approved for installation in cable tray
18. Rigid metal conduit (RMC)
19. Rigid nonmetallic conduit (RNC)

Cable and conductor installation requirements of **392.20(A)** and **(B)**, are as follows:
(A) Multiconductor cables operating at 600 volts or less can be installed in the same tray.
(B) Cables operating at over 600 volts and those operating at 600 volts or less must comply with the following:

(1) Cables operating at over 600 volts are Type MC.
(2) Cables operating at over 600 volts are separated from cables operating at 600 volts or less by a solid fixed barrier of material compatible with the cable tray.

Examples of performing fill calculations are based upon cable ratings of 2,000 volts or less, in accordance with **392.22(A)** and **392.22(B)**.

The following is a preview of **Table 392.22(A)** and **Table 392.22(B)(1)**, both used for cable tray fill. The tables look somewhat different and contain different symbols. Use **Table 392.22(A)** to follow this explanation.

When calculating the maximum raceway fill for conductors in a raceway, **Chapter 9, Table 4** is used. The most commonly used column is for 40% fill. Column 1 of **Table 392.22(A)** can be compared somewhat to the 40% column of **Chapter 9, Table 4**. The same square inch area used in Column 1 is also used in Column 2 of **Table 392.22(A)**.

The following calculations will clarify where the numbers used in **Table 392.22(A)** come from. Using a 40% fill for a cable tray measuring 3 inches of cross-sectional depth, the following comparison can be made to Column 1 of **Table 392.22(A)**. Note that Column 1 of **Table 392.22(A)** is rounded down to the next full square inch (in.^{2}).

Ventilated Tray Size	Cross-Sectional Area	Percent Fill	Total Fill	Table 392.22(A) Columns 1 and 2
6 in. × 3 in.	18 in.^{2} ×	40% =	7.2 in.^{2}	7 in.^{2}
12 in. × 3 in.	36 in.^{2} ×	40% =	14.4 in.^{2}	14 in.^{2}
18 in. × 3 in.	54 in.^{2} ×	40% =	21.6 in.^{2}	21 in.^{2}

Columns 1 and 2 of **Table 392.22(A)** are used for ladder or ventilated trough cable trays. Columns 3 and 4 are used for solid bottom cable trays. Solid bottom trays will not have as much ventilation; therefore, their fill can be expected to be less. Setting up a comparison table for solid bottom trays, using a 3 in. high tray and a 30% fill and comparing to Columns 3 and 4 of **Table 392.22(A)**, the calculated values are rounded up to the next one-half square inch (0.5 in.^{2})

Solid Bottom Tray Size	Cross-Sectional Area	Percent Fill	Total Fill	Table 392.22(A) Columns 3 and 4
6 in. × 3 in.	18 in.^{2} ×	30% =	5.4 in.^{2}	5.5 in.^{2}
12 in. × 3 in.	36 in.^{2} ×	30% =	10.8 in.^{2}	11.0 in.^{2}
18 in. × 3 in.	54 in.^{2} ×	30% =	16.2 in.^{2}	16.5 in.^{2}

392.22(A) and the accompanying **Table 392.22(A)** are used to determine the maximum number of multiconductor cables, rated 2,000 volts or less, which may be placed within a cable tray. This is commonly referred to allowable cable fill. In the following case studies, Case 1 is based upon a single layer of multiconductor cables. From a cross-sectional viewpoint, the arranged cables are then added as a "…sum of the cable diameters…" to determine the inside width of the cable tray. Case 2 is based upon 40% of the cross-sectional cable tray fill area according to Column 1 of **Table 392.22(A)**.

Case 1: Using **392.22(A)(1)(a)** for multiconductor cables in ladder, ventilated trough, and wire mesh tray, with conductor sizes 4/0 AWG or larger, the predetermined fill is based upon the "sum of diameters" for all cable placed in a single layer. For example, six 2.5 inch diameter multiconductor cables would require a 15 inch (2.5 in. x 6 cables = 15 in.) absolute minimum width cable tray. Using **Table 392.22(A)**, standard tray width results in the minimum selection of a 16 inch wide tray for practical reasons.

Case 2: Using **392.22(A)(1)(b)** and the accompanying **Table 392.22(A)** Column 1 for multiconductor cables in ladder, ventilated trough, and wire-mesh tray with all conductor sizes smaller than 4/0 AWG, the predetermined fill is based upon the Column 1 maximum allowable cable fill expressed in square inches (in.^{2}). For example, ten multiconductor cables with an individual cross-sectional area of 1.329 in.^{2} would require a 13.29 in.^{2} (1.329 in.^{2} × 10 cables = 13.29 in.^{2}) absolute minimum cross-sectional area cable tray. Again, according to **Table 392.22(A)** the next standard tray width larger than 13.29 in.^{2} is 12 inches. These and other problems will be studied in more detail later in this chapter.

Read **392.22(A)(1)(a)** through **392.22(A)(1)(c)** and carefully review **Table 392.22(A)**. Pertinent points include:

1. When 4/0 AWG and larger multiconductor cables are installed in a cable tray, they must be installed in a single layer.
2. The term "Sd" is used in **Table 392.22(A)** to denote the sum of the diameters of 4/0 AWG and larger multiconductor cables installed in the single layer area of the cable tray which also includes multiconductor cables smaller than 4/0 AWG according to **392.22(A)(1)**.
3. 1.2 Sd means 1.2 × sum of diameters.
4. This factor of 1.2 results in adding 1 inch for every 5 inches of total cable diameters. This provides 20% additional linear space to ensure these large multiconductor cables are not too crowded. In addition, it will allow a limited amount of natural cooling to occur.

Cable trays containing only multiconductor cables with a rating of 2,000 volts or less will be used in the first set of problems to determine cable tray fills. The cable tray widths listed in **Table 392.22(A)** will be used for standard cable tray widths. The following symbols and abbreviations will be used to identify cable in the cable tray illustrations.

Symbol Identification	
3/C	Number of conductors in a cable (3/C means three conductors in a multiconductor cable)
P	Power cable
L	Lighting cable
S	Signal cable
OD	Outside diameter of cable
r	*radius* = $\frac{OD}{2}$
A	Cross-sectional area $A = r \times r \times 3.1416$
Sd	Sum of the OD (Outside Diameters) of all cables (often laid adjacent in a single layer)

10.1.1 Multiconductor, Single-Layer, Vented Tray

392.22(A)(1) applies to ladder, ventilated trough, and wire mesh cable tray fill where only multiconductor cables, 2,000 volts or less, are installed. This section divides the installation into four groups:

1. All cables containing conductors 4/0 AWG or larger
2. All cables containing conductors smaller than 4/0 AWG
3. Mixing of cables containing conductors of all sizes
4. All cables that are signal or control cables

392.22(A)(1)(a) applies to the following:

- Multiconductor cables
- 2,000 volts or less
- 4/0 AWG or larger
- Installation in a single layer
- Installation in a ladder, ventilated trough, and wire mesh cable tray

392.22(A)(1)(a) requires the cables be installed in a single layer and the sum of all cable diameters not to exceed the inside cable tray width. Where this is done, then the ampacity of these cables can be determined by using **392.80(A)(1)**.

Problem 10-1

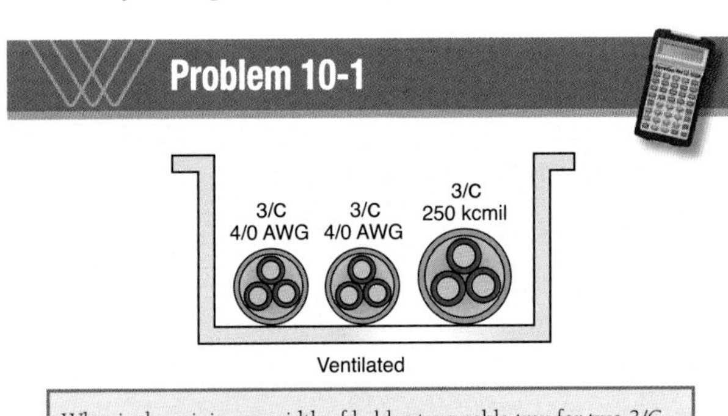

What is the minimum width of ladder-type cable tray for two 3/C, 4/0 AWG and one 3/C, 250 kcmil, 480 volt power cables? (OD 4/0 AWG = 1.875 in.; OD 250 kcmil = 2.25 in.)

Solution
392.22(A)(1)(a)
Single layer of conductors
Sd = Sum of all cable diameters
= (2 × OD 4/0 AWG) + OD 250 kcmil
= (2 × 1.875) + 2.25
= 6 in.
Answer: 6 in. minimum inside width

10.1.2 Multiconductor, More Than One Layer, Vented Tray

392.22(A)(1)(b) applies to the following:

- Multiconductor cables
- 2,000 volts or less
- All cables smaller than 4/0 AWG
- Installation in a ladder, ventilated trough or wire mesh cable tray
- Cables permitted to be stacked

392.22(A)(1)(b) requires the cross-sectional area fill not to exceed Column 1 of **Table 392.22(A)** for appropriate cable tray width.

Multiconductor cables in a ladder cable tray

Problem 10-2

What is the minimum size ladder-type cable tray needed for the installation of four 3/C, 1 AWG cables for power; three 3/C, 8 AWG cables for lighting; and one 7/C cable for signaling? (OD 1 AWG = 1.5 in.; OD 8 AWG = 0.875 in.; OD 7/C = 1 in.)

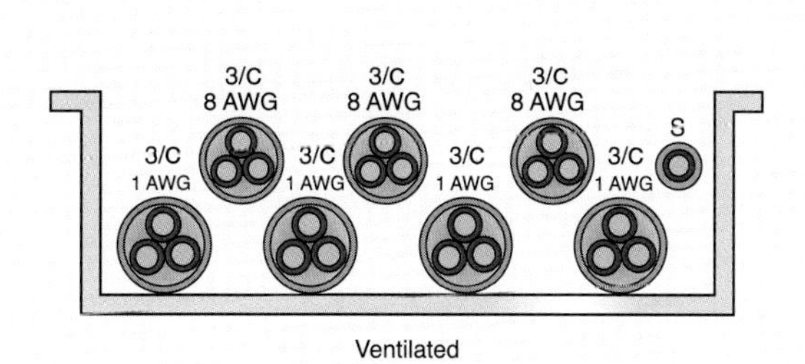

Solution

All conductors are smaller than 4/0 AWG

Fill based on cross-sectional area occupied by all cables

Step 1

Area of four 3/C 1 AWG cables

3/C 1 AWG OD = 1.5 in.

$r = \frac{OD}{2}$

$= \frac{1.5}{2}$

= 0.75 in.

A = r × r × 3.1416

= 0.75 × 0.75 × 3.1416

= 1.7672 in.^{2}

Total area for four cables = 1.7672 × 4

= 7.0688 in.^{2}

Answer: 7.0688 in.^{2}

Step 2

Area of three 3/C 8 AWG cables

3/C 8 AWG OD = 0.875

$r = \frac{OD}{2}$

$= \frac{0.875}{2}$

= 0.4375 in.

A = r × r × 3.1416

= 0.4375 × 0.4375 × 3.1416

= 0.6013 in.^{2}

Total area for three cables = 0.6013 × 3

= 1.8039 in.^{2}

Answer: 1.8039 in.^{2}

Step 3

Area of one 7/C signal cable

7/C OD = 1 in.

$r = \frac{OD}{2}$

$= \frac{1}{2}$

= 0.5 in.

A = r × r × 3.1416

= 0.5 × 0.5 × 3.1416

= 0.7854 in.^{2}

Total area for one cable = 0.7854 in.^{2}

Answer: 0.7854 in.^{2}

Calculation – Solution

Total area from all 3 types of cable

Area total = 7.0688 + 1.8039 + 0.7854

= 9.6581 in.^{2}

Table 392.22(A)

Column 1

9 in. tray max. fill = 10.5 in.^{2}

Answer: 9 in. cable tray

10.1.3 Mixing Multiconductor Single-Layer and Multilayer Vented Tray

392.22(A)(1)(c) applies to the following:

- Multiconductor cables
- 2,000 volts or less
- Cables 4/0 AWG and larger, installed with cables smaller than 4/0 AWG
- Using ladder, ventilated trough, or wire mesh cable tray

All are installed in a ladder, ventilated trough, or wire mesh cable tray and require the use of **Table 392.22(A)** Column 2, where the OD of cables 4/0 AWG and larger are added and then multiplied by 1.2.

10.1.4 Multiconductor, Signal and Control-Only Vented Tray

392.22(A)(2) applies to the following:

- Multiconductor cables
- 2,000 volts or less
- Signal and control cable only
- Installation in a ladder or ventilated trough cable tray

392.22(A)(2) requires limited use of 50% of inside cross-sectional area. Cable tray with over 6 inches of usable depth will have calculations based upon 6 inches depth only for calculating cross-sectional area.

Multiconductor Type MC cables installed in a wire mesh cable tray

For additional information, visit qr.njatcdb.org Item #1042

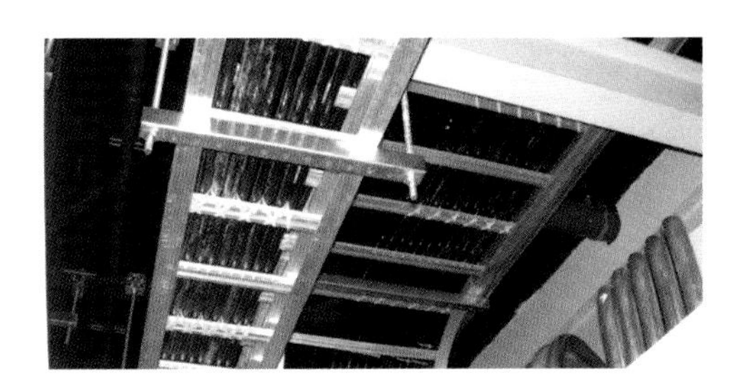

Single conductor cables installed in a ladder cable tray

Problem 10-3

A 12 in. wide cable tray contains one 3/C, 4/0 AWG cable and one 3/C, 250 kcmil cable. What is the remaining cross-sectional area which may be used by multiconductor cables less than 4/0 AWG? (OD 4/0 AWG = 2.32 in.; OD 250 kcmil = 2.96 in.)

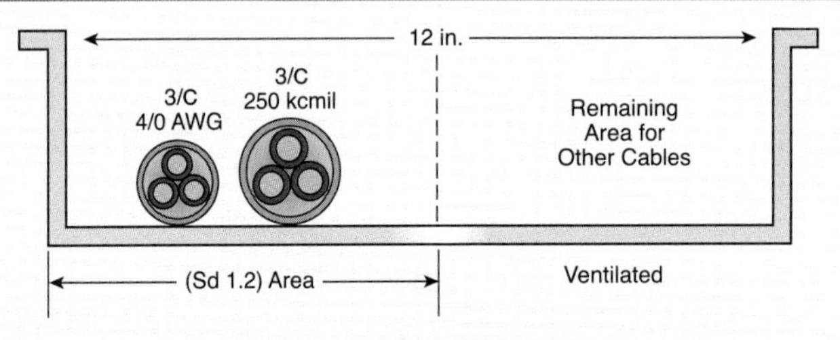

Solution

Step 1

Determine the capacity of 12 in. cable tray

Table 392.22(A)

Column 2

12 in. cable tray = 14 in.^{2}

Answer: 14 in.^{2}

Step 2

Calculate the area used by the larger multiconductor cables

A = 4/0 and larger (Sd x 1.2) or

= (OD1 + OD2) x 1.2

= (OD 4/0 AWG + OD 250 kcmil) x 1.2

= (2.32 + 2.96) x 1.2

= 6.336 in.^{2}

Answer: 6.336 in.^{2}

Solution - Calculation

Calculate the remaining area within the tray to be used for cables smaller than 4/0 AWG

Table 392.22(A)

Column 2

12 in. tray = 14 – (1.2 Sd)

= 14 – (Step 2 answer)

= 14 in.^{2} – 6.336 in.^{2}

= 7.664 in.^{2}

Answer: 7.664 in.^{2} is the remaining area for additional cables smaller than 4/0 AWG

Problem 10-4

A ladder-type cable tray has an internal width of 12 in., a depth of 8 in., and contains five 9/C signal cables with an OD of 1.27 in. each. How many 7/C control cables with an individual OD of 0.976 in. can be added without exceeding the cable tray fill limitation for a cable tray dedicated to control and signal cables?

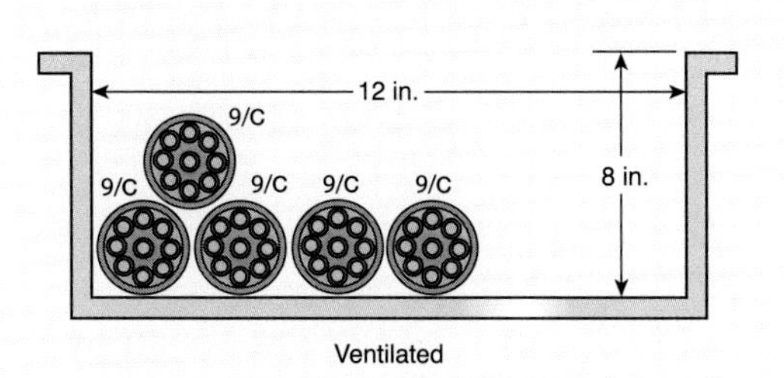

Solution

Step 1

Calculate usable cross-sectional area of the cable tray

392.22(A)(2)

Only 6 in. of the 8 in. depth may be used

Cable tray area:

A = *depth* × *width*
= 6 × 12
= 72 in.^{2}

Permitted use is limited to 50% of cross-sectional area

Usable area = 72 in.^{2} × 0.5
= 36 in.^{2}

Answer: 36 in.^{2}

Step 2

Calculate the occupied area of the cable tray

9/C cable OD = 1.27 in.

$$r = \frac{OD}{2} = \frac{1.27}{2}$$

= 0.635 in.

A = r × r × 3.1416
= 0.635 × 0.635 × 3.1416
= 1.2668 in.^{2}

Total occupied area = 1.2668 × 5 cables
= 6.334 in.^{2}

Answer: 6.334 in.^{2}

Step 3

Calculate the available area of the cable tray

Available area = *usable area* – *occupied area*
= 36 – 6.334
= 29.66 in.^{2}

Answer: 29.66 in.^{2}

Step 4

Calculate the area needed for each 7/C cable

7/C cable OD = 0.976 in.

$$r = \frac{OD}{2} = \frac{0.976}{2}$$

= 0.488 in.

A = r × r × 3.1416
= 0.488 × 0.488 × 3.1416
= 0.7482 in.^{2}

Solution - Calculation

Calculate the max. number of cables permitted in the available area of the cable tray

$$\textit{Number of conductors} = \frac{\textit{available area}}{\textit{7/C cable area}} = \frac{29.66}{0.7482}$$

= 39.6 or 39 more 7/C cables

Answer: 39 additional 7/C cables can be added before 50% fill is reached

10.1.5 Multiconductor, Single-Layer, Solid Bottom Tray

Preceding examples illustrated the use of ladder, ventilated trough and some wire mesh cable tray fills. The next few illustrations use solid bottom cable trays and calculate the maximum cable fill. Although solid bottom cable tray installations may provide additional physical damage protection to the contained cables, a solid bottom tray also presents challenges of elevated cable temperatures due to the lack of natural cooling. Because a solid bottom tray provides less ventilation, the *Code* does not permit solid bottom trays to be filled to the same capacity as ventilated trays. Due to the reduced ventilation, the ampacity of conductors within solid bottom cable tray is also reduced.

392.22(A)(3)(a) applies to solid bottom cable tray using only multiconductor cables rated 2,000 volts or less and in sizes of 4/0 AWG or larger. **392.22(A)(3)(a)** requires the cables to be installed in a single layer. Also, the sum of the diameters of all multiconductor cables must not exceed 90% of the width of the cable tray.

For additional information, visit qr.njatcdb.org Item #1043

Limitation examples:

Cable Tray Width	Limitation	Usable Width
6 in.	90%	5.4 in.
12 in.	90%	10.8 in.
18 in.	90%	16.2 in.
Other cable tray widths are reduced accordingly.		

Problem 10-5

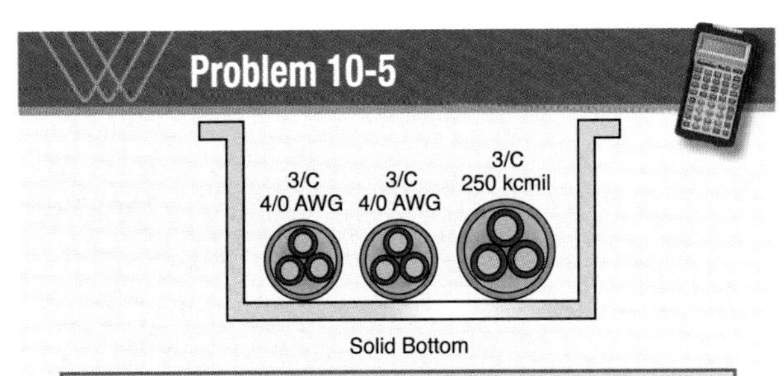

Determine the minimum standard size solid bottom cable tray to accommodate two 3/C, 4/0 AWG power cables and one 3/C 250 kcmil power cable? (OD 4/0 AWG = 1.875 in.; OD 250 kcmil = 2.25 in.)

Solution

Solid Bottom Tray

All conductors are 4/0 AWG and larger

Sd = (4/0 AWG OD × 2) + 250 kcmil OD
= (1.875 × 2) + 2.25
= 3.75 + 2.25
= 6 in.

392.22(A)(3)(a)

Cable tray width (min.) $= \frac{Sd}{90\%}$

$= \frac{6}{0.90}$

= 6.667 in.

Table 392.22(A), Column 1

Next larger standard size = 8 in.

Answer: 8 in. cable tray

10.1.6 Multiconductor, More Than One Layer, Solid Bottom

392.22(A)(3)(b) applies to solid bottom tray using only multiconductor cables, rated 2,000 volts or less, and smaller than 4/0 AWG. This section permits the multiconductor cables to be stacked and requires the sum of the cross-sectional area not to exceed cable tray fill of Column 3 in **Table 392.22(A)**.

10.1.7 Mixing Multiconductor Single-Layer and Multilayer Solid Bottom Tray

392.22(A)(3)(c) applies to solid bottom cable tray using multiconductor cables of mixed sizes all rated 2,000 volts or less. This section requires all 4/0 AWG or larger cables to be installed in a single layer with no other cables on top of cables 4/0 AWG or larger. Column 4 of **Table 392.22(A)** with "Sd" factor for maximum fill applies.

Note: The Sd factor for solid bottom cable tray is not increased. See **Table 392.22(A)** Column 4.

Problem 10-6

What is the minimum standard size solid bottom cable tray needed for two 3/C, 1/0 AWG cables; three 3/C, 2 AWG cables; and six 9/C 3 AWG control cables? (OD 1/0 AWG = 1.52 in.; OD 2 AWG = 1.14 in.; OD 9/C = 1.625 in.)

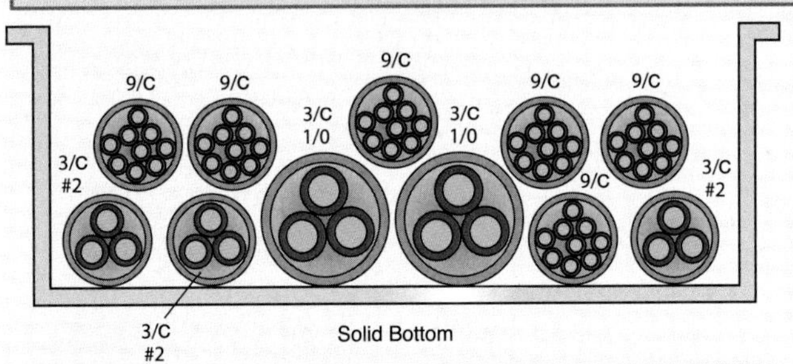

Solution

All conductors are smaller than 4/0 AWG

Fill is based upon cross-sectional area occupied

Determine the cross-sectional area for each size cable

Step 1

3/C 1/0 AWG OD = 1.52 in.

r $= \frac{OD}{2}$

$= \frac{1.52}{2}$

= 0.76 in.

A = r × r × 3.1416
= 0.76 × 0.76 × 3.1416
= 1.8146 in.^{2}

Total 1/0 AWG area = 1.8146 in.^{2} × 2
= 3.6292 in.^{2}

Answer: 3.6292 in.^{2}

Step 2

3/C 2 AWG

2 AWG OD = 1.14 in.

r $= \frac{OD}{2}$

$= \frac{1.14}{2}$

= 0.57 in.

A = r × r × 3.1416
= 0.57 × 0.57 × 3.1416
= 1.0207 in.^{2}

Total 3/C 2 AWG area = 1.0207 in.^{2} × 3
= 3.0621 in.^{2}

Answer: = 3.0621 in.^{2}

Step 3

9/C control cable OD = 1.625 in.

r $= \frac{OD}{2}$

$= \frac{1.625}{2}$

= 0.8125 in.

A = r × r × 3.1416
= 0.8125 × 0.8125 × 3.1416
= 2.0739 in.^{2}

Total 9/C area = 2.0739 in.^{2} × 6
= 12.4434 in.^{2}

Answer: 12.4434 in.^{2}

Solution - Calculation

Determine the total cross-sectional area of all cables

Total Area = 3.6292 + 3.0621 + 12.4434
= 19.1347 in.^{2}

Table 392.22(A), Column 3

Select next larger standard volume cable tray = 22 in.^{2}

Maximum fill of 22 in.^{2} = 24 in. tray

Answer: 24 in. solid bottom cable tray required

Problem 10-7

How many 3/C, 4 AWG cables can be installed in the same 12 in. solid bottom cable tray with two 4/C 500 kcmil cables?
(OD 4/C 500 kcmil = 3.12 in.; OD 3/C 4 AWG = 1.07 in.)

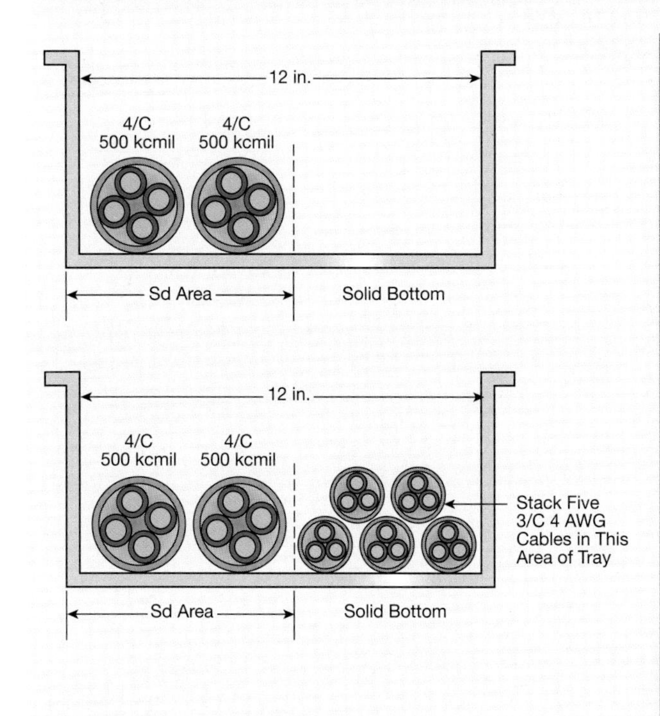

Solution

Step 1

Determine the max. capacity of a 12 in. solid bottom cable tray
Table 392.22(A), Column 4
12 in. cable tray capacity = 11 in.^{2}
Answer: 11 in.^{2}

Step 2

Determine the occupied left-side width (using a single layer fill)
Sd = OD 4/C 500 kcmil x 2
= 3.12 × 2
= 6.24 in.
(for in.^{2} units, 6.24 in. is multiplied by 1.0 in.)*
Used area (left side) = 6.24 in.^{2}
Answer: 6.24 in.^{2}

Step 3

Determine available right-side area using stacked fill
Table 392.22(A), Column 4
12 in. tray = 11.0 - Sd
= 11 – 6.24
= 4.76 in.^{2}
Answer: 4.76 in.^{2}

Step 4

Calculate the area needed for each 3/C cable

$$r = \frac{OD}{2} = \frac{1.07}{2} = 0.535 \text{ in.}$$

$$A = r \times r \times 3.1416 = 0.535 \times 0.535 \times 3.1416 = 0.8992 \text{ in.}^2$$

Answer: 0.8992 in.^{2}

Solution - Calculation

Determine the quantity of 3/C cables using stacked fill

$$\textit{Number of conductors} = \frac{\textit{available area}}{\textit{3/C cable area}} = \frac{4.76}{0.8992}$$

= 5.29 or five 3/C cables

Answer: Five 3/C 4 AWG cables

Comment
*Regarding units of measure, fill calculations from Column 2 and 4 of Table 392.22(A) are based upon square inch (in.^{2}) fill. The sum of diameters (Sd) actually is a linear measure of inches (in.) However, multiplying Sd by 1.2 in. from Column 2 or multiplying Sd by 1.0 in. from Column 4 allows the Sd units of measure to match the table units of in.^{2}.

10.1.8 Multiconductor Signal and Control-Only Solid Bottom Tray

392.22(A)(4) applies to the following:

- Multiconductor cables
- 2,000 volts or less
- Signal and control cable only
- Installation in a solid bottom cable tray

392.22(A)(4) requires limited use of 40% of inside cross-sectional area. Where the usable cable tray depth is 6 inches or less, the actual depth may be used in the calculation. Cable tray with over 6 inches of usable depth will have calculations based on a maximum depth of 6 inches only for calculating the cross-sectional area of a given tray.

Problem 10-8

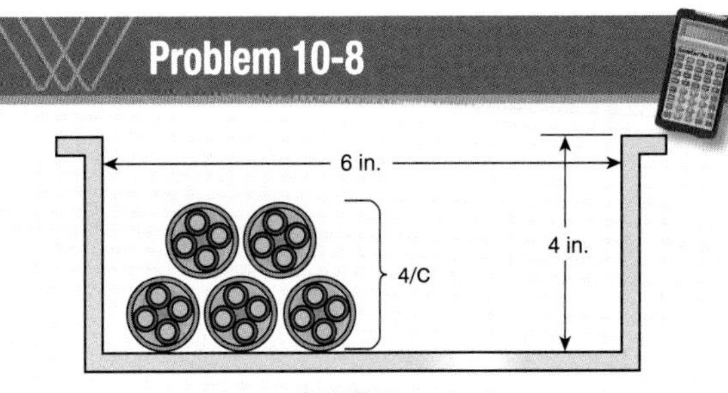

Solid Bottom

A cable tray dedicated to signal and control cables contains five 4/C control cables. How many 7/C control cables may be added to the solid bottom tray without exceeding the *Code* limitations, with internal measurements of 6 in. wide by 4 in. deep? (OD 4/C = 0.537 in.; OD 7/C = 0.975 in.)

Solution

Step 1

Calculate the usable area of the cable tray

Cable tray area $= width \times depth$
$= 6 \times 4$
$= 24 \text{ in.}^2$

392.22(A)(4)

Usable area $= total \times 40\%$
$= 24 \times 0.40$
$= 9.6 \text{ in.}^2$

Answer: 9.6 in.^2

Step 2

Calculate the occupied 4/C cable area fill

$$r = \frac{OD}{2} = \frac{0.537}{2} = 0.2685 \text{ in.}$$

$$A = r \times r \times 3.1416 = 0.2685 \times 0.2685 \times 3.1416 = 0.2265 \text{ in.}^2$$

Total occupied area $= 0.2265 \times 5$
$= 1.1325 \text{ in.}^2$

Answer: 1.1325 in.^2

Step 3

Calculate the available area

Available area $=$ *usable* area $-$ *occupied* area
$= 9.6 - 1.1325$
$= 8.4675 \text{ in.}^2$

Answer: 8.4675 in.^2

Step 4

Calculate area needed for each 7/C cable

$$r = \frac{OD}{2} = \frac{0.975}{2} = 0.4875 \text{ in.}$$

$$A = r \times r \times 3.1416 = 0.4875 \times 0.4875 \times 3.1416 = 0.7466 \text{ in.}^2$$

Answer: 0.7466 in.^2

Solution - Calculation

Calculate the max. number of 7/C cables permitted

$$\textit{Number of conductors} = \frac{\text{available area}}{\text{7/C cable area}} = \frac{5.29}{0.7466} = 11.34 \text{ or } 11 \text{ 7/C cables}$$

Answer: Eleven 7/C cables

10.1.9 Multiconductor Ventilated Channel Tray

Ventilated channel cable tray has a smaller internal width than a regular cable tray and is often used for cable tray dropouts. **392.22(A)(5)** and **Table 392.22(A)(5)** apply to the following:

- Multiconductor cable(s)
- 2,000 volts or less
- Installation in ventilated channel cables tray
- When one or more cable is installed

392.22(A)(5) and **Table 392.22(A)(5)** of the *Code* have limitations listed in Columns 1 and 2:

Table 392.22(A)(5) Column 2, More Than One Cable	
Cable Tray Width	Maximum Fill
3 in.	1.3 in.^2
4 in.	2.5 in.^2
6 in.	3.8 in.^2
Table 392.22(A)(5) Column 1, One Cable	
Cable Tray Width	Maximum Fill
3 in.	2.3 in.^2
4 in.	4.5 in.^2
6 in.	7.0 in.^2

Problem 10-9

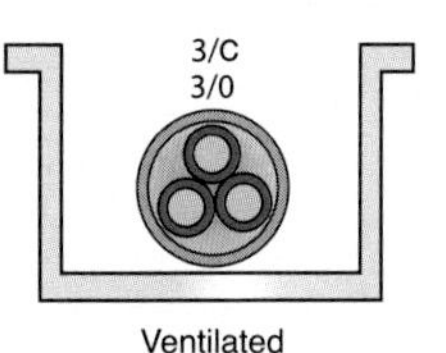

Ventilated

What width of ventilated channel cable tray would be needed for one 3/C 3/0 AWG cable? (OD 3/C 3/0 AWG = 1.7 in.)

Solution

Calculate area needed for 3/C 3/0 AWG cable

$$r = \frac{OD}{2} = \frac{1.7}{2} = 0.85 \text{ in.}$$

$$A = r \times r \times 3.1416 = 0.85 \times 0.85 \times 3.1416 = 2.27 \text{ in.}^2$$

Table 392.22(A)(5),Column 1

2.3 in.^2 for one multiconductor cable = 3 in. cable tray

Answer: 3 in. ventilated channel cable tray

Problem 10-10

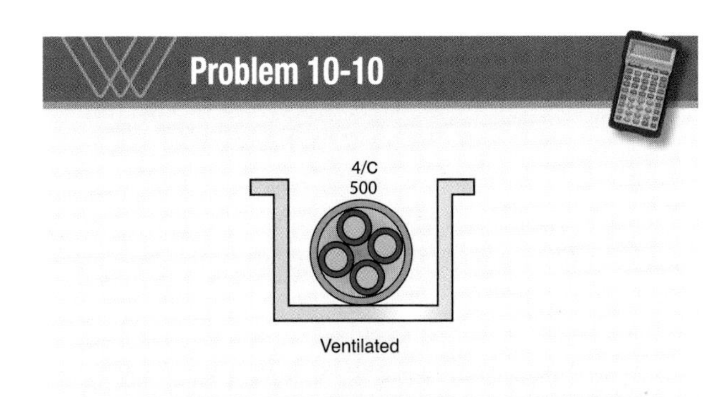

Is it permissible to install one 4/C 500 kcmil power cable with an OD of 2.5 in. in a 4 in. wide ventilated channel cable tray?

Solution

$r = \frac{OD}{2}$

$= \frac{2.5}{2}$

$= 1.25$ in.

$A = r \times r \times 3.1416$

$= 1.25 \times 1.25 \times 3.1416$

$= 4.9088$ in.^{2}

Table 392.22(A)(5),Column 1

Maximum fill one cable

4 in. cable tray, with a fill of 4.5 in.^{2} is too small

Use 6 in. cable tray with a fill of 7 in.^{2}

Answer: No. A 6 in. ventilated channel cable tray is the minimum size required.

10.2 Cable Tray Fill Calculations for Single-Conductor Installations

The next group of cable tray illustrations will be directed at the installation of single-conductors cables used in cable trays. Unless the dimensions of the insulated conductors are given, the dimensions of the diameter and cross-sectional area as given in **Chapter 9, Table 5** for copper and for aluminum and **Table 5A** for compact copper and aluminum are used. **392.10(B)(1)(a)** limits the smallest single-conductor cable installed in a cable tray to 1/0 AWG. **392.10(B)(1)(c)** permits an equipment grounding conductor as small as 4 AWG.

392.22(B) applies to the following:

- Cable 2,000 volts or less
- Where only single-conductor cable is installed
- Or where single-conductor cable assemblies (such as triplex) are installed
- When in ladder and ventilated trough and wire mesh cable tray
- Where cable is evenly distributed across the cable tray

The installations of single-conductor cables rated 2,000 volts or less are divided into five groups:

1. Where all cables are 1,000 kcmil or larger
2. Where all cables are from 250 kcmil through 900 kcmil
3. Where 1,000 kcmil cables or larger are mixed within the same tray as cables smaller than 1,000 kcmil
4. Where any 1/0 AWG through 4/0 AWG cables only in tray
5. Ventilated channel cable tray

10.2.1 Single-Conductor, 1,000 kcmil and Over Vented Tray

392.22(B)(1)(a) applies to the following:

- Single-conductor cables
- 2,000 volts or less
- Installation in a ladder, ventilated trough and wire mesh cable tray
- Conductor sizes 1,000 kcmil and over
- Single layer of cables evenly distributed

392.22(B)(1)(a) requires the sum of all cable diameters not to exceed the inside width of the cable tray.

Problem 10-11

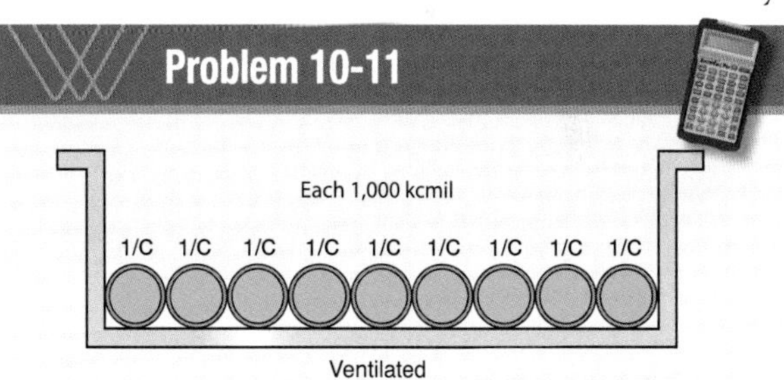

What minimum inside width of ladder-type cable tray is required for the installation of nine 480 volt, 1,000 kcmil THWN single copper conductors?

Solution

All conductors are 1,000 kcmil or larger

392.22(B)(1)(a) applies

Diameters of cables not to exceed inside width of cable tray

Chapter 9, Table 5

1,000 kcmil THWN copper OD = 1.31 in.

Min. width = *No. conductors* × OD of 1/C

= 9 × 1.31 in.

= 11.79 in.

Table 392.22(A)

Select next standard width greater than 11.79 in.

Standard cable tray width = 12 in.

Answer: 12 in. cable tray

10.2.2 Single-Conductor, From 250 kcmil Through 900 kcmil, More Than One Layer Ventilated Tray

392.22(B)(1)(b) applies to the following:

- Single-conductor cables
- 2,000 volts or less
- Installation in a ladder, ventilated trough and wire mesh cable tray
- Conductors sizes from 250 kcmil through 900 kcmil
- Conductors permitted to be stacked

392.22(B)(1)(b) requires cross-sectional area fill in square inches not to exceed cable tray fill of Column 1 of **Table 392.22(B)(1)**.

The *Code* permits single conductors from 250 kcmil through 900 kcmil to be stacked in a cable tray. This is because the larger cables are less flexible and do not get damaged or crushed as easily when other cables are placed on top of them. For smaller sizes 1/0 AWG through 4/0 AWG, cables can be easily damaged or crushed when significant weight is placed on them.

Problem 10-12

What is the minimum width of a ventilated trough cable tray needed for three 250 kcmil, THWN, copper single conductors; six 500 kcmil, THWN, copper single conductors; and three 750 kcmil, XHHW, compact aluminum single conductors? All conductors have 600 volt insulation.

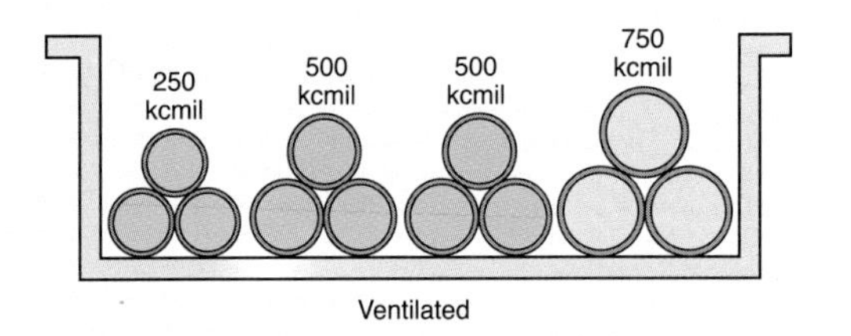

Solution
392.22(B)(1)(b)
Chapter 9, Table 5 and Table 5A
Approximate area (using in.^{2})

Table 5: 500 kcmil THWN copper =		0.7073 in.^{2}
Table 5: 250 kcmil THWN copper =		0.3970 in.^{2}
Table 5A: 750 kcmil XHHW compact alum. =		0.9331 in.^{2}
500 kcmil A =	6 × 0.7073 =	4.2438 in.^{2}
250 kcmil A =	3 × 0.3970 =	1.1910 in.^{2}
750 kcmil A =	3 × 0.9331 =	2.7993 in.^{2}
Total		8.2341 in.^{2}

Table 392.22(B)(1), Column 1
9 in. wide cable tray fill permitted to be 9.5 in.^{2}
Answer: 9 in. cable tray

Problem 10-13

How many single-conductor 500 kcmil XHHW, copper conductors can be installed with three 1,250 kcmil, XHHW copper conductors and three 1,000 kcmil, XHHW copper conductors in a 12 in. wide ventilated cable tray?

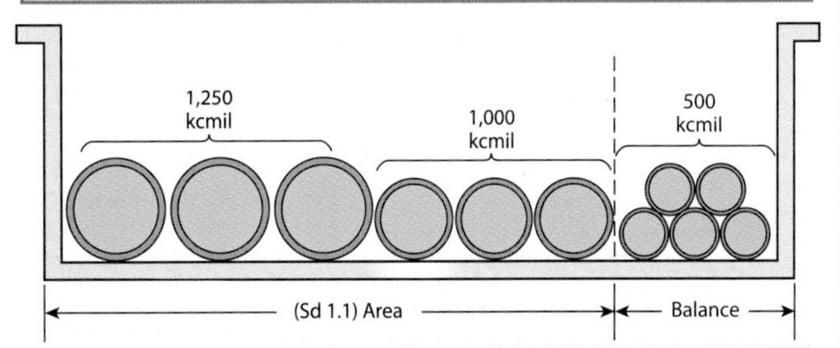

Solution
392.22(B)(1)(c) and Table 392.22(B)(1) Column 2 applies

Step 1
Occupied space (without 500 kcmil)
Find Sd for all 1,000 kcmil and over
Chapter 9, Table 5 OD of cables
3 - 1/C 1,250 kcmil XHHW = 1.479 in. × 3
= 4.437 in.
3 - 1/C 1,000 kcmil XHHW = 1.312 in. × 3
= 3.936 in.
Sd = 4.437 + 3.936
= 8.373 in.
Answer: 8.373 in.

Step 2
Remaining space (area for 500 kcmil only)
Table 392.22(B)(1), Column 2
12 in. tray = 13 – (1.1 Sd)
= 13 in.^{2} – (1.1 in. × 8.373 in.)
= 3.7897 in.^{2}
Answer: 3.7897 in.^{2}

Step 3
Quantity of 500 kcmil conductors

$$\text{No. of conductors} = \frac{\text{remaining space}}{\text{area of 500 kcmil XHHW}}$$

Chapter 9, Table 5
500 kcmil XHHW = 0.6984 in.^{2}

$$\text{No. of conductors} = \frac{3.7897 \text{ in.}^2}{0.6984 \text{ in.}^2}$$

= 5.426 or 5 conductors
Answer: Five 500 kcmil XHHW conductors

10.2.3 Single-Conductor, One Layer and Multilayer, Same Ventilated Tray

392.22(B)(1)(c) applies to the following:

- Single-conductor cables
- 2,000 volts or less
- Installation in a ladder or ventilated trough cable tray
- Some 1,000 kcmil or over
- Some from 250 kcmil through 900 kcmil

392.22(B)(1)(c) requires the use of the Sd factor of Table 392.22(B)(1) Column 2 for maximum fill.

10.2.4 Single-Conductor, 1/0 Through 4/0 AWG, One Layer Ventilated Tray

392.22(B)(1)(d) applies to the following:

- Single-conductor cables
- 2,000 volts or less
- Sizes 1/0 AWG through 4/0 AWG
- Installation in a ladder, ventilated trough and wire mesh cable tray,

392.22(B)(1)(d) requires cable to be installed in a single layer, with the total diameter of all conductors not exceeding the cable tray width.

10.2.5 Single-Conductor in Vented Channel Tray

392.22(B)(2) applies to the following:

- Single-conductor cables
- 2,000 volts or less
- Installed in 2 in., 3 in., 4 in., or 6 in. ventilated channel-type cable trays

392.22(B)(2) requires the sum of the diameter of all conductors not to exceed the inside width of the cable tray.

Problem 10-14

What is the minimum standard width of ventilated cable tray needed for four 1/0 AWG, three 2/0 AWG, and three 4/0 AWG, if all are THWN-2 copper single conductors?

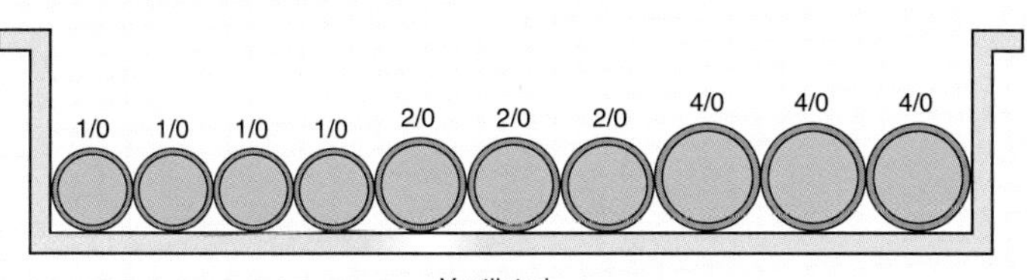

Solution
392.22(B)(1)(d)
Cable tray width not to exceed Sd of single-conductor cables
Chapter 9, Table 5
OD of cables

1/0 AWG THWN-2 =	0.486 x 4 =	1.944 in.
2/0 AWG THWN-2 =	0.532 x 3 =	1.596 in.
4/0 AWG THWN-2 =	0.642 x 3=	1.926 in.
Total		5.466 in.

Does not exceed the width of the 6 in. tray
Answer: 6 in. cable tray

Problem 10-15

What is the minimum size ventilated channel cable tray needed for four 600 kcmil, THWN, single copper conductors?

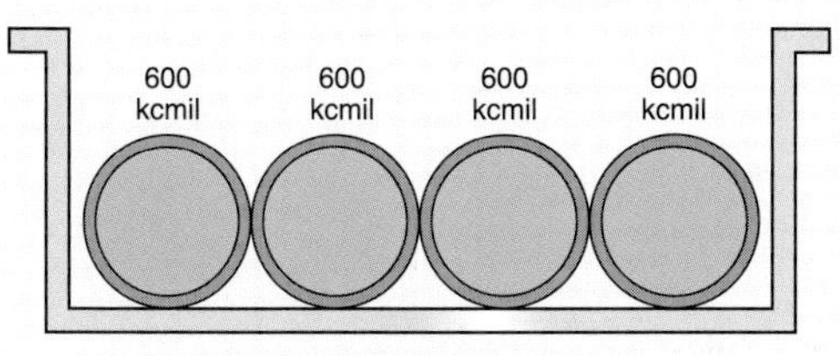

Solution
392.22(B)(2)
Cable tray width not to exceed Sd of single-conductor cables
Chapter 9, Table 5
OD of cables
600 kcmil THWN = 1.051 in.
Total width = conductors × OD
= 4 × 1.051
= 4.204 in.
Exceeds 4 in. width
Table 392.22(B)(1)
Next larger standard size = 6 in. width
Answer: 6 in. ventilated channel cable tray

10.3 Ampacity of Multiconductor Installations

The ampacity of the conductors installed in cable trays is based upon the installation of the conductors being made according to **392.80(A)(1)** for multiconductor cables, and **392.80(A)(2)** for single-conductor cables, with few variations.

The adjustment factors of **310.15(B)(3)(a)** to the allowable ampacity tables of **Article 310** are applicable to multiconductor cables installed in a cable tray. The individual adjustment factors will be based upon the number of current-carrying conductors in the cable. Each cable must be looked at individually.

For multiconductor cables, **Table 310.15(B)(16)** and **Table 310.15(B)(18)** are used. These are the tables listing the allowable ampacity of conductors when there are not more than three conductors in a raceway or cable, based upon a 30°C temperature.

For single-conductor cables, triplex, or smaller cables, **Table 310.15(B)(17)** and **Table 310.15(B)(19)** are used. These are the tables listing the ampacities of single conductors installed in open air. **Table 310.15(B)(17)** is based upon 30°C ambient temperature and **Table 310.15(B)(19)** is based upon a 40°C temperature. However, final ampacities for single-conductor cables may be less than the given values in **Table 310.15(B)(17)** and **Table 310.15(B)(19)** if the cable terminations are in equipment marked to accept only such ampacities in accordance with **110.14(C)(1)** and those found in **Table 310.15(B)(16)**.

In the following problems of this section, all conductors will be considered to be operating in an ambient area where the temperature will not exceed the temperatures listed in **Table 310.15(B)(16)** and **Table 310.15(B)(17)**. Should the installation be in an area with an ambient temperature other than the temperature listed in these allowable ampacity tables, the calculated ampacities may be different.

The allowable ampacities for conductors in different types of cable trays are the same, provided the cable tray is installed without covers. When covers over 6 feet in length are used on any type of cable tray, the ampacity of the conductors will be less than that of uncovered cable tray installations.

10.3.1 Multiconductor, Single-Layer and Multilayer, Same Uncovered Tray

Provided the installation is in accordance with **392.22(A)**, **392.80(A)(1)** is used to determine the ampacity of multiconductor cables rated 2,000 volts or less installed in cable trays. The calculation begins by applying the ampacities of **Table 310.15(B)(16)** appropriately to each multiconductor cable. The application of adjustment factors according to **310.15(B)(3)(a)** is only necessary if the number of current-carrying conductors within each multiconductor cable exceeds three current-carrying conductors. Each multiconductor cable is reviewed individually. The total sum of current-carrying conductors within all of the cables contained in a cable tray is not reviewed.

Problem 10-16

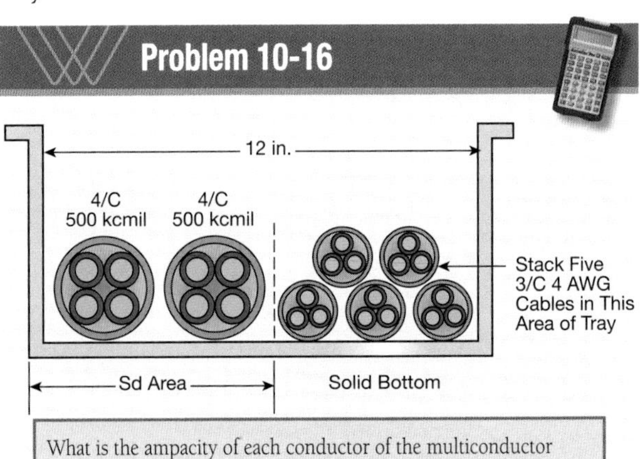

What is the ampacity of each conductor of the multiconductor cables installed in an uncovered cable tray, as shown in the following illustration? All conductors are 75°C insulated, current-carrying copper conductors.

Note: This problem is a logical extension of Problem 10-7 using the matching the cables and cable tray. The fill requirement according to 392.22(A) and associated Table 392.22(A) are satisfied.

Solution – Calculation 1

Ampacity of 4/C 500 kcmil

Table 310.15(B)(16) Ampacity

4/C 500 kcmil at 75°C = 380 amps

392.80(A)(1)(a)

More than 3 current-carrying conductors

310.15(B)(3)(a)

Table 310.15(B)(3)(a)

Adjustment factor for four conductors = 80%

500 kcmil = 380 × 0.80

= 304 amps

Answer: 304 amperes

Solution – Calculation 2

Ampacity of 3/C 4 AWG

Table 310.15(B)(16) Ampacity

3/C 4 AWG at 75°C = 85 amps

No adjustment factor necessary

4 AWG = 85 amps

Answer: 85 amperes

10.3.2 Multiconductor, Single-Layer and Multilayer, Same Solid Covered Tray

The ampacity of multiconductor cables installed in cable trays is based upon the requirements of **392.80(A)(1)**. Where a cable tray is continuously covered for more than 6 feet, **392.80(A)(1)(b)** indicates the ampacity of the conductors is required to be reduced to 95% of the ampacities of **Table 310.15(B)(16)** or **Table 310.15(B)(18)**.

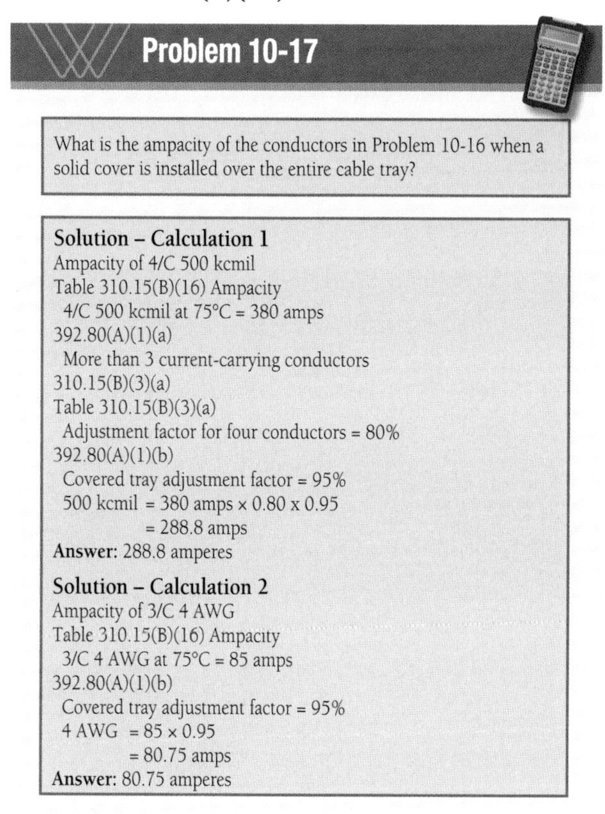

Problem 10-17

What is the ampacity of the conductors in Problem 10-16 when a solid cover is installed over the entire cable tray?

Solution – Calculation 1
Ampacity of 4/C 500 kcmil
Table 310.15(B)(16) Ampacity
4/C 500 kcmil at 75°C = 380 amps
392.80(A)(1)(a)
More than 3 current-carrying conductors
310.15(B)(3)(a)
Table 310.15(B)(3)(a)
Adjustment factor for four conductors = 80%
392.80(A)(1)(b)
Covered tray adjustment factor = 95%
500 kcmil = 380 amps x 0.80 x 0.95
= 288.8 amps
Answer: 288.8 amperes

Solution – Calculation 2
Ampacity of 3/C 4 AWG
Table 310.15(B)(16) Ampacity
3/C 4 AWG at 75°C = 85 amps
392.80(A)(1)(b)
Covered tray adjustment factor = 95%
4 AWG = 85 x 0.95
= 80.75 amps
Answer: 80.75 amperes

10.4 Ampacity of Single-Conductor Installations

The ampacity of the single-conductors cables installed in cable trays is based upon the installation of the conductors being made according to **392.80(A)(2)** with one variation used where cable separation is significant.

The adjustment factors of **310.15(B)(3)(a)** do not apply to single-conductor cables in cable tray. The allowable ampacity tables of **Article 310** are applicable to multiconductor cables installed in cable tray.

As a reminder, whenever ampacity calculations are made using tables other than **Table 310.15(B)(16)**, the final ampacity calculation for terminations may need to be adjusted according to the termination marking requirements and **110.14(C)(1)**.

10.4.1 General

The ampacities of single-conductor cables are regulated by **392.80(A)(2)**. This section is further divided into four subdivisions, three of which are discussed as follows:

a. Provided the conductors are installed in accordance with **392.22(B)**, 600 kcmil and larger, ampacities must not exceed 75% of the value given in either **Table 310.15(B)(17)** or **Table 310.15(B)(19)**. For covered tray installations, the factor of 75% is reduced to 70%.
b. Provided the conductors are installed in accordance with **392.22(B)**, 1/0 AWG through 500 kcmil and larger, ampacities must not exceed 65% of the value given in either **Table 310.15(B)(17)** or **Table 310.15(B)(19)**. For covered tray installations, the factor of 65% is reduced to 60%.
c. Where single conductors are installed with maintained spacing not less than one cable diameter between individual conductors, the ampacity of 1/0 AWG and larger must not exceed the value given in either **Table 310.15(B)(17)** or **Table 310.15(B)(19)**.

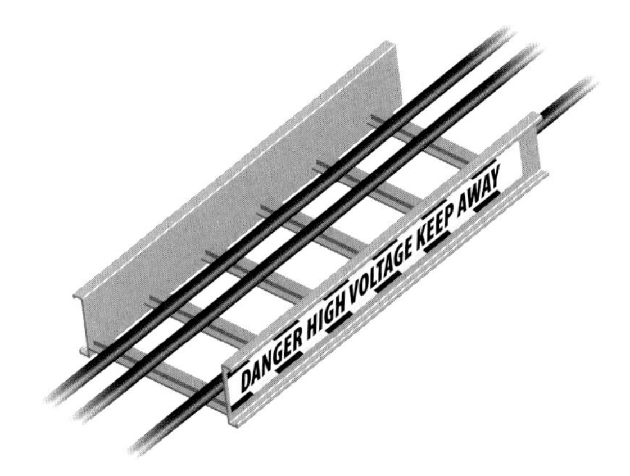

Cable trays containing conductors rated over 600 volts must be marked according to 392.18(H) and 110.21(B).

10.4.2 Single-Conductor, 600 kcmil and Larger, Uncovered Tray and Covered Tray

392.80(A)(2)(a) applies to single-conductor cables rated 2,000 volts or less and the following:

- 600 kcmil and larger
- In ventilated or solid bottom cable trays
- Requires the use of ampacity **Table 310.15(B)(17)** or **Table 310.15(B)(19)**
- Cable tray fill complies with **392.22(B)**

For these types of installations, conductors in an uncovered tray are restricted to 75% of the table ampacity. Conductors in a covered tray are restricted to 70% of table ampacity.

10.4.3 Single-Conductor, 1/0 AWG Through 500 kcmil, Uncovered and Covered Tray

392.80(A)(2)(b) applies to single-conductor cables rated 2,000 volts or less and the following:

- Sizes 1/0 AWG through 500 kcmil
- Within ventilated or solid bottom cable trays
- Requires the use of ampacity **Table 310.15(B)(17)** or **Table 310.15(B)(19)**
- Ampacity for uncovered tray = 65% of table ampacity
- Ampacity for covered tray = 60% of table ampacity
- Cable tray fill complies with **392.22(B)**

Problem 10-18

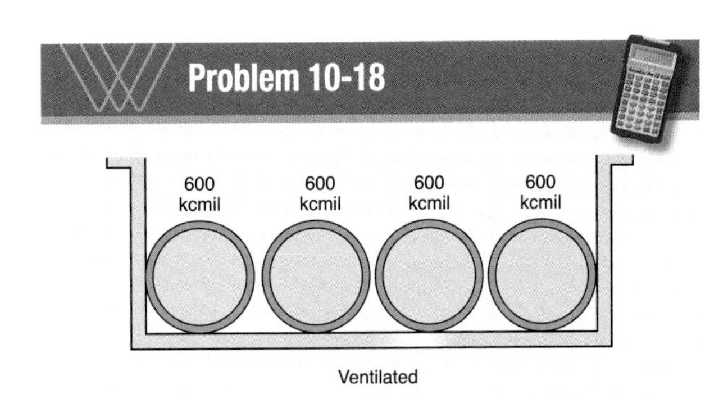

What is the ampacity of each conductor when four 600 kcmil aluminum single conductors with 75°C insulation are installed in the following:

1. An uncovered cable tray?
2. Cable tray with a full-length covering?

Solution – Calculation 1
Uncovered tray
392.80(B)(2)(a)
Table 310.15(B)(17) Ampacity
600 kcmil 75°C insulation = 545 amps
Ampacity = table value × 75%
= 545 × 0.75
= 408.75 amps
Answer: 408.75 amperes

Solution – Calculation 2
Covered tray
392.80(B)(2)(a)
Table 310.15(B)(17) Ampacity
600 kcmil 75°C insulation = 545 amps
Ampacity = table value × 70%
= 545 × 0.70
= 381.5 amps
Answer: 381.5 amperes

Problem 10-19

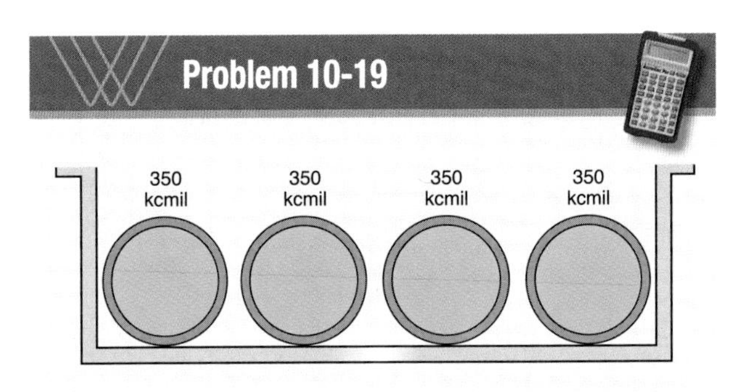

What is the ampacity of each conductor when four 350 kcmil, 75°C copper conductors are installed in both an uncovered and a covered cable tray?

Solution – Calculation 1
Uncovered tray
392.80(B)(2)(b)
Table 310.15(B)(17) Ampacity
350 kcmil 75°C = 505 amps
Ampacity = table value × 65%
= 505 × 0.65
= 328.25 amps
Answer: 328.25 amperes

Solution – Calculation 2
Covered tray
392.80(B)(2)(b)
Table 310.15(B)(17) Ampacity
350 kcmil 75°C = 505 amps
Ampacity = table value × 60%
= 505 × 0.60
= 303 amps
Answer: 303 amperes

10.4.4 Single-Conductor Uncovered, One Diameter Spacing

392.80(A)(2)(c) applies to single-conductor cables rated 2,000 volts or less and the following:

- Sizes 1/0 AWG and larger
- Single layer
- Requires conductor being spaced apart one cable diameter in an uncovered tray
- Ampacity read directly from **Table 310.15(B)(17)** or **Table 310.15(B)(19)**

392.80(A)(2)(c) provides an economical solution for cable tray circuits. For example, a 600 kcmil copper conductor using **Table 310.15(B)(16)** has a 75°C ampacity of 420 amps where as the same 600 kcmil conductor using **Table 310.15(B)(17)** has a 75°C ampacity of 690 amps. The permission to use **Table 310.15(B)(17)** allows a 64% increase in ampacity using the same conductors.

Problem 10-20

What is the ampacity of a single 500 kcmil THHN copper conductor installed in an uncovered ventilated cable tray when each conductor is separated by at least a cable diameter?

1/C 1/C 1/C 1/C

Ventilated

1 Conductor OD Separation

Solution
392.80(A)(2)(c)
Table 310.15(B)(17) Ampacity
500 kcmil THHN copper = 700 amperes
Answer: 700 amperes

Metal cable trays containing only non-powered limited conductors shall be electrically continuous or use a bonding jumper according to the requirements of 392.60(A).

Definitions and Terms

Cable Tray System - A unit or assembly of units or sections and associated fittings forming a structural system used to securely fasten or support cables and raceways.

Channel Cable Tray* - A fabricated structure consisting of a one-piece ventilated-bottom or solid-bottom channel section.

Ladder Cable Tray* - A fabricated structure consisting of two longitudinal side rails connected by individual transverse members (rungs).

Single-Rail Cable Tray* - A fabricated structure consisting of a longitudinal rail with transversely connected members (rungs) that project from one side (side-supported) or both sides (center-supported), which may be single- or multi-tier.

Solid Bottom or Nonventilated Cable Tray* - A fabricated structure consisting of a bottom without ventilation openings within integral or separate longitudinal side rails.

Trough or Ventilated Cable Tray* - A fabricated structure consisting of integral or separate longitudinal rails and a bottom having openings sufficient for the passage of air and utilizing 75% or less of the plan area of the surface to support cables where the maximum open spacings between cable support surfaces of transverse elements do not exceed 100 mm (4 in.) in the direction parallel to the tray side rails.

Wire Mesh Cable Tray* - A manufactured wire mesh tray consisting of steel wires welded at all intersections. Longitudinal wires located on the exterior of the tray are spaced at a maximum of 50 mm (2 in.) and transverse wires are spaced at a maximum of 100 mm (4 in.).

***NEMA VE 1-2009, Metal Cable Tray Systems**

Summary

- Cable tray is an economical support system for multiconductor cable-type wiring methods.
- Cable trays are available as ladder, ventilated trough, wire mesh, and solid bottom systems.
- Ventilated and wire mesh cable trays allow cables to operate at an efficient temperature.
- Cable tray may be installed with or without a cover. Where used, covers are permitted to be solid or ventilated.
- Cable tray support requirements must be in accordance with the manufacturer's instructions.
- Single-conductor installations are permitted in industrial establishments only.
- Cables operating at over 600 volts must not be placed within the same tray as cables operating at less than 600 volts unless they are separated by a barrier or the cables operating at over 600 volts are type MC cable.
- Some multiconductor cables are permitted to fill a cable tray to 40% capacity without application of adjustment factors.
- Ampacity calculations are permitted using greater ampacity than given in **Table 310.15(B)(16)**, but most important, these higher ampacities may not be used where terminations are required to be made to 75°C equipment terminals in accordance with **110.14(C)(1).**
- Conductor fill requirements of **392.22(A)** are very specific and are based upon many factors including specific type of cable tray, covered or not covered, single or multiconductor, size of conductors, sum of diameters or percent fill, conductor spacing, and, of course, voltage rating of cables.
- Conductor ampacities of **392.80(A)** are also very specific and are based upon many factors including specific type of cable tray, covered or not covered, single or multiconductor, size of conductors, and selection of ampacity using **Table 310.15(B)(16)** or other tables.
- Cable trays containing conductors rated over 600 volts shall have danger marking(s) or labels complying with **110.21(B)**
- Ampacity adjustment factors may or may not apply and conductor spacing remain an issue.
- Final ampacities for cables installed in cable trays may require stepping down to the ampacities of **Table 310.15(B)(16)**, and using the 75°C column to coordinate with temperature limitations (see **110.14(C)**) of the overcurrent protective devices, switchboards, panelboards, and other user equipment.
- Communications cables, fire alarm cables, and Class 1, 2, and 3 are also varyingly restricted from being placed within cable trays used for 600 volts or less power cables unless barriers are provided.

Electric Welders

Branch circuits and feeders supplying welders require ampacity calculations based upon those found in **Article 630 Electric Welders**. The two basic types of electric welders covered within **Article 630** are arc welders and resistance welders.

Welder loads are often substantially less than the supply current marked on the product nameplate, or as the *Code* calls it, the rating plate. Most welders operate on a duty cycle and a multiplication factor of less than 100%. Where more than one welder is placed on a branch circuit or feeder, additional demand factors may be applied which further reduces the overall calculated load.

Overcurrent protection of arc welder circuits is required to be set at not more than 200% of its rated supply current at maximum rated output (I_{1max}). The conductors which supply welders are permitted to be protected by an overcurrent protective device also set at 200%. The branch circuit overcurrent device may serve as the welder overcurrent protective device as well if it is sized at not over 200%. This 200% rating also serves as the rating of the identified disconnecting means.

Objectives

- Name the two types of electric welders and their identifying characteristics.
- Calculate the ampacity of supply conductors for individual welders of various types.
- Identify the additional calculations necessary when determining the ampacity of supply conductors for various types of electric welders when operated in groups.
- Calculate the duty cycle of various types of welders.

Chapter 11

Table of Contents

11.1 Article 630 Electric Welders

Article 630 Welders was first introduced into the *NEC* just after 1942. At that time, productivity for the war effort, especially welding, was very important. Welding machines are covered in **Article 630** of **Chapter 6, Special Equipment**. It is important to remember that according to **Section 90.3**, Chapters 1, 2, 3, and 4 apply generally and Chapters 5, 6, and 7 apply to special occupancies, special equipment, or other special conditions. These latter chapters supplement or modify the general rules. Chapters 1 through 4 apply except as amended by Chapters 5, 6, and 7 for the particular conditions. Therefore, **Article 630** may amend the requirements of Chapters 1 through 4.

Fundamental electric welding is accomplished by the use of a low voltage and an exceptionally high current.

This chapter does not provide calculations to size secondary conductors of welders.

11.1.1 Type of Welders

The *Code* contains specific requirements for the different types of welders. **Article 630** covers two basic types of welders:

1. Arc Welders

   Nonmotor Generator Type - This type of welder looks like a box and has no moving parts. It contains a transformer to step down the voltage and a bank of rectifiers to change the voltage from AC to DC, which is used for the welding process. Modern welding equipment no longer uses DC secondary circuits exclusively. This welder is used to join two pieces of metal with the use of a welding rod by properly flowing the liquid metal to bond the two pieces of metal. The AC transformer and DC rectifier welder is a nonmotor generator type welder.

   Motor Generator Type - This type of welder is often referred to as a rotary-welder because it has moving parts which rotate. As the name indicates, it is a motor generator. It has an AC motor which drives a DC generator, which is used for the actual welding. It is used in the same manner as the AC transformer and DC rectifier type.

2. Resistance Welders

   Resistance Welders - Resistance welders furnish a high current to two electrodes. When pieces of metal are placed between the electrodes, current will flow and the resistance of the metal will cause the metal to heat and fuse together. Resistance welding is often referred to as spot welding or seam welding when used in the manufacturing process.

Electric welders include portable and fixed AC arc welding machines, AC arc welding machines, DC arc welding machines, TIG welding machines, MIG/MAG welding machines, plasma arc cutting machines, plasma arc welding equipment, resistance welding machines and spot welding machines.

Type of Welder	Identification	Identification
AC Transformer and DC Rectifier	Box	Nonmotor Generator
Motor Generator Arc Welder	Rotary	Motor Generator
Resistance Spot-Welder	Resistance	

11.1.2 Duty Cycle

The term *duty cycle* used in conjunction with welders refers to the length of time there will be a demand for current flow in the circuit and the length of the time the circuit will be at rest. The duty cycle is expressed as a percentage. A duty cycle of 40% indicates that the welding circuit will be operating 40% of the time and will be at rest 60% of the time. Another way to look at it is that the current flow is heating up the conductors 40% of the time and the lack of current flow allows the conductors to cool 60% of the time. The welding current flowing can vary from full-load to partial load during the duty cycle. Since the circuit conductors are not required to carry their full ampacity at all times, the ampacity of the circuit conductors is permitted to be reduced according to the duty cycle of the welder.

11.1.3 Ampacity Multiplier

The multiplier is a number used to determine the ampacity of the circuit conductors for welding machines. The smaller the duty cycle, the less time the current will flow and the less ampacity the conductor is required to have. The multiplier is given as a decimal, which is a percentage. The multiplier given means that the ampacity of the primary conductors can have an ampacity rating less than the ampacity rating on the nameplate of the welder.

Each type of welder has its own characteristics resulting in separate multipliers for arc welders and resistance welders, based upon the duty cycle. When a welder is used 100% of the time, it has a duty cycle of 1 and the circuit conductors are calculated at 100% of the rated primary current.

11.1.4 Number of Welders

The duty cycle multiplier tables in **Article 630** are for the installation of a single welder. When more than one welder or a group of welders are installed on the same circuit, different calculations are used. These calculations are explained in **630.11(B)** and **630.31(B)** and are used to determine the ampacity of the feeder circuit conductors supplying a group of welding machines. These calculations are somewhat similar to the application of feeder demand factors of **Article 220**.

As previously noted, welders do not draw current 100% of the time, even when a group of welders are installed on the same circuit. Therefore, it is not anticipated that all the welders will be operating at full-load current, 100% of the time.

11.2 Arc Welders

Arc welders are covered by **Part II** of **Article 630**.

The following *Code* terms and symbols are applicable to arc welders.

According **Section 630.14**, the rating plate of arc welders is required to be marked with the either the I_{1max} and the I_{1eff} or the rated primary current (RPC).

I_{1eff} - This symbol is used on the rating plate of arc welders to indicate the effective input current required by the welder. This rating more accurately reflects the heating effect of the supply conductors because it considers both the current at idle as well as the current while welding. When calculating the conductor size, the I_{1eff} given on the rating plate is used. When the I_{1eff} is not given, the RPC rating is used to calculate the conductor size. For arc welders, the RPC will be used.

I_{1max} - This symbol is used on the rating plate of arc welders to indicate the current value to be used when calculating the overcurrent protective device for the welder circuit. When the I_{1max} is not given, the RPC rating is used to calculate the overcurrent protection. For arc welders, the RPC will be used.

11.2.1 AC Transformer and DC Rectifier Type Welders

AC transformer and DC rectifier welders are just one type of welder under the category of arc welder or nonmotor generator. Other nonmotor generator welders include AC arc welding machines, DC arc welding machines, TIG welding machines, MIG/MAG welding machines, plasma arc cutting machines, and plasma arc welding equipment.

AC transformer and DC rectifier welders operate more efficiently than motor generator welders because they have no rotational losses.

11.2.1.1 Load Calculation for Individual Welders - Load calculations for individual welders are the simplest of welder calculations and follow **630.11(A)** for individual welders. Using the welder machine rating plate, the demand factor of the welder is obtained from the rating plate and the multiplier is determined by using **Table 630.11(A)**, Columns 1 and 2 only. Simply match the machine duty cycle in Column 1 to the multiplier in Column 2 in the same row.

Where RPC equals the rating plate current, the following formula is used to determine the ampacity (I) of the welder branch circuit:

$$I = \text{RPC} \times \textit{multiplier}$$

But, where the rating plate contains the value for I_{1eff}, no calculations are necessary since the demand factor and the multiplier have already been accounted for and the following formula is to be used:

$$I = I_{1eff}$$

Overcurrent protection must also be provided for the welding machine as well as for the conductors which supply them. For arc welders, **Section 630.12** sets forth the maximum limit of 200% for the overcurrent protective device selection in order to provide protection against short circuits. Proper operation of the welder, together with overload circuit controls within the apparatus, will protect against overloads. However, most manufacturer installation instructions recommend an overcurrent protective device sized less than 200%. Where a welding machine is a listed or labeled product, then the manufacturer instructions become part of the listing requirements and must be followed as if it were the *Code*. **110.3(B)** indicates that listed or labeled equipment shall be installed and used in accordance with any instructions included in the listing or labeling.

For additional information, visit qr.njatcdb.org Item #1044

Problem 11-1

An AC transformer and a DC rectifier welder has a rated primary current (RPC) of 40 amperes with a duty cycle of 70%. Calculate the ampacity of the circuit conductors.

Circuit Conductors
Circuit Breaker
RPC 40 amps
Duty Cycle 70%

Solution
Table 630.11(A)
Nonmotor generator column; duty cycle 70%
Multiplier = 0.84
I = RPC × *multiplier*
= 40 × 0.84
= 33.6 amps
Answer: 33.6 amperes

Problem 11-2

An AC transformer and DC rectifier arc welder has a duty cycle of 50% and a rated primary current (RPC) of 38 amperes. Calculate the following:

1. Ampacity
2. Size of copper conductors with THWN 75°C rated insulation
3. Maximum overcurrent protective device (OCPD)

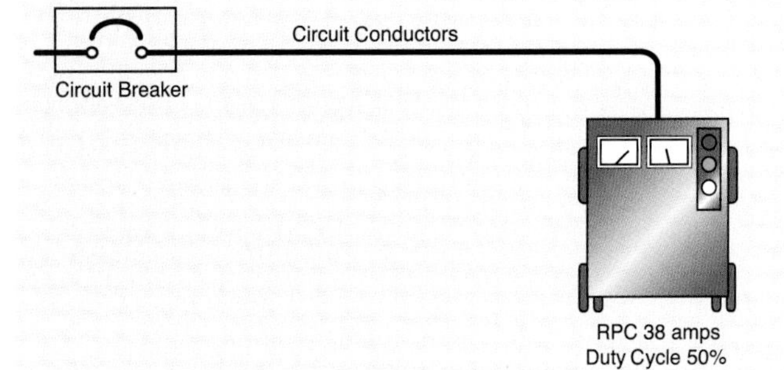

Solution – Calculation 1
Ampacity
Table 630.11(A)
Nonmotor generator column; duty cycle 50%
Multiplier = 0.71
I = RPC × *multiplier*
= 38 × 0.71
= 26.98 amps
Answer: 26.98 amperes

Solution – Calculation 2
Circuit conductors
Table 310.15(B)(16)
75°C copper column
26.98 amps = 10 AWG THWN
Double asterisk note does not apply
Answer: 10 AWG THWN

Solution – Calculation 3
Overcurrent Protective Device (OCPD)
630.12(A)
Max. OCPD = RPC × 200%
= 38 × 2.00
= 76 amps
Section 630.12, Section 240.6
76 amps is not a standard rating
Next larger standard size = 80 amps
Answer: 80 ampere max. CB

11.2.1.2 Load Calculation for Multiple Welders - 630.11(B) addresses the installation of a group of welders on the same circuit. The following conditions apply:

1. Calculate each welder using the multiplier listed in **Table 630.11(A)**.
2. Each welder is considered for its particular duty.
3. The total load is not the sum of the calculated individual currents.
4. All welders are not considered to be operating at the same time.
5. The ampacity of the circuit conductors is permitted to be a percentage of the total calculated duty load.
6. The percentage for the number of welders in the group is as follows:

Number of Welders	% of Calculated Current
Largest	100%
2nd largest	100%
3rd largest	85%
4th largest	70%
5th largest and more	60%

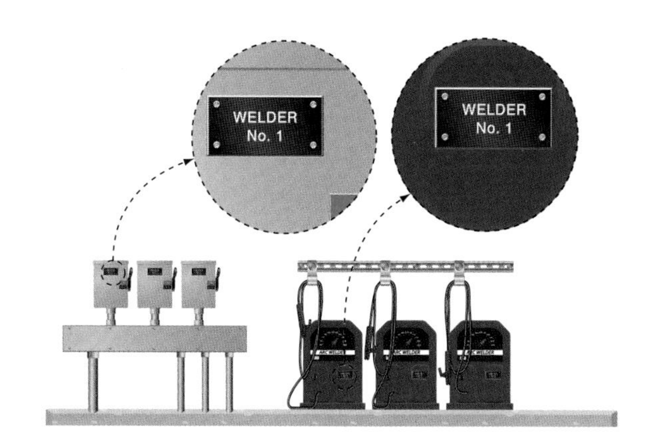

*Revised for the 2014 NEC, **Section 630.13** requires a disconnecting means must be provided in the supply circuit for each arc welder that is not equipped with a disconnect mounted as an integral part of the welder. The disconnecting means identity shall be marked in accordance with 110.22(A)*

For additional information, visit qr.njatcdb.org Item #1045

Problem 11-3

Calculate the ampacity of the circuit conductors for two welders to be installed on the same branch circuit when the welders are AC transformer and DC rectifier arc welders with the following rated primary current (RPC) ratings and duty cycles:

Welder No. 1	28 amperes	30% duty cycle
Welder No. 2	18 amperes	80% duty cycle

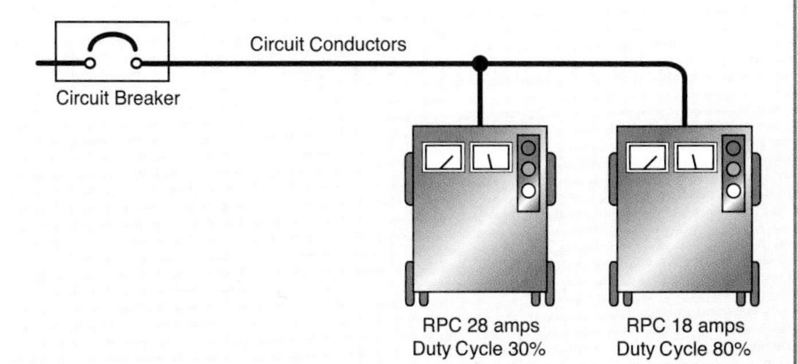

Solution

Ampacity of welders

Welder No. 1

Table 630.11(A)

Nonmotor generator column; duty cycle 30%

Multiplier = 0.55

I = RPC × *multiplier*

= 28 × 0.55

= 15.4 amps

Answer: 15.4 amperes

Welder No. 2

Table 630.11(A)

Nonmotor generator column; duty cycle 80%

Multiplier = 0.89

I = RPC × *multiplier*

= 18 × 0.89

= 16.02 amps

Answer: 16.02 amperes

Calculation

630.11(B)

Welder No. 2, largest	16.02 × 100%	16.02
Welder No. 1, 2nd largest	15.40 × 100%	15.40
Total		31.42

Answer: 31.42 amperes

Problem 11-4

Calculate the circuit conductor ampacity, the minimum size circuit conductors, and the maximum size overcurrent protection of the circuit conductors when three AC transformer and DC rectifier welders are to be installed on the same welder feeder circuit, with the following rated primary current ratings and duty cycles:

Welder No. 1	14 amperes	20% duty cycle
Welder No. 2	24 amperes	90% duty cycle
Welder No. 3	32 amperes	60% duty cycle

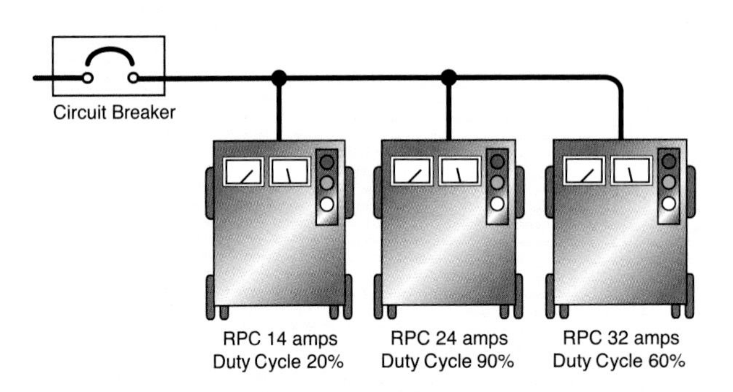

Solution– Calculation 1

Ampacity of welders

Welder No. 1

Table 630.11(A)

Nonmotor generator column; duty cycle 20%

Multiplier = 0.45

I = RPC × *multiplier*
= 14 × 0.45
= 6.3 amps

Answer: 6.3 amperes

Welder No. 2

Table 630.11(A)

Nonmotor generator column; duty cycle 90%

Multiplier = 0.95

I = RPC × *multiplier*
= 24 × 0.95
= 22.8 amps

Answer: 22.8 amperes

Welder No. 3

Table 630.11(A)

Nonmotor generator column; duty cycle 60%

Multiplier = 0.78

I = RPC × *multiplier*
= 32 amps × 0.78
= 24.96 amps

Answer: 24.96 amperes

Calculation

630.11(B)

Welder No. 3, largest	24.96 × 100%	24.96
Welder No. 2, 2nd largest	22.80 × 100%	22.80
Welder No. 1, 3rd largest	6.30 × 85%	5.36
Total		53.12

Answer: 53.12 amperes

Solution – Calculation 2

Circuit conductors

Table 310.15(B)(16)

75°C copper column

53.12 amps = 6 AWG THWN

Double asterisk note does not apply

Answer: 6 AWG THWN

Solution – Calculation 3

Overcurrent Protective Device (OCPD)

630.12(A)

Table 310.15(B)16) Ampacity

75°C copper column

6 AWG THWN = 65 amps

Max. OCPD = *conductor ampacity* x 200%
= 65 × 2.00
= 130 amps

Section 630.12, Section 240.6

130 amps not a standard rating

Next larger standard size = 150 amps

Answer: 150 ampere max. fuse or CB

Problem 11-5

Calculate the ampacity of the circuit conductors for four AC transformer and DC rectifier welders to be installed on the same circuit with the following nameplate current ratings and duty cycles:

Welder No. 1	40 amperes	20% duty cycle
Welder No. 2	28 amperes	70% duty cycle
Welder No. 3	24 amperes	50% duty cycle
Welder No. 4	16 amperes	40% duty cycle

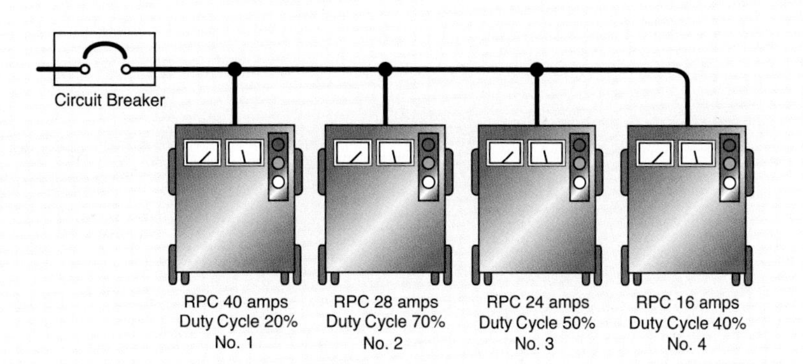

Solution

Welder No. 1

Table 630.11(A)

Nonmotor generator column; duty cycle 20%

Multiplier = 0.45

I = RPC × *multiplier*

= 40 × 0.45

= 18 amps

Answer: 18 amperes

Welder No. 2

Table 630.11(A)

Nonmotor generator column; duty cycle 70%

Multiplier = 0.84

I = RPC × *multiplier*

= 28 × 0.84

= 23.52 amps

Answer: 23.52 amperes

Welder No. 3

Table 630.11(A)

Nonmotor generator column; duty cycle 50%

Multiplier = 0.71

I = RPC × *multiplier*

= 24 × 0.71

= 17.04 amps

Answer: 17.04 amperes

Welder No. 4

Table 630.11(A)

Nonmotor generator column; duty cycle 40%

Multiplier = 0.63

I = RPC × *multiplier*

= 16 × 0.63

= 10.08 amps

Answer: 10.08 amperes

Calculation

630.11(B)

Welder No. 2, largest	23.52 × 100%	23.52
Welder No. 1, 2nd largest	18.00 × 100%	18.00
Welder No. 3, 3rd largest	17.04 × 85%	14.48
Welder No. 4, 4th largest	10.08 × 70%	7.06
Total		63.06

Answer: 63.06 amperes

Problem 11-6

Calculate the ampacity of the circuit conductors for five AC transformer and DC rectifier welders to be installed on the same circuit with the following rated primary current (RPC) ratings and duty cycles:

Welder No. 1	40 amperes	50% duty cycle
Welder No. 2	12 amperes	40% duty cycle
Welder No. 3	24 amperes	80% duty cycle
Welder No. 4	20 amperes	30% duty cycle
Welder No. 5	24 amperes	90% duty cycle

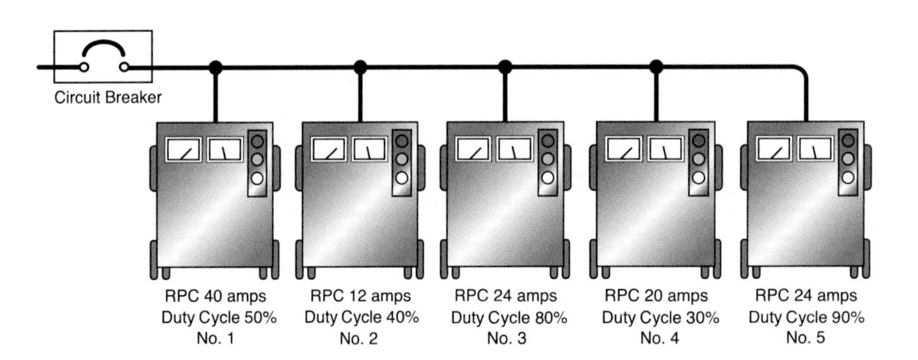

Solution

Welder No. 1

Table 630.11(A)

Nonmotor generator column; duty cycle 50%

Multiplier = 0.71

I = RPC × *multiplier*
= 40 × 0.71
= 28.4 amps

Answer: 28.4 amperes

Welder No. 2

Table 630.11(A)

Nonmotor generator column; duty cycle 40%

Multiplier = 0.63

I = RPC × *multiplier*
= 12 × 0.63
= 7.56 amps

Answer: 7.56 amperes

Welder No. 3

Table 630.11(A)

Nonmotor generator column; duty cycle 80%

Multiplier = 0.89

I = RPC × *multiplier*
= 24 × 0.89
= 21.36 amps

Answer: 21.36 amperes

Welder No. 4

Table 630.11(A)

Nonmotor generator column; duty cycle 30%

Multiplier = 0.55

I = RPC × *multiplier*
= 20 × 0.55
= 11 amps

Answer: 11 amperes

Welder No. 5

Table 630.11(A)

Nonmotor generator column; duty cycle 90%

Multiplier = 0.95

I = RPC × *multiplier*
= 24 × 0.95
= 22.8 amps

Answer: 22.8 amperes

Calculation

630.11(B)

Welder No. 1, largest	28.40 × 100%	28.40
Welder No. 5, 2nd largest	22.80 × 100%	22.80
Welder No. 3, 3rd largest	21.36 × 85%	18.16
Welder No. 4, 4th largest	11.00 × 70%	7.70
Welder No. 2, 5th largest	7.56 × 60%	4.54
Total		81.60

Answer: 81.6 amperes

11.2.2 Motor Generator Type

The motor generator arc welder uses an electric motor physically coupled to an electric generator. The electric generator actually delivers the low voltage and adjustable welding current necessary to weld. In many ways, a stand-alone fossil fuel engine-driven generator (welder) is similar to an electric motor-driven generator (welder), except that the motor-generator is powered by electricity as opposed to the fossil fuel power of the engine generator.

Motor generator arc welders are less efficiency than box-type arc welders because of the additional electrical motor. So, naturally, motor generator welders will require a somewhat larger set of supply conductors. Therefore, when sizing conductors for motor generators, the third column of **Table 630.11(A)** must be used to size these conductors.

11.2.2.1 Load Calculation for Individual Welders - 630.11(A) is used to determine the ampacity of the circuit conductors supplying a motor generator arc welder. This is similar to other arc welders. However, when using **Table 630.11(A)** to find the multiplier from the given duty cycle, the third column entitled "Motor Generator" must be used.

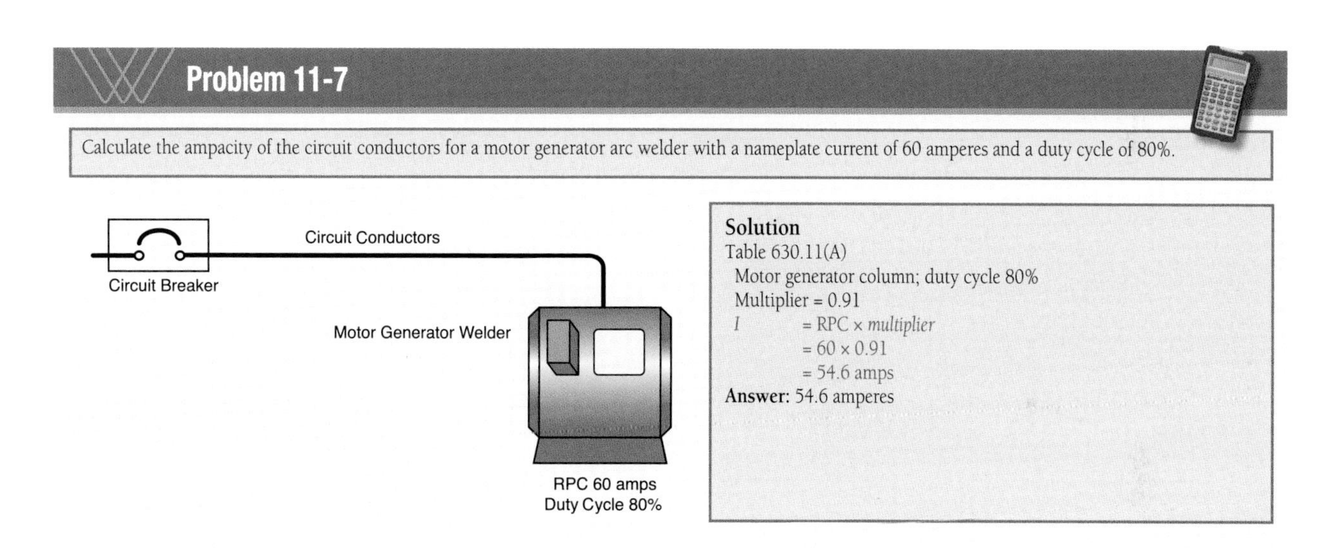

Problem 11-7
Calculate the ampacity of the circuit conductors for a motor generator arc welder with a nameplate current of 60 amperes and a duty cycle of 80%.
Circuit Conductors
Circuit Breaker
Motor Generator Welder
RPC 60 amps
Duty Cycle 80%
Solution
Table 630.11(A)
Motor generator column; duty cycle 80%
Multiplier = 0.91
I = RPC x multiplier
= 60 x 0.91
= 54.6 amps
Answer: 54.6 amperes

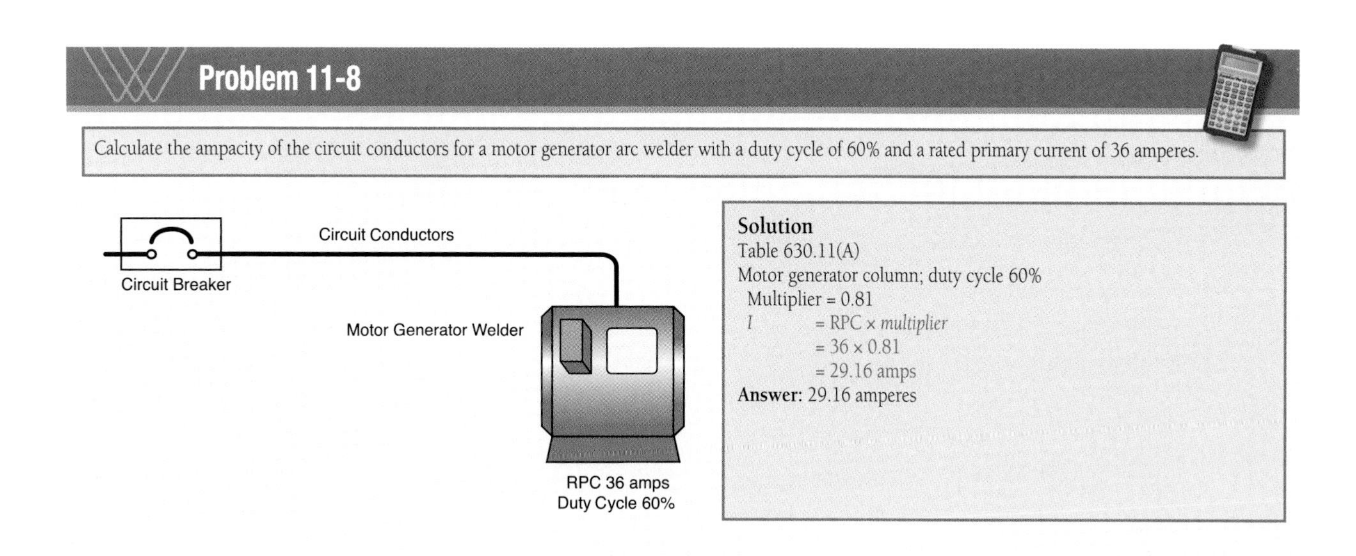

Problem 11-8
Calculate the ampacity of the circuit conductors for a motor generator arc welder with a duty cycle of 60% and a rated primary current of 36 amperes.
Circuit Conductors
Circuit Breaker
Motor Generator Welder
RPC 36 amps
Duty Cycle 60%
Solution
Table 630.11(A)
Motor generator column; duty cycle 60%
Multiplier = 0.81
I = RPC x multiplier
= 36 x 0.81
= 29.16 amps
Answer: 29.16 amperes

11.2.2.2 Load Calculation for Multiple Welders - The calculation for a group of motor generator arc welders is identical to the AC transformer and DC rectifier arc welders:

1. Calculate each welder using the multiplier listed in **Table 630.11(A)**.
2. Each welder is considered for its particular duty.
3. The total load is not the sum of the calculated individual currents.
4. All welders are not considered to be operating at the same time.
5. The ampacity of the circuit conductors is permitted to be a percentage of the total calculated duty load.
6. The percentage for the number of welders in the group is as follows:

Number of Welders	% of Calculated Current
Largest	100%
2nd Largest	100%
3rd Largest	85%
4th Largest	70%
5th Largest	60%

11.3 Resistance Welders

The next type of welders is the resistance welder. They are used for spot welding and seam welding and can be automatically or manually operated. According **Article 630, Part III**, resistance welders can be operated at different timing cycles and at less than their rated primary current, resulting in various calculations.

The following *Code* terms and symbols are applicable to resistance welders:

Nameplate - For resistance welders, the required input marking includes voltage, phase, frequency, and rated kilovolt amperes (kVA) at 50% duty cycle.

Rated primary current - The rated kilovolt-amperes (kVA) is multiplied by 1,000 and divided by the rated primary voltage using the values given on the nameplate. This rated primary current is expressed in the following formula:

$$I_{pri} = \frac{kVA \times 1{,}000}{E_{pri}}$$

Problem 11-9

Calculate the ampacity of the branch-circuit conductors for six motor generator arc welders, when all six welders have a rated primary current rating (RPC) of 21 amperes and a 60% duty cycle.

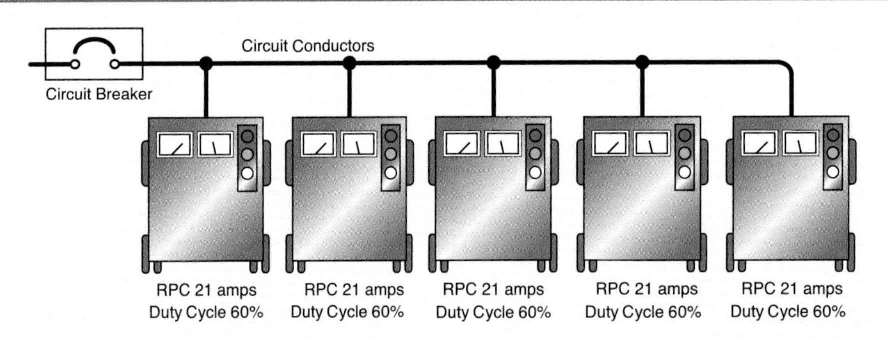

Solution
Ampacity for all 6 welders
Table 630.11(A)
Motor generator column; duty cycle 60%
Multiplier = 0.81

$I = \text{RPC} \times \textit{multiplier}$
$= 21 \times 0.81$
$= 17.01$ amps

Answer: 17.01 amperes

Calculation
630.11(B)

Largest	17.01 × 100%	17.01
2nd largest	17.01 × 100%	17.01
3rd largest	17.01 × 85%	14.46
4th largest	17.01 × 70%	11.91
5th largest	17.01 × 60%	10.21
6th largest	17.01 × 60%	10.21
Total		80.81

Answer: 80.81 amperes

Actual primary current- The current drawn from the supply circuit during each welder operation at the particular heat tap and control setting used.

Duty cycle- The percentage of time the during which the resistance welder is loaded.

11.3.1 Load Calculation for Individual Welders

The primary current can be found by using the following equation:

$$I_{pri} = \frac{\text{kVA} \times 1{,}000}{E_{pri}}$$

According to **630.31(A)(1)**, calculations depend on the duty cycle, known or unknown, and whether the welder is manually or automatically operated:

Automatic operation:

$$I_{pri} \times 70\%$$

Manual operation:

$$I_{pri} \times 50\%$$

Where a resistance type welder is designated for a specific operation, each of the following values must be known before ampacity and overcurrent calculation can be completed:

1. Actual primary current; or the actual current the primary will draw at a particular heat tap and control setting
2. The duty cycle
3. The multiplier from **Table 630.31(A)(2). See Figure 11-1.**

Problem 11-10

Calculate the ampacity of the circuit conductors for a 15 kVA resistance welder operated on a 220 volt, single-phase circuit. The duty cycle or operating time is not known and the welder is operated both automatically and manually.

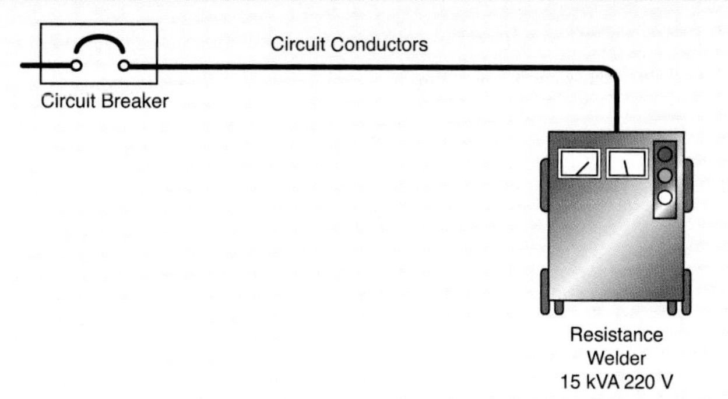

Solution – Calculation 1

Automatic operation

$$I_{pri} = \frac{kVA \times 1{,}000}{E}$$

$$= \frac{15 \times 1{,}000}{220}$$

$= 68.18$ amps

630.31(A)(1)

Conductor I = Ipri × 70%

$= 68.18 \times 0.70$

$= 47.72$ amps

Answer: 47.72 amperes

Solution – Calculation 2

Manual operation

$$I_{pri} = \frac{kVA \times 1{,}000}{E}$$

$$= \frac{15 \times 1{,}000}{220}$$

$= 68.18$ amps

630.31(A)(1)

Conductor I = Ipri × 50%

$= 68.18 \times 0.50$

$= 34.09$ amps

Answer: 34.09 amperes

Comment

If the welder circuit is used for both types of operation, then the selected ampacity must be the larger of the two.

Table 630.31(A)(2) Duty Cycle Multiplication Factors for Resistance Welders

Duty Cycle (%)	Multiplier
50	0.71
40	0.63
30	0.55
25	0.50
20	0.45
15	0.39
10	0.32
7.5	0.27
5 or less	0.22

Figure 11-1. Table 630.31(A)(2) Duty cycle multiplication factors are limited to resistance welders.

Courtesy of the Lincoln Electric Company.
A 225-amp TIG type electric welder

11.3.2 Load Calculation for Multiple Welders

When a group of resistance welders is to be supplied by the same circuit conductors, the ampacity of the circuit conductors is based the calculated conductor current in accordance with **630.31(A)** for automatic or manual operation. According to **630.31(B)**, the first welder is at 100%, and all others are at 60%.

For the largest welder:

$$Ampacity = Conductor\ I \times 100\%$$

All other welders:

$$Ampacity = Conductor\ I \times 60\%$$

Problem 11-11

Calculate the ampacity of the circuit conductors for a resistance type spot welder when the actual primary current is 40 amperes and the duty cycle is 40%.

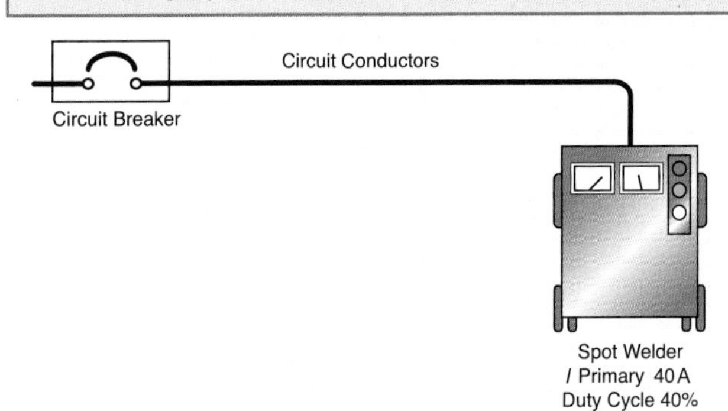

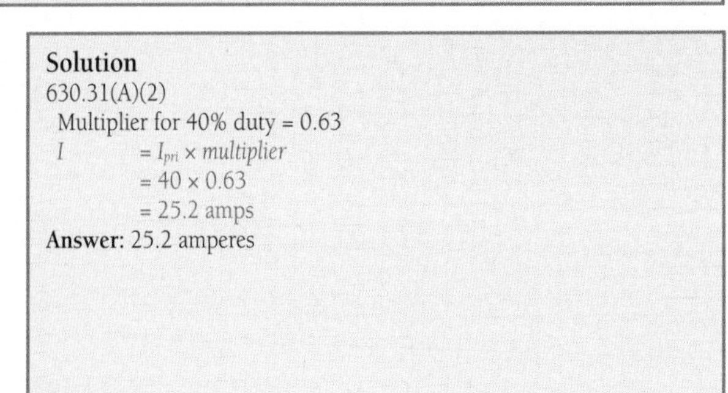

Solution

630.31(A)(2)

Multiplier for 40% duty = 0.63

$I = I_{pri} \times multiplier$

$= 40 \times 0.63$

$= 25.2$ amps

Answer: 25.2 amperes

Problem 11-12

Calculate the ampacity of the circuit conductors for two resistance welders rated 20 kVA, 240 volts, single-phase and installed on the same circuit, when no duty cycle is indicated and the welders are either automatically operated or manually operated.

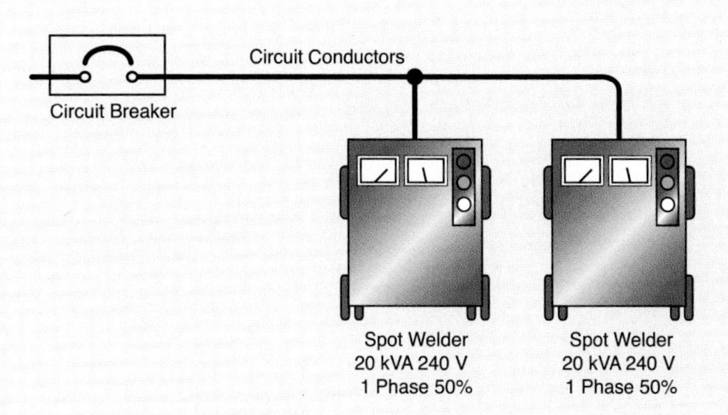

Solution – Calculation 1

Automatic operation

$I_{pri} = \frac{kVA \times 1{,}000}{E}$

$= \frac{20 \times 1{,}000}{240}$

= 83.33 amps

630.31(A)(1)

Conductor I = I_{pri} × 70%
= 83.33 x 0.70
= 58.33 amps

Welder No. 1

630.31(B)

I = *Conductor* I × 100%
= 58.33 × 1.00
= 58.33 amps

Answer: 58.33 amperes

Welder No. 2

630.31(B)

I = *Conductor* I × 60%
= 58.33 × 0.60
= 35 amps

Answer: 35 amperes

Calculation

I total = *largest + others*
= 58.33 + 35
= 93.33 amps

Answer: 93.33 amperes

Solution – Calculation 2

Manual operation

$I_{pri} = \frac{kVA \times 1{,}000}{E}$

$= \frac{20 \times 1{,}000}{240}$

= 83.33 amps

630.31(A)(1)

Conductor I = I_{pri} × 50%
= 83.33 x 0.50
= 41.67 amps

Welder No. 1

630.31(B)

I = *Conductor* I × 100%
= 41.67 × 100
= 41.67 amps

Answer: 41.67 amperes

Welder No. 2

630.31(B)

I = *Conductor* I × 60%
= 41.67 × 0.60
= 25 amps

Answer: 25 amperes

Calculation

I total = *largest + others*
= 41.67 + 25
= 66.67 amps

Answer: 66.67 amperes

11.3.3 Calculating Duty Cycle

Informational Note (3) to **630.31(B)** illustrates how to calculate the duty cycle.

Example 1

A spot welder makes 500 welds per hour on a 60-cycle AC circuit. It takes 18 cycles to make each weld.

Number of cycles in one hour:

60 cycles × 60 seconds/minute × 60 minutes/hour = 216,000 cycles

Time or number of cycles current drawn in one hour:

18 cycles × 500 welds = 9,000 cycles

$$\text{Duty cycle} = \frac{\text{time of use}}{\text{time available}}$$

$$= \frac{9{,}000}{216{,}000}$$

$$= 0.041, \text{ or } 4.1\%$$

Example 2

A spot welder is timed to be ON 6 cycles and OFF 9 cycles.

$$\text{Duty cycle} = \frac{\text{time on}}{\text{time on} + \text{time off}}$$

$$= \frac{6 \text{ cycles}}{6 \text{ cycles} + 9 \text{ cycles}}$$

$$= 0.40 \text{ or } 40\%$$

Definitions and Terms

Rating Plate - The general term for a welder nameplate used in the welding industry.

Duty cycle - The percentage of a (ten minute) cycle during which the welder is loaded or the length of time there will be a demand for current flow in the circuit, and the length of the time the circuit will be at rest.

I_{1max} - The maximum value of the rated supply current at maximum rated output.

I_{1eff} - The effective input current required by the welder that considers both welding and idle current.

Summary

- Welders are encountered more often during construction and remodeling projects than as a part of the permanent premises wiring.
- There are two basic types of electric welders: arc welders (AC transformer and DC rectifier type and motor generator type) and resistance welders.
- Actual welder loads draw substantially less current than their marked supply current. The *Code* corrects this load difference by allowing the use of duty cycles which translate into multipliers to lower the calculated load.
- Where multiple welders are connected to the same feeder, the total load is not the sum of the loads, but permits an additional demand factor to be applied to reduce the overall feeder load even further. This results in very efficient wiring; even in the temporary wiring.
- Arc welder branch circuit loads are calculated using the fundamental equation:

  $I = \text{RPC} \times \textit{multiplier}$
- Arc welder feeder loads are calculated by finding the sum of the following:

  » Largest welder load at 100%, plus
  » 2nd largest load at 100%, plus
  » 3rd largest at 85%, plus
  » 4th largest at 70%, plus
  » 5th largest and more at 60%
- Individual branch circuit loads for resistance type welders are calculated differently for automatic mode (at 70%) and manual mode (at 50%) of the I_{pri}.
- Resistance type welder loads designated for a specific purpose are calculated at:

  $I = I_{pri} \times \textit{multiplier}$
- The duty cycle of a spot welder may be calculated by finding the percentage of time the welder is "ON" or the operating per period of use (time ON + time OFF).

Solutions to Select Problems Using the *ElectriCalc® Pro*

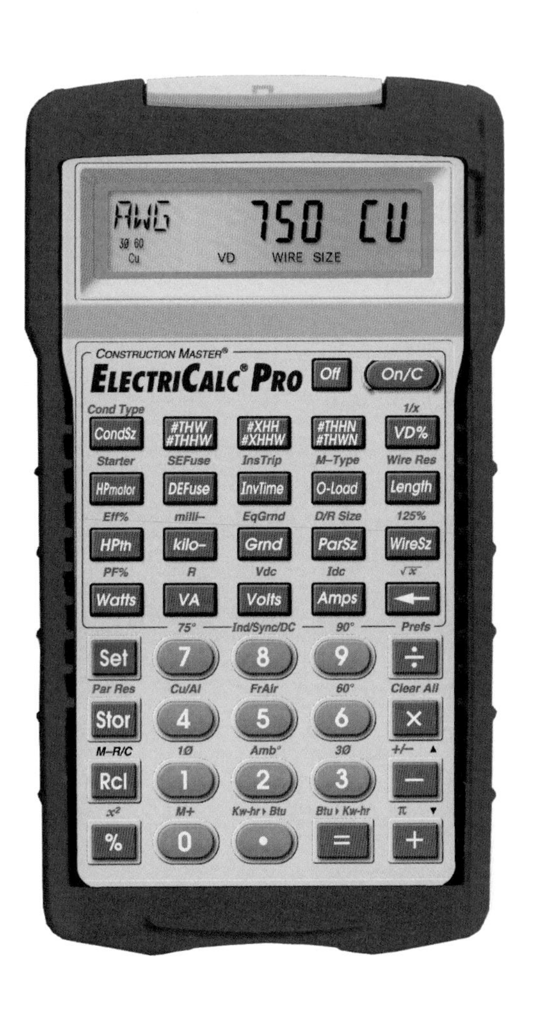

Appendix

With the 2014 edition of *Code Calculations*, our training partner Calculated Industries has provided a valuable addition feature through the use of an Electrical Industry specific calculator, the *ElectriCalc® Pro.* This calculator is preprogrammed with very accurate electrical industry and engineering data. Some calculations appearing in this Appendix may have numerical variations, especially in the third or fourth, and subsequent digits, which vary from answers in the body of the textbook. Where a variance is noticed in the Appendix and the correct entry of data has been verified, you can be assured that this calculation is correct, and the difference can be attributed to the method of rounding used throughout the textbook.

In the study of science and mathematics, the questions and decisions made regarding numerical accuracy are important. For *Code Calculations*, the NJATC does not require calculations to be performed using calculators. But, we realize that most calculations are done using some form of electronic calculator. In order to treat both manual calculations and electronic calculations fairly, the current NJATC policy is to accept an accuracy level of 3 or 4 significant digits for calculations related to the *NEC*.

For example, where look-up tables are used such as the tables of Chapter 9, it is always recommended that the same number of digits be used as are shown in the look-up tables. Most of the numbers from these tables use four significant digits, but some may use five. The full, complete, unrounded number should be used in all cases.

But for other cases, three significant digits is a simpler choice. As an example, pi (or π) has a numerical value with an infinite number of digits to the right of the decimal point. In *Code Calculations*, there is no need to work manually with ten or more significant digits for our calculations. Rather, 3.14 is used in all the textbook calculations. This also applies to the square root of three (1.7320508…), which is used as 1.73. The decision to use the simplified value in these cases is sufficient for the related problems and instruction.

Chapter 1

NOTE: If your results differ from those provided in the examples, verify that your calculator is set to the appropriate phase, wire rating, wire type and ambient temperature.

Problem 1-1

Description	Keystroke	Display
Clear Calculator	On/C On/C	0.
Toggle to Copper Wire (if necessary)	Set 4	Cu
Set Wire Rating to 75°C (THW rating)	Set 7	75
Set Ambient Temperature to 40°C*	4 0 Set 2	AMB° 40. °C
Enter Wire Size	6 WireSz	AWG 6 CU WIRE SIZE
Display Ampacity	WireSz	57.2 WIRE A

*May need to set temperature unit preference to °C via Set ÷ ÷ before storing ambient temperature.

57.2 Amps

Problem 1-2

Description	Keystroke	Display
Clear Calculator	On/C On/C	0.
Toggle to Aluminum Wire (if necessary)	Set 4	Al
Set Wire Rating to 90°C (THHN rating)	Set 9	90
Set Ambient Temperature to 75°F*	7 5 Set 2	AMB° 75. °F
Enter Wire Size	4 WireSz	AWG 4 AL WIRE SIZE
Display Ampacity	WireSz	78.0 WIRE A

*May need to set temperature unit preference to °F via Set ÷ ÷ before storing ambient temperature.

78 Amps

Problem 1-3

Description	Keystroke	Display
Clear Calculator	On/C On/C	0.
Toggle to Copper Wire (if necessary)	Set 4	Cu
Set Wire Rating to 90°C	Set 9	90
Set Ambient Temperature to 105°F*	1 0 5 Set 2	AMB° 105. °F
Enter Wire Size	1 0 WireSz	AWG 10 CU WIRE SIZE
Display Ampacity	WireSz	34.8 WIRE A

*May need to set temperature unit preference to °F via Set ÷ ÷ before storing ambient temperature.

34.8 Amps

Problem 1-4

Description	Keystroke	Display
Clear Calculator	On/C On/C	0.
Toggle to Copper Wire (if necessary)	Set 4	Cu
Set Wire Rating to 90°C	Set 9	90
Set Ambient Temperature to 45°C*	4 5 Set 2	AMB° 45. °C
Enter Wire Size	2 WireSz	AWG 2 CU WIRE SIZE
Display Ampacity	WireSz	113.1 WIRE A

*May need to set temperature unit preference to °C via Set ÷ ÷ before storing ambient temperature.

113.1 Amps

Problem 1-5

Description	Keystroke	Display
Clear Calculator	On/C On/C	0.
Toggle to Copper Wire (if necessary)	Set 4	Cu
Set Wire Rating to 90°C (THHN Rating)	Set 9	90
Set Ambient Temperature to 30°C*	3 0 Set 2	AMB° 30. °C
Enter Wire Size	4 WireSz	AWG 4 CU WIRE SIZE
Display Ampacity	WireSz	95.0 WIRE A
Store as Amps	= Amps	AMPS 95.
Enter Quantity of Wires and Find Derated Ampacity	6 Set ParSz ParSz	D/R 76.0 WIRE A

*If Ambient hasn't been changed, then entry isn't necessary, as ambient temperature is set to 30°C by default.

76 Amps

NOTE: The ECPro is also structured to compute wire size required based on amperage load, deration factors, temperature factors, etc.

Problem 1-6

Description	Keystroke	Display
Clear Calculator	On/C On/C	0.
Toggle to Aluminum Wire (if necessary)	Set 4	Al
Set Wire Rating to 75°C (THHW rating)	Set 7	75
Set Ambient Temperature to 30°C (if necessary)	3 0 Set 2	AMB° 30. °C
Enter Wire Size	1 2 WireSz	AWG 12 AL WIRE SIZE
Display Ampacity	WireSz	20.0 WIRE A
Store as Amps	= Amps	AMPS 20.
Enter Quantity of Wires and Find Derated Ampacity	3 0 Set ParSz ParSz	D/R 9.0 WIRE A

9 Amps

Problem 1-7

Description	Keystroke	Display
Clear Calculator	On/C On/C	0.
Toggle to Copper Wire (if necessary)	Set 4	Cu
Set Wire Rating to 90°C (XHHW)	Set 9	90
Set Ambient Temperature to 30°C (if necessary)	3 0 Set 2	AMB° 30. °C
Enter Wire Size	1 0 WireSz	AWG 10 CU WIRE SIZE
Display Ampacity	WireSz	40.0 WIRE A
Store as Amps	= Amps	AMPS 40.
Enter Quantity of Wires and Find Derated Ampacity	1 5 Set ParSz ParSz	D/R 20.0 WIRE A

20 Amps

Problem 1-8

Description-1	Keystroke	Display
Clear Calculator	On/C On/C	0.
Toggle to Copper Wire (if necessary)	Set 4	Cu
Set Wire Rating to 75°C (THWN Rating)	Set 7	75
Set Ambient Temperature to 30°C (if necessary)	3 0 Set 2	AMB° 30. °C
Enter Wire Size	2 WireSz	AWG 2 CU WIRE SIZE
Display Ampacity	WireSz	115.0 WIRE A
Store as Amps	= Amps	AMPS 115.
Enter Quantity of Wires and Find Derated Ampacity	9 Set ParSz ParSz	D/R 80.5 WIRE A

80.5 Amps

Description-2	Keystroke	Display
Enter Wire Size	1 WireSz	AWG 1 CU WIRE SIZE
Display Ampacity	WireSz	130.0 WIRE A
Store as Amps	= Amps	AMPS 130.
Enter Quantity of Wires and Find Derated Ampacity	9 Set ParSz ParSz	D/R 91.0 WIRE A

91 Amps

Description-3	Keystroke	Display
Enter Wire Size	1 2 WireSz	AWG 12 CU WIRE SIZE
Display Ampacity	WireSz	25.0 WIRE A
Store as Amps	= Amps	AMPS 25.
Enter Quantity of Wires and Find Derated Ampacity	9 Set ParSz ParSz	D/R 17.5 WIRE A

17.5 Amps

Problem 1-9

Description-1	Keystroke	Display
Clear Calculator	On/C On/C	0.
Toggle to Copper Wire (if necessary)	Set 4	Cu
Set Wire Rating to 90°C (THHN Rating)	Set 9	90
Set Ambient Temperature to 30°C (if necessary)	3 0 Set 2	AMB° 30. °C
Enter Wire Size	8 WireSz	AWG 8 CU WIRE SIZE
Display Ampacity	WireSz	55.0 WIRE A
Store as Amps	= Amps	AMPS 55.
Enter Quantity of Wires and Find Derated Ampacity	4 Set ParSz ParSz	D/R 44.0 WIRE A
Set Wire Rating to 60°C (TW Rating)	Set 6	60
Find Wire Ampacity	8 WireSz WireSz	40.0 WIRE A

NOTE: Use lowest value for allowable ampacity of #8 wire.

Lowest Rating: 40 Amps

Description-2	Keystroke	Display
Clear Calculator	On/C On/C	0.
Set Wire Rating to 60°C (TW Rating)	Set 6	60
Enter Wire Size	4 WireSz	AWG 4 CU WIRE SIZE
Display Ampacity	WireSz	70.0 WIRE A
Store as Amps	= Amps	AMPS 70.
Enter Quantity of Wires and Find Derated Ampacity	4 Set ParSz ParSz	D/R 56.0 WIRE A

56 Amps

Problem 1-10

Description	Keystroke	Display
Clear Calculator	On/C On/C	0.
Toggle to Copper Wire (if necessary)	Set 4	Cu
Set Wire Rating to 75°C (THWN Rating)	Set 7	75
Enter Wire Size	1 0 WireSz	AWG 10 CU WIRE SIZE
Display Ampacity	WireSz	35.0 WIRE A
Store as Amps	= Amps	AMPS 35.
Set Ambient Temperature to 45°C	4 5 Set 2	AMB° 45. °C
Enter Quantity of Wires and Find Derated Ampacity	9 Set ParSz ParSz	D/R 20.1 WIRE A

20.1 Amps

NOTE: In order for the ECPro to compute derated ampacity, the non-adjusted wire ampacity needs to be stored as Amps before ambient temperature is set. The ECPro was designed to provide a wire size that accommodates the ampacity, deration and temperature adjustments. The ECPro can provide the derated wire ampacity using the workarounds identified herein.

Problem 1-11

Description	Keystroke	Display
Reset Calculator*	Set X	NEC 2014
Change Temperature Units to °F	Set ÷ ÷ +	AMB° °F
Toggle to Copper Wire (if necessary)	Set 4	Cu
Set Wire Rating to 90°C (THHW Rating)	Set 9	90
Enter Wire Size	8 WireSz	AWG 8 CU WIRE SIZE
Display Ampacity	WireSz	55.0 WIRE A
Store as Amps	= Amps	AMPS 55.
Set Ambient Temperature to 128°F	1 2 8 Set 2	AMB° 128. °F
Enter Quantity of Wires and Find Derated Ampacity	1 2 Set ParSz ParSz	D/R 20.9 WIRE A

*Clears all values and sets calculator to 3Ø, 60°C, Copper @ 30°C Ambient Temperature.

20.9 Amps

Problem 1-12

Description	Keystroke	Display
Reset Calculator*	Set X	NEC 2014
Change Temperature Units to °C	Set ÷ ÷ +	AMB° °C
Set Wire Rating to 90°C (THHN Rating)	Set 9	90
Enter Wire Size	1 2 WireSz	AWG 12 CU WIRE SIZE
Display Ampacity	WireSz	30.0 WIRE A
Store as Amps	= Amps	AMPS 30.
Set Ambient Temperature to 50°C	5 0 Set 2	AMB° 50. °C
Enter Quantity of Wires and Find Derated Ampacity	3 6 Set ParSz ParSz	D/R 9.8 WIRE A

*Clears all values and sets calculator to 3Ø, 60°C, Copper @ 30°C Ambient Temperature.

9.8 Amps

Problem 1-13

Description	Keystroke	Display
Reset Calculator*	Set X	NEC 2014
Set Wire Rating to 75°C	Set 7	75
Enter Wire Size	0 0 0 0 WireSz	AWG 0000 CU WIRE SIZE
Display Ampacity	WireSz	230.0 WIRE A
Store as Amps	= Amps	AMPS 230.
Set Ambient Temperature to 45°C	4 5 Set 2	AMB° 45. °C
Enter Quantity of Wires and Find Derated Ampacity	4 Set ParSz ParSz	D/R 150.9 WIRE A

*Clears all values and sets calculator to 3Ø, 60°C, Copper @ 30°C Ambient Temperature.

150.9 Amps

Problem 1-14

Description	Keystroke	Display
Reset Calculator*	Set X	NEC 2014
Set Wire Rating to 90°C	Set 9	90
Find Ampacity of 12 AWG wire	1 2 WireSz WireSz	30.0 WIRE A
Store as Amps	= Amps	AMPS 30.
Set Ambient Temperature to 40°C	4 0 Set 2	AMB° 40. °C
Enter Quantity of Wires and Find Derated Ampacity	6 Set ParSz ParSz	D/R 21.8 WIRE A

*Clears all values and sets calculator to 3Ø, 60°C, Copper @ 30°C Ambient Temperature.

21.8 Amps

Problem 1-15

Description	Keystroke	Display
Reset Calculator*	Set X	NEC 2014
Set to 90°C Wire Rating	Set 9	90
Enter Wire Size	8 WireSz	AWG 8 CU WIRE SIZE
Display Ampacity	WireSz	55.0 WIRE A
Store as Amps	= Amps	AMPS 55.
Change Temperature Units to °F	Set ÷ ÷ +	AMB° °F
Enter Ambient Temperature	7 5 Set 2	AMB° 75. °F
Enter Quantity of Wires and Find Derated Ampacity	6 Set ParSz ParSz	D/R 45.8 WIRE A

*Clears all values and sets calculator to 3Ø, 60°C, Copper @ 30°C Ambient Temperature.

45.8 Amps

Problem 1-16

Description	Keystroke	Display
Reset Calculator*	Set X	NEC 2014
Set to 75°C Wire Rating	Set 7	75
Enter Wire Size	4 WireSz	AWG 4 CU WIRE SIZE
Display Ampacity	WireSz	85.0 WIRE A
Store as Amps	= Amps	AMPS 85.
Enter Quantity of Wires and Find Derated Ampacity	6 Set ParSz ParSz	D/R 68.0 WIRE A

*Clears all values and sets calculator to 3Ø, 60°C, Copper @ 30°C Ambient Temperature.

68.0 Amps

Problem 1-17

Description-1	Keystroke	Display
Clear Calculator	On/C On/C	0.
Set to 75°C Wire Rating	Set 7	75
Enter Wire Size	1 0 WireSz	AWG 10 CU WIRE SIZE
Display Ampacity	WireSz	35.0 WIRE A
Store as Amps	= Amps	AMPS 35.
Enter Quantity of Wires and Find Derated Ampacity	3 Set ParSz ParSz	D/R 35.0 WIRE A

*Clears all values and sets calculator to 3Ø, 60°C, Copper @ 30°C Ambient Temperature.

3/C: 35 Amps

Description-2	Keystroke	Display
Enter Quantity of Wires and Find Derated Ampacity	7 Set ParSz ParSz	D/R 24.5 WIRE A

7/C: 24.5 Amps

Description-3	Keystroke	Display
Enter Quantity of Wires and Find Derated Ampacity	1 5 Set ParSz ParSz	D/R 17.5 WIRE A

15/C: 17.5 Amps

Problem 1-18

Description	Keystroke	Display
Clear Calculator	On/C On/C	0.
Set to 75°C Wire Rating	Set 7	75
Enter Wire Size	0 0 0 0 WireSz	AWG 0000 CU WIRE SIZE
Display Ampacity	WireSz	230.0 WIRE A

230 Amps

Problem 1-19

Description	Keystroke	Display
Clear Calculator	On/C On/C	0.
Set to 75°C Wire Rating (THWN rating)	Set 7	75
Enter Wire Size	1 2 WireSz	AWG 12 CU WIRE SIZE
Display Ampacity	WireSz	25.0 WIRE A

25 Amps

Problem 1-20

Description	Keystroke	Display
Clear Calculator	On/C On/C	0.
Set to 75°C Wire Rating (THWN rating)	Set 7	75
Enter Wire Size	1 4 WireSz	AWG 14 CU WIRE SIZE
Display Ampacity	WireSz	20.0 WIRE A
Store as Amps	= Amps	AMPS 20.
Enter Quantity of Wires and Find Derated Ampacity	6 Set ParSz ParSz	D/R 16.0 WIRE A

16 Amps

Problem 1-21

Description	Keystroke	Display
Clear Calculator	On/C On/C	0.
Toggle to Aluminum Wire (if necessary)	Set 4	Al
Set to 90°C Wire Rating (THHN rating)	Set 9	90
Enter Wire Size	1 0 WireSz	AWG 10 AL WIRE SIZE
Display Ampacity	WireSz	35.0 WIRE A
Store as Amps	= Amps	AMPS 35.
Enter Quantity of Wires and Find Derated Ampacity	6 Set ParSz ParSz	D/R 28.0 WIRE A

28 Amps

Problem 1-22

Description	Keystroke	Display
Clear Calculator	On/C On/C	0.
Change Temperature Units to °F (if necessary)	Set ÷ ÷ +	AMB° °F
Toggle to Copper Wire (if necessary)	Set 4	Cu
Set to 90°C Wire Rating (THHN rating)	Set 9	90
Enter Wire Size	1 2 WireSz	AWG 12 CU WIRE SIZE
Display Ampacity	WireSz	30.0 WIRE A
Store as Amps	= Amps	AMPS 30.
Enter Ambient Temperature	9 0 Set 2	AMB° 90. °F
Enter Quantity of Wires and Find Derated Ampacity	4 Set ParSz ParSz	D/R 23.0 WIRE A

23 Amps

Problem 1-23

Description	Keystroke	Display
Clear Calculator	On/C On/C	0.
Change Temperature Units to °C	Set ÷ ÷ +	AMB° °C
Enter Ambient Temperature	3 0 Set 2	AMB° 30. °C
Toggle to Copper Wire (if necessary)	Set 4	Cu
Set to 90°C Wire Rating (THHN rating)	Set 9	90
Enter Wire Size	1 2 WireSz	AWG 12 CU WIRE SIZE
Display Ampacity	WireSz	30.0 WIRE A
Store as Amps	= Amps	AMPS 30.
Enter Quantity of Wires and Find Derated Ampacity	4 Set ParSz ParSz	D/R 24.0 WIRE A

24 Amps

Problem 1-24

Description	Keystroke	Display
Clear Calculator	On/C On/C	0.
Toggle to Copper Wire (if necessary)	Set 4	Cu
Set to 75°C Wire Rating (THWN rating)	Set 7	75
Enter the Amp Neutral Load	8 1 Amps	AMPS 81.
Find the Wire Size based on Ampacity	WireSz	AWG 4 CU WIRE SIZE
Enter Service Entrance Wire Size to find Minimum Ground Wire Size	1 Grnd	GRND 6 CU WIRE SIZE
Display Ground Wire Ampacity	6 WireSz WireSz	65.0 WIRE A

Grounded Conductor: 4 AWG THWN

(6 AWG doesn't handle required ampacity)

Problem 1-25

Description-1	Keystroke	Display
Reset Calculator	Set X	NEC 2014
Set to 90°C Wire Rating (THWN rating)	Set 7	75
Enter the rated Amp Load	3 8 0 Amps	AMPS 380.
Find the Wire Size based on Ampacity	WireSz	AWG 500 CU WIRE SIZE

500 kcmil THWN

Description-2	Keystroke	Display
Clear Calculator	On/C On/C	0.
Set to 75°C Wire Rating (THWN rating)	Set 7	75
Enter the Amp Neutral Load	1 2 0 Amps	AMPS 120.
Find the Wire Size based on Ampacity	WireSz	AWG 1 CU WIRE SIZE
Enter Service Entrance Wire Size to find Minimum Ground Wire Size	5 0 0 Grnd	GRND 0 CU WIRE SIZE
Display Ground Wire Ampacity	0 WireSz WireSz	150.0 WIRE A

Grounded Conductor: 1/0 AWG

(1/0 handles the 120 amp load)

Description-3	Keystroke	Display

Referencing NEC Table 310.15(B)(7); the ECPro does not provide this detail.

Description-4	Keystroke	Display
Clear Calculator	On/C On/C	0.
Enter the Amp Load	2 0 0 Amps	AMPS 200.
Find the Wire Size based on Ampacity	WireSz	AWG 000 CU WIRE SIZE

Feeder Size: 3/0 AWG

Problem 1-26

Description	Keystroke	Display
Clear Calculator	On/C On/C	0.
Toggle to Copper Wire (if necessary)	Set 4	Cu
Set to 90°C Wire Rating	Set 9	90
Enter Wire Size	1 2 WireSz	AWG 12 CU WIRE SIZE
Display Ampacity	WireSz	30.0 WIRE A
Set to 60°C Wire Rating	Set 6	60
Display Ampacity	WireSz WireSz	20.0 WIRE A

Allowable Ampacity: 20 Amps

(No deration adjustment required for 3 conductors)

Problem 1-27

Description	Keystroke	Display
Clear Calculator	On/C On/C	0.
Toggle to Aluminum Wire (if necessary)	Set 4	Al
Set to 60°C Wire Rating	Set 6	60
Find Wire Ampacity	1 0 WireSz WireSz	25.0 WIRE A
Set to 90°C Wire Rating	Set 9	90
Find Wire Ampacity	1 0 WireSz WireSz	35.0 WIRE A
Store as Amps	= Amps	AMPS 35.
Enter Ambient Temperature*	4 5 Set 2	AMB° 45. °C
Enter Quantity of Wires and Find Derated Ampacity	8 Set ParSz ParSz	D/R 21.3 WIRE A

*Change Temperature units in preference setting, if necessary.

Allowable Ampacity: 21.3 Amps

(Lower of two results)

Problem 1-28

Description	Keystroke	Display
Reset Calculator*	Set X	NEC 2014
Find Wire Ampacity	8 WireSz WireSz	40.0 WIRE A
Set to 90°C Wire Rating	Set 9	90
Find Wire Ampacity	8 WireSz WireSz	55.0 WIRE A
Store as Amps	= Amps	AMPS 55.
Enter Ambient Temperature	3 5 Set 2	AMB° 35. °C
Enter Quantity of Wires and Find Derated Ampacity	4 Set ParSz ParSz	D/R 42.2 WIRE A

*Sets calculator to 3Ø, 60°C, Copper @ 30°C Ambient Temperature.

Allowable Ampacity: 40 Amps

(Lower of two results)

Problem 1-29

Description	Keystroke	Display
Reset Calculator	Set X	NEC 2011
Set to 60°C Wire Rating	Set 6	60
Find Wire Ampacity	8 WireSz WireSz	40.0 WIRE A

Allowable Ampacity: 40 Amps

Problem 1-30

Description	Keystroke	Display
Clear Calculator	On/C On/C	0.
Set to 75°C Wire Rating (THWN rating)	Set 7	75
Find Wire Ampacity	1 0 WireSz WireSz	35.0 WIRE A

Allowable Ampacity: 35 Amps

Problem 1-31

Description	Keystroke	Display
Clear Calculator	On/C On/C	0.
Set to 75°C Wire Rating (THWN rating)	Set 7	75
Find Wire Ampacity	1 2 WireSz WireSz	25.0 WIRE A
Store as Amps	= Amps	AMPS 25.
Find Derated Ampacity	4 0 Set ParSz ParSz	D/R 10.0 WIRE A

Allowable Ampacity: 10 Amps

Problem 1-32

Description-1	Keystroke	Display
Clear Calculator	On/C On/C	0.
Set to 75°C Wire Rating (THWN rating)	Set 7	75
Find Wire Ampacity	6 WireSz WireSz	65.0 WIRE A
Store as Amps	= Amps	AMPS 65.
Find Derated Ampacity	1 8 Set ParSz ParSz	D/R 32.5 WIRE A

Allowable Ampacity (6 AWG): 32.5 Amps

Description-2	Keystroke	Display
Find Wire Ampacity	4 WireSz WireSz	85.0 WIRE A
Store as Amps	= Amps	AMPS 85.
Find Derated Ampacity	1 8 Set ParSz ParSz	D/R 42.5 WIRE A

Allowable Ampacity (4 AWG): 42.5 Amps

Chapter 2

NOTE: If your results differ from those provided in the examples, verify that your calculator is set to the appropriate phase, wire rating, wire type and ambient temperature.

Problem 2-1

Description	Keystroke	Display
Clear Calculator	On/C On/C	0.
Set to 60°C Wire Rating	Set 6	60
Find Wire Ampacity	6 WireSz WireSz	55.0 WIRE A

Allowable Ampacity: 55 Amps

Problem 2-2

Description	Keystroke	Display
Clear Calculator	On/C On/C	0.
Set to 60°C Wire Rating	Set 6	60
Toggle to Aluminum Wire	Set 4	Al
Find Wire Ampacity	2 WireSz WireSz	75.0 WIRE A

Allowable Ampacity: 75 Amps

Problem 2-3

Description	Keystroke	Display
Clear Calculator	On/C On/C	0.
Set to 75°C Wire Rating	Set 7	75
Toggle to Copper Wire	Set 4	Cu
Find Wire Ampacity	0 0 0 0 WireSz WireSz	230.0 WIRE A

Allowable Ampacity: 230 Amps

Problem 2-4

Description	Keystroke	Display
Clear Calculator	On/C On/C	0.
Set to 60°C Wire Rating	Set 6	60
Toggle to Copper Wire	Set 4	Cu
Find Wire Ampacity	6 WireSz WireSz	55.0 WIRE A
Set to 90°C Wire Rating	Set 9	90
Find Wire Ampacity	6 WireSz WireSz	75.0 WIRE A
Store as Amps	= Amps	AMPS 75.
Enter Quantity of Wires and Find Derated Ampacity	8 Set ParSz ParSz	D/R 52.5 WIRE A

Allowable Ampacity: 52.5 Amps

Problem 2-5

Description	Keystroke	Display
Clear Calculator	On/C On/C	0.
Multiply load by OCPD adjustment	2 1 X 1 2 5 %	26.25

Next Larger Size: 30 Amp Circuit Breaker

Problem 2-6

Description-1	Keystroke	Display
Clear Calculator	On/C On/C	0.
Multiply load by OCPD adjustment	4 7 X 1 2 5 %	58.75

Next Larger Size: 60 Amp Circuit Breaker

Description-2	Keystroke	Display
Clear Calculator	On/C On/C	0.
Multiply load by OCPD adjustment	4 7 X 1 0 0 %	47.

Next Larger Size: 50 Amp Circuit Breaker

Problem 2-7

Description-1	Keystroke	Display
Reset Calculator	Set X	0.
Set to 1-phase	Set 1	1 PH
Enter continuous load	6 kilo- Watts	KW 6.
Enter Voltage	2 7 7 Volts	VOLT 277.
Find Amperage	Amps	AMPS 21.66065
Multiply by 125% and store	X 1 2 5 % Stor 1	M-1 27.075812
Enter noncontinuous load	4 kilo- Watts	KW 4.
Recall Amps	Amps	AMPS 14.440433
Add to stored value	+ Rcl 1 = Amps	AMPS 41.516245

Next Larger Size: 45 Amp Circuit Breaker

Description-2	Keystroke	Display
Set to 75°C Wire Rating	Set 7	75
Find Wire Size	WireSz	AWG 8 CU WIRE SIZE

Wire Size: 8 AWG XHHW-2 Cu

Problem 2-8

Description-1	Keystroke	Display
Clear Calculator	On/C On/C	0.
Set to 1-phase	Set 1	1 PH
Enter Volts	1 2 0 Volts	VOLT 120.
Enter Load	7 • 5 kilo- Watts	KW 7.5
Find Amperage	Amps	AMPS 62.5
Adjust for continuous load	X 1 2 5 % Amps	AMPS 78.125

Next Larger Size: 80 Amp Circuit Breaker

Description-2	Keystroke	Display
Set to 75°C Wire Rating	Set 7	75
Find Wire Size	WireSz	AWG 4 CU WIRE SIZE

Wire Size: 4 AWG THWN Cu

Problem 2-9

Description-1	Keystroke	Display
Clear Calculator	On/C On/C	0.
Set to 1-phase	Set 1	1 PH
Enter Voltage	1 2 0 Volts	VOLT 120.
Enter continuous load	1 0 • 5 kilo- Watts	KW 10.5
Find Amperage	Amps	AMPS 87.5

Next Larger Size: 90 Amp Circuit Breaker

Description-2	Keystroke	Display
Set to 75°C Wire Rating	Set 7	75
Find Wire Size	WireSz	AWG 3 CU WIRE SIZE

Wire Size: 3 AWG THWN Cu

Problem 2-10

Description	Keystroke	Display
Clear Calculator	On/C On/C	0.
Multiply current load by 125%	1 7 5 X 1 2 5 %	218.75

Next Larger Size: 225 Amp Fuse or Circuit Breaker

Problem 2-11

Description-1	Keystroke	Display
Clear Calculator	On/C On/C	0.
Multiply continuous load by 125%	1 0 0 X 1 2 5 %	125.
Add noncontinuous load and store as amperage	+ 3 5 = Amps	AMPS 160.

Next Larger Size: 175 Amp Time-Delay Fuses

Description-2	Keystroke	Display
Set to 75°C Wire Rating	Set 7	75
Find Wire Size	WireSz	AWG 00 CU WIRE SIZE

Wire Size: 2/0 AWG THWN Cu

Description-3	Keystroke	Display
Clear Calculator	On/C On/C	0.
Add current loads and store as amperage	1 0 0 + 3 5 = Amps	AMPS 135.
Find Wire Size	WireSz	AWG 0 CU WIRE SIZE

Grounded Conductor: 1/0 AWG THWN Cu

Problem 2-12

Description-1	Keystroke	Display
Clear Calculator	On/C On/C	0.
Set to 3-phase	Set 3	3 PH
Enter Load	4 2 4 0 0 VA	VA 42,400.
Enter Voltage	2 0 8 Volts	VOLT 208.
Solve for current	Amps	AMPS 117.69063
Adjust for OCPD and store as Amps	X 1 2 5 % Amps	AMPS 147.11329

Next Larger Size: 150 Amp Breaker Or Fuse

Description-2	Keystroke	Display
Set to 75°C Wire Rating	Set 7	75
Find Wire Size	WireSz	AWG 0 CU WIRE SIZE

Ungrounded Conductor: 1/0 AWG THWN Cu

Description-3	Keystroke	Display
Clear Calculator	On/C On/C	0.
Enter Load	4 2 4 0 0 VA	VA 42,400.
Solve for current	Amps	AMPS 117.69063
Find Wire Size	WireSz	AWG 1 CU WIRE SIZE

Grounded Conductor: 1 AWG THWN Cu

Problem 2-13

Description-1	Keystroke	Display
Clear Calculator	On/C On/C	0.
Set to 3-phase	Set 3	3 PH
Enter Voltage	4 8 0 Volts	VOLT 480.
Enter continuous load	7 5 kilo- VA	KVA 75.
Solve for current	Amps	AMPS 90.21098
Adjust by 125% and store in memory	X 1 2 5 % Stor 1	M-1 112.76372
Enter noncontinuous load	6 0 kilo- VA	KVA 60.
Solve for current	Amps	AMPS 72.168784
Add to stored continuous load and store as Amps	+ Rcl 1 = Amps	AMPS 184.93251
Set to 75°C Wire Rating	Set 7	75
Find Wire Size	WireSz	AWG 000 CU WIRE SIZE

Ungrounded Conductor: 3/0 THWN-2 Cu

Description-2	Keystroke	Display

Service overcurrent protective device, minimum OCPD Size = 185 Amps. Section 240.4(B) and Section 240.6(A).

Next Larger Size: 200 Amp Breaker or Fuse

Chapter 3

NOTE: The ECPro does not compute or tabulate box fill/fittings, thus can only be used as a basic calculator in these instances.

Chapter 4

NOTE: If your results differ from those provided in the examples, verify that your calculator is set to the appropriate phase, wire rating, wire type and ambient temperature.

Problem 4-1

Description	Keystroke	Display
Clear Calculator	On/C On/C	0.
Set Conduit Type to RMC*	8 Set CondSz	RMC COND
Enter Wire Size	4 WireSz	AWG 4 CU WIRE SIZE
Enter Conduit Size	2 CondSz	RMC 2.00 in COND SIZE
Find Maximum Quantity of THHN wires	#THHN #THWN	THHN 16. TTL WIRES

*Conduit Type ID numbers in user's guide and on door of protective case.

16 Conductors

Problem 4-2

NOTE: The ECPro does not contain compact conductor data.

Problem 4-3

NOTE: The ECPro does not contain RHW conductor data.

Problem 4-4

NOTE: The ECPro does not contain SFF-2 conductor data.

Problem 4-5

Description	Keystroke	Display
Clear Calculator	On/C On/C	0.
Set Conduit Type to EMT	1 Set CondSz	EMT COND
Enter Wire Size	0 0 WireSz	AWG 00 CU WIRE SIZE
Enter Wire Quantity	3 #XHH #XHHW	XHHW 3. WIRES
Enter Wire Size	0 0 0 0 WireSz	AWG 0000 CU WIRE SIZE
Enter Wire Quantity	4 #XHH #XHHW	XHHW 4. WIRES
Find Conduit Size	CondSz	EMT 2.50 in COND SIZE

2.5 in. EMT

Problem 4-6

NOTE: The ECPro does not contain RHH conductor data.

Problem 4-7

Description	Keystroke	Display
Clear Calculator	On/C On/C	0.
Set Conduit Type to IMC	4 Set CondSz	IMC COND
Enter Wire Size	0 0 WireSz	AWG 00 CU WIRE SIZE
Enter Wire Quantity	3 #THHN #THWN	THHN 3. WIRES
Enter Wire Size	1 2 WireSz	AWG 12 CU WIRE SIZE
Enter Wire Quantity	3 #THHN #THWN	THHN 3. WIRES
Find Conduit Size	CondSz	IMC 1.50 in COND SIZE

1.5 in. IMC

Problem 4-8

Description	Keystroke	Display
Clear Calculator	On/C On/C	0.
Set Conduit Type to PVC-40	1 0 Set CondSz	P-4Ø COND
Enter Wire Size	0 0 0 WireSz	AWG 000 CU WIRE SIZE
Enter Wire Quantity	4 #THHN #THWN	THHN 4. WIRES
Enter Wire Size	6 WireSz	AWG 6 CU WIRE SIZE
Enter Wire Quantity	1 #THHN #THWN	THHN 1. WIRE
Find Conduit Size	CondSz	P-4Ø 2.00 in COND SIZE

2 in. PVC, Schedule 40

Problem 4-9

NOTE: The ECPro does not contain compact conductor data.

Problem 4-10

NOTE: The ECPro does not contain bare conductor data.

Problem 4-11

NOTE: The ECPro does not contain bare conductor data.

Problem 4-12

NOTE: The ECPro does not contain bare or compact conductor data.

Problem 4-13

NOTE: The ECPro does not contain bare conductor data.

Problem 4-14

NOTE: The ECPro does not compute fill areas for conduit nipple data.

Problem 4-15

NOTE: The ECPro does not compute fill areas for conduit nipple data.

Problem 4-16

NOTE: The ECPro does not allow for conductor area entries for use in conduit sizing.

Problem 4-17

NOTE: The ECPro does not allow for conductor area entries for use in conduit sizing.

Problem 4-18

Description	Keystroke	Display
Clear Calculator	On/C On/C	0.
Set Conduit Type to EMT	1 Set CondSz	EMT COND
Enter Wire Size	8 WireSz	AWG 8 CU WIRE SIZE
Enter Wire Quantity	4 #THHN #THWN	THHN 4. WIRES
Display Wire Area and Store in Memory	#THHN #THWN Stor 0	M+ 0.1464
Enter Wire Size	1 0 WireSz	AWG 10 CU WIRE SIZE
Display 1" Conduit Area	1 CondSz CondSz	EMT 0.864 COND AREA
Multiply by Fill Factor	× 4 0 %	0.3456
Subtract existing wire area	− Rcl Rcl	0.1464
Divide by New Wire Area	÷ Set #THHN #THWN =	9.4407583

9 Conductors

Problem 4-19

NOTE: The ECPro does not have TFN conductor data.

Problem 4-20

NOTE: The ECPro does not have FEP conductor data.

Problem 4-21

Description	Keystroke	Display
Clear Calculator	On/C On/C	0.
Enter Wire Size	1 2 WireSz	AWG 12 CU WIRE SIZE
Find Area	1 • 2 5 × 3 =	3.75
Multiply by Fill Factor	× 4 0 %	1.5
Divide by Wire Area	÷ Set #THW #THHW =	82.872928

82 Conductors

Problem 4-22

NOTE: The ECPro does not have Compact conductor data.

Problem 4-23

Description	Keystroke	Display
Clear Calculator	On/C On/C	0.
Enter Wire Size	1 0 WireSz	AWG 10 CU WIRE SIZE
Find Area	2 × 2 =	4.
Multiply by Fill Factor	× 4 0 %	1.6
Divide by Wire Area	÷ Set #THHN #THWN =	75.829384

75 Conductors

Problem 4-24

Description	Keystroke	Display
Clear Calculator	On/C On/C	0.
Enter Wire Size	1 0 WireSz	AWG 10 CU WIRE SIZE
Multiply Area by Fill Factor	2 • 0 5 7 × 4 0 %	0.8228
Divide by Wire Area	÷ Set #THHN #THWN	THHN 0.0211 WIRE AREA
Find Result	=	38.995261

38 Conductors

Problem 4-25

Description	Keystroke	Display
Clear Calculator	On/C On/C	0.
Enter Wire Size	8 WireSz	AWG 8 CU WIRE SIZE
Find Fill Area	1 6 #THW #THHW #THW #THHW	THW 0.6992 WIRE AREA
Store in Memory	Stor 0	M+ 0.6992
Enter Second Wire Size	8 WireSz	AWG 8 CU WIRE SIZE
Find Raceway Fill Area	3 × 4 0 %	1.2
Subtract Stored Wire Area	− Rcl Rcl =	0.5008
Divide by Wire Area	÷ Set #THHN #THWN	THHN 0.0366 WIRE AREA
Find Result	=	13.68306

13 Conductors

Problem 4-26

Description	Keystroke	Display
Clear Calculator	On/C On/C	0.
Enter Wire Size	1 WireSz	AWG 1 CU WIRE SIZE
Find Fill Area	3 × 4 × 2 0 %	2.4
Divide by Wire Area	÷ Set #THHN #THWN =	15.364917

15 Conductors

Problem 4-27

Not applicable

Problem 4-28

Description-1	Keystroke	Display
Clear Calculator	On/C On/C	0.
Find allowable fill area	5 X 5 X 2 0 %	5.
Divide by Wire Area	÷ 3 5 0 WireSz Set #THW #THHW	THW 0.5958 WIRE AREA
Find Result	=	8.3920779

8 Conductors

Description-2	Keystroke	Display

Basic calculation for 350 kcmil THWN compact stranded aluminum provided below, as the ECPro does not have compact conductor data.

Description-2	Keystroke	Display
Clear Calculator	On/C On/C	0.
Enter Fill Area from above	5	5.
Divide by Wire Area (Table 5A)	÷ • 5 2 8 1	0.5281
Find Result	=	9.4679038

9 Conductors

Problem 4-29

Description	Keystroke	Display
Clear Calculator	On/C On/C	0.
Enter Wire Size	0 WireSz	AWG 0 CU WIRE SIZE
Find Fill Area	6 #THW #THHW #THW #THHW	THW 1.3338 WIRE AREA
Store in Memory	Stor 0	M+ 1.3338
Enter Second Wire Size	4 WireSz	AWG 4 CU WIRE SIZE
Find Raceway Fill Area	3 X 3 X 2 0 %	1.8
Subtract Stored Wire Area	− Rcl Rcl =	0.4662
Divide by Wire Area	÷ Set #THW #THHW =	4.7913669

4 Conductors

Problem 4-30

Description	Keystroke	Display
Clear Calculator	On/C On/C	0.
Enter Wire Size	1 0 WireSz	AWG 10 CU WIRE SIZE
Find Raceway Fill Area	4 X 4 X 2 0 %	3.2
Divide by Wire Area	÷ Set #THHN #THWN	THHN 0.0211 WIRE AREA
Find Result	=	151.65877

151 Conductors

Problem 4-31

Description	Keystroke	Display
Clear Calculator	On/C On/C	0.
Enter Wire Size	0 0 0 0 WireSz	AWG 0000 CU WIRE SIZE
Find Max. Fill Area	2 • 5 X 3 X 7 5 %	5.625
Enter total quantity of wires (3 for splice + 2 for remaining conductors	5 #THHN #THWN	THHN 5. WIRES
Find Wire Area	#THHN #THWN	THHN 1.6185 WIRE AREA

Yes, Wire Area Less Than Max Fill

Chapter 5

NOTE: If your results differ from those provided in the examples, verify that your calculator is set to the appropriate phase, wire rating, wire type and ambient temperature.

Problem 5-1

Description	Keystroke	Display
Clear Calculator	On/C On/C	0.
Set to 3-phase (if necessary)	Set 3	3 PH
Enter Voltage	4 6 0 Volts	VOLT 460.
Enter Motor HP*	2 5 HPmotor	IND 25. HP
Find Motor FLC	Amps	FLC 34. A
Multiply by Motor Adjustment Factor	X 1 2 5 %	42.5

*If IND is not shown, select Set 8 until it is displayed in the left corner of the LCD.

Ampacity: 42.5 Amps

Problem 5-2

Description	Keystroke	Display
Clear Calculator	On/C On/C	0.
Set to 1-phase (if necessary)	Set 1	1 PH
Set to 90°C Wire Rating (THHN Rating)	Set 9	90
Enter Voltage	2 4 0 Volts	VOLT 240.
Enter Motor HP*	1 0 HPmotor	IND 10. HP
Find Motor FLC	Amps	FLC 50. A
Find 125% Wire Size	Set WireSz	AWG 6 CU WIRE SIZE 125%

*IND or SYNC can be used as a setting for single-phase motors.

Conductor Size: 6 AWG THHN Cu

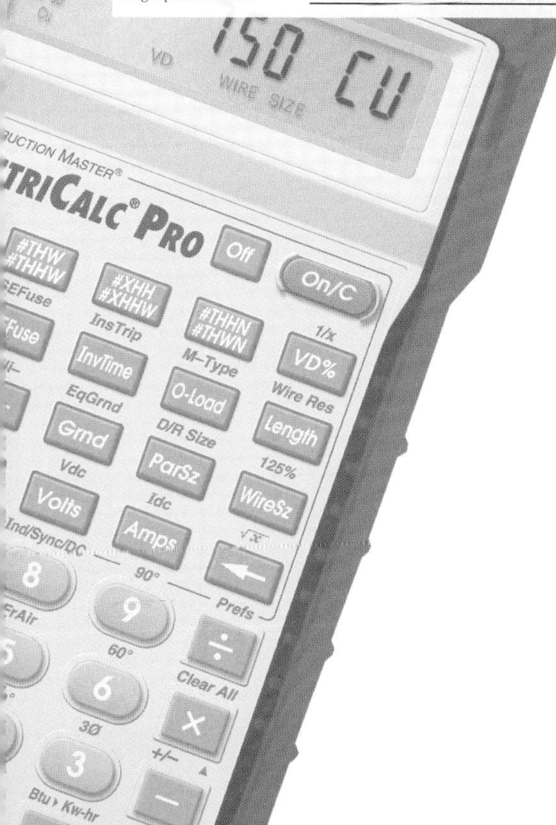

Problem 5-3

Description-1	Keystroke	Display
Clear Calculator	On/C On/C	0.
Multiply 1st nameplate FLC by 125%	6 5 X 1 2 5 %	81.25

Ampacity: 81.25 Amps

Description-2	Keystroke	Display
Multiply 2nd nameplate FLC by 125%	5 2 X 1 2 5 %	65.

Ampacity: 65 Amps

Description-3	Keystroke	Display
Multiply 3rd nameplate FLC by 125%	2 1 X 1 2 5 %	26.25

Ampacity: 26.25 Amps

Description-4	Keystroke	Display
Multiply 4th nameplate FLC by 125%	1 5 X 1 2 5 %	18.75

Ampacity: 18.75 Amps

Problem 5-4

Description	Keystroke	Display
Clear Calculator	On/C On/C	0.
Multiply highest nameplate FLC by 125%	6 5 X 1 2 5 %	81.25

Ampacity: 81.25 Amps

Problem 5-5

Description	Keystroke	Display
Clear Calculator	On/C On/C	0.
Multiply nameplate FLC by 90%*	2 4 • 5 X 9 0 %	22.05

*Factor obtained from NEC Table 430.22(E)

Ampacity: 22.05 Amps

Problem 5-6

Description	Keystroke	Display
Clear Calculator	On/C On/C	0.
Multiply nameplate FLC by 90%*	1 6 • 5 X 2 0 0 %	33.

*Factor obtained from NEC Table 430.22(E)

Ampacity: 33 Amps

Problem 5-7

Description	Keystroke	Display
Set to 75° Wire Rating (THWN rating)	Set 7	75
Enter Ampacity	3 3 Amps	AMPS 33.
Set to Copper (if necessary)	Set 4	Cu
Find Wire Size	WireSz	AWG 10 CU WIRE SIZE

Conductor Size: 10 AWG THWN Cu

Problem 5-8

Description	Keystroke	Display
Clear Calculator	On/C On/C	0.
Multiply LRC by 125%	1 0 5 X 1 2 5 %	131.25

Ampacity: 131.25 Amps

Problem 5-9

Description	Keystroke	Display
Set to 75° Wire Rating (XHHW Rating)	Set 7	75
Toggle to Copper Wire (if necessary)	Set 4	Cu
Enter Ampacity	1 3 1 • 2 5 Amps	AMPS 131.25
Find Wire Size	WireSz	AWG 0 CU WIRE SIZE

Conductor Size: 1/0 AWG XHHW Cu

Problem 5-10

Description-1	Keystroke	Display
Clear Calculator	On/C On/C	0.
Set to 3-phase (if necessary)	Set 3	3 PH
Enter Voltage	4 4 0 Volts	VOLT 440.
Enter Motor HP*	2 0 HPmotor	IND 20. HP
Find Motor FLC	Amps	FLC 27. A
Multiply by Motor Adjustment Factor	X 1 2 5 %	33.75

*If IND is not shown, select Set 8 until it is displayed in the left corner of the LCD.

Ampacity: 33.75 Amps

Description-2	Keystroke	Display
Display Motor FLC	On/C Amps	FLC 27. A
Multiply by 430.22(C) Allowable Adjustment	X 7 2 %	19.44

Ampacity: 19.44 Amps

Problem 5-11

Description-1	Keystroke	Display
Clear Calculator	On/C On/C	0.
Set to 75°C Wire Rating (THWN Rating)	Set 7	75
Enter Input Rating as Amps	3 0 Amps	AMPS 30.
Find 125% Wire Size	Set WireSz	AWG 8 CU WIRE SIZE 125%

Conductor Size: 8 AWG THWN Cu

Description-2	Keystroke	Display
Clear Calculator	On/C On/C	0.
Set to 3-phase (if necessary)	Set 3	3 PH
Enter Voltage	4 6 0 Volts	VOLT 460.
Enter Motor HP*	2 0 HPmotor	IND 20. HP
Find Motor FLC	Amps	FLC 27. A
Find 125% Wire Size	Set WireSz	AWG 10 CU WIRE SIZE 125%

*If IND is not shown, select Set 8 until it is displayed in the left corner of the LCD.

Conductor Size: 10 AWG THWN Cu

Problem 5-12

Description-1	Keystroke	Display
Clear Calculator	On/C On/C	0.
Set to 3-phase (if necessary)	Set 3	3 PH
Enter Voltage	4 6 0 Volts	VOLT 460.
Enter Motor HP*	5 0 HPmotor	IND 50. HP
Find Motor FLC	Amps	FLC 65. A
Find Branch-Circuit Ampacity	X 1 2 5 %	81.25

*If IND is not shown, select Set 8 until it is displayed in the left corner of the LCD.

Ampacity: 81.25 Amps

Description-2	Keystroke	Display
Multiply Secondary current by 125%	1 0 2 X 1 2 5 %	127.5

Ampacity: 127.5 Amps

Problem 5-13

Description-1	Keystroke	Display
Clear Calculator	On/C On/C	0.
Set to 3-phase (if necessary)	Set 3	3 PH
Enter Voltage	4 6 0 Volts	VOLT 460.
Enter Motor HP*	7 5 HPmotor	IND 75. HP
Find Motor FLC	Amps	FLC 96. A
Find Branch-Circuit Ampacity	X 1 2 5 %	120.

*If IND is not shown, select Set 8 until it is displayed in the left corner of the LCD.

Ampacity: 120 Amps

Description-2	Keystroke	Display
Multiply Secondary current by 85%*	1 5 0 X 8 5 %	127.5

*Percentage obtained from Table 430.22(E)

Ampacity: 127.5 Amps

Problem 5-14

Description-1	Keystroke	Display
Clear Calculator	On/C On/C	0.
Set to 3-phase (if necessary)	Set 3	3 PH
Enter Voltage	2 0 8 Volts	VOLT 208.
Enter Motor HP*	2 5 HPmotor	IND 25. HP
Find Motor FLC	Amps	FLC 74.8 A
Find Branch-Circuit Ampacity	X 1 2 5 %	93.5

*If IND is not shown, select Set 8 until it is displayed in the left corner of the LCD.

Ampacity: 93.5 Amps

Description-2	Keystroke	Display
Multiply Secondary current by 85%*	1 4 2 X 8 5 %	120.7

*Percentage obtained from Table 430.23(C)

Ampacity: 120.7 Amps

Problem 5-15

Description	Keystroke	Display
Clear Calculator	On/C On/C	0.
Set to 3-phase (if necessary)	Set 3	3 PH
Enter Voltage	2 0 8 Volts	VOLT 208.
Enter Motor HP*	5 HPmotor	IND 5. HP
Find Motor FLC	Amps	FLC 16.7 A
Store Value In Memory	Stor 1	M-1 16.7
Find Branch-Circuit Ampacity	X 1 2 5 %	20.875
Add Second Motor FLC	+ Rcl 1 =	37.575

*If IND is not shown, select Set 8 until it is displayed in the left corner of the LCD.

Ampacity: 37.575 Amps

Problem 5-16

Description	Keystroke	Display
Clear Calculator	On/C On/C	0.
Set to 1-phase (if necessary)	Set 1	1 PH
Enter Voltage	2 3 0 Volts	VOLT 230.
Enter Largest Motor HP*	5 HPmotor	IND 5. HP
Find Motor FLC	Amps	FLC 28. A
Find Branch-Circuit Ampacity	X 1 2 5 %	35.
Store value in memory	Stor 0	M+ 35.
Enter Next Motor HP*	2 HPmotor	IND 2. HP
Find Motor FLC	Amps	FLC 12. A
Add to M+ Value and Display Result	+ Rcl Rcl =	47.

*IND or SYNC can be used as a setting for single-phase motors.

Ampacity: 47 Amps

Problem 5-17

Description	Keystroke	Display
Clear Calculator	On/C On/C	0.
Set to 3-phase (if necessary)	Set 3	3 PH
Enter Voltage	4 6 0 Volts	VOLT 460.
Enter First Induction Motor HP*	1 0 HPmotor	IND 10. HP
Find Motor FLC	Amps	FLC 14. A
Store in Memory Register 1	Stor 1	M-1 14.
Enter Second Induction Motor HP*	2 5 HPmotor	IND 25. HP
Find Motor FLC	Amps	FLC 34. A
Store in Memory Register 2	Stor 2	M-2 34.
Toggle to Sync Motors	Set 8 (repeat until SYNC displayed)	SYNC
Enter Synchronous Motor HP	3 0 HPmotor	SYNC 30. HP
Find Motor FLC	Amps	FLC 32. A
Store in Memory Register 3	Stor 3	M-3 32.
Recall largest FLC and adjust ampacity	Rcl 2 X 1 2 5 %	42.5
Add remaining FLCs and find total	+ Rcl 1 + Rcl 3 =	88.5

*If IND is not shown, select Set 8 until it is displayed in the left corner of the LCD.

Ampacity: 88.5 Amps

Problem 5-18

Description	Keystroke	Display
Clear Calculator	On/C On/C	0.
Set to 1-phase (if necessary)	Set 1	1 PH
Enter Voltage	1 2 0 Volts	VOLT 120.
Enter Largest Motor HP*	1 HPmotor	IND 1. HP
Find Motor FLC	Amps	FLC 16. A
Find Branch-Circuit Ampacity	X 1 2 5 %	20.
Store value in memory	Stor 0	M+ 20.
Enter Lighting Watts	6 0 0 Watts	WATT 600.
Find Amperage	Amps	AMPS 5.
Add to M+ Value and Display Result	+ Rcl Rcl =	25.

*IND or SYNC can be used as a setting for single-phase motors.

Ampacity: 25 Amps

Problem 5-19

Description	Keystroke	Display
Clear Calculator	On/C On/C	0.
Set to 1-phase (if necessary)	Set 1	1 PH
Enter Voltage	2 4 0 Volts	VOLT 240.
Enter Largest Motor HP*	3 HPmotor	IND 3. HP
Find Motor FLC	Amps	FLC 17. A
Store value in memory	Stor 0	M+ 17.
Enter Heater Watts	1 2 0 0 Watts	WATT 1,200.
Find Amperage	Amps	AMPS 5.
Add to M+ Value and Display Result	+ Rcl Rcl =	22.
Multiply by Adjustment factors	X 1 2 5 %	27.5

*IND or SYNC can be used as a setting for single-phase motors.

Ampacity: 27.5 Amps

Problem 5-20

Description	Keystroke	Display
Clear Calculator	On/C On/C	0.
Set to 1-phase (if necessary)	Set 1	1 PH
Enter Voltage	2 4 0 Volts	VOLT 240.
Enter First Motor HP*	1 0 HPmotor	IND 10. HP
Find Motor FLC	Amps	FLC 50. A
Multiply by Adjustment factor	X 1 2 5 %	62.5
Store value in memory	Stor 0	M+ 62.5
Recall FLC	Rcl Amps	FLC 50. A
Store value in memory	Stor 0	M+ 50.
Determine Periodic Motor FLC	2 7 X 9 0 %	24.3
Add to M+ Value and Display Result	+ Rcl Rcl =	136.8

*IND or SYNC can be used as a setting for single-phase motors.

Ampacity: 136.8 Amps

Problem 5-21

Description	Keystroke	Display
Clear Calculator	On/C On/C	0.
Set to 3-phase (if necessary)	Set 3	3 PH
Enter Voltage	4 6 0 Volts	VOLT 460.
Enter Largest Motor HP*	4 0 HPmotor	IND 40. HP
Find Motor FLC	Amps	FLC 52. A
Find branch circuit ampacity	X 1 2 5 %	65.

*If IND is not shown, select Set 8 until it is displayed in the left corner of the LCD.

Ampacity: 65 Amps

Example

Description-1	Keystroke	Display
Clear Calculator	On/C On/C	0.
Set to 3-phase (if necessary)	Set 3	3 PH
Enter Voltage	4 6 0 Volts	VOLT 460.
Enter Largest Motor HP*	2 0 HPmotor	IND 20. HP
Find Motor FLC	Amps	FLC 27. A
Find Time-Delay Fuse Size	DEFuse	AMPS 47.25 dE

*If IND is not shown, select Set 8 until it is displayed in the left corner of the LCD.

Next Larger Rating: 50 Amps

Description-2	Keystroke	Display
Find Non Time-Delay Fuse Size	Set DEFuse	AMPS 81. SE

Next Larger Rating: 90 Amps

Description-3	Keystroke	Display
Find Inverse-Time Breaker Size	InvTime	AMPS 67.5 b2

Next Larger Rating: 70 Amps

Problem 5-22

Description	Keystroke	Display
Clear Calculator	On/C On/C	0.
Set to 3-phase (if necessary)	Set 3	3 PH
Enter Voltage	2 4 0 Volts	VOLT 240.
Enter Motor HP*	2 5 HPmotor	IND 25. HP
Find Motor FLC	Amps	FLC 68. A
Find Time-Delay Fuse Size	DEFuse	AMPS 119. dE

*If IND is not shown, select Set 8 until it is displayed in the left corner of the LCD.

Next Larger Rating: 125 Amps

Problem 5-23

Description	Keystroke	Display
Clear Calculator	On/C On/C	0.
Set to 3-phase (if necessary)	Set 3	3 PH
Enter Voltage	2 4 0 Volts	VOLT 240.
Enter Motor HP*	2 5 HPmotor	IND 25. HP
Find Motor FLC	Amps	FLC 68. A
Find Non Time-Delay Fuse Size	Set DEFuse	AMPS 204. SE

*If IND is not shown, select Set 8 until it is displayed in the left corner of the LCD.

Next Larger Rating: 225 Amps

Problem 5-24

Description	Keystroke	Display
Clear Calculator	On/C On/C	0.
Set to 3-phase (if necessary)	Set 3	3 PH
Enter Voltage	4 6 0 Volts	VOLT 460.
Enter Motor HP*	5 0 HPmotor	IND 50. HP
Find Motor FLC	Amps	FLC 65. A
Multiply by Adjustment Factor (430.52(C)(1), Note 2b)	X 2 2 5 %	146.25

*If IND is not shown, select Set 8 until it is displayed in the left corner of the LCD.

Next Smaller Rating: 125 Amps

Problem 5-25

Description-1	Keystroke	Display
Clear Calculator	On/C On/C	0.
Set to 3-phase (if necessary)	Set 3	3 PH
Enter Voltage	4 6 0 Volts	VOLT 460.
Enter Motor HP*	1 5 HPmotor	IND 15. HP
Find Motor FLC	Amps	FLC 21. A
Find Fuse Size	Set DEFuse	AMPS 63. SE

*If IND is not shown, select Set 8 until it is displayed in the left corner of the LCD.

Next Larger Rating: 70 Amps

Description-2	Keystroke	Display
Show FLC	Rcl Amps	FLC 21. A
Multiply by Adjustment Factor (430.52(C)(1), Note 2a)	X 4 0 0 %	84.

Next Smaller Rating: 80 Amps

Problem 5-26

Description-1	Keystroke	Display
Clear Calculator	On/C On/C	0.
Set to 3-phase (if necessary)	Set 3	3 PH
Enter Voltage	4 4 0 Volts	VOLT 440.
Enter Motor HP*	2 5 HPmotor	IND 25. HP
Find Motor FLC	Amps	FLC 34. A
Find Fuse Size	DEFuse	AMPS 59.5 dE

*If IND is not shown, select Set 8 until it is displayed in the left corner of the LCD.

Next Larger Rating: 60 Amps

Description-2	Keystroke	Display
Show FLC	Rcl Amps	FLC 34. A
Multiply by Adjustment Factor (430.52(C)(1), Note 2a)	X 2 2 5 %	76.5

Next Smaller Rating: 70 Amps

Problem 5-27

Description-1	Keystroke	Display
Clear Calculator	On/C On/C	0.
Set to 3-phase (if necessary)	Set 3	3 PH
Enter Voltage	2 4 0 Volts	VOLT 240.
Enter Motor HP*	3 0 HPmotor	IND 30. HP
Find Motor FLC	Amps	FLC 80. A
Find Breaker Size	InvTime	AMPS 200. b2

*If IND is not shown, select Set 8 until it is displayed in the left corner of the LCD.

Rating: 200 Amps

Description-2	Keystroke	Display
Show FLC	Rcl Amps	FLC 80. A
Multiply by Adjustment Factor (430.52(C)(1), Note 2c)	X 4 0 0 %	320.

Next Smaller Rating: 300 Amps

Problem 5-28

Description-1	Keystroke	Display
Clear Calculator	On/C On/C	0.
Set to 1-phase (if necessary)	Set 1	1 PH
Enter Voltage	2 4 0 Volts	VOLT 240.
Enter Motor HP*	5 HPmotor	IND 5. HP
Find Motor FLC	Amps	FLC 28. A
Find Fuse Size	DEFuse	AMPS 49. dE

*If IND is not shown, select Set 8 until it is displayed in the left corner of the LCD.

Next Larger Rating: 50 Amps

Description-2	Keystroke	Display
Show FLC	Rcl Amps	FLC 28. A
Multiply by Adjustment Factor (430.52(C)(1), Note 2b)	X 2 2 5 %	63.

Next Smaller Rating: 60 Amps

Problem 5-29

Description	Keystroke	Display
Clear Calculator	On/C On/C	0.
Set to 3-phase (if necessary)	Set 3	3 PH
Enter Voltage	2 4 0 Volts	VOLT 240.
Enter Motor HP*	6 0 HPmotor	IND 60. HP
Find Motor FLC	Amps	FLC 154. A
Multiply by Adjustment Factor (430.52)**	X 1 5 0 %	231.

*If IND is not shown, select Set 8 until it is displayed in the left corner of the LCD.

Next Larger Rating: 250 Amps

**The ECPro does not utilize wound-rotor setting for determining protection percentages.

Problem 5-30

Description-1	Keystroke	Display
Clear Calculator	On/C On/C	0.
Set to 3-phase (if necessary)	Set 3	3 PH
Enter Voltage	2 0 8 Volts	VOLT 208.
Enter Motor HP*	2 5 HPmotor	IND 25. HP
Find Motor FLC	Amps	FLC 74.8 A
Find Breaker Size	Set InvTime	AMPS 598.4 b1

*If IND is not shown, select Set 8 until it is displayed in the left corner of the LCD.

Rating: 598.4 Amps

Description-2	Keystroke	Display
Show FLC	Rcl Amps	FLC 74.8 A
Multiply by Adjustment Factor (430.52(C)(3), Note 1)	X 1 3 0 0 %	972.4

Rating: 972.4 Amps

Problem 5-31

Description-1	Keystroke	Display
Clear Calculator	On/C On/C	0.
Set to 3-phase (if necessary)	Set 3	3 PH
Enter Voltage	4 6 0 Volts	VOLT 460.
Enter Motor HP*	2 5 HPmotor	IND 25. HP
Find Motor FLC	Amps	FLC 34. A
Multiply by Adjustment Factor (430.52)**	X 1 1 0 0 %	374.

*If IND is not shown, select Set 8 until it is displayed in the left corner of the LCD.

Rating: 374 Amps

**The ECPro does not utilize data for Energy-Efficient motors.

Description-2	Keystroke	Display
Show FLC	Rcl Amps	FLC 34. A
Multiply by Adjustment Factor (430.52(C)(3), Note 1)	X 1 7 0 0 %	578.

Rating: 578 Amps

Problem 5-32

Description	Keystroke	Display
Clear Calculator	On/C On/C	0.
Set to 3-phase (if necessary)	Set 3	3 PH
Enter Voltage	4 6 0 Volts	VOLT 460.
Enter Motor HP*	5 0 HPmotor	SYNC 50. HP
Find Motor FLC	Amps	FLC 52. A
Find Fuse Size	Set DEFuse	AMPS 156. SE

*If SYNC is not shown, select Set 8 until it is displayed in the left corner of the LCD.

Next Larger Rating: 175 Amps

Problem 5-33

Description	Keystroke	Display
Clear Calculator	On/C On/C	0.
Set to 3-phase (if necessary)	Set 3	3 PH
Enter Voltage	4 6 0 Volts	VOLT 460.
Enter Motor HP*	6 0 HPmotor	SYNC 60. HP
Find Motor FLC	Amps	FLC 61. A
Adjust for Power Factor	X 1 2 5 % Amps	AMPS 76.25
Find Fuse Size	DEFuse	AMPS 133.44 dE

*If SYNC is not shown, select Set 8 until it is displayed in the left corner of the LCD.

Next Larger Rating: 150 Amps

Problem 5-34

Description	Keystroke	Display
Clear Calculator	On/C On/C	0.
Compute Ampacity and Store as Amps	9 1 X 1 2 5 % Amps	AMPS 113.75
Set to 75°C Wire Rating (THW)	Set 7	75
Find Wire Size	WireSz	AWG 2 CU WIRE SIZE
Find Wire Ampacity	WireSz	115.0 WIRE A

Next Larger Rating: 125 Amps

Problem 5-35

Description	Keystroke	Display
Clear Calculator	On/C On/C	0.
Multiply Ampacity by adjustment*	7 5 X 1 2 5 %	93.75

*Reference 430.32(A)(1)

Rating: 93.75 Amps

Problem 5-36

Description	Keystroke	Display
Clear Calculator	On/C On/C	0.
Multiply Ampacity by adjustment*	7 5 × 1 4 0 %	105.

*Reference 430.32(C)

Rating: 105 Amps

Problem 5-37

Description-1	Keystroke	Display
Clear Calculator	On/C On/C	0.
Multiply Ampacity by adjustment*	1 9 × 1 2 5 %	23.75

*Reference 430.32(A)(1)

Rating: 23.75 Amps

Description-2	Keystroke	Display
Clear Calculator	On/C On/C	0.
Multiply Ampacity by adjustment*	1 9 × 1 4 0 %	26.6

*Reference 430.32(C)

Rating: 26.6 Amps

Problem 5-38

Description-1	Keystroke	Display
Clear Calculator	On/C On/C	0.
Multiply Ampacity by adjustment*	3 1 • 5 × 1 1 5 %	36.225

*Reference 430.32(A)(1)

Rating: 36.225 Amps

Description-2	Keystroke	Display
Clear Calculator	On/C On/C	0.
Multiply Ampacity by adjustment*	3 1 • 5 × 1 3 0 %	40.95

*Reference 430.32(C)

Rating: 40.95 Amps

Problem 5-39

Description	Keystroke	Display
Clear Calculator	On/C On/C	0.
Set to 3-phase (if necessary)	Set 3	3 PH
Enter Voltage	4 4 0 Volts	VOLT 440.
Enter Motor HP*	5 HPmotor	IND 5. HP
Find FLC	Amps	FLC 7.6 A
Multiply FLC by adjustment**	× 1 7 0 %	12.92

*If IND is not shown, select Set 8 until it is displayed in the left corner of the LCD.
**Reference 430.32(A)(2)

Rating: 12.92 Amps

Problem 5-40

Description	Keystroke	Display
Clear Calculator	On/C On/C	0.
Multiply Ampacity by adjustment*	1 5 × 1 4 0 %	21.

*Reference 430.32(C)

Next Smaller Rating: 20 Amps

Problem 5-41

Description	Keystroke	Display
Clear Calculator	On/C On/C	0.
Multiply Ampacity by adjustment*	2 5 × 1 3 0 %	32.5

*Reference 430.32(C)

Next Smaller Rating: 30 Amps

Problem 5-42

Description	Keystroke	Display
Clear Calculator	On/C On/C	0.
Multiply Ampacity by adjustment*	2 9 × 1 2 5 %	36.25

*Reference 430.32(A)(1)

Rating: 36.25 Amps

Problem 5-43

Description-1	Keystroke	Display
Clear Calculator	On/C On/C	0.
Enter HP	2 5	25.
Multiply by 1000	× 1 0 0 0	1,000.
Divide by Voltage	÷ 4 6 0	460.
Divide by 1.73 and store result for later recall	÷ 1 • 7 3 = Stor 0	M+ 31.414928
Multiply by minimum kVA/hp factor*	× 8	8.
Find Result	=	251.31943

*Reference Table 430.7(B)

LRC lowest: 251.32 Amps

Description-2	Keystroke	Display
Clear Calculator	On/C On/C	0.
Determine average kVA/hp factor*	8 + 8 • 9 9 ÷ 2 =	8.495
Multiply by stored value and display result	× Rcl 0 =	266.86982

*Reference Table 430.7(B)

LRC average: 266.87 Amps

Description-3	Keystroke	Display
Clear Calculator	On/C On/C	0.
Enter max kVA/hp factor*	8 • 9 9	8.99
Multiply by stored value and display result	× Rcl Rcl =	282.42021

*Reference Table 430.7(B)

LRC max: 282.42 Amps

Problem 5-44

Description-1	Keystroke	Display
The ECPro does not compute LRC. Using table 430.251(B),		LRC: 183 Amps

Description-2	Keystroke	Display
Clear Calculator	On/C On/C	0.
Set to 3-phase (if necessary)	Set 3	3 PH
Enter Voltage	4 6 0 Volts	VOLT 460.
Enter Motor HP*	2 5 HPmotor	IND 25. HP
Find Motor FLC	Amps	FLC 34. A
Multiply by six and display result	× 6 =	204.

*If IND is not shown, select Set 8 until it is displayed in the left corner of the LCD.

LRC: 204 Amps

Description-3	Keystroke	Display
Clear Calculator	On/C On/C	0.
Enter HP	2 5	25.
Multiply by 1000	× 1 0 0 0	1,000.
Divide by Voltage	÷ 4 6 0	460.
Divide by 1.73	÷ 1 • 7 3	1.73
Multiply by max kVA/hp factor* and display result	× 5 • 5 9 =	175.60945

*Reference Table 430.7(B)

LRC max: 175.61 Amps

Problem 5-45

Description	Keystroke	Display
Clear Calculator	On/C On/C	0.
Enter HP	1 0	10.
Multiply by 1000	× 1 0 0 0	1,000.
Divide by Voltage	÷ 2 4 0	240.
Divide by 1.73	÷ 1 • 7 3	1.73
Multiply by max kVA/hp factor and display result	× 3 • 1 4 =	75.626204

NOTE: Reference Tables 430.7(B) and 430.251(B)

HP Switch Size: 5 HP

Problem 5-46

Description	Keystroke	Display
Clear Calculator	On/C On/C	0.
Enter HP	1 0	10.
Multiply by 1000	× 1 0 0 0	1,000.
Divide by Voltage	÷ 2 4 0	240.
Divide by 1.73	÷ 1 • 7 3	1.73
Multiply by max kVA/hp factor and display result	× 5 • 5 9 =	134.63391

NOTE: Reference Tables 430.7(B) and 430.251(B)

HP Switch Size: 10 HP

Problem 5-47

Description	Keystroke	Display
Clear Calculator	On/C On/C	0.
Enter HP	1 0	10.
Multiply by 1000	× 1 0 0 0	1,000.
Divide by Voltage	÷ 2 4 0	240.
Divide by 1.73	÷ 1 • 7 3	1.73
Multiply by max kVA/hp factor and display result	× 1 7 • 9 9 =	433.28516

NOTE: Reference Tables 430.7(B) and 430.251(B)

HP Switch Size: 30 HP

Problem 5-48

Description	Keystroke	Display
Clear Calculator	On/C On/C	0.
Add both LRC values from Table 430.251(B)	2 3 2 + 2 9 0 =	522.

HP Switch Size: 40 HP

Problem 5-49

Description	Keystroke	Display
Clear Calculator	On/C On/C	0.
Set to 1-phase	Set 1	1 PH
Enter Wattage	1 6 0 0 0 Watts	WATT 16,000.
Enter Voltage	2 4 0 Volts	VOLT 240.
Find Amperage	Amps	AMPS 66.666667
Add LRC rating and display result*	+ 1 6 8 =	234.66667

*Reference Table 430.251(A)

HP Switch Size: 7.5 HP

Problem 5-50

Description	Keystroke	Display
Clear Calculator	On/C On/C	0.
Set to 3-phase (if necessary)	Set 3	3 PH
Enter Voltage	4 6 0 Volts	VOLT 460.
Enter Motor HP*	2 0 HPmotor	IND 20. HP
Find FLC	Amps	FLC 27. A
Multiply FLC by adjustment**	× 1 1 5 %	31.05

*If IND is not shown, select Set 8 until it is displayed in the left corner of the LCD.
**Reference 430.110(A)

Next Larger Rating: 35 Amps

Problem 5-51

Description	Keystroke	Display
Clear Calculator	On/C On/C	0.
Set to 1-phase (if necessary)	Set 1	1 PH
Enter Voltage	2 4 0 Volts	VOLT 240.
Enter Motor HP*	5 HPmotor	IND 5. HP
Find FLC	Amps	FLC 28. A
Multiply FLC by adjustment**	X 1 1 5 %	32.2

*If IND is not shown, select Set 8 until it is displayed in the left corner of the LCD.
**Reference 430.110(A)

Next Larger Rating: 35 Amps

Problem 5-52

Description	Keystroke	Display
Clear Calculator	On/C On/C	0.
Set to 3-phase (if necessary)	Set 3	3 PH
Enter Voltage	4 6 0 Volts	VOLT 460.
Enter 1st Motor HP*	1 0 HPmotor	IND 10. HP
Find FLC	Amps	FLC 14. A
Store in Memory	Stor 0	M+ 14.
Enter 2nd Motor HP	1 5 HPmotor	IND 15. HP
Find FLC	Amps	FLC 21. A
Add Stored Memory value	+ Rcl Rcl =	35.
Multiply FLC by adjustment**	X 1 1 5 %	40.25

*If IND is not shown, select Set 8 until it is displayed in the left corner of the LCD.
**Reference 430.110(C)(2)

Next Larger Rating: 45 Amps

Problem 5-53

Description	Keystroke	Display
Clear Calculator	On/C On/C	0.
Set to 1-phase (if necessary)	Set 1	1 PH
Enter Voltage	2 4 0 Volts	VOLT 240.
Enter 1st Motor HP*	5 HPmotor	IND 5. HP
Find FLC	Amps	FLC 28. A
Store in Memory	Stor 0	M+ 28.
Enter Wattage	5 0 0 0 Watts	WATT 5,000.
Find Amperage	Amps	AMPS 20.833333
Add Stored Memory value	+ Rcl Rcl =	48.833333
Multiply FLC by adjustment**	X 1 1 5 %	56.158333

*If IND is not shown, select Set 8 until it is displayed in the left corner of the LCD.
**Reference 430.110(C)(2)

Next Larger Rating: 60 Amps

Problem 5-54

Description	Keystroke	Display
Clear Calculator	On/C On/C	0.
Set to 3-phase (if necessary)	Set 3	3 PH
Enter Voltage	2 4 0 Volts	VOLT 240.
Enter Motor HP*	2 5 HPmotor	IND 25. HP
Find FLC	Amps	FLC 68. A
Multiply FLC by adjustment**	X 1 1 5 %	78.2

*If IND is not shown, select Set 8 until it is displayed in the left corner of the LCD.
**Reference 430.110(A)

Rating: 78.2 Amps

Problem 5-55

Description	Keystroke	Display
Clear Calculator	On/C On/C	0.
Set to 3-phase (if necessary)	Set 3	3 PH
Enter Voltage	2 4 0 Volts	VOLT 240.
Enter Motor HP*	3 0 HPmotor	IND 30. HP
Find FLC	Amps	FLC 80. A
Multiply FLC by adjustment**	X 1 1 5 %	92.

*If IND is not shown, select Set 8 until it is displayed in the left corner of the LCD.
**Reference 430.110(A)

Rating: 92 Amps

Problem 5-56

Description	Keystroke	Display
Clear Calculator	On/C On/C	0.
Set to 75°C Wire Rating (THWN)	Set 7	75
Multiply BCSC by adjustment factor and store as Amps	2 5 X 1 2 5 % Amps	AMPS 31.25
Find Wire Size	WireSz	AWG 10 CU WIRE SIZE

10 AWG THWN Cu

Problem 5-57

Description	Keystroke	Display
Clear Calculator	On/C On/C	0.
Set to 75°C Wire Rating (THWN)	Set 7	75
Multiply BCSC by adjustment factor and store as Amps	4 3 X 1 2 5 % Amps	AMPS 53.75
Find Wire Size	WireSz	AWG 6 CU WIRE SIZE

6 AWG THWN Cu

Problem 5-58

Description	Keystroke	Display
Clear Calculator	On/C On/C	0.
Set to 75°C Wire Rating (XHHW)	Set 7	75
Multiply largest BCSC by adjustment factor	5 6 X 1 2 5 %	70.
Add remaining BCSC values and store result as amps	+ 5 6 = Amps	AMPS 126.
Find Wire Size	WireSz	AWG 1 CU WIRE SIZE

1 AWG XHHW-2 Cu

Problem 5-59

Description	Keystroke	Display
Clear Calculator	On/C On/C	0.
Multiply RLC by adjustment factor*	2 7 X 1 4 0 %	37.8

*Reference 440.52(A)(1)

Rating: 37.8 Amps

Problem 5-60

Description	Keystroke	Display
Clear Calculator	On/C On/C	0.
Multiply RLC by adjustment factor*	2 7 X 1 2 5 %	33.75

*Reference 440.52(B)(3)

Next Smaller Rating: 30 Amps

Problem 5-61

Description-1	Keystroke	Display
Clear Calculator	On/C On/C	0.
Multiply BCSC by adjustment factor*	5 6 X 1 7 5 %	98.

*Reference 440.22(A)

Next Smaller Rating: 90 Amps

Description-2	Keystroke	Display
Clear Calculator	On/C On/C	0.
Multiply BCSC by adjustment factor*	5 6 X 2 2 5 %	126.

*Reference 440.22(A)

Next Smaller Rating: 125 Amps

Chapter 6

NOTE: If your results differ from those provided in the examples, verify that your calculator is set to the appropriate phase, wire rating, wire type and ambient temperature.

Problem 6-1

Description	Keystroke	Display
Clear Calculator	On/C On/C	0.
Divide length by 1000	1 7 5 ÷ 1 0 0 0	1,000.
Multiply by Resistance value* and show result	× 3 • 0 7 =	0.53725

*NEC Ch. 9, Table 8 (the ECPro utilizes stranded uncoated copper wires only)

0.53725 Ohms

Problem 6-2

Description	Keystroke	Display
Clear Calculator	On/C On/C	0.
Divide length by 1000	2 0 0 ÷ 1 0 0 0	1,000.
Multiply by Resistance value* and show result	× 3 • 0 7 =	0.614

*NEC Ch. 9, Table 8 (the ECPro utilizes stranded uncoated copper wires only)

0.614 Ohms

Problem 6-3

Description-1	Keystroke	Display
Clear Calculator	On/C On/C	0.
Divide length by 1000	1 0 0 0 ÷ 1 0 0 0	1,000.
Multiply by Resistance value*	× • 3 0 8	0.308
Divide by Temperature Adjustment Factor	÷ 1 • 0 5 =	0.2933333

*NEC Ch. 9, Table 8

0.2933 Ohms

ECPro Operation*

Description-2	Keystroke	Display
Clear Calculator	On/C On/C	0.
Set to 60°C Wire Rating	Set 6	60
Enter Wire Size	4 WireSz	AWG 4 CU WIRE SIZE
Display Resistance	Set Length	OHMS 0.2930774 WIRE
Multiply by length and divide by 1000	× 1 0 0 0 ÷ 1 0 0 0 =	0.2930774

*The ECPro adjusts for temperature per Note 2 under Table 8, resulting in slight differences.

0.2931 Ohms

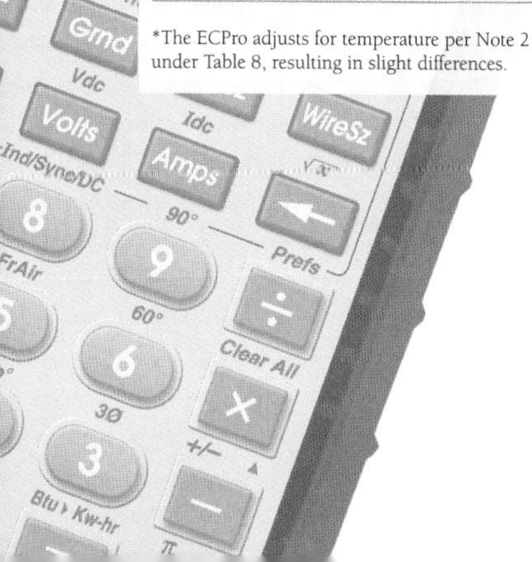

Problem 6-4

Description-1	Keystroke	Display
Clear Calculator	On/C On/C	0.
Divide length by 1000	1 0 0 0 ÷ 1 0 0 0	1,000.
Multiply by Resistance value*	× • 3 1 9	0.319
Multiply by Temperature Adjustment Factor	× 1 • 0 5 =	0.33495

*NEC Ch. 9, Table 8

0.33495 Ohms

ECPro Operation*

Description-2	Keystroke	Display
Clear Calculator	On/C On/C	0.
Set to 90°C Wire Rating	Set 9	90
Toggle to Aluminum Wire	Set 4	Al
Enter Wire Size	2 WireSz	AWG 2 AL WIRE SIZE
Display Resistance	Set Length	OHMS 0.3347905 WIRE
Multiply by length and divide by 1000	× 1 0 0 0 ÷ 1 0 0 0 =	0.3347905

*The ECPro adjusts for temperature per Note 2 under Table 8, resulting in slight differences.

0.33479 Ohms

Problem 6-5

Description-1	Keystroke	Display
Clear Calculator	On/C On/C	0.
Divide length by 1000	4 0 0 ÷ 1 0 0 0	1,000.
Multiply by Resistance value*	× • 4 9 1	0.491
Multiply by Temperature Adjustment Factor	× 1 • 0 5 =	0.20622

*NEC Ch. 9, Table 8

0.20622 Ohms

ECPro Operation*

Description-2	Keystroke	Display
Clear Calculator	On/C On/C	0.
Set to 90°C Wire Rating	Set 9	90
Toggle to Copper Wire	Set 4	Cu
Enter Wire Size	6 WireSz	AWG 6 CU WIRE SIZE
Display Resistance	Set Length	OHMS 0.514789 WIRE
Multiply by length and divide by 1000	× 4 0 0 ÷ 1 0 0 0 =	0.2059156

*The ECPro adjusts for temperature per Note 2 under Table 8, resulting in slight differences.

0.20592 Ohms

Problem 6-6

Description-1	Keystroke	Display
Clear Calculator	On/C On/C	0.
Divide length by 1000	3 0 0 ÷ 1 0 0 0	1,000.
Multiply by Resistance value*	× • 5 0 8	0.508
Multiply by Temperature Adjustment Factor	× 1 • 0 5 =	0.16002

*NEC Ch. 9, Table 8

0.16002 Ohms

ECPro Operation* Description-2	Keystroke	Display
Clear Calculator	On/C On/C	0.
Set to 90°C Wire (THHN)	Set 9	90
Toggle to Aluminum Wire	Set 4	Al
Enter Wire Size	4 WireSz	AWG 4 AL WIRE SIZE
Display Resistance	Set Length	OHMS 0.533146 WIRE
Multiply by length and divide by 1000	× 3 0 0 ÷ 1 0 0 0 =	0.1599438

*The ECPro adjusts for temperature per Note 2 under Table 8, resulting in slight differences.

0.15994 Ohms

Problem 6-7

Description-1	Keystroke	Display
Clear Calculator	On/C On/C	0.
Divide length by 1000	6 0 ÷ 1 0 0 0	1,000.
Multiply by Resistance value*	× 3 • 0 7	3.07
Multiply by current	× 1 2	12.
Multiply by 2 (single-phase)	× 2 =	4.4208

*NEC Ch. 9, Table 8

4.42 Volts

ECPro Operation* Description-2	Keystroke	Display
Clear Calculator	On/C On/C	0.
Set to 1-phase	Set 1	1 PH
Set to 75°C Wire (THWN)	Set 7	75
Toggle to Copper Wire (if necessary)	Set 4	Cu
Enter Voltage	1 2 0 Volts	VOLT 120.
Enter Amperage	1 2 Amps	AMPS 12.
Enter Length	6 0 Length	FEET 60.
Enter Wire Size	1 4 WireSz	AWG 14 CU WIRE SIZE
Find Voltage Drop	VD%	DROP 4.5 V

*The ECPro uses uncoated stranded copper, thus slight differences may occur.

4.5 Volts

Problem 6-8

Description-1	Keystroke	Display
Multiply result by amperage	4 • 4 2 × 1 2 =	53.04

53.04 Watts

ECPro Operation* Description-2	Keystroke	Display
Multiply result by amperage	4 • 5 × 1 2 =	54.

*The ECPro uses uncoated stranded copper, thus slight differences may occur.

54 Watts

Problem 6-9

Description-1	Keystroke	Display
Clear Calculator	On/C On/C	0.
Divide length by 1000	1 0 0 ÷ 1 0 0 0	1,000.
Multiply by Resistance value*	× • 4 9 1	0.491
Multiply by current	× 6 0	60.
Multiply by 1.73 (3-phase)	× 1 • 7 3 =	5.09658

*NEC Ch. 9, Table 8

5.097 Volts

ECPro Operation* Description-2	Keystroke	Display
Clear Calculator	On/C On/C	0.
Set to 3-phase	Set 3	3 PH
Set to 75°C Wire (THWN)	Set 7	75
Enter Voltage	4 8 0 Volts	VOLT 480.
Enter Amperage	6 0 Amps	AMPS 60.
Enter Length	1 0 0 Length	FEET 100.
Enter Wire Size	6 WireSz	AWG 6 CU WIRE SIZE
Find Voltage Drop	VD%	DROP 5.1 V

*The ECPro uses uncoated stranded copper, thus slight differences may occur.

5.1 Volts

Problem 6-10

Description-1	Keystroke	Display
Multiply result by amperage and adjust for 3-phase	5 • 0 9 7 × 6 0 × 1 • 7 3 =	529.0686

529.07 Watts

ECPro Operation* Description-2	Keystroke	Display
Multiply result by amperage and adjust for 3-phase	5 • 1 × 6 0 × 1 • 7 3 =	529.38

*The ECPro uses uncoated stranded copper, thus slight differences may occur.

529.38 Watts

Problem 6-11

Description-1	Keystroke	Display
Clear Calculator	On/C On/C	0.
Divide length by 1000	1 5 0 ÷ 1 0 0 0	1,000.
Multiply by Resistance value*	X • 2 5 3	0.253
Multiply by current	X 9 0	90.
Multiply by 1.73 (3-phase)	X 1 • 7 3 =	5.908815

*NEC Ch. 9, Table 8

5.91 Volts

ECPro Operation Description-2	Keystroke	Display
Clear Calculator	On/C On/C	0.
Set to 3-phase	Set 3	3 PH
Toggle to Aluminum wire	Set 4	Al
Set to 75°C Wire (THWN)	Set 7	75
Enter Voltage	2 4 0 Volts	VOLT 240.
Enter Amperage	9 0 Amps	AMPS 90.
Enter Length	1 5 0 Length	FEET 150.
Enter Wire Size	1 WireSz	AWG 1 AL WIRE SIZE
Find Voltage Drop	VD%	DROP 5.9 V

5.9 Volts

Problem 6-12

Description-1	Keystroke	Display
Clear Calculator	On/C On/C	0.
Divide length by 1000	1 6 0 ÷ 1 0 0 0	1,000.
Multiply by Resistance value*	X • 1 9 4	0.194
Multiply by current	X 1 2 5	125.
Multiply by 2 (1-phase) and adjust for temp.	X 2 X 1 • 0 5 =	8.148

*NEC Ch. 9, Table 8

8.148 Volts

ECPro Operation Description-2	Keystroke	Display
Clear Calculator	On/C On/C	0.
Set to 1-phase	Set 1	1 PH
Toggle to Copper wire	Set 4	Cu
Set to 90°C Wire (THHN)	Set 9	90
Enter Amperage	1 2 5 Amps	AMPS 125.
Enter Length	1 6 0 Length	FEET 160.
Enter Wire Size	2 WireSz	AWG 2 CU WIRE SIZE
Find Voltage Drop	VD%	DROP 8.1 V

8.1 Volts

Problem 6-13

Description-1	Keystroke	Display
Clear Calculator	On/C On/C	0.
Determine Voltage Drop and store in memory	2 4 0 X 3 % Stor 1	M-1 7.2
Multiply k-factor by current and length	1 2 • 9 X 2 0 0 X 1 5 0 =	387,000.
Multiply by 2 (1-phase) and divide by Voltage Drop	X 2 ÷ Rcl 1 =	107,500.

NOTE: NEC Ch. 9, Table 8 – Next larger cmil wire size **2/0 AWG THWN Cu**

ECPro Operation Description-2	Keystroke	Display
Clear Calculator	On/C On/C	0.
Set to 1-phase	Set 1	1 PH
Toggle to Copper wire (if necessary)	Set 4	Cu
Set to 75°C Wire (THWN)	Set 7	75
Enter Allowable Voltage Drop Percentage	3 VD%	DROP 3.0 V %
Enter Amperage	2 0 0 Amps	AMPS 200.
Enter Voltage	2 4 0 Volts	VOLT 240.
Enter Length	1 5 0 Length	FEET 150.
Find Wire Size*	WireSz	AWG 000 CU WIRE SIZE
Find Secondary Wire Size*	WireSz	AWG 00 CU VD WIRE SIZE

*The ECPro displays the largest wire size between ampacity and VD calculations. VD is lit on the display for VD wire size.

3/0 AWG THWN Cu

NOTE: In the example above, ampacity requirements would have required a 3/0 wire size.

Problem 6-14

Description-1	Keystroke	Display
Clear Calculator	On/C On/C	0.
Determine Voltage Drop and store in memory	2 0 8 X 3 % Stor 1	M-1 6.24
Multiply k-factor by current and length	2 1 • 2 X 1 1 5 X 1 3 0 =	316,940.
Multiply by 1.73 (3-phase) and divide by Voltage Drop	X 1 • 7 3 ÷ Rcl 1 =	87,869.583

NOTE: NEC Ch. 9, Table 8 – Next larger cmil wire size **1/0 AWG THWN Al**

ECPro Operation Description-2	Keystroke	Display
Clear Calculator	On/C On/C	0.
Set to 3-phase	Set 3	3 PH
Toggle to Aluminum wire (if necessary)	Set 4	Al
Set to 75°C Wire (THWN)	Set 7	75
Enter Allowable Voltage Drop Percentage*	3 VD%	DROP 3.0 V %
Enter Amperage	1 1 5 Amps	AMPS 115.
Enter Voltage	2 0 8 Volts	VOLT 208.
Enter Length	1 3 0 Length	FEET 130.
Find Wire Size**	WireSz	AWG 0 AL WIRE SIZE
Find Secondary Wire Size**	WireSz	AWG 0 AL VD WIRE SIZE

*VD% is set to 3% by default.
**The ECPro displays the largest wire size between ampacity and VD calculations. VD is lit on the display for VD wire size.

1/0 AWG THWN Al

Problem 6-15

Description-1	Keystroke	Display
Clear Calculator	On/C On/C	0.
Multiply k-factor by current	2 1 • 2 X 5 0 =	1,060.
Multiply by 2 (1-phase) and store result	X 2 = Stor 1	M-1 2,120.
Determine cmil* and multiply by Voltage Drop	2 6 2 4 0 X 1 2 0 X 3 %	94,464.
Divide by stored value and display result	÷ Rcl 1 =	44.558491

*NEC Ch. 9, Table 8

44.56 Feet

ECPro Operation

Description-2	Keystroke	Display
Clear Calculator	On/C On/C	0.
Set to 1-phase	Set 1	1 PH
Toggle to Aluminum wire (if necessary)	Set 4	Al
Set to 75°C Wire (THWN)	Set 7	75
Enter Allowable Voltage Drop Percentage*	3 VD%	DROP 3.0 V %
Enter Amperage	5 0 Amps	AMPS 50.
Enter Voltage	1 2 0 Volts	VOLT 120.
Enter Wire Size	6 WireSz	AWG 6 AL WIRE SIZE
Find Length	Length	FEET 44.554455

*VD% is set to 3% by default.

44.55 Feet

Problem 6-16

Description-1	Keystroke	Display
Clear Calculator	On/C On/C	0.
Multiply k-factor by current	1 2 • 9 X 1 4 0 =	1,806.
Multiply by 1.73 (3-phase) and store result	X 1 • 7 3 = Stor 1	M-1 3,124.38
Determine cmil* and multiply by Voltage Drop	1 0 5 6 0 0 X 4 8 0 X 3 %	1,520,640.
Divide by stored value and display result	÷ Rcl 1 =	486.70136

*NEC Ch. 9, Table 8

487 Feet

ECPro Operation

Description-2	Keystroke	Display
Clear Calculator	On/C On/C	0.
Set to 3-phase	Set 3	3 PH
Toggle to Copper wire (if necessary)	Set 4	Cu
Set to 75°C Wire (THWN)	Set 7	75
Enter Allowable Voltage Drop Percentage*	3 VD%	DROP 3.0 V %
Enter Amperage	1 4 0 Amps	AMPS 140.
Enter Voltage	4 8 0 Volts	VOLT 480.
Enter Wire Size	0 WireSz	AWG 0 CU WIRE SIZE
Find Length	Length	FEET 486.75901

*VD% is set to 3% by default.

487 Feet

Problem 6-17

Description	Keystroke	Display
Clear Calculator	On/C On/C	0.
Multiply k-factor by distance	1 2 • 9 X 1 6 0 =	2,064.
Multiply by 1.73 (3-phase) and store result	X 1 • 7 3 = Stor 1	M-1 3,570.72
Determine cmil* and multiply by Voltage Drop	1 0 3 8 0 X 2 2 0 X 3 %	68,508.
Divide by stored value and display result	÷ Rcl 1 =	19.186047

*NEC Ch. 9, Table 8

19.19 Amps

NOTE: The ECPro does not compute maximum current using voltage drop data.

Problem 6-18

Description	Keystroke	Display
Clear Calculator	On/C On/C	0.
Multiply k-factor by distance	2 1 • 2 X 1 2 0 =	2,544.
Multiply by 2 (1-phase) and store result	X 2 = Stor 1	M-1 5,088.
Determine cmil* and multiply by Voltage Drop	1 6 5 1 0 X 1 1 5 X 3 %	56,959.5
Divide by stored value and display result	÷ Rcl 1 =	11.19487

*NEC Ch. 9, Table 8

11.19 Amps

NOTE: The ECPro does not compute maximum current using voltage drop data.

Problem 6-19

Description-1	Keystroke	Display
Clear Calculator	On/C On/C	0.
Set to 3-phase	Set 3	3 PH
Toggle to Copper wire (if necessary)	Set 4	Cu
Set to 75°C Wire (THWN)	Set 7	75
Enter Amperage	2 2 5 Amps	AMPS 225.
Enter Voltage	4 8 0 Volts	VOLT 480.
Enter Wire Size	0 0 0 0 WireSz	AWG 0000 CU WIRE SIZE
Enter Length	3 0 0 Length	FEET 300.
Find Voltage Drop	VD%	DROP 7.1 V

7.1 Volts

Description-2	Keystroke	Display
Clear Calculator	On/C On/C	0.
Enter Amperage	1 0 5 Amps	AMPS 105.
Enter Wire Size	2 WireSz	AWG 2 CU WIRE SIZE
Enter Length	1 2 5 Length	FEET 125.
Find Voltage Drop	VD%	DROP 4.4 V

4.4 Volts

Description-3	Keystroke	Display
Add Voltage Drops Together	7 • 1 + 4 • 4 =	11.5

11.5 Volts

Description-4	Keystroke	Display
Divide Voltage Drop result by Total Voltage	÷ Volts X 1 0 0 =	2.3958333

2.4 %

Problem 6-20

Description	Keystroke	Display
Clear Calculator	On/C On/C	0.
Set to 1-phase	Set 1	1 PH
Enter Voltage	2 4 0 Volts	VOLT 240.
Enter Amperage	8 0 Amps	AMPS 80.
Enter Wire Size	4 WireSz	AWG 4 CU WIRE SIZE
Enter Length	2 0 0 Length	FEET 200.
Find Voltage Drop*	VD%	DROP 9.9 V

*The ECPro does not internally utilize the k-value method for computing voltage drop.

9.9 Volts

Problem 6-21

Description	Keystroke	Display
Clear Calculator	On/C On/C	0.
Set to 3-phase	Set 3	3 PH
Toggle to Aluminum wire	Set 4	Al
Enter Voltage	2 4 0 Volts	VOLT 240.
Enter Amperage	4 5 Amps	AMPS 45.
Enter Wire Size	8 WireSz	AWG 8 AL WIRE SIZE
Enter Length	1 7 5 Length	FEET 175.
Find Voltage Drop*	VD%	DROP 17.5 V

*The ECPro does not internally utilize the k-value method for computing voltage drop.

17.5 Volts

Problem 6-22

Description	Keystroke	Display
Clear Calculator	On/C On/C	0.
Set to 1-phase	Set 1	1 PH
Toggle to Copper wire (if necessary)	Set 4	Cu
Set to 75°C Wire Rating (THWN)	Set 7	75
Enter Voltage	2 4 0 Volts	VOLT 240.
Enter Amperage	6 0 Amps	AMPS 60.
Enter Length	2 0 0 Length	FEET 200.
Find Wire Size	WireSz	AWG 3 CU VD WIRE SIZE

NOTE: The ECPro does not internally utilize the k-value method for computing voltage drop.
NOTE: The ECPro uses 3% as the maximum Voltage Drop percentage as a default value.
NOTE: The ECPro computes both Ampacity derived (310.16/17) and Voltage Drop wire sizes, displaying the largest size first.

3 AWG THWN Cu

Problem 6-23

Description-1	Keystroke	Display
Clear Calculator	On/C On/C	0.
Find line-to-neutral circuit impedance	3 0 0 X • 0 5 ÷ 1 0 0 0 =	0.015

0.015 Ohms

Description-2	Keystroke	Display
Find line-to-neutral voltage drop	X 3 2 0 =	4.8

4.8 Volts

Description-3	Keystroke	Display
Find line-to-line voltage drop	X 1 • 7 3 2 =	8.3136

8.31 Volts

Description-4	Keystroke	Display
Find voltage drop percentage of circuit voltage	X 1 0 0 ÷ 4 8 0 =	1.732

1.73 %

Description-5	Keystroke	Display
Find actual voltage present at the load	4 8 0 = 8 • 3 1 =	471.69

471.69 Volts

Problem 6-24

Description-1	Keystroke	Display
Clear Calculator	On/C On/C	0.
Find line-to-neutral circuit impedance	1 5 0 X • 0 9 4 ÷ 1 0 0 0 =	0.0141

0.0141 Ohms

Description-2	Keystroke	Display
Find line-to-neutral voltage drop	X 1 4 3 =	2.0163

2.016 Volts

Description-3	Keystroke	Display
Find line-to-line voltage drop	X 1 • 7 3 2 =	3.4922316

3.49 Voltage Drop

Description-4	Keystroke	Display
Find voltage drop percentage of circuit voltage	X 1 0 0 ÷ 2 0 8 =	1.6789575

1.68%

Description-5	Keystroke	Display
Find actual voltage present at the load	2 0 8 = 3 • 4 9 =	204.51

204.51 Volts

Chapter 7

NOTE: The ECPro does not have specific functions for Appliance calculations, thus it is only useful here as a standard calculator or for looking up ampacity-derived wire sizes.

NOTE: If your results differ from those provided in the examples, verify that your calculator is set to the appropriate phase, wire rating, wire type and ambient temperature.

Problem 7-1

Description	Keystroke	Display
Clear Calculator	On/C On/C	0.
Multiply Table 220.55 value by 70%	8 0 0 0 X 7 0 %	5,600.

5,600 VA Neutral

Problem 7-2

Description	Keystroke	Display
Clear Calculator	On/C On/C	0.
Multiply Table 220.55 value by 70% for neutral	1 1 0 0 0 X 7 0 %	7,700.

7,700 VA Neutral

Problem 7-3

Description	Keystroke	Display
Clear Calculator	On/C On/C	0.
Multiply Table 220.55 value* by 70% for neutral	1 1 0 0 0 X 7 0 %	7,700.

*Derived from Table 220.55 using largest range rating.

7,700 VA Neutral

Problem 7-4

Description-1	Keystroke	Display
Clear Calculator	On/C On/C	0.
Add Wattage Rating and multiply by demand factor*	2 • 5 + 3 = X 7 5 0 =	4,125.

*Derived from Table 220.55

4,125 VA

Description-2	Keystroke	Display
Find neutral wire size	X 7 0 %	2,887.5

2,888 VA Neutral

Problem 7-5

Description-1	Keystroke	Display
Clear Calculator	On/C On/C	0.
Add Wattage Rating and multiply by demand factor*	3 X 8 X 5 5 0 =	13,200.

*Derived from Table 220.55

13,200 VA

Description-2	Keystroke	Display
Find neutral wire size	X 7 0 %	9,240.

9,240 VA Neutral

Problem 7-6

Description-1	Keystroke	Display
Clear Calculator	On/C On/C	0.
Total 1st Wattage Rating, multiply by demand factor* and store in memory	3 X 3 X 7 0 0 = Stor 0	M+ 6,300.
Total 2nd Wattage Rating and multiply by demand factor*	3 X 8 X 5 5 0 =	13,200.
Add value from memory	+ Rcl Rcl =	19,500.

*Derived from Table 220.55

19,500 VA

Description-2	Keystroke	Display
Find neutral wire size	X 7 0 %	13,650.

13,650 VA Neutral

Problem 7-7

Description-1	Keystroke	Display
Clear Calculator	On/C On/C	0.
Compute Line Demand*	1 5 + 2 8 = X 1 0 0 0 =	43,000.

*Derived from Table 220.55

43,000 VA

Description-2	Keystroke	Display
Find neutral wire size	X 7 0 %	30,100.

30,100 VA Neutral

Problem 7-8

Description-1	Keystroke	Display
Clear Calculator	On/C On/C	0.
Compute Line Demand*	• 7 5 X 4 8 = + 2 5 = X 1 0 0 0 =	61,000.

*Derived from Table 220.55

61,000 VA

Description-2	Keystroke	Display
Find neutral wire size	X 7 0 %	42,700.

42,700 VA Neutral

Problem 7-9

Description-1	Keystroke	Display
Clear Calculator	On/C On/C	0.
Compute Demand Adjustment*	1 5 − 1 2 = X 5 %	0.15
Compute Line Demand*	+ 1 X 8 0 0 0 =	9,200.

*Derived from Table 220.55

9,200 VA

Description-2	Keystroke	Display
Find neutral wire size	X 7 0 %	6,440.

6,440 VA Neutral

Problem 7-10

Description-1	Keystroke	Display
Clear Calculator	On/C On/C	0.
Compute Demand Adjustment*	1 6 − 1 2 = × 5 %	0.2
Compute Line Demand*	+ 1 × 3 9 0 0 0 =	46,800.

*Derived from Table 220.55

46,800 VA

Description-2	Keystroke	Display
Find neutral wire size	× 7 0 %	32,760.

32,760 VA Neutral

Problem 7-11

Description-1	Keystroke	Display
Clear Calculator	On/C On/C	0.
Add ranges and find average rating	1 2 + 1 4 + 1 6 × 4 ÷ 1 2 =	14.
Compute Demand Adjustment*	− 1 2 = × 5 %	0.1
Compute Line Demand*	+ 1 × 2 7 0 0 0 =	29,700.

*Derived from Table 220.55 using twelve 14-kW ranges.

29,700 VA

Description-2	Keystroke	Display
Find neutral wire size	× 7 0 %	20,790.

20,790 VA Neutral

Problem 7-12

Description-1	Keystroke	Display
Clear Calculator	On/C On/C	0.
Evenly divide ranges	3 0 ÷ 3 =	10.
Multiply by number of ranges on adjacent phases	× 2 =	20.
Compute Line Demand*	3 5 0 0 0 × 3 ÷ 2 =	52,500.

*Derived from Table 220.55 using twenty 12-kW ranges

52,500 VA

Description-2	Keystroke	Display
Find neutral wire size	× 7 0 %	36,750.

36,750 VA Neutral

Problem 7-13

Description-1	Keystroke	Display
Clear Calculator	On/C On/C	0.
Compute Demand Adjustment*	1 6 − 1 2 = × 5 %	0.2
Compute Line Demand*	+ 1 × 8 0 0 0 =	9,600.

*Derived from Table 220.55

9,600 VA

Description-2	Keystroke	Display
Find neutral wire size	× 7 0 %	6,720.

6,720 VA Neutral

Problem 7-14

No calculation necessary per NEC 210.19(A)(3) Exception #2

Problem 7-15

Description	Keystroke	Display
Clear Calculator	On/C On/C	0.
Add all loads together	4 + 4 + 5 =	13.
Find Adjustment for being over 12 kW*	− 1 2 × 5 %	0.05
Determine Max Demand*	+ 1 × 8 0 0 0 =	8,400.

* Derived from Table 220.55

8,400 VA Neutral

Problem 7-16

Description-1	Keystroke	Display
Clear Calculator	On/C On/C	0.
Set to 1-phase	Set 1	1 PH
Set to 75°C Wire Rating	Set 7	75
Enter Demand* as kW	8 kilo- Watts	KW 8.
Enter Voltage	2 4 0 Volts	VOLT 240.
Find Amperage	Amps	AMPS 33.333333
Below minimum, enter 40 amps	4 0 Amps	AMPS 40.
Determine Minimum Wire Size	WireSz	AWG 8 CU WIRE SIZE

*Derived from Table 220.55

8 AWG THWN Cu

Description-2	Keystroke	Display
Determine Neutral Ampacity	Amps × 7 0 % Amps	AMPS 28.
Determine Minimum Neutral Wire Size	WireSz	AWG 10 CU WIRE SIZE

10 AWG THWN Cu

Problem 7-17

Description-1	Keystroke	Display
Clear Calculator	On/C On/C	0.
Set to 1-phase	Set 1	1 PH
Set to 75°C Wire Rating	Set 7	75
Find Demand Adjustment*	1 8 − 1 2 × 5 %	0.3
Compute Demand*	+ 1 × 8 0 0 0 =	10,400.
Enter Demand as kW	Watts	WATT 10,400.
Enter Voltage	2 4 0 Volts	VOLT 240.
Find Amperage	Amps	AMPS 43.333333
Determine Minimum Wire Size	WireSz	AWG 8 CU WIRE SIZE

*Derived from Table 220.55

8 AWG THHN Cu

Description-2	Keystroke	Display
Determine Neutral Ampacity	Amps × 7 0 % Amps	AMPS 30.333333
Determine Minimum Neutral Wire Size	WireSz	AWG 10 CU WIRE SIZE

10 AWG THHN Cu

Problem 7-18

Description-1	Keystroke	Display
Clear Calculator	On/C On/C	0.
Compute Demand Adjustment*	1 6 − 1 2 × 5 %	0.2
Compute Demand*	+ 1 × 2 1 0 0 0 =	25,200.

*Derived from Table 220.55

25,200 VA

Description-2	Keystroke	Display
Determine Neutral Ampacity	× 7 0 %	17,640.

17,640 VA Neutral

Problem 7-19

Description-1	Keystroke	Display
Enter Demand*	2 3 0 0 0	23,000.

*Derived from Table 220.55

23,000 VA

Description-2	Keystroke	Display
Determine Neutral Ampacity	× 7 0 %	16,100.

16,100 VA Neutral

Problem 7-20

Description	Keystroke	Display
Clear Calculator	On/C On/C	0.
Add Appliance loads together	3 × 1 6 + 7 • 5 + 4 =	59.5
Multiply by Demand Factor*	× 7 0 %	41.65

* Derived from Table 220.56

41,650 VA

Problem 7-21

Description	Keystroke	Display
Clear Calculator	On/C On/C	0.
Add Appliance loads together	2 × 2 4 + 1 0 + 3 + 6 =	67.
Multiply by Demand Factor*	× 6 5 %	43.55

*Derived from Table 220.56
NOTE: Use sum of two largest loads, since it is greater.

48,000 VA

Problem 7-22

Description	Keystrokes	Display
Set to 1-phase	Set 1	1 PH
Set to 90° Wire Rating	Set 9	90
Enter Load	5 kilo- Watts	KW 5.
Enter Voltage	2 4 0 Volts	VOLT 240.
Find Amperage	Amps	AMPS 20.833333
Find Wire Size	WireSz	AWG 14 CU WIRE SIZE

NOTE: Per 240.4(D)(7), 10 AWG copper must be used due to restrictions of smaller conductor sizes.
NOTE: Per 240.4(D)(7), 10 AWG copper limited to 30 Amps Protection Rating: 30 Amps.

10 AWG Cu

Problem 7-23

Description	Keystroke	Display
Clear Calculator	On/C On/C	0.
Set to 1-phase	Set 1	1 PH
Set to 60° Wire Rating	Set 6	60
Enter Load	4 • 5 kilo- Watts	KW 4.5
Enter Voltage	2 4 0 Volts	VOLT 240.
Find Amperage	Amps	AMPS 18.75
Find Wire Size	WireSz	AWG 12 CU WIRE SIZE

NOTE: Per 240.4(D)(7), 10 AWG copper must be used due to restrictions of smaller conductor size.
NOTE: Per 240.4(D)(7), 10 AWG copper limited to 30 Amps Protection Rating: 30 Amps.

10 AWG Cu

Problem 7-24

Description-1	Keystroke	Display
Clear Calculator	On/C On/C	0.
Set to 60°C Wire Rating (per 334.80)	Set 6	60
Enter Amperage	8 • 1 Amps	AMPS 8.1
Find Wire Size	WireSz	AWG 14 CU WIRE SIZE

NOTE: Per 240.4(D)(3), 14 AWG copper limited to 15 Amps, thus acceptable.
NOTE: Per 240.4(D)(3), 10 AWG copper limited to 30 Amps Protection Rating: 15 Amps.

14 AWG Cu

Description-2	Keystroke	Display
Clear Calculator	On/C On/C	0.
Enter Max Wire Rating	1 5	15.
Multiply by 80%	× 8 0 % =	12.

NOTE: 8.1 amps less than 12 amps, thus use the 15 amp switch.

Protection Rating: 15 Amps

Problem 7-25

Description-1	Keystroke	Display
Clear Calculator	On/C On/C	0.
Set to 75°C Wire Rating	Set 7	75
Multiply nameplate current by 125% and enter as Amperage	2 6 • 5 × 1 2 5 % Amps	AMPS 33.125
Find Wire Size	WireSz	AWG 10 CU WIRE SIZE

NOTE: Per 240.4(D)(7), 10 AWG copper limited to 30 Amps, thus next size is needed.

8 AWG THWN Cu

Description-2	Keystroke	Display
Find Wire Ampacity Rating	8 WireSz WireSz	50.0 WIRE A

Protection Rating: 50 Amps

Description-3	Keystroke	Display
Find the Equipment Ground Wire Size	Set Grnd	EQPG 10 CU WIRE SIZE

Equipment Ground Size: 10 AWG Cu

Chapter 8

NOTE: The ECPro does not have specific functions for load calculations, thus it is only useful here as a standard calculator or for looking up ampacity derived wire sizes.

NOTE: If your results differ from those provided in the examples, verify that your calculator is set to the appropriate phase, wire rating, wire type and ambient temperature.

Problem 8-1

Description	Keystroke	Display
Clear Calculator	On/C On/C	0.
Find Area	5 5 X 6 5 =	3,575.
Determine Lighting Load*	X 3 =	10,725.
		10,725 VA

* Derived from Table 220.12

Problem 8-2

Description	Keystroke	Display
Clear Calculator	On/C On/C	0.
Set to 1-phase	Set 1	1 PH
Enter Load	1 0 7 2 5 VA	VA 10,725.
Enter Voltage	1 2 0 Volts	VOLT 120.
Solve for Amperage	Amps	AMPS 89.375
Divide by Circuit Size	÷ 1 5 =	5.9583333
		6 circuits

Problem 8-3

Description	Keystroke	Display
Clear Calculator	On/C On/C	0.
Add all loads together*	1 0 7 2 5 + 3 0 0 0 + 1 5 0 0 =	15,225.
Subtract 3,000	− 3 0 0 0 =	12,225.
Multiply result by demand factor**	X • 3 5 =	4,278.75
Add 3,000 and display demand	+ 3 0 0 0 =	7,278.75.
		7,279 VA

* Includes 2 small appliances and 1 laundry load per 220.52 (A) & (B).
** Derived from Table 220.42

Problem 8-4

Description	Keystroke	Display
Clear Calculator	On/C On/C	0.
Determine neutral line size	5 0 0 0 X • 7 =	3,500.
		3,500 VA

NOTE: Per 220.54, minimum feeder size is the larger of 5,000 VA or dryer nameplate rating.

Problem 8-5

Description	Keystroke	Display
Clear Calculator	On/C On/C	0.
Set to 1-phase	Set 1	1 PH
Enter Voltage	1 2 0 Volts	VOLT 120.
Enter Disposal Amperage	6 Amps	AMPS 6.
Find VA and add to memory	VA Stor 0	M+ 720.
Enter Dishwasher Wattage	1 5 0 0 Watts	WATT 1,500.
Find VA and add to memory	VA Stor 0	M+ 1,500.
Enter Water Heater Wattage	5 0 0 0 Watts	WATT 5,000.
Find VA and add to memory	VA Stor 0	M+ 5,000.
Enter Hood Amperage	4 • 4 Amps	AMPS 4.4
Find VA and add to memory	VA Stor 0	M+ 528.
Enter Trash Compactor Wattage	9 6 0 Watts	WATT 960.
Find VA and add to memory	VA Stor 0	M+ 960.
Recall Total of the Loads	Rcl Rcl	8,708.
Adjust for demand factor*	X • 7 5 =	6,531.
		6,531 VA

* Reference 220.53

Problem 8-6

Description	Keystroke	Display
Clear Calculator	On/C On/C	0.
Compute neutral line*	8 0 0 0 X • 7 =	5,600.
		5,600 VA

* Reference 220.61(B)(1)

Problem 8-7

Description-1	Keystroke	Display
Clear Calculator	On/C On/C	0.
Compute Total Strip Heater Load	4 X 5 0 0 = Stor 0	M+ 2,000.
Compute Total Bathroom Heater Load	2 X 7 5 0 =	1,500.
Add stored value for total heating load	+ Rcl Rcl =	3,500.
		3,500 VA

Description-2	Keystroke	Display
Enter Voltage	2 4 0 Volts	VOLT 240.
Enter motor hp (should be in 1Ø)	7 • 5 HPmotor	IND 7.5 HP
Find Motor FLC	Amps	FLC 40. A
Compute VA	VA	9,600.
		9,600 VA

NOTE: Per 220.60, largest load for noncoincidental load is used.

Problem 8-8

Description	Keystroke	Display
Clear Calculator	On/C On/C	0.
Compute Result	9 6 0 0 X . 2 5 =	2,400.

NOTE: Per 220.50, largest load for noncoincidental load is used. **2,400 VA**

Problem 8-9

Summarize the total feeder demand for the Smith house using the standard calculations. (Straight addition not needing example results: Feeder: 38,810, Neutral: 22,910)

Problem 8-10

Description-1	Keystroke	Display
Clear Calculator	On/C On/C	0.
Enter Voltage	2 4 0 Volts	VOLT 240.
Enter Ungrounded Conductor Load	3 8 8 1 0 VA	VA 38,810.
Find Amperage	Amps	AMPS 161.70833

161.71 Amps

Description-2	Keystroke	Display
Enter Neutral Conductor Load	2 2 9 1 0 VA	VA 22,910.
Find Amperage	Amps	AMPS 95.458333

95.46 Amps

Problem 8-11

Description-1	Keystroke	Display
Clear Calculator	On/C On/C	0.
Enter Service Rating	1 7 5	175.
Multiply by 310.15 (B) (7) Factor	X . 8 3 Amps	AMPS 145.25
Set to 75°C (THWN)	Set 7	75
Find Conductor Size	WireSz	0 CU

1/0 AWG Cu

Description-2	Keystroke	Display
Clear Calculator	On/C On/C	0.
Set to 75°C Wire Rating	Set 7	75
Enter Neutral Conductor Amperage	9 6 Amps	AMPS 96.
Find Wire Size	WireSz	AWG 3 CU WIRE SIZE

3 AWG Cu

Problem 8-12

Description	Keystroke	Display
Find Grounded Wire Size	0 Grnd	GRND 6 CU WIRE SIZE

6 AWG Cu

Problem 8-13

Calculate the feeder demand for the Smith house, using optional method (straight addition of all loads resulting in 31,773 VA).

Problem 8-14

Description-1	Keystroke	Display
Clear Calculator	On/C On/C	0.
Enter Voltage	2 4 0 Volts	VOLT 240.
Enter Ungrounded Conductor Load	3 1 7 7 3 VA	VA 31,773.
Find Amperage	Amps	AMPS 132.3875

132 Amps

Description-4	Keystroke	Display
Clear Calculator	On/C On/C	0.
Enter Neutral Conductor Load	2 0 0 8 7 VA	VA 20,087.
Find Amperage	Amps	AMPS 83.695833
Find Wire Size	WireSz	AWG 4 CU WIRE SIZE

4 AWG Cu

Description-5	Keystroke	Display
Enter Feeder Wire Size and find Ground Wire Size	1 Grnd	GRND 6 CU WIRE SIZE

6 AWG Cu

Problem 8-15

Description	Keystroke	Display
Clear Calculator	On/C On/C	0.
Find Total Area	1 4 0 0 X 1 8 =	25,200.
Determine Lighting Load*	X 3 =	75,600.

* Derived from Table 220.12 **75,600 VA**

Problem 8-16

Calculate the feeder demand for the lighting, small appliance and laundry loads for the Homestead Apartments (simple calculations computing to a total of 53,100 VA).

Problem 8-17

Description	Keystroke	Display
Clear Calculator	On/C On/C	0.
Set to 1-phase	Set 1	1 PH
Enter Voltage	1 2 0 Volts	VOLT 120.
Enter Disposal Amperage	6 • 2 Amps	AMPS 6.2
Find VA and add to memory	VA Stor 0	M+ 744.
Enter Dishwasher Wattage	2 0 0 0 Watts	WATT 2,000.
Find VA and add to memory	VA Stor 0	M+ 2,000.
Enter Water Heater Wattage	3 0 0 0 Watts	WATT 3,000.
Find VA and add to memory	VA Stor 0	M+ 3,000.
Recall Total of the Loads	Rcl Rcl	5,744.
Multiply by number of apartments	X 1 8 =	103,392.
Adjust for demand factor*	X 7 5 %	77,544.

*Reference 220.53

77,544 VA

Problem 8-18

Description-1	Keystroke	Display
Clear Calculator	On/C On/C	0.
Compute Total Load	1 8 X 5 0 0 0 =	90,000.
Multiply by Demand Factor*	X 4 0 %	36,000.

*Reference 220.54

36,000 VA

Description-2	Keystroke	Display
Find Neutral Conductor Load*	X 7 0 %	25,200.

*Reference 220.61(B)

25,200 VA

Problem 8-19

Description	Keystroke	Display
Clear Calculator	On/C On/C	0.
Find Neutral Conductor Load	3 3 0 0 0* X 7 0 %**	23,100.

* Reference 220.55
** Reference 220.61(B)

23,100 VA

Problem 8-20

Description	Keystroke	Display
Clear Calculator	On/C On/C	0.
Enter Voltage	2 4 0 Volts	VOLT 240.
Enter Motor HP	3 HPmotor	IND 3. HP
Find FLC	Amps	FLC 17. A
Find VA	VA	VA 4,080.
Multiply by number of units	X 1 8 =	73,440.

73,440 VA

Problem 8-21

Description	Keystroke	Display
Clear Calculator	On/C	0.
Multiply VA by 25%	Rcl VA X 2 5 %	1,020.

NOTE: References Problem 8-20

1,020 VA

Problem 8-22

Description	Keystroke	Display
Clear Calculator	On/C On/C	0.
Multiply load by quantity	5 0 0 X 6 =	3,000.

3,000 VA

Problem 8-23

Summing all loads results in: Feeder: 277,104; Neutral: 181,944.

Problem 8-24

Description-1	Keystroke	Display
Clear Calculator	On/C On/C	0.
Set to 1-phase	Set 1	1 PH
Toggle to Aluminum Wire	Set 4	Al
Toggle to 75°C Wire	Set 7	75
Enter Voltage	2 4 0 Volts	VOLT 240.
Enter Ungrounded Conductor Load	2 7 7 1 0 4 VA	VA 277,104.
Find Amperage	Amps	AMPS 1,154.6

1,154.6 Amps

Description-2	Keystroke	Display
Enter Neutral Conductor Load	1 8 1 9 4 4 VA	VA 181,944.
Find Amperage	Amps	AMPS 758.1
Less first 200 amps at 100%	− 2 0 0 =	558.1
Calculate remaining amps at 70%	X 7 0 %	390.67
Calculate total neutral ampacity	+ 2 0 0 =	590.67

591 Amps

Problem 8-25

Description-1	Keystroke	Display
Clear Calculator	On/C On/C	0.
Set to 1-phase	Set 1	1 PH
Toggle to Aluminum Wire (if necessary)	Set 4	Al
Toggle to 75°C Wire	Set 7	75
Enter Ungrounded Conductor Load	2 7 7 1 0 4 VA	VA 277,104.
Find Amperage	Amps	AMPS 1,154.6
Enter amount of Parallel Wires	3 ParSz	PAR 750 AL WIRE SIZE

750 kcmil

Description-2	Keystroke	Display
Enter amperage of Neutral Load*	5 9 1 Amps	AMPS 591.
Enter amount of Parallel Wires	3 ParSz	PAR 250 AL WIRE SIZE

250 kcmil

Problem 8-26

Description	Keystroke	Display
Clear Calculator	On/C On/C	0.
Set to RMC	8 Set CondSz	RMC COND
Enter Feeder Wire Size	7 5 0 WireSz	AWG 750 AL WIRE SIZE
Enter Quantity of Wires	2 #THHN #THWN	THHN 2. WIRES
Enter Neutral Wire Size	2 5 0 WireSz	AWG 250 AL WIRE SIZE
Enter Quantity of Wires	1 #THHN #THWN	THHN 1. WIRE
Show Conduit Size	CondSz	RMC 3.00 in COND SIZE

3 inch RMC

Problem 8-27

Simple calculations computing to a total of 225,464 VA.

Problem 8-28

Description-1	Keystroke	Display
Clear Calculator	On/C On/C	0.
Set to 1-phase (if necessary)	Set 1	1 PH
Enter VA	2 2 5 4 6 4 VA	VA 225,464.
Enter Voltage	2 4 0 Volts	VOLT 240.
Find Amperage	Amps	AMPS 939.43333

939.43 Amps

Description-2	Keystroke	Display
Enter Neutral Conductor Load	1 8 1 9 4 4 VA	VA 181,944.
Find Amperage	Amps	AMPS 758.1
Less first 200 amps at 100%	− 2 0 0 =	558.1
Calculate remaining amps at 70%	× 7 0 %	390.67
Calculate total neutral ampacity	+ 2 0 0 =	590.67

591 Amps

Problem 8-29

Description 1	Keystroke	Display
Clear Calculator	On/C On/C	0.
Compute Store Area	1 5 0 × 6 0 =	9,000.
Compute Lighting Load* and store in memory	× 3 = Stor 0	M+ 27,000.

Description 2	Keystroke	Display
Compute Storage Area Lighting Load*	5 0 0 0 × • 2 5 =	1,250.

Description 3	Keystroke	Display
Find Total Load	+ Rcl Rcl =	28,250.

* Reference Table 220.12

28,250

Problem 8-30

Description	Keystroke	Display
Clear Calculator	On/C On/C	0.
Compute Receptacle Load*	3 6 × 1 8 0 =	6,480.

*Reference 220.14(I) and Table 220.44

6,480 VA

Problem 8-31

Section 600.5(A) requires a 20 Ampere branch circuit. Section 220.14(F) requires 1,200 VA for sign load.

Problem 8-32

Description	Keystroke	Display
Clear Calculator	On/C On/C	0.
Compute Load*	2 × 2 0 × 2 0 0 =	8,000.

*Reference 220.43(A)

8,000 VA

Problem 8-33

Description	Keystroke	Display
Clear Calculator	On/C On/C	0.
Compute number of units	3 0 ÷ 5 =	6.
Compute Load*	6 × 1 8 0 =	1,080.

* Reference 220.14(H)

1,080 VA

Problem 8-34

Description	Keystroke	Display
Clear Calculator	On/C On/C	0.
Enter kilo Watts	1 0 kilo- Watts	KW 10.
Convert to Volt-Amps	VA	VA 10,000.

10,000 VA

Problem 8-35

Description	Keystroke	Display
Clear Calculator	On/C On/C	0.
Set to 1-phase (if necessary)	Set 1	1 PH
Enter Voltage	1 2 0 Volts	VOLT 120.
Enter Motor HP	1 ÷ 3 HPmotor	IND 0.33 HP
Find FLC	Amps	FLC 7.2 A
Multiply by number of motors and store as Amperage	× 3 = Amps	AMPS 21.6
Compute Load	VA	VA 2,592.

2,592 VA

Problem 8-36

Description	Keystroke	Display
Clear Calculator	On/C On/C	0.
Set to 3-phase (if necessary)	Set 3	3 PH
Enter Voltage	2 0 8 Volts	VOLT 208.
Enter Motor HP	2 0 HPmotor	IND 20. HP
Find FLC	Amps	FLC 59.4 A
Compute Load*	VA	21,399.834

* Answer based on actual square root of 3 (1.7320508).

21,400 VA

Problem 8-37

Description	Keystroke	Display
Clear Calculator	On/C On/C	0.
Compute Load*	2 1 4 0 0 × 2 5 %	5,350.

* Answer based on actual square root of 3 (1.7320508).

5,350 VA

Problem 8-38

Simple addition resulting in: Noncontinuous: 48,120 VA; Continuous: 47,602 VA.

Problem 8-39

Description-1	Keystroke	Display
Clear Calculator	On/C On/C	0.
Set to 3-phase (if necessary)	Set 3	3 PH
Enter Voltage	2 0 8 Volts	VOLT 208.
Enter Ungrounded Load	9 3 3 7 0 VA	VA 93,370.
Find Ampacity	Amps	AMPS 259.1692

Next Larger OCPD: 300 Amps

Description-2	Keystroke	Display
Adjust for wire deration*	÷ 8 0 %	323.96151

Ungrounded: 324 Amps

Description-3	Keystroke	Display
Enter Neutral Load	4 7 6 0 2 VA	VA 47,602.
Find Ampacity	Amps	AMPS 132.12994
Adjust for wire deration*	÷ 8 0 %	165.16242

*Reference Tables 310.15(B)(5)(c) and 310.15(B)(3)(a)

Neutral: 165 Amps

Problem 8-40

Note: The ECPro does not account for bare conductors for conduit size calculations.

Description-1	Keystroke	Display
Clear Calculator	On/C On/C	0.
Set to Copper (if necessary)	Set 4	Cu
Set to 75°C Wire Rating	Set 7	75
Set to RMC conduit	8 Set CondSz	RMC COND
Enter Ungrounded Feeder Ampacity	3 2 4 Amps	AMPS 324.
Find Wire Size	WireSz	AWG 400 CU WIRE SIZE
Enter number of wires	3 #THHN #THWN	THHN 3. WIRES

400 kcmil

Description-2	Keystroke	Display
Enter Neutral Ampacity	1 6 5 Amps	AMPS 165.
Find Wire Size	WireSz	AWG 00 CU WIRE SIZE
Enter number of wires	1 #THHN #THWN	THHN 1. WIRE

2/0 AWG

Problem 8-41

Description	Keystroke	Display
Find Conduit Size	CondSz	RMC 3.00 in COND SIZE

NOTE: The ECPro does not utilize bare conductors for sizing.

3 inch RMC

Problem 8-42

Description	Keystroke	Display
Clear Calculator	On/C On/C	0.
Compute Area	1 2 5 X 8 8 X 5 =	55,000.
Multiply by lighting load factor*	X 3 • 5 VA	VA 192,500.

*Reference Table 220.12.

192,500 VA

Problem 8-43

Description	Keystroke	Display
Set to 1-phase	Set 1	1 PH
Adjust 8-42 Computed Load by 125%	VA X 1 2 5 % VA	VA 240,625.
Enter Voltage	2 7 7 Volts	VOLT 277.
Compute total Ampacity	Amps	AMPS 868.68231
Divide by circuit rating size	÷ 2 0 =	43.434116

44 circuits

Problem 8-44

Description	Keystroke	Display
Clear Calculator	On/C On/C	0.
Multiply area by load factor* and enter as VA	5 5 0 0 0 X 4 • 5 VA	VA 247,500.

*Reference Tables 220.12 and 220.14(K)

247,500 VA

Problem 8-45

Description	Keystroke	Display
Clear Calculator	On/C On/C	0.
Multiply receptacle load by quantity/floor and # of floors	1 8 0 X 5 0 X 5 =	45,000.
Subtract 10,000	= 1 0 0 0 0 =	35,000.
Adjust for load requirements	X 5 0 %	17,500.
Add back first 10,000	+ 1 0 0 0 0 VA	VA 27,500.

NOTE: Reference 220.14(I) and 220.14(K)

Per 220.14(K), Larger Load needs to be used: 55,000 VA

Problem 8-46

Minimum receptacle load is 55,000 VA, thus minimum transformer rating is 55 kVA.

Problem 8-47

Description	Keystroke	Display
Clear Calculator	On/C On/C	0.
Set to 3-phase	Set 3	3 PH
Enter Voltage	4 6 0 Volts	VOLT 460.
Enter first motor size and find FLC	5 HPmotor Amps	FLC 7.6 A
Find VA, multiply by quantity and store in memory	VA X 2 Stor 0	M+ 12,110.499
Enter second motor size and find FLC	1 0 HPmotor Amps	FLC 14. A
Find VA, multiply by quantity and store in memory	VA X 2 Stor 0	M+ 22,308.814
Enter third motor size and find FLC	5 0 HPmotor Amps	FLC 65. A
Find VA, multiply by quantity and store in memory	VA X 2 Stor 0	M+ 103,576.64
Add 25% of largest VA (last computation above)	VA X 2 5 % Stor 0	M+ 12,947.08
Recall and Display Total Load	Rcl Rcl	150,943.03

NOTE: VA results based on actual square root of 3.

150,943 VA

Problem 8-48

Adding values from previous solutions, Noncontinuous: 205,794 VA; Continuous: 192,500 VA.

Problem 8-49

Simple addition after adjustments from 220.61(B)(2) resulting in 1,104 amps.

Chapter 9

NOTE: If your results differ from those provided in the examples, verify that your calculator is set to the appropriate phase, wire rating, wire type and ambient temperature.

Problem 9-1

Description-1	Keystroke	Display
Clear Calculator	On/C On/C	0.
Set to 3-phase	Set 3	3 PH
Enter Voltage	4 8 0 Volts	VOLT 480.
Enter Load	1 5 0 0 0 VA	VA 15,000.
Solve for Amps	Amps	AMPS 18.042196
Apply adjustment factor*	X 1 2 5 %	22.552745

*Reference 450.3(B) **Next Larger Size: 25 Amps**

Description-2	Keystroke	Display
Clear Calculator	On/C	0.
Solve for Amps	Amps	AMPS 18.042196
Apply adjustment factor*	X 2 5 0 %	45.10549

*Reference 450.3(B) **Next Smaller Size: 45 Amps**

Problem 9-2

Description	Keystroke	Display
Clear Calculator	On/C On/C	0.
Set to 1-phase	Set 1	1 PH
Enter Voltage	4 8 0 Volts	VOLT 480.
Enter Load	2 5 0 0 VA	VA 2,500.
Solve for Amps	Amps	AMPS 5.2083333
Apply adjustment factor*	X 1 6 7 %	8.6979167

*Reference 450.3(B) **8.69 Amps or less**

Problem 9-3

Description	Keystroke	Display
Clear Calculator	On/C On/C	0.
Set to 1-phase	Set 1	1 PH
Enter Voltage	1 2 0 Volts	VOLT 120.
Enter Load	2 0 0 VA	VA 200.
Solve for Amps	Amps	AMPS 1.6666667
Apply adjustment factor*	X 3 0 0 %	5.

* Reference 450.3(B) **5 Amps or less**

Problem 9-4

Description-1	Keystroke	Display
Clear Calculator	On/C On/C	0.
Set to 3-phase	Set 3	3 PH
Set to 75°C Wire Rating (if necessary)	Set 7	75
Enter Voltage	4 4 0 Volts	VOLT 440.
Enter Load	4 5 0 0 0 VA	VA 45,000.
Solve for Amps	Amps	AMPS 59.047187
Apply adjustment factor*	X 1 2 5 %	73.808983
*Reference 450.3(B)	**Next Larger Size: 80 Amps**	
Find Wire Size	= Amps WireSz	AWG 4 CU WIRE SIZE

Description-2	Keystroke	Display
Enter Load	4 5 0 0 0 VA	VA 45,000.
Solve for Amps	Amps	AMPS 59.047187
Find Wire Size	WireSz	AWG 6 CU WIRE SIZE
Show Wire Ampacity	WireSz	65.0 WIRE A

NOTE: Reference 240.4(A) and 240.6(A) **Next Larger Size: 70 Amps**

Description-3	Keystroke	Display
Show Amps	Amps	AMPS 59.047187
Adjust for loading*	X 1 2 5 %	73.808983
* Reference 450.3(B)	**Next Larger Size: 80 Amps**	
Find Wire Size	8 0 Amps WireSz	AWG 4 CU WIRE SIZE

Problem 9-5

Description	Keystroke	Display
Clear Calculator	On/C On/C	0.
Set to 3-phase	Set 3	3 PH
Enter Voltage	2 4 0 Volts	VOLT 240.
Enter Load	4 0 0 0 0 VA	VA 40,000.
Solve for Amps	Amps	AMPS 96.225045
Apply adjustment factor*	X 1 2 5 %	120.28131

*Reference 450.3(B) **Next Larger Size: 125 Amps**

Problem 9-6

Description	Keystroke	Display
Clear Calculator	On/C On/C	0.
Set to 1-phase	Set 1	1 PH
Enter Voltage	2 4 0 Volts	VOLT 240.
Enter Load	2 0 0 0 VA	VA 2,000.
Solve for Amps	Amps	AMPS 8.3333333
Apply adjustment factor*	X 1 6 7 %	13.916667

* Reference 450.3(B)

<u>13.9 Amps or less</u>

Problem 9-7

Description-1	Keystroke	Display
Clear Calculator	On/C On/C	0.
Enter Secondary Current and adjust for load*	9 6 • 3 X 1 2 5 %	120.375

<u>Next Larger Size: 125 Amps</u>

Description-2	Keystroke	Display
Enter Primary Current and adjust for load*	4 8 • 2 X 2 5 0 %	120.5

*Reference 450.3(B)

<u>Next Smaller Size: 110 Amps</u>

Description-3	Keystroke	Display
Enter primary OPCD amperage	1 1 0 Amps	AMPS 110.
Set to 75°C Wire Rating	Set 7	75
Solve for Wire Size	WireSz	AWG 2 CU WIRE SIZE

<u>2 AWG THWN Cu</u>

Problem 9-8

Description-1	Keystroke	Display
Clear Calculator	On/C On/C	0.
Set to 3-phase	Set 3	3 PH
Enter Voltage	2 0 8 Volts	VOLT 208.
Enter Load	7 5 0 0 0 VA	VA 75,000.
Solve for Amps	Amps	AMPS 208.17918
Apply adjustment factor*	X 1 2 5 %	260.22398

*Reference 450.3(B)

<u>Next Larger Size: 300 Amps</u>

Description-2	Keystroke	Display
Enter breaker size and divide by desired breaker size	3 0 0 ÷ 5 0 =	6.

<u>Six 50 Amp circuit breakers</u>

Problem 9-9

Description	Keystroke	Display
Clear Calculator	On/C On/C	0.
Enter Primary Current (from illustration)	5 4 • 2	54.2
Multiply by adjustment factor*	X 6 =	325.2

*Reference 450.3(B)

<u>300 Amps or less</u>

Problem 9-10

Description	Keystroke	Display
Clear Calculator	On/C On/C	0.
Enter Primary Current (from illustration)	5 4 • 2	54.2
Multiply by adjustment factor*	X 4 =	216.8

*Reference 450.3(B)

<u>200 Amps or less</u>

Problem 9-11

Description-1	Keystroke	Display
Clear Calculator	On/C On/C	0.
Enter Primary Current	2 0 • 8 Amps	AMPS 20.8
Find OCPD @ 125%*	1 2 5 O-Load	AMPS 26. ol

*Reference 450.3(B) and 240.6(A)

Next Larger Size: 30 Amps

Description-2	Keystroke	Display
Clear Calculator	On/C On/C	0.
Enter Secondary Current (from illustration)	8 3 • 3 Amps	AMPS 83.3
Show Wire size	WireSz	AWG 4 CU WIRE SIZE
Show Wire Ampacity	WireSz	85.0 WIRE A

85 Amps

Find Secondary to Primary Voltage Ratio (reference illustration for voltages)	1 2 0 ÷ 4 8 0 =	0.25

Verification No. 1

Description	Keystroke	Display
Multiply with Secondary OCPD value	× 8 5 =	21.25

Verification No. 1 Fails rating (25/30 OCPD rating exceeds 21.25 Amps)

Verification No. 2

Description	Keystroke	Display
Clear Calculator	On/C On/C	0.
Increase to 3 THWN wires	3 WireSz WireSz	100.0 WIRE A
Multiply by voltage ratio	× • 2 5 =	25.

Verification No. 2 Acceptable Rating does not exceed 25 Amps

Verification No. 3

Description	Keystroke	Display
Clear Calculator	On/C On/C	0.
Increase size of secondary conductors	1 WireSz WireSz	130.0 WIRE A
Multiply by voltage ratio	× • 2 5 =	32.5

Use Next Smaller Size: OCPD 30 Amps

Primary OCPD: 25 Amps
Primary Conductor: 10 AWG Cu
Secondary Conductor: 3 AWG Cu
Primary OCPD: 30 Amps
Primary Conductor: 10 AWG Cu
Secondary Conductor: 1 AWG Cu

Problem 9-12

Description-1	Keystroke	Display
Clear Calculator	On/C On/C	0.
Determine adjusted ampacity*	1 2 × 1 2 5 % Amps	AMPS 15.

*Reference 450.3(B)

15 Amps

Description-2	Keystroke	Display
Determine conductor size (75°C Setting)	WireSz	AWG 14 CU WIRE SIZE

14 AWG THWN Copper

Description-3	Keystroke	Display
Show conductor Wire Rating	WireSz	20.0 WIRE A
Determine max Feeder OCPD*	× 1 0 =	200.

*Reference 240.21(B)(1)(4)

Cannot exceed 200 Amps

Problem 9-13

Description-1	Keystroke	Display
Clear Calculator	On/C On/C	0.
Determine Primary OCPD (Table 450.3(B) & illustration)*	1 3 5 X 2 5 0 %	337.5

Next Smaller Size: 300 Amps

Description-2	Keystroke	Display
Enter OCPD Rating as amperage	3 0 0 Amps	AMPS 300.
Determine Conductor Size (75°C Rating)	WireSz	AWG 350 CU WIRE SIZE

Wire Size: 350 kcmil per phase

Description-2a	Keystroke	Display
Determine Secondary OCPD (Table 450.3(B) & illustration)*	3 1 2 X 1 2 5 %	390.

*Reference 240.6(A)

Next Larger Size: 400 Amps

Description-3	Keystroke	Display
Enter OCPD Rating as amperage	3 0 0 Amps	AMPS 300.
Determine Conductor Size (75°C Rating)	WireSz	AWG 350 CU WIRE SIZE
Show Ampacity	WireSz	310.0 WIRE A

Wire Size: 350 kcmil per phase and neutral

Description-4	Keystroke	Display
Secondary overcurrent protection (60 amp circuit) – Section 215.2(A)(1) Max OCPD,		60 amps

Description-5	Keystroke	Display
Enter OCPD Rating as amperage	6 0 Amps	AMPS 60.
Determine Conductor Size (75°C Rating)	WireSz	AWG 6 CU WIRE SIZE
Display Ampacity	WireSz	65.0 WIRE A
Multiply by 10 and show result	X 1 0 =	650.
Determine Voltage Ratio	4 8 0 ÷ 2 0 8 =	2.3076923
Multiply by Primary OCPD	X 3 0 0 =	692.30769
Ratio Failed, so use next larger conductor	4 WireSz WireSz	85.0 WIRE A

Wire Size: 4 AWG THWN Cu

Description-6	Keystroke	Display
Find OCPD secondary conductors	3 1 2 X 1 2 5 % Amps	AMPS 390.

Next Larger Size: 400 Amps

	Keystroke	Display
Add both OCPD's together and verify that it doesn't exceed 400 amps	3 0 0 + 6 0 =	360.

OCPD max not exceeded

Problem 9-14

Description-1	Keystroke	Display
Clear Calculator	On/C On/C	0.
Determine Primary OCPD*	9 0 X 1 2 5 %	112.5

*Reference 450.3(B) and 240.6(A)

Next Larger Size: 125 Amps

Description-2	Keystroke	Display
Enter Computed Primary OCPD as amperage	= Amps	AMPS 112.5
Determine Conductor Size (75°C Rating)	WireSz	AWG 2 CU WIRE SIZE

Wire Size: 2 AWG THWN Cu

Description-3	Keystroke	Display
Determine Secondary OCPD (Table 450.3(B) & illustration)*	2 0 8 X 1 2 5 %	260.

*Reference 240.6(A)

Next Larger Size: 300 Amps, but restrict to 200 amp due to panel

Description-4	Keystroke	Display
Enter Secondary Ampacity	2 0 0 Amps	AMPS 200.
Determine Wire Size	WireSz	AWG 000 CU WIRE SIZE

Wire Size: 3/0 THWN AWG Cu

Problem 9-15

Description-1	Keystroke	Display
Determine Primary OCPD*	9 0 X 1 2 5 %	112.5

*Reference 450.3(B) and 240.6(A)

Next Larger Size: 125 amps

Description-2	Keystroke	Display
Enter Primary Current as Amperage	1 2 5 Amps	AMPS 125.
Determine Conductor Size (75°C Rating)	WireSz	AWG 1 CU WIRE SIZE
Show Ampacity	WireSz	130.0 WIRE A

NOTE: Exceeds OCPD, use next smaller wire size.

Next Smaller Wire Size: 2 AWG THWN Cu

Description-3	Keystroke	Display
Determine Feeder Ampacity	3 0 0 ÷ 3 = Amps	AMPS 100.

NOTE: Reference 240.21(B)(2)(1)

Wire Size: 2 AWG THWN Copper is sufficient to carry 100 Amps

Problem 9-16

Description-1	Keystroke	Display
Set to 3-phase	Set 3	3 PH
Set to 75°C Wire Rating (if necessary)	Set 7	75
Enter Voltage	2 4 0 Volts	VOLT 240.
Enter Transformer Rating	8 0 0 0 0 VA	VA 80,000.
Find Amperage	Amps	AMPS 192.45009
Determine OCPD*	1 9 3 X 1 2 5 %	241.25

*Reference 450.3(B) and 240.6(A) **Next Larger Size: 250 Amps**

Description-2	Keystroke	Display
Enter OCPD Ampacity	2 5 0 Amps	AMPS 250.
Determine Conductor Size (75°C Rating)	WireSz	AWG 250 CU WIRE SIZE
Verify Ampacity	WireSz	255.0 WIRE A

Secondary Conductor: 250 kcmil THWN copper

Problem 9-17

Description-1	Keystroke	Display
Set to 3-phase	Set 3	3 PH
Enter Voltage	2 0 8 Volts	VOLT 208.
Enter Transformer Rating	4 5 0 0 0 VA	VA 45,000.
Find Amperage	Amps	AMPS 124.90751
Determine OCPD*	X 1 2 5 %	156.13439

*Reference 450.3(B) and 240.6(A) **Next Larger Size: 175 Amps**

Description-2	Keystroke	Display
Enter OCPD Ampacity	1 7 5 Amps	AMPS 175.
Determine Conductor Size (75°C Rating)	WireSz	AWG 00 CU WIRE SIZE
Verify Ampacity	WireSz	175.0 WIRE A

Secondary Conductor: 2/0 AWG THWN Cu

Problem 9-18

Description	Keystroke	Display
Sum of motor Locked-Rotor Currents	3 6 5 + 9 2 + 6 0 =	517.
Multiply result by voltage ratio	X 2 4 0 ÷ 4 8 0 =	258.5

NOTE: Reference 430.251(B) and 240.4(B) **Primary OCPD: 300 Amps**

Chapter 10

NOTE: If your results differ from those provided in the examples, verify that your calculator is set to the appropriate phase, wire rating, wire type and ambient temperature.

Problem 10-1

Description	Keystroke	Display
Clear Calculator	On/C On/C	0.
Compute Minimum Width	2 X 1 • 8 7 5 + 2 • 2 5 =	6.

NOTE: Reference 392.22(A)(1)(a)

Tray Size: 6 in.

Problem 10-2

Description	Keystroke	Display
Clear Calculator	On/C On/C	0.
Compute Area of one 1 AWG Cable	1 • 5 X = X Set + ÷ 4 =	1.7671459
Multiply by number of cables and store in memory	X 4 = Stor 0	M+ 7.0685835
Compute Area of one 8 AWG cable	• 8 7 5 X = X Set + ÷ 4 =	0.6013205
Multiply by number of cables and store in memory	X 3 = Stor 0	M+ 1.8039614
Compute Area of 7/C Cable	1 X = X Set + ÷ 4 =	0.7853982
Add to Recalled Memory Values and Show Total	+ Rcl Rcl =	9.657943

NOTE: Reference 392.22(A), Column 1

Tray Size: 9 in.

Problem 10-3

Description	Keystroke	Display
Clear Calculator	On/C On/C	0.
Add OD of cables	2 • 3 2 + 2 • 9 6 =	5.28
Multiply by Adjustment Factor	X 1 • 2 =	6.336
Determine capacity for 12 in tray	– 1 4 =	-7.664

NOTE: Reference 392.22(A), Column 2

Remaining Space: 7.664 in² for cables smaller than 4/0 AWG

Problem 10-4

Description-1	Keystroke	Display
Clear Calculator	On/C On/C	0.
Compute allowable tray area*	6 X 1 2 X 5 0 %	36.

*Reference 392.22(A)(2)

36 in²

Description-2	Keystroke	Display
Compute 9/C area	1 • 2 7 ÷ 2 X = X Set + =	1.2667687
Multiply by number of cables and store result	X 5 = Stor 1	M-1 6.3338435

6.334 in²

Description-3	Keystroke	Display
Compute 7/C area	• 9 7 6 ÷ 2 X = X Set + =	0.7481514
Store in second memory register for later recall	Stor 2	M-2 0.7481514
Subtract Stored 9/C area from allowable tray area	3 6 – Rcl 1 =	29.666157

29.66 in²

Description-4	Keystroke	Display
Divide result by stored 7/C area	÷ Rcl 2 =	39.652609

NOTE: Reference 392.22(A)(2)

39 7/C cables

Problem 10-5

Description	Keystroke	Display
Clear Calculator	On/C On/C	0.
Compute Total OD	2 X 1 • 8 7 5 + 2 • 2 5 =	6.
Compute Allowable Fill for 6 inch tray	÷ 9 0 %	6.6666667

NOTE: Reference 392.22(A)(3)(a) and 392.22(A), Column 1

Next Larger Standard Size: 8 in

Problem 10-6

Description	Keystroke	Display
Clear Calculator	On/C On/C	0.
Compute 1/0 AWG Total Area	1 • 5 2 X = X Set + ÷ 4 X 2 =	3.6291678
Store in Accumulative Memory (M+)	Stor 0	M+ 3.6291678
Compute 250 kcmil Total Area	1 • 1 4 X = X Set + ÷ 4 X 3 =	3.0621104
Store in Accumulative Memory (M+)	Stor 0	M+ 3.0621104
Compute 3 AWG Total Area	1 • 6 2 5 X = X Set + ÷ 4 X 6 =	12.443652
Store in Accumulative Memory (M+)	Stor 0	M+ 12.443652
Display Memory Total	Rcl Rcl	19.13493

NOTE: Reference 392.22(A), Column 3

24 inch tray required

Problem 10-7

Description	Keystroke	Display
Clear Calculator	On/C On/C	0.
Compute 500 kcmil Total Space	3 • 1 2 X 2 =	6.24
Compute Remaining Tray Area	1 1 − 6 • 2 4 =	4.76
Compute 4 AWG Area	1 • 0 7 X = X Set + ÷ 4 =	0.8992024
Store in Memory for later recall	Stor 1	M-1 0.8992024
Compute Number of Cables	4 • 7 6 ÷ Rcl 1 =	5.2935804

NOTE: Reference 392.22(A), Column 4

5 cables

Problem 10-8

Description	Keystroke	Display
Clear Calculator	On/C On/C	0.
Compute Allowable Tray Area	6 X 4 X 4 0 %	9.6
Compute 4/C Total Area	• 5 3 7 X = X Set + ÷ 4 X 5 =	1.1324224
Store in Memory for later recall	Stor 1	M-1 1.1324224
Compute Remaining Area	9 • 6 − Rcl 1 =	8.4675776
Store in Memory for later recall	Stor 1	M 1 8.4675776
Compute 7/C Total Area	• 9 7 5 X = X Set + ÷ 4 =	0.7466191
Store in Memory 2 for later recall	Stor 2	M-2 0.7466191
Compute Number of Cables	Rcl 1 ÷ Rcl 2 =	11.341228

NOTE: Reference 392.22(A)(4)

11 cables

Problem 10-9

Description	Keystroke	Display
Compute Cable Area	1 • 7 X = X Set + ÷ 4 =	2.2698007

NOTE: Reference 392.22(A)(5), Column 1

3 inch tray

Problem 10-10

Description	Keystroke	Display
Compute Cable Area	2 • 5 X = X Set + ÷ 4 =	4.9087385

NOTE: Reference 392.22(A)(5), Column 1

No. A 6 in. ventilated channel cable tray is the minimum size required

Problem 10-11

Description	Keystroke	Display
Multiply wire diameter by wire quantity	1 • 3 1 X 9 =	11.79

NOTE: Reference 392.22(B)(1) and 392.22(A)

Tray Size: 12 in.

Problem 10-12

Description	Keystroke	Display
Find 250 kcmil area	2 5 0 WireSz Set #THHN #THWN	THHN 0.397 WIRE AREA
Multiply by quantity and add to M+	X 3 = Stor 0	M+ 1.191
Find 500 kcmil area	5 0 0 WireSz Set #THHN #THWN	THHN 0.7073 WIRE AREA
Multiply by quantity and add to M+	X 6 = Stor 0	M+ 4.2438
Multiply Compact 750 kcmil area by quantity	• 9 3 3 1 X 3 =	2.7993
Add stored value to find total fill area	+ Rcl Rcl =	8.2341

NOTE: Reference 392.22(B)(1)(b) and 392.22(B)(1), Column 1

Tray Size: 9 in.

Problem 10-13

Description-1	Keystroke	Display
Compute 1250 kcmil total length and store in M+	3 X 1 • 4 7 9 = Stor 0	M+ 4.437
Compute 1000 kcmil total length and store in M+	3 X 1 • 3 1 2 = Stor 0	M+ 3.936
Multiply total by Sd factor and store in Memory 1	Rcl Rcl X 1 • 1 = Stor 1	M-1 9.2103
Find remaining area	1 3 − Rcl 1 =	3.7897

3.7897 in.²

Description-2	Keystroke	Display
Divide result by 500 kcmil area	÷ • 6 9 8 4 =	5.42626

NOTE: Reference 392.22(B)(1)(c) and 392.22(B)(1), Column 2
NOTE: Wire diameter not supported by the ECPro.

5 Conductors

Problem 10-14

Description	Keystroke	Display
Compute 1/0 AWG total length and store in M+	4 X • 4 8 6 = Stor 0	M+ 1.944
Compute 2/0 AWG total length and store in M+	3 X • 5 3 2 = Stor 0	M+ 1.596
Compute 4/0 AWG total length and store in M+	3 X • 6 4 2 = Stor 0	M+ 1.926
Display resulting total	Rcl Rcl	5.466

6 inch tray

Problem 10-15

Description	Keystroke	Display
Compute total length	4 X 1 • 0 5 1 =	4.204

NOTE: Reference 392.22(B)(1)

6 inch tray

Problem 10-16

Description-1	Keystroke	Display
Set to 75°C Wire Rating (if necessary)	Set 7	75
Show Ampacity of 500 kcmil copper	5 0 0 WireSz WireSz	380.0 WIRE A
Store as Amps	= Amps	AMPS 380.
Find Derated Ampacity	4 Set ParSz ParSz	D/R 304.0 WIRE A

500 kcmil: 304 Amps

Description-2	Keystroke	Display
Set to 75°C Wire Rating (if necessary)	Set 7	75
Show Ampacity of 4 AWG copper	4 WireSz WireSz	85.0 WIRE A
Store as Amps	= Amps	AMPS 85.

4 AWG: 85 Amps

Problem 10-17

Description-1	Keystroke	Display
Set to 75°C Wire Rating (if necessary)	Set 7	75
Show Ampacity of 500 kcmil copper	5 0 0 WireSz WireSz	380.0 WIRE A
Adjust per 392.80(A)(1)(b)	X 9 5 % Amps	AMPS 361.
Find Derated Size	4 Set ParSz ParSz	D/R 288.8 WIRE A

500 kcmil: 288.8 Amps

Description-2	Keystroke	Display
Show Ampacity of 4 AWG copper (be sure calculator set to 75°C Cu)	4 WireSz WireSz	85.0 WIRE A
Adjust per 392.80(A)(1)(b)	X 9 5 % Amps	AMPS 80.75

4 AWG: 80.75 Amps

Problem 10-18

Description-1	Keystroke	Display
Set to Aluminum	Set 4	Al
Set to Free Air	Set 5	FrAir
Show Ampacity of 600 kcmil aluminum*	6 0 0 WireSz WireSz	545.0 WIRE A
Adjust per 392.80(B)(2)(a)	X 7 5 %	408.75

*Be sure wire rating is set to 75°C

408.75 Amps

Description-2	Keystroke	Display
Clear Calculator	On/C	0.
Show Ampacity of 600 kcmil aluminum*	6 0 0 WireSz WireSz	545.0 WIRE A
Adjust per 392.80(B)(2)(a)	X 7 0 %	381.5

*Be sure wire rating is set to 75°C

381.5 Amps

Problem 10-19

Description-1	Keystroke	Display
Set to Copper	Set 4	Cu
Set to Free Air (if necessary)	Set 5	FrAir
Show Ampacity of 350 kcmil copper (be sure calculator set to 75°C)	3 5 0 WireSz WireSz	505.0 WIRE A
Adjust per 392.80(B)(2)(b)	X 6 5 %	328.25

328.25 Amps

Description-2	Keystroke	Display
Show Ampacity of 350 kcmil copper (be sure calculator set to 75°C Cu)	3 5 0 WireSz WireSz	505.0 WIRE A
Adjust per 392.80(B)(2)(b)	X 6 0 %	303.

303 Amps

Problem 10-20

Description	Keystroke	Display
Set to 90°C Wire Rating (if necessary)	Set 9	90
Set to Free Air (if necessary)	Set 5	FrAir
Find Ampacity	5 0 0 WireSz WireSz	700.0 WIRE A
		<u>700 Amps</u>

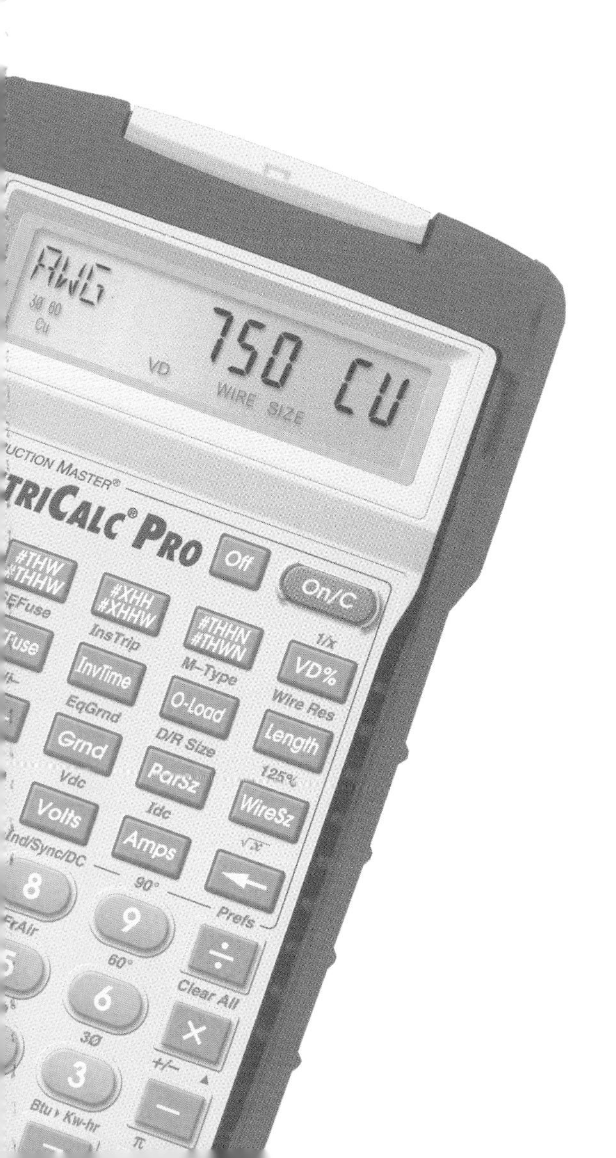

Chapter 11

NOTE: If your results differ from those provided in the examples, verify that your calculator is set to the appropriate phase, wire rating, wire type and ambient temperature.

Problem 11-1

Description	Keystroke	Display
Reset Calculator (clears previous settings)	Set X	NEC 2014
Multiply RPC by Table 630.11(A) Factor	4 0 X • 8 4 =	33.6

33.6 Amps

Problem 11-2

Description-1	Keystroke	Display
Multiply RPC by Table 630.11(A) Factor	3 8 X • 7 1 =	26.98

26.98 Amps

Description-2	Keystroke	Display
Enter previous result as Amps	Amps	AMPS 26.98
Find Wire Size	WireSz	AWG 10 CU WIRE SIZE

10 AWG THWN Copper

Description-3	Keystroke	Display
Multiply RPC by 200% (Section 630.12(A))	3 8 X 2 0 0 %	76.

Next Larger Standard Size: 80 Amps

Problem 11-3

Description-1	Keystroke	Display
Multiply RPC by Table 630.11(A) Factor	2 8 X • 5 5 =	15.4

15.4 Amps

Description-2	Keystroke	Display
Multiply RPC by Table 630.11(A) Factor	1 8 X • 8 9 =	16.02

16.02 Amps

Description-3	Keystroke	Display
Add Welder 1 ampacity for total ampacity	+ 1 5 • 4 =	31.42

31.42 Amps

Problem 11-4

Description-1	Keystroke	Display
Multiply Largest RPC by Table 630.11(A) Factor and store in M+	3 2 X • 7 8 = Stor 0	M+ 24.96
Multiply 2nd Largest RPC by Table 630.11(A) Factor and store in M+	2 4 X • 9 5 = Stor 0	M+ 22.8
Multiply 3rd Largest RPC by Table 630.11(A) Factor and adjust per 630.11(B)	1 4 X • 4 5 X • 8 5 =	5.355
Add Stored M+ value and display total	+ Rcl Rcl =	53.115

53.115 Amps

Description-2	Keystroke	Display
Enter previous result as Amps	Amps	AMPS 53.115
Set to 75°C insulation rating	Set 7	75
Find Wire Size	WireSz	AWG 6 CU WIRE SIZE

6 AWG THWN Copper

Description-3	Keystroke	Display
Find Ampacity of Circuit Conductor	WireSz	65.0 WIRE A
Multiply by 200% (Section 630.12(A))	X 2 0 0 %	130.

Next Larger Standard Size: 150 Amps

Problem 11-5

Description-1	Keystroke	Display
Determine Ampacity of Welder 1	4 0 X • 4 5 =	18.
		18 Amps

Description-2	Keystroke	Display
Determine Ampacity of Welder 2	2 8 X • 8 4 =	23.52
		23.52 Amps

Description-3	Keystroke	Display
Determine Ampacity of Welder 3	2 4 X • 7 1 =	17.04
		17.04 Amps

Description-4	Keystroke	Display
Determine Ampacity of Welder 4	1 6 X • 6 3 =	10.08
		10.08 Amps

Description-5	Keystroke	Display
Enter Largest Ampacity into M+	2 3 • 5 2 Stor 0	M+ 23.52
Enter 2nd Largest Ampacity into M+	1 8 Stor 0	M+ 18.
Adjust 3rd Largest ampacity* and store in M+	1 7 • 0 4 X • 8 5 = Stor 0	M+ 14.484
Adjust 4th Largest ampacity* and store in M+	1 0 • 0 8 X • 7 = Stor 0	M+ 7.056
Display Resulting Total	Rcl Rcl	63.06
*Reference Table 630.11(A) and 630.11(B) for adjustment factors.		**63.06 Amps**

Problem 11-6

Description-1	Keystroke	Display
Determine Ampacity of Welder 1	4 0 X • 7 1 =	28.4
		28.4 Amps

Description-2	Keystroke	Display
Determine Ampacity of Welder 2	1 2 X • 6 3 =	7.56
		7.56 Amps

Description-3	Keystroke	Display
Determine Ampacity of Welder 3	2 4 X • 8 9 =	21.36
		21.36 Amps

Description-4	Keystroke	Display
Determine Ampacity of Welder 4	2 0 X • 5 5 =	11.
		11 Amps

Description-5	Keystroke	Display
Determine Ampacity of Welder 5	2 4 X • 9 5 =	22.8
		22.8 Amps

Description-6	Keystroke	Display
Enter Largest Ampacity into M+	2 8 • 4 Stor 0	M+ 28.4
Enter 2nd Largest Ampacity into M+	2 2 • 8 Stor 0	M+ 22.8
Adjust 3rd Largest ampacity* and store in M+	2 1 • 3 6 X • 8 5 = Stor 0	M+ 18.156
Adjust 4th Largest ampacity* and store in M+	1 1 X • 7 = Stor 0	M+ 7.7
Adjust 5th Largest ampacity* and store in M+	7 • 5 6 X • 6 = Stor 0	M+ 4.536
Display Resulting Total	Rcl Rcl	81.592
*Reference Table 630.11(A) and 630.11(B) for adjustment factors.		**81.6 Amps**

Problem 11-7

Description	Keystroke	Display
Determine Ampacity	6 0 X • 9 1 =	54.6
NOTE: Reference Table 630.11(A)		**54.6 Amps**

Problem 11-8

Description	Keystroke	Display
Determine Ampacity	3 6 X • 8 1 =	29.16
NOTE: Reference Table 630.11(A)		**29.16 Amps**

Problem 11-9

Description	Keystroke	Display
Determine Ampacity*	2 1 X • 8 1 =	17.01
Add all 630.11(B) adjustment factors together	1 + 1 + • 8 5 + • 7 + • 6 + • 6 =	4.75
Multiply by ampacity	X 1 7 • 0 1 =	80.7975

*Reference Table 630.11(A)

80.8 Amps

NOTE: Method above only works when all welders are the same rating and duty cycle.

Problem 11-10

Description-1	Keystroke	Display
Set to 1-phase	Set 1	1 PH
Enter Voltage	2 2 0 Volts	VOLT 220.
Enter Power Rating	1 5 0 0 0 VA	VA 15,000.
Find Amperage	Amps	AMPS 68.181818
Adjust per 630.31(A)(1)	X • 7 =	47.727273

Automatic: 47.72 Amps

Description-2	Keystroke	Display
Recall Amperage	Rcl Amps	AMPS 68.181818
Adjust per 630.31(A)(1)	X • 5 =	34.090909

Manual: 34.09 Amps

Problem 11-11

Description	Keystroke	Display
Multiply amps by 40% duty*	4 0 X • 6 3 =	25.2

*Reference 630.31(A)(2)

25.2 Amps

Problem 11-12

Description-1	Keystroke	Display
Set to 1-phase	Set 1	1 PH
Enter Voltage	2 4 0 Volts	VOLT 240.
Enter Power Rating	2 0 0 0 0 VA	VA 20,000.
Find Amperage	Amps	AMPS 83.333333
Adjust per 630.31(A)(1)	X • 7 =	58.333333
Store in M+	Stor 0	M+ 58.333333
Multiply by 630.31(B) factor	X • 6 =	35.
Add to recalled M+ value	+ Rcl Rcl =	93.333333

Automatic: 93.33 Amps

Description-2	Keystroke	Display
Set to 1-phase	Set 1	1 PH
Enter Voltage	2 4 0 Volts	VOLT 240.
Enter Power Rating	2 0 0 0 0 VA	VA 20,000.
Find Amperage	Amps	AMPS 83.333333
Adjust per 630.31(A)(1)	X • 5 =	41.666667
Store in M+	Stor 0	M+ 41.666667
Multiply by 630.31(B) factor	X • 6 =	25.
Add to recalled M+ value	+ Rcl Rcl =	66.666667

Manual: 66.67 Amps

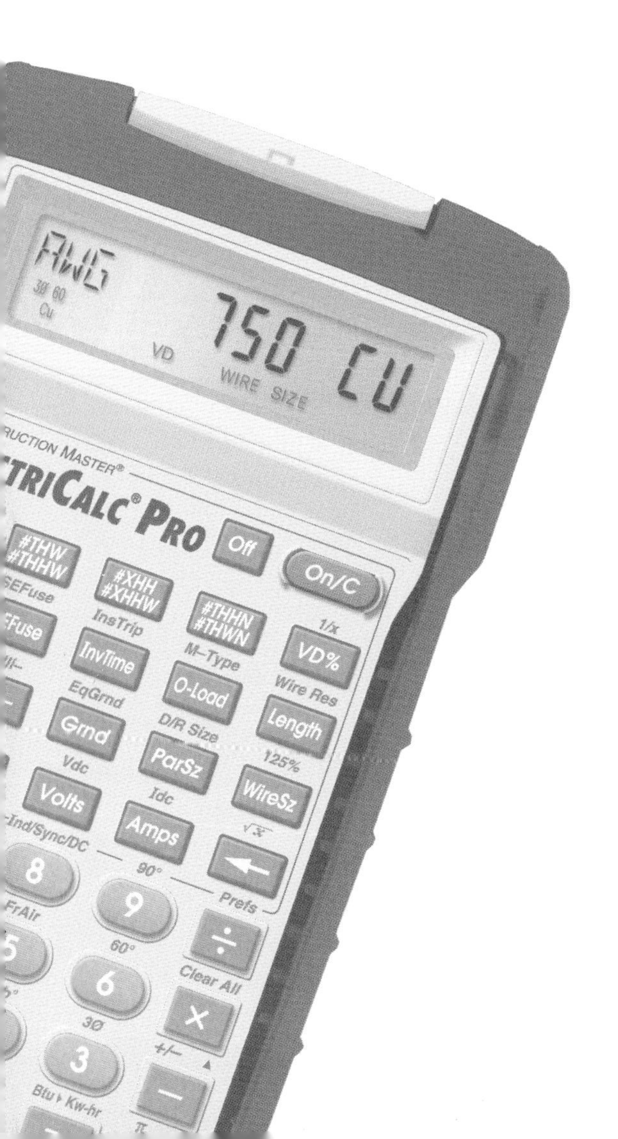
AWG
750 CU
WIRE SIZE
VD
Off
On/C
VD%
Length
WireSz
O-Load
ParSz
Amps
Volts
Grnd
InvTime
Clear All
Prefs